ELECTROMAGNETIC COMPATIBILITY HANDBOOK

ELECTROMAGNETIC COMPATIBILITY HANDBOOK

J. L. Norman Violette, Ph.D.
Donald R. J. White, MSEE
Michael F. Violette, BSEE

 VAN NOSTRAND REINHOLD COMPANY
—————————————————————— New York

Copyright © 1987 by Van Nostrand Reinhold Company Inc.
Library of Congress Catalog Card Number 85-29475
ISBN 0-442-28903-0

Printed in the United States of America

Van Nostrand Reinhold Company Inc.
115 Fifth Avenue
New York, New York 10003

Van Nostrand Reinhold Company Limited
Molly Millars Lane
Wokingham, Berkshire RG11 2PY, England

Van Nostrand Reinhold
480 La Trobe Street
Melbourne, Victoria 3000, Australia

Macmillan of Canada
Division of Canada Publishing Corporation
164 Commander Boulevard
Agincourt, Ontario M1S 3C7, Canada

16   15   14   13   12   11   10   9   8   7   6   5   4   3   2   1

Library of Congress Cataloging-in-Publication Data

Violette, J. L. Norman.
  Electromagnetic compatibility handbook.

  Includes bibliographies and index.
  1. Electromagnetic compatibility. I. White,
Donald R. J. II. Violette, Michael F. III. Title.
TX6553.V53 1985   621.38'0436   85-29475
ISBN 0-442-28903-0

PREFACE

The growth of the electrical industry has resulted in the utilization of essentially the entire electromagnetic spectrum to perform many of the tasks associated with various industries: communication services; air, sea, and ground radio navigation systems; military operations; scientific research; medical treatment; computer-related operations; and space explorations. The proliferation of solid-state electronic products associated with all of these has created an electromagnetic environment wherein the probability of mutual electromagnetic interference (EMI) from man-made sources continues to increase at a rapid pace. In addition to man-made interference sources, the effects due to lightning and other natural EMI sources must be considered.

The capability of electrical/electronic systems to operate within a given environment without unacceptable performance degradation of any system due to EMI is referred to as Electromagnetic Compatibility (EMC). As EMI becomes more pronounced, so does the requirement for EMC Engineering at the system, subsystem, circuit, and component level. As a discipline, EMC engineering is generally not taught as part of standard curriculum; its required skills, instead, are acquired from practical experience "in the field."

This Handbook features relevant developments of appropriate EMC topics; mathematical formulations to support analytical quantitative data presented in tabular and graphical format, and empirical data derived from several years of "hands-on" experience by the authors and others within the EMC community. Rigorous mathematical formulations are generally not presented in their entirety, but the results are provided to facilitate understanding of EMC situations and solutions in terms of basic scientific and engineering principles.

The Handbook emphasizes:

1. Identification of EMI sources and methods for quantifying resultant ambient electromagnetic fields, whenever this is possible;
2. Identification of EMI receptors, or victims, and quantitative determination of susceptibility;
3. Identification of coupling paths between sources and receptors and quantification of the degree of coupling;

4. Development of the most cost-effective approaches to eliminate EMI problems.

This Handbook is intended for engineers and technicians whose activities include the design, development, fabrication, retrofit, and maintenance of electrical and electronic equipment for communication, control, computer, power distribution, military, automotive, and space systems. The Handbook addresses problems associated with a broad range of the frequency spectrum, from DC to microwaves. EMC principles applicable to digital circuits are presented, with time-to-frequency domain transformations developed. Digital designers will find the Handbook useful whether their task be the development of electronic games or highly sensitive military and medical diagnostic equipment. Many topics from this Handbook are applicable to the design of automotive systems to assist in improving fuel economy and pollution control. Managers will also find this Handbook useful, as will students of electrical engineering.

J. L. NORMAN VIOLETTE
DONALD R. J. WHITE
MICHAEL F. VIOLETTE

ACKNOWLEDGMENTS

The authors would like to extend their appreciation to the people who worked to make this Handbook a reality. Since the concept of this book took root some five years ago, many individuals have been involved in the project: some actively gathering, sorting, and typing material; some profing the many pages of text; while others gently prodded the project to completion with words of encouragement.

The authors would like to thank their wives: Bette Violette, who spent long hours typing and editing many pages of text for the manuscript; Colleen White, who helped prepare original material for the book; and Liv Violette, who was instrumental in preparing and proofing much of the artwork and figures for the manuscript.

No one could have done more to actually get the first manuscript out the door than Mr. Michel Mardiguian, to whom we gratefully extend our appreciation. Mr. Ulf Nilsson reviewed various versions of the manuscript until an acceptable work was hammered out. Others involved in the book over the years include Mr. John D. Osburn, who worked during the conceptual stage; Mr. John Brown, who provided essential material to the authors; Mr. Joseph Violette, who performed typing and editing; and Messrs. Erik Syvrud and Steven Strickland, who provided technical guidance to the authors.

CONTENTS

Chapter 1
AN INTRODUCTION TO
ELECTROMAGNETIC COMPATIBILITY

Electrical, electromechanical, and electronic equipment all must comply with specifications intended to assure electromagnetic compatibility (EMC), which is the ability of systems, subsystems, circuits, and components to function as designed, without malfunction or unacceptable degradation of performance due to electromagnetic interference (EMI), within their intended operational environment. Essentially, any equipment or system should not adversely affect the operation of any other equipment or system as a result of radiated or conducted EMI, and, in turn, it should not be affected by same. This chapter defines EMI and EMC, discusses use of the electromagnetic frequency spectrum, describes basic elements of all EMI situations, and introduces the concepts of intersystem and intrasystem interference.

ELECTROMAGNETIC INTERFERENCE

Electromagnetic interference (EMI) consists of any unwanted, spurious, conducted, and/or radiated signal(s) of electrical origin that can cause unacceptable degradation of system or equipment performance. The effects of EMI can range from minor nuisance to catastrophic consequences. A minor nuisance is judged to result, for example, when someone is watching a television program for entertainment, and interference ("snow") appears momentarily on the screen due to the operation of an electrical appliance such as a hair dryer or an electric shaver. On the other hand, a serious consequence could result should an interfering signal disturb the normal operation of medical electronic equipment being used to monitor the condition of a patient under intensive care in a hospital. Another serious consequence could result should unwanted interference disrupt critical communications during a time of emergency, such as during a rescue operation. These examples are indicative of the nature and potential consequences of EMI and the importance of effective control to achieve EMC.

The Nature of EMI

The origins of EMI are basically electrical with the unwanted emissions being either conducted (voltages and/or currents) or radiated (electric and/or magnetic fields). In the time domain, EMI can be transient, impulsive, or steady-state. Within the frequency domain, EMI can contain components with frequencies ranging from the lower power frequencies of 50, 60, and 400 Hz, on up into the microwave region. Also within the frequency domain, EMI signals can be either narrowband or broadband, and coherent or noncoherent. The sources of EMI can be classified as man-made or natural. Within the man-made class, a further delineation is made to identify intentional versus unintentional (or incidental) sources of EMI. The classifications of EMI are designed to facilitate recognition of sources, determine receptor susceptibility, identify coupling paths between source and receptor, and assist in determining means of control.

The Frequency Spectrum

Frequency plays a significant role in EMC. Performance specifications, such as insertion loss and attenuation characteristics for filters and shielding effectiveness for materials, are generally provided as a function of frequency. The worldwide legal utilization of the electromagnetic spectrum is determined by international treaty between the nations of the International Telecommunications Union (ITU). The frequency band allocation in the United States by the Federal Communications Commission (FCC) is based upon the prevention of mutual interference between the various users of electromagnetic radiations within the United States. The emissions from licensed communication-electronic (C-E) transmitters are often found to be "culprit" emissions when they interfere inadvertently with other systems, although C-E transmitter emissions are designed to provide desirable services elsewhere.

The Frequency–Wavelength Relationship. The relationship between the frequency (f), wavelength (λ), and propagation velocity (c) of an electromagnetic wave is:

$$c = f\lambda \qquad (1\text{-}1)$$

where $c = 3 \times 10^8$ meters/second is the velocity at which electromagnetic waves propagate through free space, f is the frequency in Hertz, and λ is the wavelength in meters. Many EMC-related analyses and calculations are based upon the frequency–wavelength relationship of electromagnetic waves. The

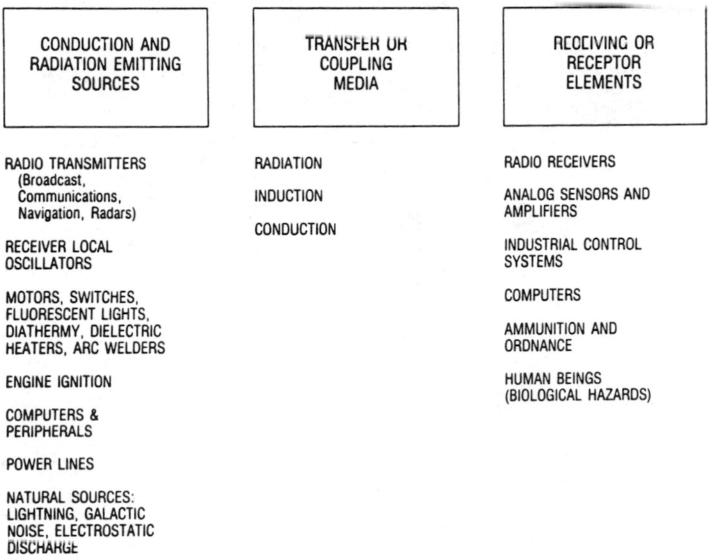

CONDUCTION AND RADIATION EMITTING SOURCES	TRANSFER OR COUPLING MEDIA	RECEIVING OR RECEPTOR ELEMENTS

RADIO TRANSMITTERS (Broadcast, Communications, Navigation, Radars)	RADIATION	RADIO RECEIVERS
	INDUCTION	ANALOG SENSORS AND AMPLIFIERS
RECEIVER LOCAL OSCILLATORS	CONDUCTION	INDUSTRIAL CONTROL SYSTEMS
MOTORS, SWITCHES, FLUORESCENT LIGHTS, DIATHERMY, DIELECTRIC HEATERS, ARC WELDERS		COMPUTERS
		AMMUNITION AND ORDNANCE
ENGINE IGNITION		HUMAN BEINGS (BIOLOGICAL HAZARDS)
COMPUTERS & PERIPHERALS		
POWER LINES		
NATURAL SOURCES: LIGHTNING, GALACTIC NOISE, ELECTROSTATIC DISCHARGE		

Fig. 1-1. Three basic elements of an emitting-susceptibility situation. (*Courtesy of Interference Control Technologies*)

following variations of equation (1-1) are often useful (1):

$$\lambda \text{ (meters)} = \frac{300}{f \text{ (megahertz)}} \qquad (1\text{-}2)$$

$$\lambda \text{ (centimeters)} = \frac{30}{f \text{ (gigahertz)}} \qquad (1\text{-}3)$$

$$\lambda \text{ (feet)} = \frac{984}{f \text{ (megahertz)}} \qquad (1\text{-}4)$$

Basic Elements of EMI Situations

Figure 1-1 defines the three essential elements for an EMI situation to exist: an electrical noise (EMI) source, a coupling path, and a victim receptor. The noise source emission can be either a conducted voltage or current, or an electric or magnetic field propagated through space. Some representative EMI sources are indicated in Fig. 1-1, as are some receptors. It can be seen that some equipment or systems can serve as both sources and receptors. Methods of coupling between sources and receptors can be divided into two

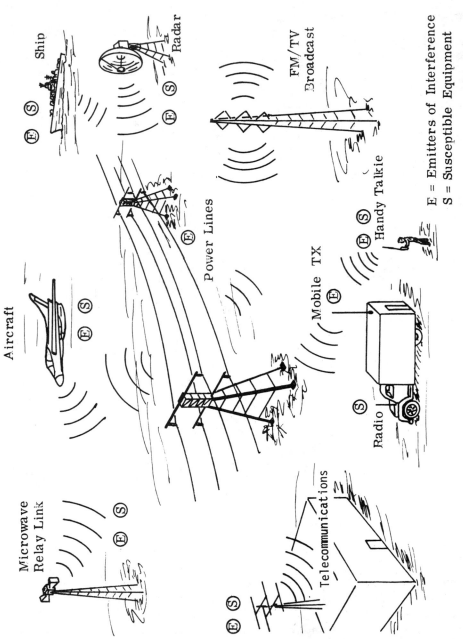

Fig. 1-2. Illustration of potential EMI situation between discrete systems (Inter-System EMI). *(Courtesy of Interference Control*

basic groups: radiation or field coupling by electromagnetic wave propagation through space or materials, and coupling via conducting paths through which currents can flow.

Intersystem and Intrasystem EMI Situations

The ability to anticipate and identify potential and existing EMI situations often depends upon whether the situation exists between two or more discrete systems (intersystem EMI) or between elements within the same system (intrasystem EMI).

Intersystem EMI.   Figure 1-2 illustrates potential EMI situations between discrete systems operating within a wide frequency range from power to microwave frequencies (50/60 Hz to several GHz). The objective of good frequency management is the prevention of interference by assigning frequencies such that a proper spread is maintained between the frequencies of the respective electromagnetic spectrum users. However, the generation of harmonics and unpredictable equipment/system susceptibilities and mal functions often result in frequency-related EMI situations. The level of power associated with each system varies from a few watts (for the hand-held transceiver) to several megawatts (for the radar systems). Also, no single operating agency (i.e., no single "manager") has jurisdiction over all systems to dictate actions necessary to achieve EMC. For the situation depicted, EMC usually is achieved by industrial association, voluntary regulation, government-enforced regulation, and negotiated agreements between the affected parties (such as frequency management, prescribed antenna radiation patterns, etc.). In most intersystem EMI situations, coupling usually results from radiation, such as via antenna-to-antenna or antenna-to-cable coupling paths. Engineering methods and techniques to control intersystem EMI are discussed in Chapter 4 of this *Handbook* and in reference 2.

Intrasystem EMI.   Intrasystem EMI occurs as a result of "self-jamming" or undesirable emission coupling within a system, as illustrated in Fig. 1-3. In this illustration, interference can develop as a result of transient voltage and/or current spikes that appear on power cables and wiring harnesses. These spikes can be electrically (capacitively) and/or magnetically (inductively) coupled into adjoining sensitive cables via cable-to-cable mutual capacitance and/or mutual inductance, thereby causing undesirable signals to appear. Intrasystem EMI can also be coupled from one part of a system to another as a result of voltage drops induced across common ground impedances by ground currents. In contrast to intersystem EMI, intrasystem EMI generally can be controlled by a single "manager" of the system, who could be the system or equipment design engineer. Design methods and techniques to control

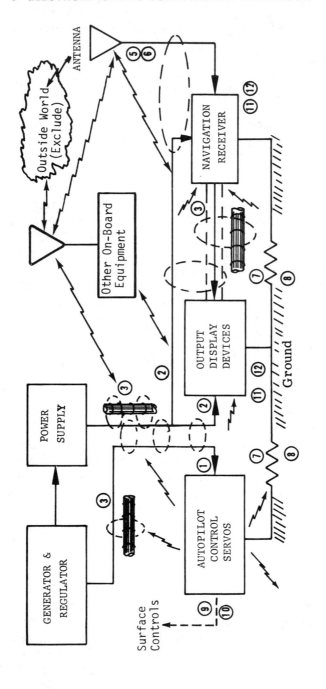

1. Power Cable Conducted Emission
2. Power Cable Conducted Susceptibility
3. Interconnecting Cable Conducted Emission
4. Interconnecting Cable Conducted Susceptibility
5. Antenna Lead Conducted Emission
6. Antenna Lead Conducted Susceptibility
7. Common Ground Impedance Emission Coupling
8. Common Ground Impedance Suscept. Coupling
9. H-Field Radiation
10. E-Field Radiation
11. H-Field Susceptibility
12. E-Field Susceptibility

Fig. 1-3. Examples of Intra-System EMI. (*Courtesy of Interference Control Technologies*)

intrasystem EMI are developed throughout this *Handbook* starting with Chapter 5, and also within the references (3).

AN OVERVIEW OF EMI CONTROL TECHNIQUES

The control of EMI involves the implementation of various methods and procedures associated with the different phases of system development, such as the conceptual, design, development/prototype, production, test/evaluation, and operational phases. Each of these phases provides certain opportunities for implementing EMC engineering, with the conceptual phase providing the most cost-effective opportunities and the operational phase usually offering the least number of possibilities. A truism based upon experience can be stated at this point: *The longer one waits to implement EMC engineering during a system's life cycle, the faster the opportunities for optimization vanish, and the higher the cost is to implement those opportunities that remain.* Proper system planning and programming include EMC methods, procedures, and techniques at the outset of system design. The success of achieving EMC is predicated upon the capability to perform emission and susceptibility analysis for the prediction of EMI and upon effective diagnostic procedures to support test and evaluation of systems undergoing development and production testing.

Intersystem EMI Control Techniques

Figure 1-4 illustrates basic elements of concern in an intersystem EMI control problem. This form of EMI is distinguished by interference between two or more discrete systems that are generally under the control of separate users. EMI control techniques are grouped into four categories: (1) frequency management, (2) time management, (3) location management, and (4) direction management. System isolation/separation, antenna polarization, and other control features appear in Fig. 1-4 as subcategories. These categories of intersystem EMI control are discussed in later chapters of this *Handbook* and in reference 2.

Intrasystem EMI Control Techniques

Figure 1-5 illustrates the basic elements of concern in an intrasystem EMI control problem. With respect to the classification of intrasystem EMI as suggested in Fig. 1-5, the interest is mainly in performance degradation due to self-jamming. Other potential problems also must be considered from a systems approach: (1) problems associated with susceptibility to externally produced conducted and/or radiated emissions, and (2) undesirable effects of conducted and/or radiated emissions produced by the given system on

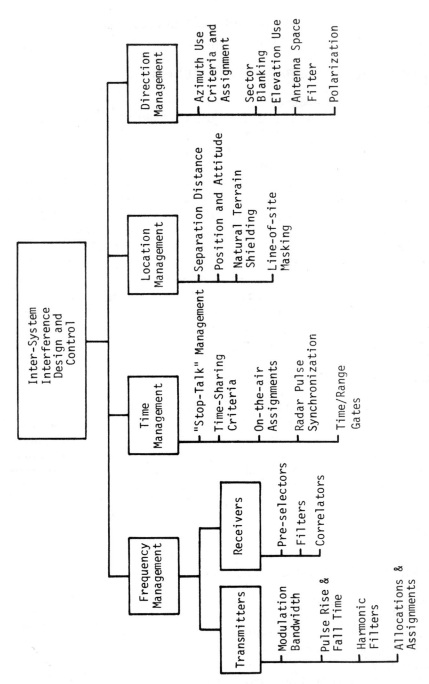

Fig. 1-4. Inter-System EMI control techniques. (*Courtesy of Interference Control Technologies*)

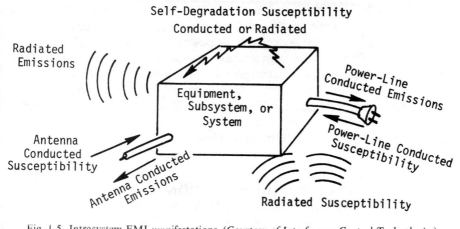

Fig. 1-5. Intrasystem EMI manifestations. (*Courtesy of Interference Control Technologies*)

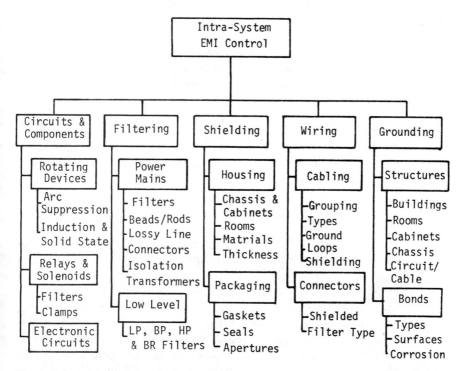

Fig. 1-6. Intrasystem EMI control organization tree. (*Courtesy of Interference Control Technologies*)

adjacent systems. Thus, intra-system EMI control considers both self-jamming and susceptibility to external emissions, and the generation of excessive levels of emissions capable of posing a threat to nearby systems.

Figure 1-6 illustrates an organization tree that groups intrasystem EMI control techniques into five categories: (1) circuits and components, (2) filtering, (3) shielding, (4) wiring, and (5) grounding. Bonding, connector considerations, packaging, and other devices and methods for EMI control appear as subcategories. The functions of the techniques illustrated in Fig. 1-6 in the control of EMI are treated in later chapters of this *Handbook* and in reference 3.

AN OVERVIEW OF INSTRUMENTATION AND MEASUREMENT TECHNIQUES

The prediction, analysis, diagnostics, and eventual control of EMI could not be satisfactorily accomplished without proper instrumentation and the ability to perform measurements during system development, quality assurance, and operational testing. Several sections of this *Handbook* present EMI-related instrumentation and measurement techniques.

Chapters 18 through 20 discuss instrument terminology, calibration, potential errors peculiar to instruments and errors associated with applications and procedures, and other related topics. Measurement techniques and procedures pertaining to CISPR, the FCC, and the VDE are also presented. Additional instrumentation and measurement details are available in references 4, 5, and 6.

EMI tests and measurements are performed for the following reasons: (1) component performance testing, (2) design/developmental testing, (3) operational field testing, (4) electromagnetic ambient site surveys, and (5) system-level EMC checkout and quality assurance performance testing. A brief overview of each topic follows.

Component Performance Testing

Part of the EMC design or field service engineer's mission is to select appropriate EMI control components. This requires the availability of time and frequency domain performance data for many components, such as filters, shields, gaskets, cable shields, surge suppressors, optical isolators, isolation transformers, ferrite beads and rods, and suppressant tubing. Larger items, such as cabinet racks and shielded rooms, are also considered to be EMI-control components in the general sense, whose performance (i.e., shielding effectiveness) must be determined. Measurement and test techniques are required to determine component performance, usually attenuation or shielding effectiveness as a function of frequency, with the data usually published by and available from the manufacturers of the specific components.

Design/Developmental Testing

EMC testing should be performed as early as is practical during system development, such as during the "brassboard" or prototype stage. The adequacy of EMC design should not be delayed until the late stages of a program, such as during quality assurance, when the cost of EMC redesign generally exceeds the cost of earlier EMC implementation. The emphasis again is on identifying problems as early as possible during system testing.

Operational Field Testing

The most pressing EMI situation occurs when an installed operational system malfunctions due to EMI, such as a computer "crashing" because of an unforeseen electromagnetic ambient. Such a situation often results in costly delays in production, loss of administrative time, and so on. Field problems often occur as a result of improper attention to EMC during system development, as was mentioned previously, or in spite of previous efforts, although the probability of malfunction is reduced by early EMC actions. Diagnostic testing in the field permits the identification of the causes of system failure due to EMI and the determination of remedial actions.

Electromagnetic Ambient Site Surveys

The solution to many EMI problems is greatly facilitated if the ambient conducted and/or radiated emissions can be determined, that is, if the "threat" can be identified and quantified. Whenever possible, an EMI site survey is highly desirable prior to the design of new equipment or the retrofitting of existing equipment designated for installation within a site area. A typical site survey procedure may include: (1) the determination of conducted and radiated parameters to be measured and the range of frequencies for which parameter values are required, (2) the selection of appropriate measuring equipment, (3) the determination of appropriate formats to gather and display data, (4) data analysis and display of results, and (5) the selection of an appropriate site among alternate sites (if applicable) based upon EMI ambient levels and other system/program considerations. In instances where the site in question has been determined, the site survey provides information on the levels of ambient EMI signals as a function of frequency. These can be compared to equipment susceptibilities using an amplitude and frequency culling procedure to determine the required EMC hardening for equipment design or retrofit. A major consideration for a site survey is the range of frequencies for which the measured data are required because frequency determines equipment requirements, time required to conduct the survey, time and complexity for data analysis and display, and report preparation if required.

System-Level EMC Checkout and Product Quality Assurance

The increasing concern for the control of EMI emissions has resulted in the establishment of EMI specifications on the international and national levels. International limits are specified for voluntary, cooperative compliance by the CISPR member countries, with the degree of enforcement varying with each member nation. In West Germany, the VDE has established VDE 0871 and VDE 0875 limits for intentional and unintentional emitters, respectively. The VDE limits basically use the CISPR recommendations with added requirements peculiar to West Germany, which strongly enforces compliance when compared to most of the other nations. In the United States, the FCC *Rule and Regulations* specify the limits, with Part 15J specifically dedicated to digital devices, heretofore a significant source of interfering emissions. Other countries also have emission limits that are enforced to varying degrees. Similarly, any equipment sold to the U.S. military must comply with appropriate military specifications, such as MIL-STD-461 and MIL-E-6051.

All equipment to be marketed in countries with enforced emission limits or to the military must be tested to verify compliance. The testing requirements vary from subsystem ("box-level") testing, such as that required for emission and susceptibility levels by MIL-STD-461, to full system EMC checkout, such as that required by MIL-E-6051. Consumer product testing usually involves testing an end product to determine emission levels without regard to susceptibility. Details of instrumentation and measurement procedures required to determine product EMC compliance are discussed in later chapters of this *Handbook* and in references 4 and 5.

REFERENCES

1. *Reference Data for Radio Engineers*, Sixth Edition, Third Printing. New York: Howard W. Sams and Company, 1979.
2. Duff, William G., and White, Donald R. J. *EMI Prediction and Analysis Techniques*, A Handbook Series on Electromagnetic Interference and Compatibility, Volume 5. Germantown, MD: Don White Consultants, Inc., 1972.
3. White, Donald R. J. *EMI Control Methods and Techniques*, A Handbook Series on Electromagnetic Interference and Compatibility, Volume 3. Germantown, MD: Don White Consultants, Inc., 1973.
4. White, Donald R. J. *EMI Test Methods and Procedures*, A Handbook Series on Electromagnetic Interference and Compatibility, Volume 2. Gainsesville, VA: Don White Consultants, Inc., 1974.
5. White, Donald R. J. *EMI Test Instrumentation and Systems*, A Handbook Series on Electromagnetic Interference and Compatibility, Volume 4, Second Printing. Germantown, MD: 1976.
6. Hill, James S., and White, Donald R. J. *EMI Specifications, Standards, and Regulations*, A Handbook Series on Electromagnetic Interference and Compatibility, Volume 6. Germantown, MD: Don White Consultants, Inc., 1975.

Chapter 2
Sources of
Electromagnetic Interference

CLASSIFICATION OF EMI AND SOURCES

Any device or apparatus that transmits, distributes, processes, or otherwise utilizes any form of electrical energy can be a source of EMI if any aspect of its operation generates conducted and/or radiated electromagnetic signals that can cause a degradation of performance of any other equipment or system that shares the same environment. Signals that are traceable to an electrical origin are essentially found everywhere: on the earth's surface, in underground locations such as mines and tunnels, under water, in the atmosphere, and in outer space (1, 2).

Figure 2-1 represents a classification of EMI sources into several categories. The first major breakout classifies sources as either of natural origin or man-made. The variation of EMI signal amplitude as a function of frequency is the next criterion, which specifies sources as being either broadband or narrowband. Broadband sources can be further identified as either coherent or incoherent. Sources can also be categorized into conducted or radiated signals. The final criterion is whether an EMI signal is generated intentionally or unintentionally (incidentally generated). Also shown is a category of restricted radiators including industrial, scientific, and medical (ISM) equipment. Brief descriptions of terms illustrated in Fig. 2-1 follow.

Natural EMI sources: Sources of electromagnetic noise associated with natural phenomena including atmospheric charge/discharge phenomena, such as lightning and precipitation static; and extraterrestrial sources including radiation from the sun and galactic sources such as radio stars, galaxies, and other cosmic sources (1, 3–5).

Man-made EMI sources: Sources of electromagnetic noise associated with man-made electrical devices such as power lines, auto ignition, fluorescent lights, and so on.

Broadband EMI: electromagnetic conducted and radiated signals whose amplitude variation as a function of frequency (spectral density function)

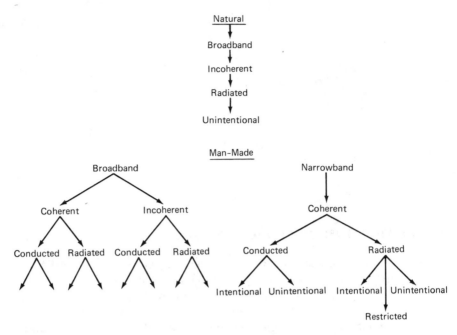

Fig. 2-1. Classification of EMI sources.

extends over a frequency range that is greater than the bandwidth of a specified receptor. Within a broadband noise environment, the response of a receptor is proportional to its frequency bandwidth for coherent noise signals, and proportional to the square root of its frequency bandwidth for incoherent noise. The spectral density amplitude function of a broadband signal is expressed in terms of a specified bandwidth in addition to being a function of frequency. Broadband noise may be defined mathematically and otherwise as a function whose spectral density is a continuous function of frequency over the frequency range of interest.

Narrowband EMI: electromagnetic conducted and radiated signals whose amplitude variation as a function of frequency (spectral density function) extends over a frequency range that is narrower than the bandwidth of a specified receptor. Within a narrowband noise environment, the response of a receptor is independent of its bandwidth once the latter is greater than the frequency range of the noise signal. Narrowband noise may be defined mathematically and otherwise as a function whose spectral density consists of a line spectrum as a function of frequency over the frequency range of interest.

Coherent and incoherent broadband signals: A signal is said to be coherent when neighboring components of the signal (within the frequency domain) have a well-defined relationship of amplitude, frequency, and phase. Conversely, a signal is said to be incoherent when neighboring components of the signal (within the frequency domain) are random or pseudo-random (bandwidth limited) in phase or/and amplitude.

Conducted EMI: Conducted noise signals are those transmitted via electrical conduction paths, such as wires, ground planes, and so forth. Quantities are expressed in terms of voltages and currents.

Radiated EMI: Radiated signals are identified as electric and magnetic fields transmitted through space from source to receptor. A further differentiation is made between the near and far fields. Whenever the source-to-observer distance is less than $(\lambda/2\pi)$, the situation is considered to be that of a near field; and for a distance greater than $(\lambda/2\pi)$, a far field situation is considered to exist. The parameter λ is the free space wavelength determined in accordance with equations (1-2) through (1-4).

Intentional radiating emitters: a category of emitters whose primary function depends upon radiated emissions. Communication-electronic (C-E) licensed systems, including communications, navigation, and radar systems, fall within this category.

Restricted radiating devices: devices that intentionally use electromagnetic radiation for other than communication or data transfer purposes, such as garage door operating systems and radio-controlled models; and low-power communication devices that do not require licensing, such as wireless microphones. Also included are devices that utilize electromagnetic waves for industrial, scientific, medical (ISM), and other purposes that include the use of electromagnetic radiation for purposes other than communications (3).

Unintentional (incidental) radiating devices: devices that radiate radio frequency signals but whose primary function is not to generate or utilize these signals. An extensive class of emitters is included within this category.

Representative sources of electromagnetic interference for various natural and man-made categories are illustrated in Fig. 2-2. Table 2-1 provides a list of representative EMI sources within the categories presented in Fig. 2-2. This table is not all-inclusive, and additional sources can be identified within references 1–4.

DATA REPORTING FORMATS

Engineering design and analysis requires that quantitative data be available to determine useful circuit and other parameters. Analysis of the effects of the noise sources discussed and illustrated in Fig. 2-2 and Table 2-1 requires that the noise levels produced be quantified in useful formats, often referred to as spectrum signatures. A spectrum signature is the identification, in terms of an amplitude versus frequency, of the emissions from one or more electromagnetic sources. A spectrum signature, or frequency–amplitude profile, is also dependent upon other variables such as the source-to-observer distance (near and far fields), directional properties and polarization of the emissions,

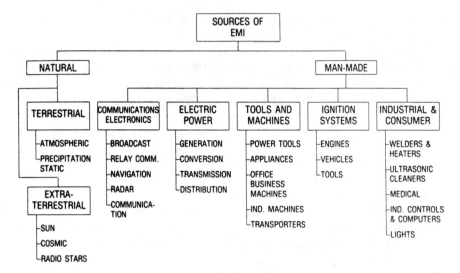

Fig. 2-2. Sources of electromagnetic interference (*Courtesy of Interference Control Technologies*)

and the effects of terrain and other obstacles along the propagation paths. Finally, a system of units must be standardized to express the desired quantities.

The Decibel

The decibel (dB) is the unit most often applied to EMC engineering. Basically, the unit is defined as a common logarithm (base 10) function:

$$dB = 10 \log (P_1/P_2) \tag{2-1}$$

where P_1 and P_2 are levels of power. In equation (2-1), P_2 is considered to be the reference level. When P_2 is expressed in watts, then the decibel abbreviation becomes dBW; if P_2 is in milliwatts, then the expression becomes dBm; and so on. The relationship between power and voltage is $P = V^2/R$, where R is the resistance between the terminals where V is measured. Substituting the voltage relationship for power into equation (2-1):

$$dB = 10 \log \frac{V_1^2/R_1}{V_2^2/R_2} \tag{2-2}$$

The relationship in equation (2-2) becomes:

$$dB = 10 \log \frac{V_1^2}{V_2^2} + 10 \log \frac{R_2}{R_1} \tag{2-3}$$

If $R_1 - R_2$, equation (2-3) may be expressed as follows:

$$\text{dBV} = 20 \log \frac{V_1}{V_2} \tag{2-4}$$

where V_2 is the reference of the ratio V_1/V_2 and equal to 1 volt. If V_2 is 1 microvolt, then equation (2-4) is expressed in "dB above 1 microvolt," or

Table 2-1. Examples of sources of electromagnetic interference.

(Courtesy of Interference Control Technologies.)

NATURAL SOURCES

1.1 Terrestrial Sources:
  a. Atmospherics (thunderstorms around the world)
  b. Lightning discharges (local storms)
  c. Precipitation Static
  d. Whistlers

1.2 Extra-Terrestrial Sources:
  a. Cosmic Noise
  b. Radio Stars
  c. Sun:
      Disturbed
      Quiet

MAN-MADE SOURCES

2.1 Electric-Power:
  a. Conversion (Step Up/Down)
      Faulty/Dirty Insulators
      Faulty Transformers
  b. Distribution
      Faulty/Dirty Insulators
      Faulty Transformers
      Faulty Wiring
      Pick-up and Re-radiation
      Poor Grounding
  c. Generators
  d. Transmission Lines
      Faulty/Dirty Insulators
      Pick-up and Re-radiation

2.2 Communications-Electronics (C-E):
  a. Broadcast
      MF Amplitude Modulation
      VHF/FM

VHF/FM
VHF/UHF TV
  b. Communications (non relay)
      Aeronautical Mobile
      Amateurs (Hams)
      Citizens radio
      Facsimile
      HF Telegraphy
      HF Telephony
      Land Mobile
      Maritime mobile
      Radio-Control Devices
      Telemetry
      Telephone Circuits
      Wireless Microphone
  c. Navigation (non radar)
      Aircraft Beacons
      Instrument Landing Systems
      Loran
      Marker Beacons
      Omega
      VOR/TACAN/VORTAC
  d. Radar
      Air Search
      Air Surface Detection
      Air Traffic Control
      Harbor
      Mapping
      Police Speed Monitor
      Surface Search
      Tracking/Fire Control
      Weather
  e. Relay Communications
      Ionospheric Scatter
      Microwave Relay Links
      Satellite Relay
      Tropospheric Scatter

Table 2.1 - (Continued)

2.3 Tools and Machines
   a. Appliances
       Air Conditioners
       Blenders
       Deep Freezes
       Fans
       Lawn Mowers, electric
       Mix Masters
       Ovens, electric
       Ovens, Microwave
       Refrigerators
       Sewing Machines
       Vacuum Cleaners
       Water Pumps
   b. Industrial Machines
       Electric Cranes
       Fork-lift Trucks
       Lathes
       Milling Machines
       Printing Presses
       Punch Presses
       Rotary Punches
       Screw Machines
   c. Office/Business Machines
       Adding Machines
       Calculators
       Cash Registers
       Electric Typewriters
       Reproduction Equipment
   d. Power Tools
       Band Saws
       Drill Press
       Electric Drills
       Electric Hand Saws
       Electric Grinders
       Electric Sanders
       Hobby Tools
       Routers/Joiners
       Table Saws
   e. Transporters
       Converyor Belts
       Elevators
       Escalators
       Moving Sidewalks

2.4 Ignition Systems
   a. Engines
   b. Tools
       Auxiliary Generators

       Lawn Mowers
       Portable Saws
   c. Vehicles
       Aircraft
       Automobiles
       Farm Machinery
       In-board Motors
       Mini-bikes
       Motorcycles
       Out-board Motors
       Tanks
       Tractors
       Trucks
2.5 Industrial & Consumer
       (non motor/engines)
   a. Heaters and Gluers
       Diaelectric Heaters
       Plastic Pre-Heaters
       Wood Gluers
   b. Industrial Controls &
       Computers
       Card Punches
       Card Readers
       Computers
       Machine Controllers
       Peripheral Equipment
       Process Controllers
       Silicon-Controlled
         Rectifiers
       Teletypewriters
   c. Lights
       Faulty Incadescent
       Fluorescent Lamps
       Light Dimmers
       Neon Lights
       R-F Excited, Gas-Display
         Signs
       R-F Excited, Gas Laser
   d. Medical Equipment
       Defibrillators
       Diathermy
       X-Ray Machines
   e. Ultra-sonic Cleaners
   f. Welders & Heaters
       Arc Welders
       Heli-Arc Welders
       Induction Heaters
       Plastic Welders
       R-F Stabilized Welders

Table 2-2. Conversion of dB to voltage, current, and power ratio.

(Courtesy of Interference Control Technologies.)

dB	Voltage or Current Ratio	Power Ratio	dB	Voltage or Current Ratio	Power Ratio
0	1.00	1.00	0	1.00	1.00
.5	1.06	1.12	-.5	.94	.89
1	1.12	1.26	-1	.89	.79
2	1.26	1.58	-2	.79	.63
3	1.41	2.00	-3	.71	.50
4	1.58	2.51	-4	.63	.40
5	1.78	3.16	-5	.56	.32
6	2.00	4.00	-6	.50	.25
7	2.24	5.01	-7	.45	.20
8	2.51	6.31	-8	.40	.16
9	2.82	7.94	-9	.35	.13
10	3.16	10.0	-10	.32	.10
11	3.55	12.6	-11	.28	.079
12	4.00	16.0	-12	.25	.063
13	4.47	20.0	-13	.22	.050
14	5.01	25.1	-14	.20	.040
15	5.62	31.6	-15	.18	.032
16	6.31	39.8	-16	.16	.025
17	7.08	50.1	-17	.14	.020
18	7.94	63.1	-18	.13	.016
19	8.91	79.4	-19	.11	.013
20	10.0	100	-20	.10	.010
25	17.8	316	-25	.056	.0032
30	31.6	1000	-30	.032	.0010
35	56.2	3.16×10^3	-35	.018	.00032
40	100	10^4	-40	.010	.00010
45	178	3.16×10^4	-45	.0056	3.16×10^{-5}
50	316	10^5	-50	.0032	10^{-5}
55	562	3.16×10^5	-55	.0018	3.16×10^{-6}
60	1000	10^6	-60	10^{-3}	10^{-6}
65	1.78×10^3	3.16×10^6	-65	5.62×10^{-4}	3.16×10^{-7}
70	3.16×10^3	10^7	-70	3.16×10^{-4}	10^{-7}
75	5.62×10^3	3.16×10^7	-75	1.78×10^{-4}	3.16×10^{-8}
80	10^4	10^8	-80	10^{-4}	10^{-8}
85	1.78×10^4	3.16×10^8	-85	5.62×10^{-5}	3.16×10^{-9}
90	3.16×10^4	10^9	-90	3.16×10^{-5}	10^{-9}
95	5.62×10^4	3.16×10^9	-95	1.78×10^{-5}	3.16×10^{-10}
100	10^5	10^{10}	-100	10^{-5}	10^{-10}
105	1.78×10^5	3.16×10^{10}	-105	5.62×10^{-6}	3.16×10^{-11}
110	3.16×10^5	10^{11}	-110	3.16×10^{-6}	10^{-11}
115	5.62×10^5	3.16×10^{11}	-115	1.78×10^{-6}	3.16×10^{-12}
120	10^6	10^{12}	-120	10^{-6}	10^{-12}
130	3.16×10^6	10^{13}	-130	3.16×10^{-7}	10^{-13}
140	10^7	10^{14}	-140	10^{-7}	10^{-14}
150	3.16×10^7	10^{15}	-150	10^{-8}	10^{-15}
160	10^8	10^{16}	-160	10^{-8}	10^{-16}
170	3.16×10^8	10^{17}	-170	3.16×10^{-9}	10^{-17}
180	10^9	10^{18}	-180	10^{-9}	10^{-18}
190	3.16×10^9	10^{19}	-190	3.16×10^{-10}	10^{-19}
200	10^{10}	10^{20}	-200	10^{-10}	10^{-20}

dBμV, etc. Similar expressions may be derived for units of current, to obtain:

$$dBA = 20 \log \frac{I_1}{I_2} \qquad (2\text{-}5)$$

where dBA stands for "dB above 1 ampere," with the reference I_2 equal to 1 ampere. If I_2 is 1 microampere, then equation (2-5) becomes "dB above 1 microampere," or dBμA, and so on.

The above quantities were derived for and are applied to narrowband conducted quantities expressed in units of voltage, current, and power. Similar decibel expressions are derived and applied to narrowband radiated quantities, specifically radiated power density, in watts per square meter (W/m²) or milliwatts per square meter (mW/m²); electric field strength in volts/meter (V/m) or microvolts per meter (μV/m); and magnetic field strength in amperes per meter (A/m) or microamperes per meter (μA/m). In units of decibels, radiated quantities are expressed in typical units such as dBV/m, dBμV/m, dBA/m, dBμA/m, dBmW/m², and so on depending upon the reference unit of the field ratios.

Broadband quantities must take bandwidth into account and incorporate it into the decibel expression. For broadband quantities, the signal strength is a function of a specified unit bandwidth with typical units such as volts per kilohertz (V/kHz) and amperes per megahertz (A/MHz) for broadband conducted quantities, and volts per meter per megahertz (V/m/MHz) and amperes per meter per kilohertz (A/m/kHz) for radiated field quantities. Tables 2-2 and 2-3 provide a comparison of decibel-to-ratio conversions, and vice versa, that can be applied to appropriate quantities.

Table 2-4 provides broadband and narrowband equivalent terms for given receptor (receiver) impulse bandwidth, which is equal to the area of the voltage response selectivity curve of the receiver as a function of frequency divided

Table 2-3. Conversion of voltage, current, and power ratio to dB. (Courtesy of Interference Control Technologies.)

Voltage or current ratio	Power ratio	dB	Voltage or current ratio	Power ratio	dB
1.00	1.00	.0	1.00	1.00	0
1.02	1.04	.2	.98	.96	−.2
1.04	1.08	.3	.96	.92	−.4
1.07	1.14	.6	.93	.86	−.6
1.10	1.21	.8	.90	.81	−.9
1.15	1.32	1.2	.87	.76	−1.2
1.20	1.44	1.6	.85	.72	−1.4
1.30	1.69	2.3	.80	.64	−1.9
1.40	1.96	2.9	.75	.56	−2.5

Table 2-3. Conversion of voltage, current, and power ratio to dB. (Courtesy of Interference Control Technologies.)

Voltage or current ratio	Power ratio	dB	Voltage or current ratio	Power ratio	dB
1.50	2.25	3.5	.70	.49	−3.1
1.60	2.56	4.1	.65	.42	−3.7
1.70	2.89	4.6	.60	.36	−4.4
1.80	3.24	5.1	.55	.30	−5.2
1.90	3.61	5.6	.50	.25	−6.0
2.00	4.00	6.0	.48	.23	−6.4
2.20	4.84	6.8	.45	.20	−6.9
2.40	5.76	7.6	.43	.18	−7.3
2.50	6.25	8.0	.40	.16	−8.0
2.60	6.76	8.3	.38	.14	−8.4
2.80	7.84	8.9	.35	.12	−9.1
3.00	9.00	9.5	.33	.11	−9.6
3.25	10.6	10.2	.30	.090	−10.5
3.50	12.3	10.9	.28	.078	−11.1
3.75	14.1	11.5	.26	.068	−11.7
4.00	16.0	12.0	.24	.058	−12.4
4.50	20.3	13.1	.22	.048	−13.2
5.00	25.0	14.0	.20	.040	−14.0
5.50	30.0	14.8	.18	.032	−14.9
6.00	36.0	15.6	.17	.029	−15.4
6.50	42.3	16.3	.15	.023	−16.5
7.00	49.0	16.9	.14	.020	−17.1
7.50	56.3	17.5	.13	.017	−17.7
8.00	64.0	18.1	.12	.014	−18.4
9.00	81.0	19.1	.11	.012	−19.2
10	100	20.0	.10	.010	−20.0
30	900	29.5	.03	9×10^{-4}	−30.5
100	10^4	40	.01	10^4	−40
300	9×10^4	49.5	.003	9×10^{-6}	−50.5
1000	10^6	60	.001	10^{-6}	−60
3×10^3	9×10^6	69.5	3×10^{-4}	9×10^{-8}	−70.5
10^4	10^8	80	10^{-4}	10^{-8}	−80
3×10^4	9×10^8	89.5	3×10^{-5}	9×10^{-10}	−90.5
10^6	10^{10}	100	10^{-5}	10^{-10}	−100
10^6	10^{12}	120	10^{-6}	10^{-12}	−120
10^7	10^{14}	140	10^{-7}	10^{-14}	−140
10^8	10^{16}	160	10^{-8}	10^{-16}	−160
10^9	10^{18}	180	10^{-9}	10^{-18}	−180
10^{10}	10^{20}	200	10^{-10}	10^{-20}	−200

Table 2-4. RFI/EMI broad- and narrowband equivalent terms.

(Courtesy of Interference Control Technologies.)

dBm	dBµV	Volts	Broadband Equivalent Signal Level* in dBµV/MHz for Indicated Receiver Impulse Bandwidth*						
			10 Hz	100 Hz	1 kHz	10 kHz	100 kHz	1 MHz	10 MHz
+53	160	100	260	240	220	200	180	160	140
+48	155	56	255	235	215	195	175	155	135
+43	150	32	250	230	210	190	170	150	130
+38	145	18	245	225	205	185	165	145	125
+33	140	10	240	220	200	180	160	140	120
+28	135	5.6	235	215	195	175	155	135	115
+23	130	3.2	230	210	190	170	150	130	110
+18	125	1.8	225	205	185	165	145	125	105
+13	120	1.0	220	200	180	160	140	120	100
+8	115	.56	215	195	175	155	135	115	95
+3	110	.32	210	190	170	150	130	110	90
-2	105	.18	205	185	165	145	125	105	85
-7	100	.10	200	180	160	140	120	100	80
-8	99	89 mV	199	179	159	139	119	99	79
-9	98	79 mV	198	178	158	138	118	98	78
-10	97	71 mV	197	177	157	137	117	97	77
-11	96	63 mV	196	176	156	136	116	96	76
-12	95	56 mV	195	175	155	135	115	95	75
-13	94	50 mV	194	174	154	134	114	94	74
-14	93	45 mV	193	173	153	133	113	93	73
-15	92	40 mV	192	172	152	132	112	92	72
-16	91	35 mV	191	171	151	131	111	91	71
-17	90	32 mV	190	170	150	130	110	90	70
-18	89	28 mV	189	169	149	129	109	89	69
-19	88	25 mV	188	168	148	128	108	88	68
-20	87	22 mV	187	167	147	127	107	87	67
-21	86	20 mV	186	166	146	126	106	86	66
-22	85	18 mV	185	165	145	125	105	85	65
-23	84	16 mV	184	164	144	124	104	84	64
-24	83	14 mV	183	163	143	123	103	83	63
-25	82	13 mV	182	162	142	122	102	82	62
-26	81	11 mV	181	161	141	121	101	81	61
-27	80	10 mV	180	160	140	120	100	80	60
-28	79	8.9 mV	179	159	139	119	99	79	59
-29	78	7.9 mV	178	158	138	118	98	78	58

							mV		dB
57	77	97	117	137	157	177	7.1 mV	77	-30
56	76	96	116	136	156	176	6.3 mV	76	-31
55	75	95	115	135	155	175	5.6 mV	75	-32
54	74	94	114	134	154	174	5.0 mV	74	-33
53	73	93	113	133	153	173	4.5 mV	73	-34
52	72	92	112	132	152	172	4.0 mV	72	-35
51	71	91	111	131	151	171	3.5 mV	71	-36
50	70	90	110	130	150	170	3.2 mV	70	-37
49	69	89	109	129	149	169	2.8 mV	69	-38
48	68	88	108	128	148	168	2.5 mV	68	-39
47	67	87	107	127	147	167	2.2 mV	67	-40
46	66	86	106	126	146	166	2.0 mV	66	-41
45	65	85	105	125	145	165	1.8 mV	65	-42
44	64	84	104	124	144	164	1.6 mV	64	-43
43	63	83	103	123	143	163	1.4 mV	63	-44
42	62	82	102	122	142	162	1.3 mV	62	-45
41	61	81	101	121	141	161	1.1 mV	61	-46
40	60	80	100	120	140	160	1.0 mV	60	-47
39	59	79	99	119	139	159	.89 mV	59	-48
38	58	78	98	118	138	158	.79 mV	58	-49
37	57	77	97	117	137	157	.71 mV	57	-50
36	56	76	96	116	136	156	.63 mV	56	-51
35	55	75	95	115	135	155	.56 mV	55	-52
34	54	74	94	114	134	154	.50 mV	54	-53
33	53	73	93	113	133	153	.45 mV	53	-54
32	52	72	92	112	132	152	.40 mV	52	-55
31	51	71	91	111	131	151	.35 mV	51	-56
30	50	70	90	110	130	150	.32 mV	50	-57
29	49	69	89	109	129	149	.28 mV	49	-58
28	48	68	88	108	128	148	.25 mV	48	-59
27	47	67	87	107	127	147	.22 mV	47	-60
26	46	66	86	106	126	146	.20 mV	46	-61
25	45	65	85	105	125	145	.18 mV	45	-62
24	44	64	84	104	124	144	.16 mV	44	-63
23	43	63	83	103	123	143	.14 mV	43	-64
22	42	62	82	102	122	142	.13 mV	42	-65
21	41	61	81	101	121	141	.11 mV	41	-66
20	40	60	80	100	120	140	.10 mV	40	-67

(CONTINUED OVER)

Broadband Equivalent Signal Level* in dBμV/MHz for Indicated Receiver Impulse Bandwidth*

dBm	dBμV	Microvolts	10 Hz	100 Hz	1 kHz	10 kHz	100 kHz	1 MHz	10 MHz
-68	39	89	139	119	99	79	59	39	19
-69	38	79	138	118	98	78	58	38	18
-70	37	71	137	117	97	77	57	37	17
-71	36	63	136	116	96	76	56	36	16
-72	35	56	135	115	95	75	55	35	15
-73	34	50	134	114	94	74	54	34	14
-74	33	45	133	113	93	73	53	33	13
-75	32	40	132	112	92	72	52	32	12
-76	31	35	131	111	91	71	51	31	11
-77	30	32	130	110	90	70	50	30	10
-78	29	28	129	109	88	69	49	29	9
-79	28	25	128	108	87	68	48	28	8
-80	27	22	127	107	86	67	47	27	7
-81	26	20	126	106	86	66	46	26	6
-82	25	18	125	105	85	65	45	25	5
-83	24	16	124	104	84	64	44	24	4
-84	23	14	123	103	83	63	43	23	3
-85	22	13	122	102	82	62	42	22	2
-86	21	11	121	101	81	61	41	21	1
-87	20	10	120	100	80	60	40	20	0
-88	19	8.9	119	99	79	59	39	19	-1
-89	18	7.9	118	98	78	58	38	18	-2
-90	17	7.1	117	97	77	57	37	17	-3
-91	16	6.3	116	96	76	56	36	16	-4
-92	15	5.6	115	95	75	55	35	15	-5
-93	14	5.0	114	94	74	54	34	14	-6
-94	13	4.5	113	93	73	53	33	13	-7
-95	12	4.0	112	92	72	52	32	12	-8
-96	11	3.5	111	91	71	51	31	11	-9
-97	10	3.2	110	90	70	50	30	10	-10
-98	9	2.8	109	89	69	49	29	9	-11
-99	8	2.5	108	88	68	48	28	8	-12
-100	7	2.2	107	87	67	47	27	7	-13
-101	6	2.0	106	86	66	46	26	6	-14
-102	5	1.8	105	85	65	45	25	5	-15
-103	4	1.6	104	84	64	44	24	4	-16
-104	3	1.4	103	83	63	43	23	3	-17
-105	2	1.3	102	82	62	42	22	2	-18

Level	0	1.0	100	80	60	40	20	0	-20
-107	0	1.0	99	79	60	40	20	0	-19
-108	-1	.89	98	78	59	39	18	-1	-20
-109	-2	.79	97	77	58	38	17	-2	-21
-110	-3	.71	96	76	57	37	16	-3	-22
-111	-4	.63	95	75	56	36	15	-4	-23
-112	-5	.56	95	75	55	35	15	-5	-24
-113	-6	.50	94	74	54	34	14	-6	-25
-114	-7	.45	93	73	53	33	13	-7	-26
-115	-8	.40	92	72	52	32	12	-8	-27
-116	-9	.35	91	71	51	31	11	-9	-28
-117	-10	.32	90	70	50	30	10	-10	-29
-118	-11	.28	89	69	49	29	9	-11	-30
-119	-12	.25	88	68	48	28	8	-12	-31
-120	-13	.22	87	67	47	27	7	-13	-32
-121	-14	.20	86	66	46	26	6	-14	-33
-122	-15	.18	85	65	45	25	5	-15	-34
-123	-16	.16	84	64	44	24	4	-16	-35
-124	-17	.14	83	63	43	23	3	-17	-36
-125	-18	.13	82	62	42	22	2	-18	-37
-126	-19	.11	81	61	41	21	1	-19	-38
-127	-20	.10	80	60	40	20	0	-20	-39
-132	-25	.056	79	59	39	19	-1	-21	-40
-137	-30	.032	78	58	38	18	-2	-22	-41
-142	-35	.018	77	57	37	17	-3	-23	-42
-147	-40	.010	76	56	36	16	-4	-24	-43
-152	-45	.006	75	55	35	15	-5	-25	-44

*To obtain broadband level corresponding to other impulse bandwidths, choose the next lower bandwidth and subtract the following dB therefrom:

Bandwidth Multiplier	Subtract dB
1.1	1
1.3	2
1.4	3
1.6	4
1.8	5
2.0	6

Bandwidth Multiplier	Subtract dB
2.2	7
2.5	8
2.8	9
3.2	10
3.5	11
4.0	12

Bandwidth Multiplier	Subtract dB
4.5	13
5.0	14
5.6	15
6.3	16
7.1	17
7.9	18
8.9	19

EXAMPLE: Determine equivalent broadband signal level corresponding to a 10 mV signal and a bandwidth of 20 kHz. From the chart, it is seen that a 10 mV (80 dBµV) signal and a 10 kHz bandwidth yields an equivalent broadband level of 120 dBµV/MHz. Since 20 kHz is 2 times 10 kHz, the value to be subtracted from the above table is 6 dB. Thus, the answer is 120 dBµV/MHz - 6 dB or 114 dBµV/MHz.

Table 2-5. Electric field strength to magnetic fields to power densities.
(Courtesy of Interference Control Technologies.)

dBµV/m	V/m	dBµA/m	A/m	dBpT	pT & mT	nT & T	Gauss	Tesla	dBW/m²	dBm/m²	mW/cm²
280	10^8	228.5	2.65×10^5	230.5	3.33×10^{11}	3.33×10^8	3.33×10^3	.333	134.3	164.3	2.67×10^{12}
260	10^7	208.5	2.65×10^4	210.5	3.33×10^{10}	3.33×10^7	.333	.033	114.3	144.3	2.67×10^{10}
240	10^6	188.5	2.65×10^3	190.5	3.33×10^9	3.33×10^6	33.3	.003	94.3	124.3	2.67×10^8
220	10^5	168.5	265	170.5	3.33×10^8	3.33×10^5	3.33	3.33×10^{-4}	74.3	104.3	2.67×10^6
200	10^4	148.5	26.5	150.5	3.33×10^7	3.33×10^4	.333	3.33×10^{-5}	54.3	84.3	2.67×10^4
180	10^3	128.5	2.65	130.5	3.33×10^6	3.33×10^3	.0333	3.33×10^{-6}	34.3	64.3	267
160	100	108.5	.265	110.5	3.33×10^5	.333	3.33×10^{-3}	3.33×10^{-7}	14.3	44.3	2.67
140	10	88.5	.0265	90.5	3.33×10^4	33.3	3.33×10^{-4}	3.33×10^{-8}	−5.7	24.3	.0267
120	1	68.5	2.65×10^{-3}	70.5	3.33×10^3	3.33	3.33×10^{-5}	3.33×10^{-9}	−25.7	4.3	2.67×10^{-4}
100	.10	48.5	2.65×10^{-4}	50.5	.333	.333	3.33×10^{-6}	3.33×10^{-10}	−45.7	−15.7	2.67×10^{-6}
80	.01	28.5	2.65×10^{-5}	30.5	33.3	.0333	3.33×10^{-7}	3.33×10^{-11}	−65.7	−35.7	2.67×10^{-8}
60	10^{-3}	8.5	2.65×10^{-6}	10.5	3.33	3.33×10^{-3}	3.33×10^{-8}	3.33×10^{-12}	−85.7	−55.7	2.67×10^{-10}
40	10^{-4}	−11.5	2.65×10^{-7}	−9.5	.333	3.33×10^{-4}	3.33×10^{-9}	3.33×10^{-13}	−105.7	−75.7	2.67×10^{-12}
20	10^{-5}	−31.5	2.65×10^{-8}	−29.5	.0333	3.33×10^{-5}	3.33×10^{-10}	3.33×10^{-14}	−125.7	−95.7	2.67×10^{-14}
19	8.91×10^{-6}	−32.5	2.36×10^{-8}	−30.5	.0297	2.97×10^{-5}	2.97×10^{-14}	2.97×10^{-10}	−126.7	−96.7	2.12×10^{-14}
18	7.94×10^{-6}	−33.5	2.10×10^{-8}	−31.5	.0265	2.65×10^{-5}	2.65×10^{-10}	2.65×10^{-14}	−127.7	−97.7	1.68×10^{-14}
17	7.08×10^{-6}	−34.5	1.87×10^{-8}	−32.5	.0236	2.36×10^{-5}	2.36×10^{-10}	2.36×10^{-14}	−128.7	−98.7	1.34×10^{-14}

16	6.31×10^{-6}	-35.5	1.67×10^{-8}	-33.5	$.0210$	2.10×10^{-5}	2.10×10^{-10}	2.10×10^{-14}	-129.7	-99.7	1.06×10^{-14}
15	5.62×10^{-6}	-36.5	1.49×10^{-8}	-34.5	$.0187$	1.87×10^{-5}	1.87×10^{-10}	1.87×10^{-14}	-130.7	-100.7	8.43×10^{-15}
14	5.01×10^{-6}	-37.5	1.33×10^{-8}	-35.5	$.0167$	1.67×10^{-5}	1.67×10^{-10}	1.67×10^{-14}	-131.7	-101.7	6.70×10^{-15}
13	4.47×10^{-6}	-38.5	1.18×10^{-8}	-36.5	$.0149$	1.49×10^{-5}	1.49×10^{-10}	1.49×10^{-14}	-132.7	-102.7	5.32×10^{-15}
12	3.98×10^{-6}	-39.5	1.05×10^{-8}	-37.5	$.0133$	1.33×10^{-5}	1.33×10^{-10}	1.33×10^{-14}	-133.7	-103.7	4.23×10^{-15}
11	3.55×10^{-6}	-40.5	9.40×10^{-9}	-38.5	$.0118$	1.18×10^{-5}	1.18×10^{-10}	1.18×10^{-14}	-134.7	-104.7	3.36×10^{-15}
10	3.16×10^{-6}	-41.5	8.38×10^{-9}	-39.5	$.0105$	1.05×10^{-5}	1.05×10^{-10}	1.05×10^{-14}	-135.7	-105.7	2.67×10^{-15}
9	2.82×10^{-6}	-42.5	7.46×10^{-9}	-40.5	9.40×10^{-3}	9.40×10^{-6}	9.40×10^{-11}	9.40×10^{-15}	-136.7	-106.7	2.12×10^{-15}
8	2.51×10^{-6}	-43.5	6.65×10^{-9}	-41.5	8.38×10^{-3}	8.38×10^{-6}	8.38×10^{-11}	8.38×10^{-15}	-137.7	-107.7	1.68×10^{-15}
7	2.24×10^{-6}	-44.5	5.93×10^{-9}	-42.5	7.46×10^{-3}	7.46×10^{-6}	7.46×10^{-11}	7.46×10^{-15}	-138.7	-108.7	1.34×10^{-15}
6	2.00×10^{-6}	-45.5	5.28×10^{-9}	-43.5	6.65×10^{-3}	6.65×10^{-6}	6.65×10^{-11}	6.65×10^{-15}	-139.7	-109.7	1.06×10^{-15}
5	1.78×10^{-6}	-46.5	4.71×10^{-9}	-44.5	5.93×10^{-3}	5.93×10^{-6}	5.93×10^{-11}	5.93×10^{-15}	-140.7	-110.7	8.43×10^{-16}
4	1.58×10^{-6}	-47.5	4.20×10^{-9}	-45.5	5.28×10^{-3}	5.28×10^{-6}	5.28×10^{-11}	5.28×10^{-15}	-141.7	-111.7	6.70×10^{-16}
3	1.41×10^{-6}	-48.5	3.74×10^{-9}	-46.5	4.71×10^{-3}	4.71×10^{-6}	4.71×10^{-11}	4.71×10^{-15}	-142.7	-112.7	5.32×10^{-16}
2	1.26×10^{-6}	-49.5	3.33×10^{-9}	-47.5	4.20×10^{-3}	4.20×10^{-6}	4.20×10^{-11}	4.20×10^{-15}	-143.7	-113.7	4.23×10^{-16}
1	1.12×10^{-6}	-50.5	2.97×10^{-9}	-48.5	3.74×10^{-3}	3.74×10^{-6}	3.74×10^{-11}	3.74×10^{-15}	-144.7	-114.7	1.36×10^{-16}
0	10^{-6}	-51.5	2.65×10^{-9}	-49.5	3.33×10^{-3}	3.33×10^{-6}	3.33×10^{-11}	3.33×10^{-15}	-146.7	-116.7	2.67×10^{-16}
-20	10^{-7}	-71.5	2.65×10^{-10}	-69.5	3.33×10^{-4}	3.33×10^{-7}	3.33×10^{-12}	3.33×10^{-16}	-165.7	-135.7	2.67×10^{-18}
-40	10^{-8}	-91.5	2.65×10^{-11}	-89.5	3.33×10^{-5}	3.33×10^{-8}	3.33×10^{-13}	3.33×10^{-17}	-185.7	-155.7	2.67×10^{-20}
-60	10^{-9}	-111.5	2.65×10^{-12}	-109.5	3.33×10^{-6}	3.33×10^{-9}	3.33×10^{-14}	3.33×10^{-18}	-205.7	-175.7	2.67×10^{-22}

Table 2-5 (cont'd). Magnetic flux density to field strength to power densities.

dBpT	pT/mΓ	nT/Γ	Gauss	Tesla	dBµA/m	A/m	dBµV/m	V/m	dBW/m²	dBm/m²	mW/cm²
240	10^{12}	10^9	10^4	1	238	8×10^5	289.5	3×10^8	143.8	173.8	2.4×10^{11}
220	10^{11}	10^8	10^3	.1	218	8×10^4	269.5	3×10^7	123.8	153.8	2.4×10^{11}
200	10^{10}	10^7	10^2	.01	198	8×10^3	249.5	3×10^6	103.8	133.8	2.4×10^9
180	10^9	10^6	10	.001	178	796	229.5	3×10^5	83.8	113.8	2.4×10^7
160	10^8	10^5	1	10^{-4}	158	80	209.5	3×10^4	63.8	93.8	2.4×10^5
140	10^7	10^4	.1	10^{-5}	138	8	189.5	3×10^3	43.8	73.8	2.4×10^1
120	10^6	10^3	.01	10^{-6}	118	.8	169.5	300	23.8	53.8	23.9
100	10^5	10^2	.001	10^{-7}	98	.08	149.5	30	3.8	33.8	.24
80	10^4	10	10^{-4}	10^{-8}	78	.008	129.5	3	−16.2	13.8	.0024
60	10^3	1	10^{-5}	10^{-9}	58	8×10^{-4}	109.5	.3	−36.2	−6.2	2.39×10^{-4}
40	100	.1	10^{-6}	10^{-10}	38	8×10^{-5}	89.5	.03	−56.2	−26.2	2.39×10^{-7}
20	10.000	.01	10^{-7}	10^{-11}	18	7.96×10^{-6}	69.5	.003	−76.2	−46.2	2.39×10^{-9}
19	8.913	8.91×10^{-3}	8.91×10^{-8}	8.91×10^{-12}	17	7.09×10^{-6}	68.5	2.67×10^{-3}	−77.2	−47.2	1.90×10^{-9}
18	7.943	7.94×10^{-3}	7.94×10^{-8}	7.94×10^{-12}	16	6.32×10^{-6}	67.5	2.38×10^{-3}	−78.2	−48.2	1.51×10^{-9}
17	7.079	7.08×10^{-3}	7.08×10^{-8}	7.08×10^{-12}	15	5.63×10^{-6}	66.5	2.12×10^{-3}	−79.2	−49.2	1.20×10^{-9}
16	6.310	6.31×10^{-3}	6.31×10^{-8}	6.31×10^{-12}	14	5.02×10^{-6}	65.5	1.89×10^{-3}	−80.2	−50.2	9.50×10^{-10}
15	5.623	5.62×10^{-3}	5.62×10^{-8}	5.62×10^{-12}	13	4.47×10^{-6}	64.5	1.69×10^{-3}	−81.2	−51.2	7.55×10^{-10}

14	5.012	5.01×10^{-3}	5.01×10^{-8}	5.01×10^{-12}	12	3.99×10^{-6}	63.5	1.50×10^{-3}	-82.2	-52.2	6.00×10^{-10}
13	4.467	4.47×10^{-3}	4.47×10^{-8}	4.47×10^{-12}	11	3.55×10^{-6}	62.5	1.34×10^{-3}	-83.2	-53.2	4.76×10^{-10}
12	3.981	3.98×10^{-3}	3.98×10^{-8}	3.98×10^{-12}	10	3.17×10^{-6}	61.5	1.19×10^{-3}	-84.2	-54.2	3.78×10^{-10}
11	3.548	3.55×10^{-3}	3.55×10^{-8}	3.55×10^{-12}	9	2.82×10^{-6}	60.5	1.06×10^{-3}	-85.2	-55.2	3.01×10^{-10}
10	3.162	3.16×10^{-3}	3.16×10^{-8}	3.16×10^{-12}	8	2.52×10^{-6}	59.5	9.49×10^{-4}	-86.2	-56.2	2.39×10^{-10}
9	2.818	2.82×10^{-3}	2.82×10^{-8}	2.82×10^{-12}	7	2.24×10^{-6}	58.5	8.46×10^{-4}	-87.2	-57.2	1.90×10^{-10}
8	2.512	2.51×10^{-3}	2.51×10^{-8}	2.51×10^{-12}	6	2.00×10^{-6}	57.5	7.54×10^{-4}	-88.2	-58.2	1.51×10^{-10}
7	2.239	2.24×10^{-3}	2.24×10^{-8}	2.24×10^{-12}	5	1.78×10^{-6}	56.5	6.72×10^{-4}	-89.2	-59.2	1.20×10^{-10}
6	1.995	2.00×10^{-3}	2.00×10^{-8}	2.00×10^{-12}	4	1.59×10^{-6}	55.5	5.99×10^{-4}	-90.2	-60.2	9.50×10^{-11}
5	1.778	1.78×10^{-3}	1.78×10^{-8}	1.78×10^{-12}	3	1.42×10^{-6}	54.5	5.33×10^{-4}	-91.2	-61.2	7.55×10^{-11}
4	1.585	1.58×10^{-3}	1.58×10^{-8}	1.58×10^{-12}	2	1.26×10^{-6}	53.5	4.75×10^{-4}	-92.2	-62.2	6.00×10^{-11}
3	1.413	1.41×10^{-3}	1.41×10^{-8}	1.41×10^{-12}	1	1.12×10^{-6}	52.5	4.24×10^{-4}	-93.2	-63.2	4.76×10^{-11}
2	1.259	1.26×10^{-3}	1.26×10^{-8}	1.26×10^{-12}	0	1.00×10^{-6}	51.5	3.78×10^{-4}	-94.2	-64.2	3.78×10^{-11}
1	1.122	1.12×10^{-3}	1.12×10^{-8}	1.12×10^{-12}	-1	8.93×10^{-7}	50.5	3.37×10^{-4}	-95.2	-65.2	3.01×10^{-11}
0	1.000	10^{-3}	10^{-8}	10^{-12}	-2	7.96×10^{-7}	49.5	3.00×10^{-4}	-96.2	-66.2	2.39×10^{-11}
-20	.1	10^{-4}	10^{-9}	10^{-13}	-22	8×10^{-8}	29.5	3×10^{-5}	-116.2	-86.2	2.4×10^{-13}
-40	.01	10^{-5}	10^{-10}	10^{-14}	-42	8×10^{-9}	9.5	3×10^{-6}	-136.2	-106.2	2.4×10^{-15}
-60	.001	10^{-6}	10^{-11}	10^{-15}	-62	8×10^{-10}	-10.5	3×10^{-7}	-156.2	-126.2	2.4×10^{-17}

by the height of the curve. The impulse bandwidth is a measure of the receptor/receiver response to a broadband signal. An illustrative example is provided on the use of Table 2-4.

Table 2-5 provides factors for conversion of electric field strength and magnetic field strength to power density.

Emission Spectra

Emission spectra have been measured for many known noise sources, and are available in the form of experimental spectral data and envelope statistics (1–4). In any relationship between parameters, the question arises regarding useful ways to format and present data. This is especially important when empirical relationships are involved that include many variables that are often inseparable in practice. In presenting emission spectra, principal effects are displayed, and the more significant variables are identified. For spectrum–amplitude profiles of emission sources, the problem is compounded by such considerations as near versus far field distances, the distance between emitter and measuring receiver, terrain and other propagation path effects, emitter radiator directivity (or, radiating element "gain"), and polarization.

An emitter spectrum signature is identified in terms of an amplitude versus frequency portrayal of emissions from one or more sources. Amplitude units are determined by whether the emissions are conducted or radiated, narrowband or broadband.

Narrowband Conducted Emission Units. Conducted narrowband signals are measured and expressed in terms of decibel (dB) voltage units: dBV, dBmV, dBμV, corresponding to the measured voltage level referenced to 1 volt, 1 millivolt, or 1 microvolt, respectively; or decibel current units: dBA, dBmA, or dBμA, corresponding to the measured current level referenced to 1 ampere, 1 milliampere, or 1 microampere, respectively. Also, conducted power levels are measured in decibel units referenced to 1 watt (dBW), 1 milliwatt (dBm), 1 microwatt (dBμW), or 1 picowatt (dBpW).

Broadband Conducted Emission Units. Conducted broadband signals are measured and expressed in terms of voltage and current decibel units per specified bandwidth, such as dBV/MHz, dBV/kHz, dBμV/MHz, dBA/Hz, dBμA/MHz, and so on.

Narrowband Radiated Emission Units. Radiated narrowband signals are measured and expressed in decibel electric field units: dBV/m and dBμV/m, corresponding to the measured electric field level referenced to 1 volt/meter and 1 microvolt/meter, respectively; decibel magnetic field intensity units: dBA/m and dBμA/m, corresponding to the measured magnetic field intensity (H) level referenced to 1 ampere/meter and 1 microampere/meter, respectively; decibel magnetic flux density units: dBT, dBpT, and dBG, corresponding to the measured magnetic flux density (B) level referenced to 1 Tesla, 1 picoTesla, and 1 gauss, respectively.

Broadband Radiated Emission Units. Radiated broadband signals are measured and expressed in terms of radiated electric and magnetic field decibel units per specified bandwidth, such as dBV/m/MHz, dBμV/m/MHz; dBA/m/kHz, to cite a few typical examples.

Radiated Emission Power Density. A unit of radiated power frequently encountered is the decibel of broadband power density referenced to 1 milliwatt per square meter per kilohertz bandwidth, abbreviated dBm/m^2/kHz. The capture area of the intercept antenna in square meters and the receiver bandwidth must be known to convert this radiation amplitude unit to a received signal in dBm (decibel power unit referred to 1 milliwatt) (3).

Far-Field Electric Strength and Equivalent Power Density. Table 2-6 provides conversion units for power density and equivalent electric field strength in the far field (essentially plane waves) where the wave impedance is equal to $120\pi = 377$ ohms (Ω). The relationship between the electric field strength and power density is (3):

$$P_D = \frac{E^2}{Z_0} \tag{2-6}$$

where E is the electric field strength in volts/meter, and Z_0 is the free space (plane wave) impedance equal to 377 ohms. The conversion from basic units to decibels is performed by using equation (2-1) for power density, and equation (2-4) for electric field, with appropriate substitutions of power density and electric field reference units.

Amplitude Probability Distributions (APDs). An emission may be formatted as a function of time, or number of discrete events, as it is observed during a given time interval, T (2, 3). The probability that a given level of the function will be exceeded in value during a given time interval is referred to as the amplitude probability distribution of the function, such as a noise emission function. For example (2), consider a noise signal voltage $V(t)$ observed at the output of a receiving antenna circuit, tuned to a carrier frequency f_0, represented as a function of time by the relationship:

$$V(t) = V_0(t) \cos \left[2\pi f_0 t + \phi(t) \right] \tag{2-7}$$

Table 2-6. Field intensity and power density relationships (related by free space impedance = 377 ohms).

(Courtesy of Interference Control Technologies.)

Volts/m	dBμV/m	Watts/m²	dBW/m²	Watts/cm²	dBW/cm²	mW/cm²	dBm/cm²
10,000	200	265,000	+54	27	+14	26,500	+44
7,000	197	130,000	+51	13	+11	13,000	+41
5,000	194	66,300	+48	6.6	+8	6,630	+38
3,000	190	23,900	+44	2.4	+4	2,390	+34
2,000	186	10,600	+40	1.1	0	1,060	+30
1,000	180	2,650	+34	.27	−6	265	+24
700	177	1,300	+31	.13	−9	130	+21
500	174	663	+28	.066	−12	66	+18
300	170	239	+24	.024	−16	24	+14
200	166	106	+20	.011	−20	11	+10
100	160	27	+14	27×10^{-4}	−26	2.7	+4
70	157	13	+11	13×10^{-4}	−29	1.3	+1
50	154	6.6	+8	6.6×10^{-4}	−32	.66	−2
30	150	2.4	+4	2.4×10^{-4}	−36	.24	−6
20	146	1.1	0	1.1×10^{-4}	−40	.11	−10
10	140	.27	−6	27×10^{-5}	−46	.027	−16
7	137	.13	−9	13×10^{-6}	−49	.013	−19
5	134	.066	−12	6.6×10^{-6}	−52	66×10^{-4}	−22
3	130	.024	−16	2.4×10^{-6}	−56	24×10^{-4}	−26
2	126	.011	−20	1.1×10^{-6}	−60	11×10^{-4}	−30
1	120	27×10^{-4}	−26	27×10^{-8}	−66	2.7×10^{-4}	−36
0.7	117	13×10^{-4}	−29	13×10^{-8}	−69	1.3×10^{-4}	−39
0.5	114	6.6×10^{-4}	−32	6.6×10^{-8}	−72	66×10^{-4}	−42
0.3	110	2.4×10^{-4}	−36	2.4×10^{-8}	−76	24×10^{-6}	−46
0.2	106	1.1×10^{-4}	−40	1.1×10^{-8}	−80	11×10^{-6}	−50
0.1	100	27×10^{-6}	−46	27×10^{-10}	−86	2.7×10^{-6}	−56
70×10^{-3}	97	13×10^{-6}	−49	13×10^{-10}	−89	1.3×10^{-6}	−59
50×10^{-3}	94	6.6×10^{-6}	−52	6.6×10^{-10}	−92	66×10^{-8}	−62
30×10^{-3}	90	2.4×10^{-6}	−56	2.4×10^{-10}	−96	24×10^{-8}	−66
20×10^{-3}	86	1.1×10^{-6}	−60	1.1×10^{-10}	−100	11×10^{-8}	−70
10×10^{-3}	80	27×10^{-8}	−66	27×10^{-12}	−106	2.7×10^{-8}	−76
7×10^{-3}	77	13×10^{-8}	−69	13×10^{-12}	−109	1.3×10^{-8}	−79
5×10^{-3}	74	6.6×10^{-8}	−72	6.6×10^{-12}	−112	66×10^{-10}	−82
3×10^{-3}	70	2.4×10^{-8}	−76	2.4×10^{-12}	−116	24×10^{-10}	−86
2×10^{-3}	66	1.1×10^{-8}	−80	1.1×10^{-12}	−120	11×10^{-10}	−90
1×10^{-3}	60	27×10^{-10}	−86	27×10^{-14}	−126	2.7×10^{-10}	−96
700×10^{-6}	57	13×10^{-10}	−89	13×10^{-14}	−129	1.3×10^{-10}	−99
500×10^{-6}	54	6.6×10^{-10}	−92	6.6×10^{-14}	−132	66×10^{-12}	−102
300×10^{-6}	50	2.4×10^{-10}	−96	2.4×10^{-14}	−136	24×10^{-12}	−106
200×10^{-6}	46	1.1×10^{-10}	−100	1.1×10^{-14}	−140	11×10^{-12}	−110
100×10^{-6}	40	27×10^{-12}	−106	27×10^{-16}	−146	2.7×10^{-12}	−116
70×10^{-6}	37	13×10^{-12}	−109	13×10^{-16}	−149	1.3×10^{-12}	−119
50×10^{-6}	34	6.6×10^{-12}	−112	6.6×10^{-16}	−152	66×10^{-14}	−122
30×10^{-6}	30	2.4×10^{-12}	−116	2.4×10^{-16}	−156	24×10^{-14}	−126
20×10^{-6}	26	1.1×10^{-12}	−120	1.1×10^{-16}	−160	11×10^{-14}	−130
10×10^{-6}	20	27×10^{-14}	−126	27×10^{-18}	−166	2.7×10^{-14}	−136
7×10^{-6}	17	13×10^{-14}	−129	13×10^{-18}	−169	1.3×10^{-14}	−139
5×10^{-6}	14	6.6×10^{-14}	−132	6.6×10^{-18}	−172	66×10^{-16}	−142
3×10^{-6}	10	2.4×10^{-14}	−136	2.4×10^{-18}	−176	24×10^{-16}	−146
2×10^{-6}	6	1.1×10^{-14}	−140	1.1×10^{-18}	−180	11×10^{-16}	−150
1×10^{-6}	0	27×10^{-16}	−146	27×10^{-20}	−186	2.7×10^{-16}	−156

where the amplitude $V_0(t)$ and the phase $\phi(t)$ are functions of time. The cumulative probability that the instantaneous value of $V_0(t)$ exceeds a specified level, say V_k, during a period of observation T, is called the amplitude probability distribution (APD) of the function $V(t)$. Several such noise distribution functions are presented in this chapter.

Thermal Noise Power Reference. Received noise power may be referred to thermal noise power at a reference temperature, say 290° Kelvin (K)(5). Atmospheric noise level maps are published (for example, CCIR [International Consultative Radio Committee] Report 322) to display noise levels at various frequencies in decibels above kT_0B, where k is the Boltzmann constant equal to 1.38×10^{-23} joules per degree Kelvin, T_0 is the reference temperature in degrees Kelvin (usually 290°K), and B is the effective receiver noise bandwidth in Hertz. Atmospheric, galactic, solar, and other noise data are presented using the kT_0B reference (2, 3, 5).

Time-Domain to Frequency-Domain Conversion

Digital and transient phenomena are often represented (approximated) in the time-domain by pulse-type waveforms such as are illustrated in the sketches of Fig. 2-3a. Other such waveforms not shown include the cosine squared, gaussian, triangular, and sawtooth waveforms (6, 7). Frequency-domain representations of the time-domain waveforms may be obtained by well-known Fourier analysis mathematical techniques. However, these techniques involve the solution of somewhat complex integrals to obtain the continuous amplitude spectrum of nonperiodic waveforms, or the amplitude coefficients of the sinusoidal components of the resulting Fourier series representation of periodic waveforms. An approximation technique has been developed to quickly and simply obtain an upper bound approximation envelope for the amplitude function of continuous Fourier spectra for nonperiodic waveforms, or of the discrete line spectra for periodic waveforms (8). Examples of these envelope approximations are shown in Fig. 2-3b for the corresponding waveforms of Fig. 2-3a. Also shown are the relationships between the time and frequency-domain representation parameters. In the time-domain, A is the pulse amplitude (usually of a voltage or current pulse), τ is the pulse duration, which is the time measured between the pulse leading and trailing edges at the $0.5A$ (mid-amplitude) level, τ_r is the pulse rise-time, measured as the time required for the pulse leading edge to rise from 0.1A to 0.9A, and τ_f is the pulse fall-time, measured as the time required for the pulse trailing edge to fall from $0.9A$ to $0.1A$. In many (perhaps most) applications, it is assumed that the rise and fall-times are equal.

The frequency-domain spectrum amplitude envelopes of Fig. 2-3b are determined from the parameters of the corresponding time-domain waveforms and utilized to approximate the emissions of pulse-type waveforms such as those

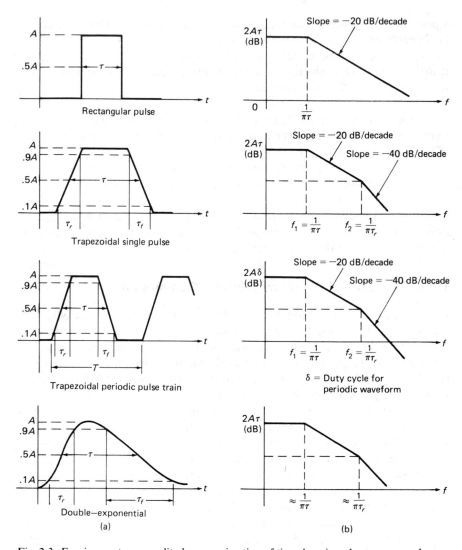

Fig. 2-3. Fourier spectrum amplitude approximation of time-domain pulse-type wave-shapes.

associated with digital devices. An examination of the spectrum envelopes indicates that pulse rise-time determines the spread of frequency components contained in the corresponding time-domain waveform. The upper corner frequency, $f_2 = 1/\pi\tau_r$, is referred to as the bandwidth of the spectra shown. Note that for the rectangular pulse, the rise and fall-times are zero, and the frequency f_2 extends out to infinity. This illustrates that pulses with short rise-times (such as those for high-speed logic) contain a broad frequency spectrum.

It must be noted that if the waveform fall-time is less than the rise-time, then the upper corner frequency f_2 is determined by $f_2 = 1/\pi\tau_f$. The spectrum amplitude for the nonperiodic waveform is continuous and represents a spectrum density in amplitude units per unit frequency bandwidth (such as volts/Hz, amps/kHz, etc.), depending on the units of the amplitude A as it represents a voltage or current. The spectral density unit of $2A\tau$ arises from the time units of τ, in seconds, with time being the reciprocal unit of frequency. For the periodic waveforms, the amplitude units of the frequency spectrum are those of A (i.e., volts or amps, etc.). The units of $2A\delta$ are those of A, since the duty cycle (δ) of the periodic waveform is dimensionless.

The frequency spectrum envelope approximations of Fig. 2-3b provide an upper bound (or "worst case") of the actual Fourier spectrum amplitude of the time-domain waveforms. For the periodic waveforms, frequency components present must be integral multiples (harmonics) of the fundamental frequency, $f_0 = 1/T$, where T is the period of the time-domain waveform. Note also that the frequencies f_1 and f_2 determine the break points of the slope of the envelope segments, and they are generally unrelated to the frequency content of the periodic waveform frequency spectrum.

NATURAL SOURCES OF EMI

Natural sources of EMI are divided into two groups based upon their different physical properties. The first group includes those sources located in the atmosphere (terrestrial sources), and the second group is comprised of those sources associated with emissions originating from regions beyond the earth's atmosphere (extraterrestrial sources). Terrestrial emissions tend to be of a transient nature, whereas the emissions originating from the sun, stars, galaxies, and other outer-space regions tend to be more like band-limited white noise. These natural emissions are often the limiting factor in determining the receiver sensitivity obtainable.

Atmospheric (Terrestrial) Natural EMI Sources

Atmospheric Radio-Frequency EMI.   The dominant natural radio-noise below 30 MHz is atmospheric noise produced by electrical discharges during thunderstorms. This noise has a moderately broad emission spectrum with the largest amplitude frequency components generated occurring between 2 and 30 kHz. For frequencies below ionospheric cutoff, most of the atmospheric noise detected within the temperate zones is produced by local thunderstorms during the summer and by tropical region thunderstorms during winter. Noise signals from tropical disturbances are propagated by sky waves supported by the ionosphere and transmitted over distances of several thousand kilometers.

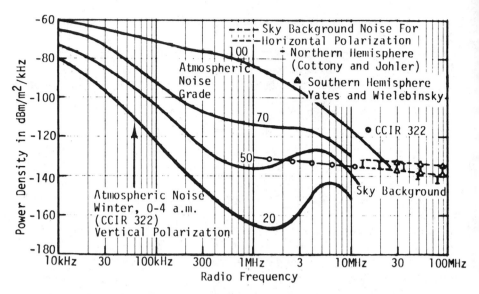

Fig. 2-4. Spectral distribution of atmospheric noise. (*Courtesy of Interference Control Technologies*)

At frequencies above that of ionospheric cutoff, local lightning discharges account for producing most of the atmospheric noise. Remote location from tropical thunderstorms and the decreasing amplitude of the spectral density with frequency cause atmospheric noise to become insignificant at frequencies above 30 MHz in temperate and polar regions. Figure 2-4 illustrates the spectral distribution of atmospheric noise (3) by providing curves of four CCIR grades of atmospheric noise, numbered 20, 50, 70, and 100, respectively.

Each noise-grade curve displays a variation of average atmospheric noise power as a function of frequency for a geographical location whose observed relative noise level is proportional to the noise grade number.

Within the frequency range from 300 kHz to 30 MHz, atmospheric noise is not always the dominant natural noise source. Sky-background (galactic) noise often dominates in temperate and tropical regions and may be dominant in polar regions. Over subpolar continents and oceans, atmospheric noise clearly dominates only during the daily 8 P.M. to 4 A.M. period, an interval that is extended somewhat beyond both limits during the summer and autumn periods. For the remainder of the day, atmospheric noise levels are usually lower than the corresponding levels of sky-background noise.

The diurnal increase in atmospheric noise begins in the afternoon and rises in level throughout the evening as a consequence of reduced ionospheric D-region absorption. Electrical discharge activity is at a maximum during the afternoon within the sunlit tropical regions of the continents and decreases after sunset, attaining a minimum level sometime after midnight. Because of long-range, moderate-loss sky-wave propagation during the evening, transpolar and transequatorial propagation paths from sunlit tropical regions contribute large atmospheric noise signals at locations within the Northern Hemisphere.

Precipitation Static. Precipitation static (known as "P-static") refers to the transfer of charge, and eventual discharge, produced by the relative motion of dust, rain, hail, and snow as these atmosphere-borne particles move past electrically insulated surfaces located on structures, vehicles, aircraft, spacecraft, and so forth. When the potential builds to a level of 40 to 1000 kilovolts, a corona or spark discharge occurs, thereby generating broadband static noise with a frequency spectrum extending into the VHF and UHF bands (9).

Lightning Strike Transients

In addition to generating radio-frequency noise, lightning strikes frequently cause injury to personnel and damage to structures and equipment. Injury and damage can be caused by direct strikes that result in high-amplitude voltage and current transients that appear directly at locations where there is direct attachment of the lightning discharge. Also, transients can be induced into circuits by the coupling of the high-amplitude, time-changing electric and magnetic fields produced by a lightning discharge. Lightning characteristics, parameters, and effects are determined experimentally, by simulation, and displayed statistically (9–11).

Galactic (Extraterrestrial) Natural EMI Sources

Sources of galactic noise at radio frequencies are located beyond the earth's atmosphere. Principal noise sources include the sun, radio sky-background radiation, and cosmic sources distributed within the galactic plane (3, 5).

Figure 2-5 provides composite data on solar, planetary, and stellar noise (3). Also, radio-sky maps are published showing brightness temperature contours in degrees Kelvin at 136 MHz and 400 MHz (5).

MAN-MADE EMI SOURCES

In addition to natural EMI sources, numerous sources associated with man-made devices exist and contribute undesirable conducted and radiated emissions. Man-made EMI can originate from terrestrial and extraterrestrial

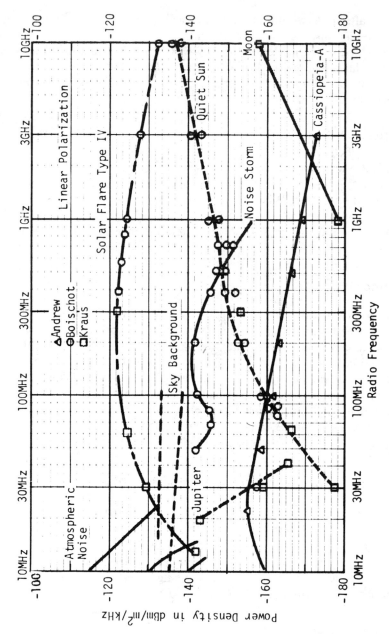

Fig. 2-5. Solar, planetary, and stellar noise. (*Courtesy of Interference Control Technologies*)

sources, the latter being earth satellites or spacecraft. With possible exceptions, extraterrestrial man-made sources are generally less significant than earth-bound sources, since they represent a relatively small fraction of the total EMI-source population. However, the fields-of-view of transmitters and receivers on space platforms cover vast areas of the earth's surface, and the number and significance of these platforms are likely to increase.

The general classification of man-made (and other) EMI sources is illustrated by Fig. 2-1. Various classes of emitters and their characteristics are described in the following sections.

Communications-Electronics (C-E) Emitters

The emissions from this class of man-made emitters are generally narrow-band, coherent, intentionally radiated signals. Also present, however, are spurious radiations, usually in the form of harmonics and sidebands. This class of emitters can be subdivided into the following five categories:

(1) Common Broadcast C-E Emitters (with allocated frequency bands). They include:

- Amplitude modulation (AM) band: 535–1605 kHz
- Frequency modulation (FM) band: 88–108 MHz
- VHF television:
    Lower bands: 54–72 MHz, 76–88 MHz (Channels 2–6)
    Upper bands: 174–216 MHz (Channels 7–13)
- UHF television: 470–806 MHz
- International broadcasting: selected bands between 5950 kHz and 26.1 MHz

To illustrate spectrum amplitude profiles of C-E emitters, Fig. 2-6 presents two typical spectrum signatures: one that of an FM broadcast transmitter, and the other the signature of a typical VHF TV transmitter. Note for the FM transmitter the nominal 180-kHz, 3-dB emission bandwidth around the fundamental FM frequency of 92 MHz and the relatively low (approximately −90 dB) out-of-band noise levels as controlled by the FCC *Rules and Regulations*. For the VHF TV signal, the nominal 6-MHz transmission bandwidth levels are readily observed. To determine the emission signatures of both signals in Fig. 2-6, the bandwidth of the measuring receiver must be much less than the broadcast transmitter bandwidth. The amplitude is presented in relative units of dBμV/m/kHz for a prescribed distance between the transmitter and the receiver.

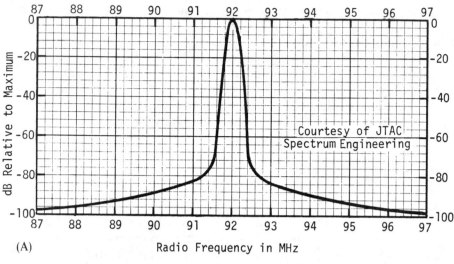

(A)

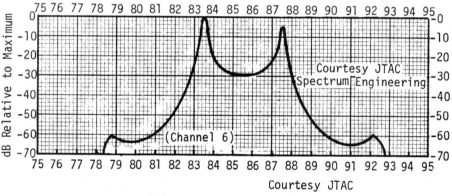

(B)

Fig. 2-6. Typical spectrum signatures of (a) FM and (b) VHF TV broadcast transmitters. (*Courtesy of Interference Control Technologies*)
(a) Typical spectrum of an FM broadcast transmitter.
(b) Typical spectrum of a VHF TV broadcast transmitter.

(2) Non-Relay Communications C-E Emitters.

Non-relay communications systems and equipment comprise the largest and most varied category of C-E emitters. They occupy allocated frequency bands interspersed between other services ranging from 14 kHz to about 960 MHz. For frequencies above 1 GHz, point-to-point communications are generally of the relay type.

Figure 2-7 illustrates a typical spectrum signature of a VHF land-mobile transmitter (3). The 25-kHz transmitter bandwidth is evident, and the emis-

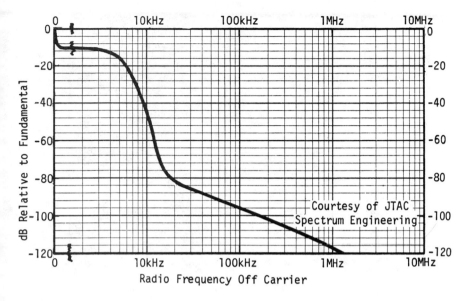

Fig. 2-7. Typical spectrum signature of VHF land-mobile transmitter. (*Courtesy of Interference Control Technologies*)

sion spectrum indicates that the out-of-band emission levels are 100 dB below the fundamental at a frequency displacement of ±150 kHz on either side of the fundamental (carrier) frequency (3).

(3) Relay Communication C-E Emitters. Relay communications C-E emitters can generally be placed in one of the following four subdivisions:

- Common-carrier, microwave relay: interspersed frequencies between 2.1 and 12.2 GHz
- Satellite relay: interspersed frequencies between 399.9 MHz and 275 GHz
- Ionospheric scatter: 400 to 500 MHz
- Tropospheric scatter: interspersed frequencies between 1.8 and 5.6 GHz

Figure 2-8 illustrates a typical spectrum signature of a 1-watt, C-band, microwave relay transmitter. Note the rapid amplitude fall-off with frequency. This is typical of most relay-communications systems that have a low spectrum pollution level and that operate under severe adjacent-channel requirements.

(4) Navigational C-E Emitters. The following represent typical navigational aid transmitters included within this category (exclusive of radar

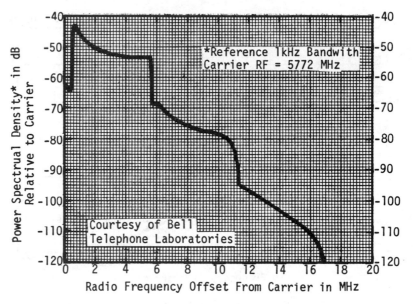

Fig. 2-8. Typical spectrum signature of a C-band microwave relay transmitter. (*Courtesy of Interference Control Technologies*)

systems):

- VHF omni-range (VOR): 108–118 MHz
- Tactical air navigation (TACAN): 960–1215 MHz
- Marker beacons: 74.6–75.4 MHz
- Instrument landing systems (ILS):
  Localizer: 108–112 MHz
  Glide path: 328.6–335.4 MHz
- Altimeters: 4.2–4.4 GHz
- Direction finders: 405–415 kHz
- Loran C: 90–110 kHz
- Loran A: 1.8–2.0 MHz
- Maritime radio navigation: 285–325 kHz
  2.9–3.1 GHz
  5.47–5.65 GHz
- Land radio navigation: 1605–1800 kHz

Figure 2-9 illustrates a typical spectrum signature of a LORAN C, low-frequency (100 kHz) transmitter (3). Note the 20-kHz, 99.5% power bandwidth centered about 100 kHz, the sideband emission fall-off, and the harmonic

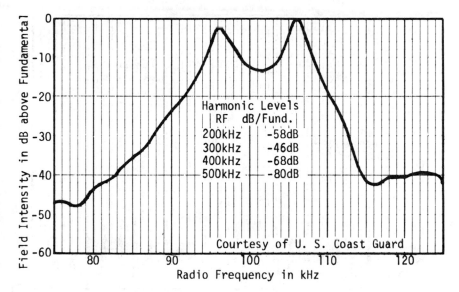

Harmonic Levels	
RF	dB/Fund.
200kHz	-58dB
300kHz	-46dB
400kHz	-68dB
500kHz	-80dB

Courtesy of U. S. Coast Guard

Fig. 2-9. Typical spectrum signature of a 100-kHz LORAN C transmitter. (*Courtesy of Interference Control Technologies*)

levels. Although relatively high emission levels exist out-of-band, the system transmitter is also operated within a comparatively uncongested portion of the frequency spectrum.

Figure 2-10 illustrates the spectrum signature of a VHF Doppler VOR transmitter operated by the Federal Aviation Administration (FAA). The 10-kHz, frequency-modulated, variable-phase subcarrier components are readily observed.

(5) Radar C-E Emitters. Among C-E emitters, radars are perhaps the greatest offenders with respect to the production of EMI, for these reasons: (1) the high-level peak pulse powers (up to several megawatts); (2) the attendant spectrum spread due to short pulses occupying broad basebands; (3) the relatively high levels of harmonic radiation; and (4) the high carrier frequencies at which radars operate for efficient coupling of signals.

Radars occupy intermittent portions of the frequency spectrum from 225 MHz to several GHz. They are employed in many capacities, including air traffic control, air and surface search and rescue, harbor surveillance, mapping, tracking and fire control, highway speed monitoring, and weather prediction. The Department of Defense operates the highest-powered radars.

Figures 2-11 and 2-12 illustrate typical spectrum signatures of P-band (VHF) long-range search radars, and FAA S-band air traffic control radars, respectively. Note the wide emission sidebands and high-level harmonics.

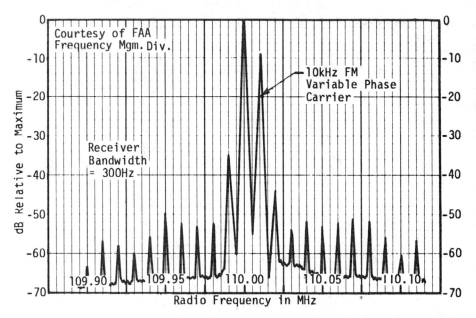

Fig. 2-10. Spectrum signature of FAA, VHF Doppler VOR. (*Courtesy of Interference Control Technologies*)

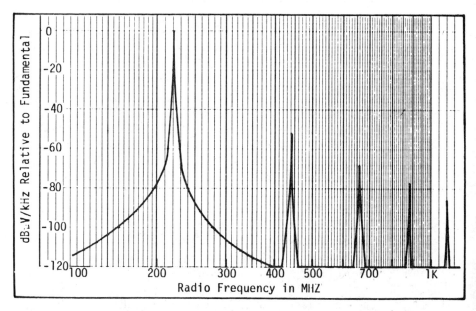

Fig. 2-11. Typical spectrum signatures of a P-Band (VHF) long-range search radar. (*Courtesy of Interference Control Technologies*)

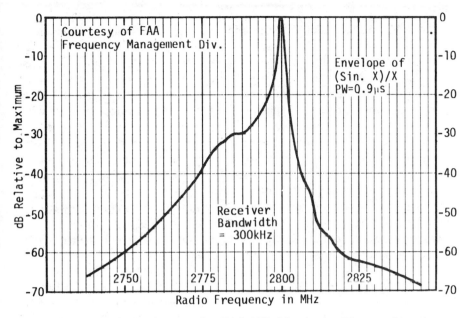

Fig. 2-12. Typical spectrum signature of an FAA ASR, S-band radar. (*Courtesy of Interference Control Technologies*)

Restricted Emitters

According to FCC *Rules and Regulations* Part 15.4, a restricted radiation device basically is a device that intentionally generates radio-frequency energy to perform a given function other than electrical communication. Such devices include emitters referred to as industrial, scientific, and medical (ISM) equipment, such as ultrasonic, industrial heating, and medical diathermy equipment, RF-stabilized arc welders, and induction cooking ranges. Other restricted emitters include low power communication devices, such as wireless microphones, phonograph oscillators, radio-controlled garage door operators, and radio-controlled models (12). Typically, the output power and operating frequency of these devices are restricted; however, in the presence of sensitive receptors, the emission levels could be significant and cause interference.

Incidental EMI Emitters

An incidental radiating device is one that radiates radio-frequency energy during the course of its operation although the device is not intentionally designed to radiate this energy (12) (i.e., the production of electromagnetic emissions is not inherent in the intended function of the source). Except for the C-E equipment and restricted radiators discussed above, all of the man-made EMI sources identified in Fig. 2-2 are unintentional (incidental) emitters.

An incidental emission may be random in time and thus can be stochastically represented as a function of time and/or frequency; or it may be deterministic (and thus appears as a relatively narrowband spectrum). Within a metropolitan area, many types of unintentional generators of EMI exist (1–3). Dominant sources include automotive ignition systems, electric power transmission and distribution lines, fluorescent lamps, and industrial equipment such as RF-stabilized arc welders. Other urban incidental noise sources include electric trains and buses, dielectric heaters, plastic welders and heaters, and wood-gluing equipment. Within industrial and residential areas, numerous incidental noise sources exist that can cause localized conducted and/or radiated interference. Examples include electrical switching equipment, electric motors, household appliances, electrical tools, and digital devices (computers, games, microprocessor-controlled equipment). As a result of the noise-producing aspect of digital devices, the FCC published Part 15J of the *Rules and Regulations* (R & R). The objective of Part 15J is to specify allowable upper limits for the conducted and radiated emissions produced by digital devices.

Interest in the analysis and control of EMI generated by industrial, commercial, and consumer equipment is the result of manufacturers and consumers being obligated to respond to directives and other pressures from regulatory agencies as well as general public concern. Until the late 1970s and early 1980s, efforts to control EMI were almost purely a reaction to public pressure. However, with the publication of FCC R & R Part 15J, at least some standard exists for the suppression of emissions. The next question appears to be one of how to specify reasonable levels of immunity, that is, susceptibility levels for test equipment to assure that no unacceptable degradation of performance results when the equipment is subjected to a specified electromagnetic environment.

Efforts directed toward achieving a general understanding of the intensity and geographical extent of incidental EMI in a metropolitan area have, for the most part, consisted of the performance of area noise surveys in and above several cities and towns of the world. All of these surveys have been considerably limited in scope, only one having examined the variation of metropolitan-area EMI over an extended period of one year. Most surveys to date have covered only a small portion of the frequency spectrum, and all have been geographically restricted.

Surface incidental EMI surveys comprise the major portion of the area noise investigations that have been performed and reported. Aerial surveys have been systematically performed only during the last few years, and the reported data represent a small body of information. Both surface and aerial incidental EMI surveys have employed techniques to develop descriptions (i.e., signatures) of the noise-producing mechanisms and also statistical processes to describe the spectral and other characteristics of EMI sources. Upon these foundations, many analytical models have been developed that accurately represent several aspects of the range and frequency dependence of metropolitan-area EMI. Many of these models appear in reference 2.

Automotive Ignition Noise

Automotive ignition noise represents the composite radiation from metropolitan-area vehicular traffic. Individual vehicles radiate pulse trains consisting of periodic, narrow pulses of 1 to 5 nanoseconds time duration. The peak pulse amplitudes from single vehicles vary as a function of type of ignition, vehicle speed, and mechanical loading. Over long periods of time, the radiation from a single vehicle will also change in response to aging and wear. The large number of different engine and drive-train designs occurring in the vehicle mix that constitutes urban traffic adds to the diversity of radiated ignition system waveforms, which, in total, constitute the impulsive noise of automotive traffic.

The results of studies of automotive ignition noise, representing the measurements made by four investigators working on three continents, are presented in Fig. 2-13. The ordinate axis of the figure gives the broadband radiated electric field strength in units of dBμV/m/kHz.

The measurements by Critchlow were made in Washington, D.C., and those of George were made on Long Island, New York. Both sets of data represent automotive traffic noise levels during business hours in the vicinity of major roadways at a distance of approximately 10 meters from the road centerline. The measurements by George characterize automotive interference from vehicles manufactured prior to World War II. Ignition noise suppression techniques used in the vehicles observed by Critchlow in 1960 were relatively advanced over those encountered by George. This difference in vehicle ignition design partially accounts for the higher intensities of radiated noise observed by George.

Ellis, working in Melbourne and other cities of eastern Australia, observed business-day automotive noise levels at VHF and UHF frequencies that were comparable to the earlier results of George. Traffic observed by Ellis was essentially comprised of British and American vehicles. Ellis also recorded traffic density, along with noise-level spectra, and a strong correlation can be seen in his results between automotive ignition interference level and traffic density.

The most extensive statistical accumulation of electric field intensity spectra produced by automotive ignition noise was made in three Japanese cities by Suzuki. His data were accumulated from an observation point located

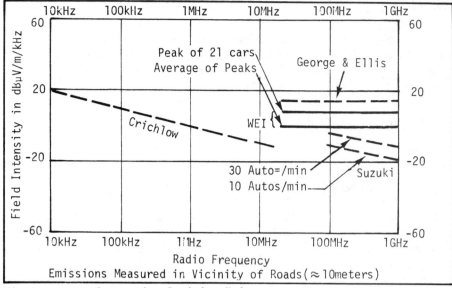

Figure   - Automotive Ignition Noise

Fig. 2-13. Automotive ignition noise. (*Courtesy of Interference Control Technologies*)

by the roadside. The large variation in flow rate of vehicle traffic passing Suzuki's observation point provides a perspective on the quantitative dependence of noise level on traffic density. The median traffic noise level increases by 17 dB for a tenfold increase in traffic flow. Field intensity levels are approximately normally distributed for all but the few values where data were insufficient to establish the form of the distribution function.

A mathematical model can be formulated for the median automotive noise field intensity from Suzuki's data as follows:

$$E \text{ dB}\mu\text{V/m} = -11 + 10 \log (B) + 17 \log (C) - 20 \log (R/10) \qquad (2\text{-}8)$$

where B = receiver bandwidth in kHz, C = vehicular traffic rate in cars/minute, R = measuring distance to road in meters, and log is the common logarithm to the base 10.

More recent measurements of the ignition system radiation from American automobiles have raised the question of the validity of some previously reported data. In 1970, White Electromagnetics, Inc. (WEI) reported the results of a survey performed for the Automobile Manufacturers Association on various 1969 and 1970 car models. All of the automobiles (which represented several manufacturers) included in this survey had been tested previously for compliance with the Society of Automotive Engineers (SAE) J55/a specification limit. Tests were performed and observations made in the frequency range from 20 MHz to 1 GHz, at a distance of 33 feet from the test vehicles, and the measuring antennas were placed 10 feet above the ground. The receiver impulse bandwidth was 16 kHz for the receiver tuned below 30 MHz, and 350 kHz for the receiver tuned above 30 MHz.

Figure 2-13 illustrates the general results corresponding to 21 individual vehicles and also the results for the same vehicles taken as a simultaneously operating group. The following general conclusions can be reached when reference is made to the plotted results of the WEI tests (Fig. 2-13):

1. Radiation below about 100 MHz shows a greater tendency toward vertical polarization. The effect is more pronounced (0 to 20-dB variations) with a 10-dB average) when the automobiles are considered individually than when they are considered as a simultaneously operating group.
2. Peak emission levels from automobiles considered as a group (0 to 10 dBμV/m/kHz) show no enhancement over that of the most offending individual automobile within the group (5 to 10 dBμV/m/kHz).
3. The average of the peak emission from automobiles considered as a simultaneously operating group (-10 dBμV/m/kHz) is approximately equal to that of an automobile from within the group considered individually.
4. Above 1 GHz, radiation levels decrease at a rate of approximately 20 dB per octave.

It is significant to note that the total number of pulses per second coming from all of the automobiles never exceeded the measuring instrument bandwidth; that is, the pulse repetition frequency of the ignition sources is smaller than the measuring receiver bandwidth. Thus, although both the average signal level and the duty cycle will rise as more automobiles are simultaneously operated, there will not necessarily be a change in peak emissions unless an offending automobile with high peaks is operated.

Additional developments and illustrations of automotive noise are presented in reference 2.

Power Transmission Line EMI

After automotive ignition noise, the next source of significant incidental EMI radiation is comprised of high-voltage generation and transmission equipment (13). In the case of power transmission lines, the noise levels reach maximum intensity during conditions of rain, fog, snow, or high relative humidity. Whenever power transmission lines and components deteriorate with age or otherwise become damaged, their impulsive noise emissions are observed to increase appreciably at frequencies above 50 MHz. These emissions are readily detectable in the absence of ignition noise.

Emissions arising from electric power production, conversion, and transport facilities occur at frequencies ranging from the fundamental power frequencies of 50/60 Hz, and into the UHF region. Two types of noise sources are identified as associated with power line systems: gap breakdown and corona discharge (2).

Noise due to gap discharge is produced by the flow of discharge currents in the air gap between two points of unequal potential in electric power equipment. The breakdown may be caused by mechanical damage, aging fractures, and accumulation of conducting-surface contamination. Radiation components due to gap discharge extend into the UHF range of frequencies (2).

Corona discharge requires a single body, charged at a sufficiently high positive or negative potential with respect to some other body. As the potential increases, the high-mobility free electrons in the vicinity are accelerated by the electric field either toward or away from the charged body. The accelerated electrons produce excited molecules, positive ions, and other free electrons through the process of inelastic collisions (2). The process results in the emission of visible radiation generally within the bluish part of the spectrum, and also RF radiation extending into the UHF range (2).

Power transmission lines are capable of supporting long-distance, low-attenuation propagation of high frequency (HF) transients and other signals by virtue of their ability to function as either coaxial waveguides, with the high frequency currents being confined between the inner conductor and the sheath, or as a single line above ground. In addition, a transmission line exhibits unknown resonances, the resonances arising from the periodic mechanical construction of the line and supporting members, which thereby form resonant circuits elements. These resonances may then somewhat enhance noise spectra peaks.

The time-domain waveform of power transmission line noise is random and impulsive. It generally has a greater pulse width but occurs less frequently than automotive ignition noise. Whenever bursts of transmission line interference are observed, the typical pulse duration is several milliseconds. The fine structure of the burst pulse corresponds to short-duration, fast rise-time, distorted square waves, often occurring at high repetition rates. Pulse rise-times and durations of approximately 2 and 10 nanoseconds, respectively, have been recorded using very-wideband receivers (3, 4). These emissions have been identified as resulting from actuations of inductive loads on power lines.

The field strength generated by any radiating source is a function of the distance from the source to the point of observation, d. The field components exhibit a dependency proportional to d^{-n}, with values of n equal to 1, 2, and 3, respectively. At distance d very close to a radiating source, all three components of the electric and/or magnetic fields are generally present; that is, those components that vary, or are proportional to d^{-1}, d^{-2}, and d^{-3}. As the distance from the source increases, the significant electric and magnetic field components are those that vary as d^{-1}, with the other two components decreasing rapidly with increasing d (2).

Figure 2-14 depicts noise from high voltage transmission lines measured by several observers at different distances from the transmission lines. The broadband electric field strength (in dBμV/m/kHz) is given as a function of frequency, line voltage, and lateral distance of the observation point from the line. The line voltage values (in kilovolts) are indicated by the numbers in parentheses, while the distance (in feet) from the outer conductor of the transmission line to the point of observation is given by the numbers in angular brackets.

All of the data used to construct Fig. 2-14 were obtained under dry weather conditions. The noise level can increase an additional 10 to 20 dB if the relative humidity is high, or if fog, mist, snow, or rain is present (2, 3). Other power line noise distribution data may be found in reference 12.

Arc Welders, Heaters, and Gluers

Several primary types of industrial material processing and assembly equipment are known to be major sources of incidental radio noise. Extensive radio-noise field strength data exist for some, including RF-stabilized arc welders, plastic welders and heaters, and wood-gluing equipment. Of these, only arc welders emit a broadband spark spectrum. Emissions from the others are manifested in the form of line spectra at harmonics of the radio-excitation frequency.

The most extensive compilation of data on the interference produced by RF-stabilized arc welders was published by Garlan, who studied more than 100 operational installations. His data are sufficiently detailed to allow the determination of the dominant emission frequencies and the variation of interference with distance from the source to point of observation (d), whereby the electric field strength is proportional to $d^{-1.5}$ for values of d varying from 1000 feet to 1 mile (3).

The left side of Fig. 2-15 represents a compilation of Garlan's data on 72 extensively investigated welder units at an observation distance of 1,000 feet. Based upon these data, dominant radiation bands for the broadband electric field emission spectra are identified at 750 kHz, 3 MHz, and 20 MHz. Individual welders usually do not radiate equally within each frequency interval, with the emissions generally concentrated within one of the bands identified above (3).

Pearce has examined the radiated field strength of several types of industrial processing and assembly equipment such as plastic welders and heaters

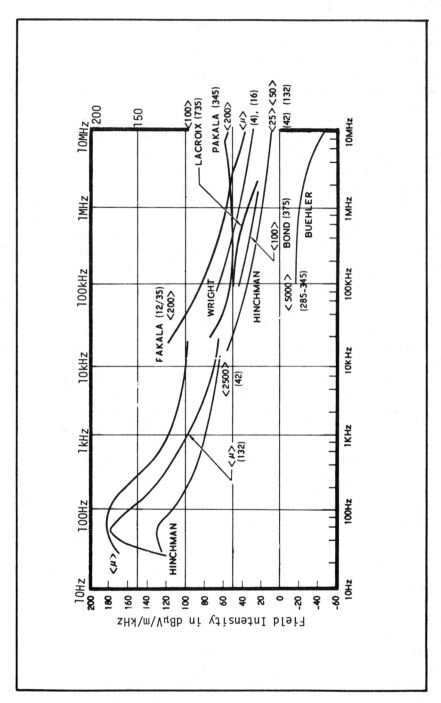

Fig. 2-14. Representative transmission line noise. (*Courtesy of Interference Control Technologies*)

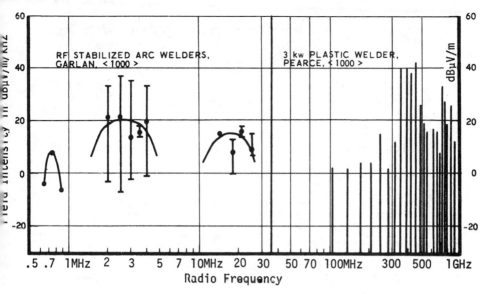

Fig. 2-15. RF-stabilized arc welder and plastic welder noise. (*Courtesy of Interference Control Technologies*)

and wood gluers. The equipment tested was powered by RF sources that radiated appreciable amounts of energy (in the form of line spectra) via coupling from leads connecting the power supply to the working unit or via leakage from power supply enclosures. The right side of Fig. 2-15 presents radiated field strength based upon Pearce's data on plastic welders (3-kW power output). The level of radiated electric-field interference produced by the plastic welders is indicated in units of dBμV/m, which is essentially a narrowband unit independent of receptor bandwidth. The plastic welder emits at a fundamental interference frequency of 35 MHz and exhibits most of the harmonics up to the limit of Pearce's measurements (approximately 1 GHz). Several other presentations of noise from industrial welders are available in reference 2.

Fluorescent Lamps

Clark has determined the intensity of radiated noise for several configurations of fluorescent lighting fixtures. His reported results were obtained on unmodified (i.e., as received from the manufacturer) fluorescent assemblies as well as on assemblies that had been specially modified to reduce noise emissions covering a broad portion of the radio-frequency spectrum. Both hot-cathode and cold-cathode fluorescent light assemblies were included in the study. It

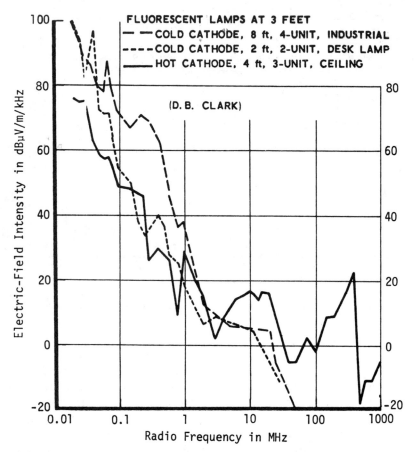

Fig. 2-16. Radiated noise from three fluorescent lamp assemblies. (*Courtesy of Interference Control Technologies*)

was determined that the cold-cathode variety emits noise of markedly decreased intensity within and above the high frequency (HF) band.

Figure 2-16 illustrates the maximum electric field intensities of broadband radiated noise for three unmodified fluorescent assemblies, with one assembly using hot-cathode units and the other two cold-cathode units. The legend at the top of Fig. 2-16 indicates the length and the number of fluorescent bulbs included within each assembly. In all three cases, the intercept antenna was located 3 feet from the installation with no measurements at greater distances reported (3). The distinctive, high frequency emissions of the hot-cathode assembly are clearly evident in the VHF and UHF bands.

Other Incidental EMI Emitters

Electric trains and buses represent two additional noise sources found in metropolitan areas. In the case of electric trains, arc-discharge noise is generated as a result of sporadic contact between the electrical power supply bus and the pantograph. The pulse width of the radiating spark is usually a fraction of a microsecond, the average repetition frequency being less than or equal to one pulse per second. Although the emitted noise spectrum is usually confined to below 30 MHz, there is evidence indicating that high-speed electric trains produce a radiated noise spectrum that extends into the VHF band.

The drive motors and on-board equipment of electric trains and buses can also be significant noise sources, with the radiated energy from both of these types of source concentrated in the lower-frequency bands. Very little quantitative information on the radiated noise from electric trains and buses has been reported, Amamiya's work being one of the few references (3). However, the characteristics of the measuring instruments used for this work were unreported, and the emphasis was placed on conducted interference.

DC electric motors and some types of AC motors of both multiple and fractional horsepower are also significant noise sources. The heavy DC drive motors used in older elevator installations emit detectable noise, and AC motors of certain small appliances produce intense noise over a broad portion of the frequency spectrum extending into the UHF band.

Overall Mathematical Models of Incidental Emitters

The composite surface, incidental, radio-noise environment of a metropolitan area is a product of the distribution of the individual noise sources discussed in the preceding sections. Composite metropolitan-area noise has been the subject of several experimental investigations during the past 20 years (2). The motivation for these investigations and also for the more recent theoretical studies has been provided by the increasing requirements of radio communication and information transmission systems, specifically the effects of noise levels on the performance of urban radio-telephone, aircraft navigation, radio broadcast, and television systems.

It is apparent even from limited observations that the level of man-made radio noise is subject to wide geographic and short-term time variations. These variations make it difficult to specify an exact interference level for any given time and/or geographic location without the benefit of actual measurements; however, it is possible to suggest typical levels of unintentional noise. The adopted convention is to classify the levels of unintentional, man-made interference as a function of the level of urbanization within the area where measurements are actually made, or to develop actual noise-level contour maps. As more data become available and data correlation methods are introduced, incidental noise classification could be accomplished on the basis of local industrial and/or commercial activity.

Figure 2-17 illustrates typical (median) levels of incidental man-made radio noise at three arbitrarily defined levels of urbanization density: urban (0 to 10 miles from a major city center), suburban (10 to 30 miles from a major city center), and rural (beyond 30 miles from a major city center). Urban area measurements were made within the business areas of New York City, New York; Baltimore, Maryland; Washington, D.C.; Denver, Colorado; Melbourne, Australia; and Tel Aviv, Haifa, and Jerusalem in Israel. The suburban curve was obtained from data taken in Boulder, Colorado; near Washington, D.C. Melbourne, Tel Aviv, and Haifa; and in suburban areas of England. Finally, rural measurements were made at locations that were considered to be as free of man-made noise as possible (the stations given in CCIR Report 322 and the experimental radio facilities near Boulder, Colorado).

At each defined noise level in Fig. 2-17, the curves are discontinuous within the frequency interval of 10 MHz to 20 MHz. The data used to construct Fig. 2-17 do not include measurements within this frequency interval. For frequencies at and below 10 MHz, data supplied by ITSA/ESSA were used; and for frequencies at and above 20 MHz, data obtained by Young, Simpson, Hamer, Ellis, FCC ACLMRS, and ITSA/ESSA were used to derive the given curves. These studies usually involved measurements at two to three specific frequencies. In the case of the measurements at or above 20 MHz, the specific frequencies selected generally varied from location to location (i.e., from investigator to investigator). In addition, individual investigators either concentrated their measurements near 20 MHz or at the upper end of the spectral region studied. To minimize possible frequency-location bias, the best fit for each set of observations was taken, and equal-frequency points from these curves were then used to obtain a best fit to all the data.

The reason for the disparity between the data across the 10 MHz to 20 MHz interface is not clear, although below 10 MHz all measurements were made with a vertically polarized whip or rod antenna, while most measurements above 20 MHz (especially above 100 MHz) were made using a horizontally polarized, tuned-dipole antenna. However, polarization per se does not explain the disparity because measurements that were made with both polarizations of the same antenna evidenced only a 2-dB average difference. Accordingly, Fig. 2-17 suggests a graphically constructed interpolation model from about 1 MHz to 200 MHz.

Next to each curve segment in Fig. 2-17 the root-mean-square (RMS) standard deviation (in decibels) of the median values used to obtain that segment

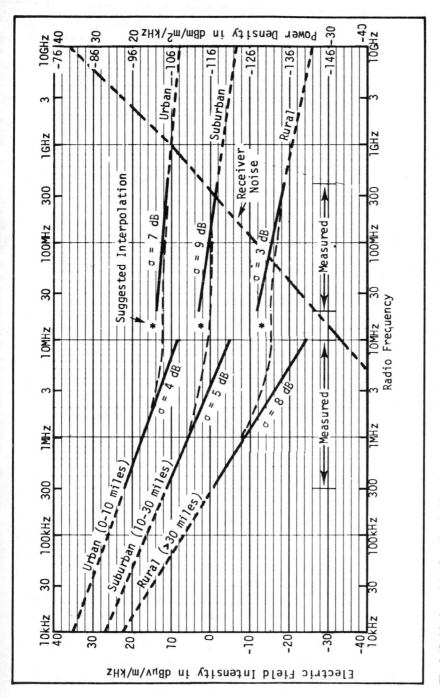

Fig. 2-17. Math models of median incidental man-made noise based on lossless omnidirectional antenna near surface. (*Courtesy of Interference Control Technologies*)

is given. Note that this deviation is found from the average power values for each location and frequency and is not the instantaneous (peak) variation, which is considerably larger than the values shown.

In estimating the noise level at the receiver due to external sources, the gain, polarization, and orientation of the receiving antenna should be considered. The median operating field strengths and equivalent power densities given in Fig. 2-17 represent the values to be expected from an omnidirectional, short, lossless, vertical antenna near the surface of the earth. For convenient reference, Fig. 2-17 depicts the typical receiver noise based on a noise figure of 10 dB at 1 MHz that rises to 15 dB at 1 GHz. Many government and commercial receivers would be more sensitive than this, while many consumer receivers would be less so.

COMPONENT SOURCES OF EMI

A component source of EMI (component emitter) is one from which EMI emanates from a single element (component) rather than an ensemble of components such as equipment and devices discussed in the previous sections. Examples of component noise sources would include wires and cables; connectors, motors, generators, and other rotating devices; switches, relays, and solenoids; and vacuum tubes, transistors, diodes, gas lamps, thyristors, and similar units. Actually, this distinction is somewhat academic because these components require energy and connecting wires from other sources to perform their function. Thus, they are not true sources of EMI, but are EMI transducers in that they convert electrical energy to electrical noise.

Wires, Cables, and Connectors

Wires and cables, while not generating EMI by themselves, provide an induction or radiation "medium" for coupling undesired energy into or out of other wires, cables, circuits, or equipment. Magnetic and electric field coupling together with the corresponding EMI-control techniques are discussed in detail in Chapter 6 of this *Handbook*.

While not a direct source of electrical noise, connectors may indirectly generate EMI as a result of poor contacts. In effect, the connector acts like a variable-impedance switch that can be environment-sensitive (e.g., sensitive to shock or vibration). The result is impedance-modulation of a current or voltage source, which can then emit EMI. Other attributes of connectors that contribute to the generation of EMI are the voltage standing wave ratio (VSWR), inadequate circumferential shielding (poor backshells), and/or contact potential. This topic together with the corresponding EMI-control techniques are discussed in detail in Chapters 6 and 7 or this *Handbook*.

Motors and Generators

Motors and generators, which use brushes and commutators to operate, are inherent sources of broadband, transient electrical noise. Transients develop as a result of an arc discharge upon separation of the rotating brush–commutator interface. An arc is generated as a result of the rapidly collapsing magnetic field in the armature winding of the motor or generator. The EMI-coupling mode may be either conduction or radiation, and significant transients can exist with resultant frequency components up to 100 MHz or higher. Interference resulting from magnetic induction can be significant at frequencies below 100 kHz. These component sources of EMI together with the corresponding EMI-control techniques are discussed in detail in Chapter 16 of this *Handbook*.

Switches, Relays, and Solenoids

Up to 300 MHz (or higher), electromagnetic switches, relays, and solenoids are capable of generating EMI in sensitive equipment. For example, upon the de-energizing of a relay (i.e., opening a switch to the source of relay power), the stored magnetic energy can develop an induced voltage that is 10 to 20 times greater than the supply voltage. Thus, arcing will develop at the switch contacts, which may conduct and/or radiate broadband transient EMI. These component sources of EMI together with the corresponding EMI-control techniques are discussed in detail in Chapter 16 of this *Handbook*.

Active Devices

Active devices (i.e., components that are capable of amplifying the current or voltage in a circuit) such as vacuum tubes, transistors, diodes, gas lamps, and thyristors can also generate EMI. Vacuum tubes generate Johnson (thermal) and shot noise, may develop microphonics at low frequencies, or can produce parasitic oscillations at high frequencies. Diodes may produce transients as a result of reverse recovery periods that develop transient spikes from AC-supplied sources. Transistors generate thermal, shot, and flicker noise, which limits their sensitivity with respect to low-level amplification; parasitic oscillations may also occur at high frequencies. These component sources of EMI are reviewed in detail in Chapter 16 of this *Handbook*.

Energy and Signal Sources

Virtually any device that consumes power or generates electromagnetic energy or control signals represents a potential source of EMI. In general, higher-energy-level sources constitute the greater EMI threat. Typical examples include AC power-main sources, switching-regulator power supplies, modulators, servo-control circuits, computers, and base-band digital data transmitters. Figure 2-18 shows the conducted EMI signature of a typical open frame switching power supply and the radiated EMI profiles from a typical piece of digital equipment (at 10 meters). Superimposed on the emissions profile in Fig. 2-18a are the regulatory limits (FCC and VDE) on conducted noise. The VDE 0871 Class B radiated limits at 10 meters are shown superimposed on the radiated emissions profiles in Fig. 2-18b.

These and additional sources of EMI and the corresponding EMI-control techniques are discussed in later chapters of this *Handbook* (consult the index).

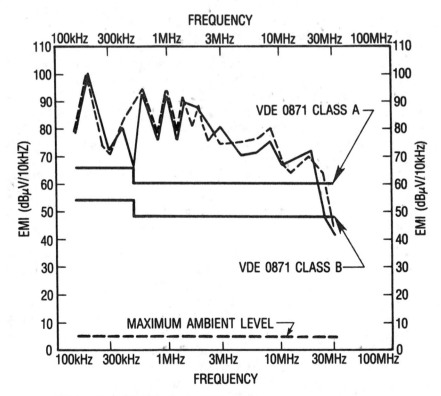

—100 WATTS SWITCHING POWER SUPPLY
—OPERATING FREQUENCY 160 kHz
—MEASURED EACH LINE TO GROUND WITH 50Ω ARTIFICIAL NETWORK

Fig. 2-18A. Typical conducted emissions of an open-frame, unfiltered switching power supply. (*Courtesy of Interference Control Technologies*)

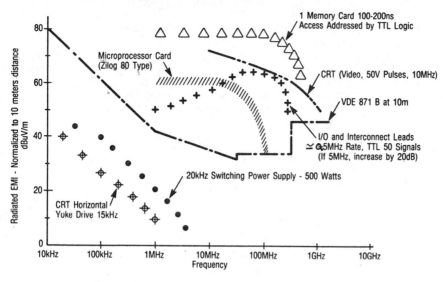

Fig. 2-18B. Typical/EDP devices radiated profiles. (Unshielded hardware; no shielded cables.)
(*Courtesy of Interference Control Technologies*)

REFERENCES

1. Herman, John R. *Electromagnetic Ambients and Man-Made Noise.* Gainesville, VA: Don White Consultants, Inc., 1979.
2. Skomal, Edward N. *Man-Made Radio Noise.* New York: Van Nostrand Reinhold Company, 1978.
3. White, Donald R. J. *EMI Control Methods and Techniques,* A Handbook Series on Electromagnetic Interference and Compatibility, Volume 3. Germantown, MD: Don White Consultants, Inc., 1973
4. Duff, William G., and White, Donald R. J. *EMI Prediction and Analysis Techniques,* A Handbook Series on Electromagnetic Interference and Compatibility, Volume 5. Gremantown, MD: Don White Consultants, Inc., 1972.
5. *Reference Data for Radio Engineers,* Sixth Edition, Third Printing. New York: Howard W. Sams and Company, 1979.
6. White, Donald R. J. *EMI Control Methodology and Procedures.* Third Edition, Second Printing. Gainesville, VA: Don White Consultants, Inc., 1982.
7. Freeman, Ernest R., and Sachs, Michael. "Electromagnetic Compatibility Design Guide." *NAVAIR AD 1115.* Dedham, MA: Artech House, 1982.
8. Rehkopf, Harold, L. "Prediction of Pulse Spectral Levels." *Proceedings of the Fourth National IRE Symposium of RFI,* 1962.

9. Air Force Systems Command. *Design Handbook 1–4 (AFSC DH 1–4), Electromagnetic Compatibility*, Third Edition, Revisions 3 and 4. Andrews AFB, MD: Department of the Air Force, Headquarters AFSC, 5 January 1977.

10. Hart, William C., and Malone, Edgar W. *Lightning and Lightning Protection.* Gainesville, VA: Don White Consultants, Inc., 1979.

11. Uman, Martin A., and Krider, E. P. "A Review of Natural Lightning: Experimental Data and Modeling." *IEEE Transactions on Electromagnetic Compatibility*, Volume EMC-24, Number 2, May 1982.

12. Federal Communications Commission. "Industrial, Scientific, and Medical Equipment." *Rules and Regulations*, Volume II, Part 18. Washington, D.C.: Government Printing Office, July 1981.

13. Sheikh, A. U. H., and Parsons, J. D. "Statistics of Electromagnetic Noise Due to High-Voltage Power Lines." *IEEE Transactions of Electromagnetic Compatibility*, Volume EMC-23, Number 4, November 1981.

Chapter 3
ELECTROMAGNETIC INTERFERENCE RECEPTORS AND SUSCEPTIBILITY CRITERIA

The previous chapter surveyed sources of electromagnetic interference (EMI). Since it takes both an emission source and a susceptible receptor (i.e., a victim) to create an EMI situation, the topic of receptors and susceptibility forms the subject of this chapter. The term "receptor" refers to the generic class of devices, equipment, and/or systems that, when exposed to conducted and/or radiated electromagnetic energy from emitting sources, will either degrade or malfunction in performance. Receptors may be classified as summarized in Fig. 3-1.

Many devices and systems can be emission sources and susceptible receptors simultaneously. Examples include most communications-electronics (C-E) systems, since they contain both transmitters and receivers, and digital devices, including several types of computers and peripherals. Figure 3-1 shows that receptors of EMI may be divided into natural and man-made categories. This parallels the classification for EMI emitters previously illustrated in Fig. 2-1. The problem most commonly encountered is interference in man-made devices; so this chapter emphasizes this receptor class. A discussion of natural receptors is also included.

This chapter surveys several types of EMI receptors, including C-E receivers, low-level sensors, I-F amplifiers, video and audio amplifiers, digital devices, computers, and status monitors and indicators. Associated susceptibility criteria of these receptors are also summarized. Among others, these criteria include voice intelligibility, digital bit error rate, radar and visual displays, and other outputs.

NATURAL RECEPTORS OF EMI

Natural receptors of EMI include humans, animals, and plants. Research emphasis is on the effects of nonionizing radiation absorption by humans. The term "nonionizing radiation" refers to electromagnetic radiation within the

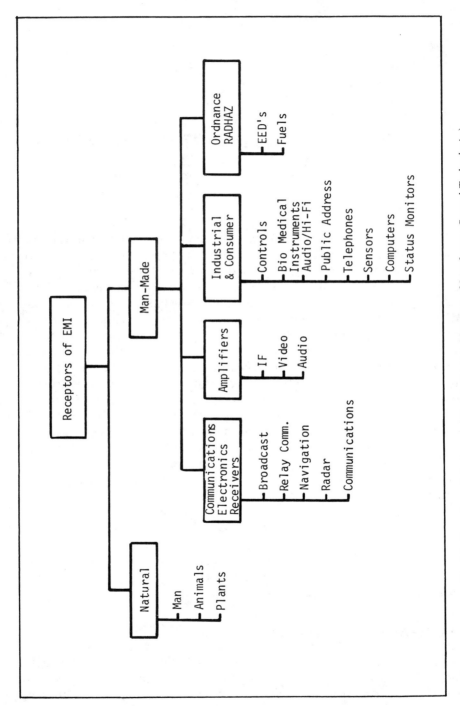

Fig. 3-1. Receptors of electromagnetic interference. (*Courtesy of Interference Control Technologies*)

frequency range from power frequencies (50/60 Hz) up to microwave frequencies (to 100 GHz), and excludes X rays, gamma rays, and other ionizing radiation associated with nuclear and similar phenomena.

Radiation Hazards to Humans

This section is based upon information contained in the American National Standard *ANSI C95.1—1982*, entitled "Safety Levels with Respect to Human Exposure to Radio Frequency Electromagnetic Fields, 300 kHz to 100 GHz" (1). The limits recommended by this document are intended for the prevention of possible harmful effects in humans exposed to electromagnetic fields within the frequency range specified above. The specified limits apply to nonoccupational as well as to occupational exposures, but they are not intended for application to intentional radiation exposure of patients in a medical capacity.

The limits established for maximum field strength and equivalent plane-wave power densities are based upon well-established findings that the body as a whole exhibits frequency dependency for absorbing RF electromagnetic energy. Maximum absorption rates occur when the long axis of the body is parallel to the electric (**E**) field vector and is 0.4 × wavelength of the incident field. At frequencies that result in maximum absorption, which define whole-body resonance, the electrical cross section of an exposed body increases in area. This increase occurs at a frequency near 70 MHz for Standard Man (long axis of 175 centimeters). Recommended maximum limits are based upon this frequency dependency (i.e., susceptibility) with the limits reduced appropriately across the range of frequencies at which human bodies from small infants to large adults exhibit whole-body resonance.

Radio Frequency Protection Guide (RFPG). This refers to electric and magnetic field strengths or equivalent plane-wave power densities that should not be exceeded except in cases where careful consideration has shown that: (1) there is a good reason for doing so, (2) the effects of the increased energy absorbed by the body have been carefully estimated and determined to be safe, and (3) the increased risk of unwanted biological effects has been carefully considered and determined to be acceptable.

Specific Absorption Rate (SAR). This refers to the time rate at which RF electromagnetic energy is imparted to (absorbed by) an element of mass of a biological body.

Table 3-1 provides the protection guides for human exposure to electromagnetic energy at radio frequencies (RFPG) in the frequency range from 300 kHz to 100 GHz. The limits are provided in terms of the mean squared electric (E^2) and magnetic (H^2) field strengths and in terms of the equivalent

Table 3-1. Radio frequency electric field (E^2) and magnetic field (H^2) squared and equivalent plane-wave power density limits for human exposure (from _ANSI C95.1—1982_).

1	2	3	4
Frequency range (MHz)	Electric field squared (E^2) (V^2/m^2)	Magnetic field squared (H^2) (A^2/m^2)	Power density (mW/cm^2)
0.3–3.0	400,000	2.5	100
3.0–30	$4000\,(900/f_{MHz}^2)$	$0.025\,(900/f_{MHz}^2)$	$900/f_{mHz}^2$
30–300	40,000	0.025	1.0
300–1500	$4000\,(f_{MHz}/300)$	$0.025\,(f_{MHz}/300)$	$f_{MHz}/300$
1500–100,000	20,000	0.125	5.0

plane-wave, free-space power density, as a function of frequency. In the near-field, the only applicable limits are the mean-squared electric (E^2) and magnetic (H^2) field strengths (columns 2 and 3). For convenience, these limits may be expressed in terms of the equivalent plane-wave power density (column 4). Column 4 values are based upon a free-space wave impedance of 377 ohms. Where mixed or broadband fields are present at a number of freqeuncies, the fraction of the RFPG incurred within each frequency range indicated in Table 3-1 should be determined, and the sum of all the fractions should not exceed unity (1).

Figure 3-2 illustrates the electric and magnetic field squared and equivalent plane-wave power density values of Table 3-1 plotted as functions of frequency. Note that the minimum values for the limits occur within the frequency range between 30 MHz and 300 MHz, which corresponds to the frequencies where the whole-body resonance is likely to occur.

Exclusions. The limits provided by Table 3-1 may be exceeded in the frequency range of 300 kHz to 100 GHz if it can be shown by laboratory procedures that the exposure conditions produce specific absorption rates (SARs) below 0.4 watt/kilogram as averaged over the whole body, and spatial peak SAR values are below 8 watts/kilogram as averaged over any 1 gram of tissue. At frequencies between 300 kHz and 1 GHz, the limits may be exceeded if the RF input power of the radiating device is 7 watts or less (1).

Measurements and Time-Averaging. Measurements to determine radiated field and power density levels shall be made at distances of 5 centimeters or more from any object. Measurement techniques and instrumentation are specified in _ANSI C95.3—1979_, entitled "Techniques and Instrumentation for the Measurement of Potentially Hazardous Electromagnetic Radiation at Microwave Frequencies."

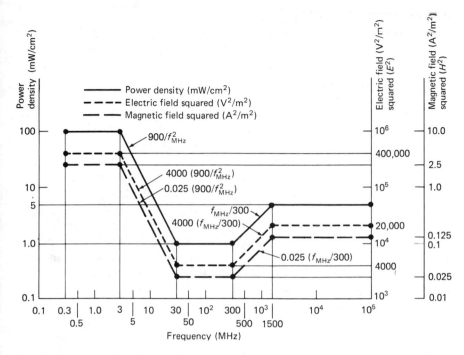

Fig. 3-2. Radio frequency electric field squared (E^2) magnetic field squared (H^2) and equivalent plane-wave power density limits for whole body exposure of human beings.

For pulsed and nonpulsed fields, the power density, the squares of the field strengths, and the SAR values of input power are averaged over any 6-minute period. The time-averaged values should not exceed the values given in Table 3-1 or the exclusions specified above.

MAN-MADE RECEPTORS OF EMI

Figure 3-1 showed that receptors of EMI may be divided into natural and man-made classes. This section discusses the latter. As shown in Fig. 3-1, man-made receptors are classified into communications-electronics receivers, amplifiers and digital circuits, industrial and consumer devices, ordnance, and fuels.

It is helpful to establish some measure of latent susceptibility of receptors to electromagnetic exposure in order to rate them for classification purposes. This then requires that a scoring system be used to calculate and assess relative vulnerability to EMI environments. Such a scoring technique has been developed and is discussed below.

Many receptors, especially those using wideband amplifiers, are susceptible to intermodulation and audio rectification. Spurious responses are usually developed from the heterodyne process in superheterodyne-type receivers. Cross modulation is an adjacent channel phenomenon in receivers. Both spurious responses and cross modulation are discussed in Chapter 4.

Audio Rectification

Except perhaps for in-band EMI, audio rectification is a principal cause of interference in analog amplifiers and digital devices. It is often the reason why a radar causes a computer to malfunction, a physician's paging system causes an EKG to produce an erroneous output, or a citizens band (CB) transmitter signal is detected by stereo equipment. A common factor is that a high-amplitude, high-frequency, radiated interfering transmitter signal finds entry (i.e., is coupled) into a victim amplifier or digital circuit via power and/or signal cables, is enhanced by parasitic circuits resonant at or near the interfering signal frequency, and thence is demodulated by a nonlinear element such as the detector of an AM radio or the base-emitter junction of an input transistor to a digital gate. The demodulated interfering signal then becomes part of the detector output signal or causes changes in the digital gate bias voltage, thereby causing errors in the gate output.

Susceptibility to audio rectification thus results from a combination of high-amplitude, high-frequency radiated signals and nonlinear circuit phenomena whereby a radiated signal is coupled into the input signal and/or power leads of an amplifier or digital circuit. The high-level emission would be rejected within the stopband of an ideal amplifier frequency response curve, as shown by the dashed line in Fig. 3-3. However, because of internal parasitic inductances and capacitances associated with circuit wiring, components, and so forth, resonances and anti-resonances are developed at various frequencies within the stopband, thereby causing deviations from the ideal amplifier response and developing many RF leakage windows that become regions of EMI susceptibility at high frequencies, as shown in Fig. 3-3. The term "audio rectification" is approximately 50 years old, and may be something of a misnomer. The phenomenon was originally associated with audio amplifiers; however, victims also include video and IF amplifiers, carrier-operated devices (receivers), and digital logic. The word "rectification" could be replaced by the more generalized term "demodulation." Other suggestions include replacing the term "audio rectification" with "carrier demodulation" or "interference demodulation" to describe this out-of-band response phenomenon, which is the cause of many EMI problems (2, 3).

The complex out-of-band response indicated in Fig. 3-3 can only be ascertained deterministically by measurement of each unit of equipment (amplifier, other receptor) to determine the out-of-band response characteristics, which will vary from unit to unit for models of equipment of presumably identical design, fabrication, and so on. Audio rectification occurs whenever a high

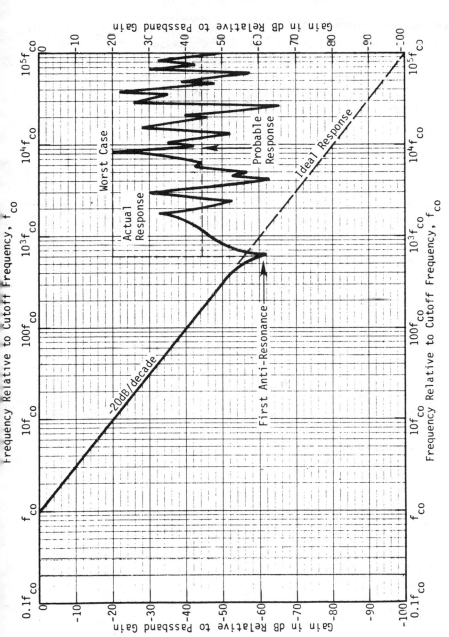

Fig. 3-3. Frequency response of amplifier to low-level interference (ideal) and high-level audio-rectification interference. (*Courtesy of Interference Control Technologies*)

frequency signal of sufficient amplitude is coupled into the cables of a receptor and the signal frequency falls close to one of the resonant peaks of the out-of-band response shown in Fig. 3-3. For example, should a signal be present at a frequency corresponding to the "worst case" peak indicated, then the amplifier attenuation at that frequency would be 20 dB instead of the expected ideal attenuation of approximately 78 dB.

For audio rectification to occur, the carrier need not be amplitude-modulated such as the signals associated with AM radio, TV, or pulsed-radar. The out-of-band response can develop slope-detection of interfering FM carriers such as the signals of FM broadcasts and TV audio. In general, little quantitative information appears in the literature, which contains numerous articles on audio rectification, because manufacturers generally do not perform the necessary measurements on equipment unless it is specifically required (2).

Audio Rectification Coupling Paths

Figure 3-4 depicts an electronic device whose input/output power and signal cables are subject to high-frequency radiated signals, thus illustrating the possible audio rectification coupling paths. Path A represents radiation pickup onto the input signal and control leads going into the device. Path B represents interconnecting cables between the device and another box (peripheral). The cables labeled C represent the power input cables into the device. Path D represents any radiation that may penetrate the housing or enclosure of the device; however, this path is usually less significant than the cable coupling paths (2).

All cable radiation pickup illustrated in Fig. 3-4 results in the generation of common-mode (CM) and differential-mode (DM) signals into the cable

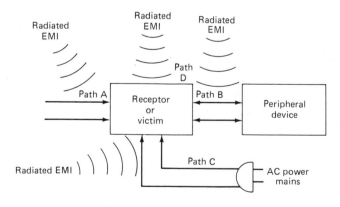

Fig. 3-4. Possible audio rectification coupling path.

circuitry. The CM component contains an interference signal on each lead, with each signal usually of different amplitude, but with all signals being in phase with respect to time. The amplitude of each CM signal is determined relative to some CM reference, most likely the reference ground plane of the system. The electronic device response is a function of the differential mode signal at the input terminals. Therefore, the effect of the CM interference is a function of the amount of CM-to-DM conversion that results from circuit impedance unbalance. The DM components of interference appear directly at the device input terminals (3).

The degree of the effects of interference due to audio rectification depends upon the level of the interfering signal with respect to the desired signal and system susceptibility. (It should be noted that this criterion applies regardless of the form of the interference.) As illustrated in Fig. 3-4, the input cables (path A) will likely carry the lowest-level signals and therefore represent the highest susceptibility. The intentional signals on the interconnecting cables (path B) are likely to be of higher amplitude than the signals on the input cables and therefore present a less susceptible situation. The power cables are generally decoupled from signal leads and tend to be less susceptible to audio rectification (exceptions are acknowledged). Radiation that penetrates the equipment housing can couple into loops and wiring formed by internal circuitry, such as printed circuit boards, interconnecting cables, and so on. The internal wiring and trace lengths are generally of smaller dimensions than the external cables, so if the enclosure housing is metallized to some degree to provide some shielding, audio rectification via this coupling path is less likely to occur (2). (Again, exceptions are acknowledged.)

Eliminating Audio Rectification

To eliminate, or, at least reduce, the effects of audio rectification, it appears highly desirable to minimize the pickup of electromagnetic radiation by external cables. The possible approaches include the following:

1. *Inserting ferrite beads onto the cables.* Since the cables can contain both CM and DM interfering signals, the beads should be placed as close as possible to the point of entry to the victim box. For this type of bead insertion, the bandwidth of the intentional signal should be below the frequency for which the bead first becomes effective. The effectiveness of a ferrite bead is a function of both frequency and circuit impedance, with the specific numbers depending upon the type of material used to construct the bead (4). Here, reference to manufacturer's data is recommended to determine bead type, configuration, and so forth. For entering or exiting cables, ferrite beads or yokes could be placed around entire harnesses which would reduce CM only and not appreciably interfere with intentional signals (2). Thus, the ferrite bead

configured as a common mode choke can be used even within the intended signal bandwidth.

2. *Inserting absorbing jacket sheath over cable harness.* The insertion of a ferrite-impregnated, absorbing sheath over a cable harness will reduce the CM coupling. Again, the bandwidth of the intentional signal should be below the frequency at which the absorbing sheath first becomes effective. The effectiveness of an absorbing sheath becomes significant at frequencies above 10 MHz (2, 3).

3. *Twisting the wires of every cable pair.* Twisting each cable supply and return pair will provide a relatively inexpensive reduction of the DM coupling.

4. *On ribbon cable, connecting every other lead (ground return) together as a zero-signal reference at both ends of the cable connectors.* This will maximize the capacitance of each signal lead to ground, and effectively provide shielding for the signal leads. Note that this is generally achievable in practice only when the signal lines are unbalanced.

5. *Using filters at point of entry or exit.* To be most effective, the filters should be hybrid types to suppress both CM and DM interference. Filter-pin connectors could be effective for harnesses containing several cable leads.

6. *Shielding cables or harnesses.* To reduce radiation pickup, one or more of the victim cable-pairs can be shielded with the shield(s) properly grounded. If the shield is braided and the shield length is less than 1/30th of a wavelength, then it is recommended that the shield be grounded at one end (generally, the receiving end) to prevent possible enhancement of the common mode. If the shield length is equal to or greater than 1/30th of a wavelength, or the shield is homogeneous (i.e., a solid shell), then the shield can be grounded at both ends to provide additional suppression of the DM without enhancing the CM (2, 3).

7. *Connecting high frequency capacitors across victim input terminals.* Where only one or a few low-level amplifiers are involved, connecting a high-frequency ceramic disk capacitor directly across the input terminals of the victim(s) can provide a quick-fix to an audio rectification problem. The value of the capacitor should not compromise the necessary bandwidth of the intentional signal which places a limit on the maximum value of the capacitance, while the self-resonant frequency of the capacitor due to its lead inductance should be above the frequency of the interference signal.

The above EMI fixes can be determined either by prediction or measurement. The objective is to isolate the effects of the possible coupling paths, determine which of the coupling paths dominate, and determine the most cost-effective way to achieve a solution. In some relatively simple situations, such as when only one or two cable pairs are present, a trial-and-error approach may be used. However, this approach may rapidly become nonproductive as system complexity increases (2).

Receptor Susceptibility Index

The measure of receptor in-band susceptibility presents a unique method for computing vulnerability to a variety of electromagnetic ambients. Sensitivity, N, and bandwidth, B, are suggested as pertinent parameters to be used to specify susceptibility. The greater the sensitivity, that is, the lower the inherent system analog noise level or the digital noise margin, and the greater the system bandwidth, the greater the system tendency toward being affected by EMI. In short, susceptibility to EMI is directly proportional to system bandwidth (B) and inversely proportional to the sensitivity (N).

A *coherent* signal or EMI emission has a specified and well-defined amplitude and phase relationship between adjacent frequency components, whereas *incoherent* signals or EMI emissions have a random amplitude and phase relationship between adjacent frequency components.

If an interfering signal is coherent (e.g., a broadband transient or pulse emission), then a receptor susceptibility *voltage-to-noise* ratio, RS_c, where the subscript c stands for coherent, may be expressed as follows:

$$RS_c = \frac{kB}{N_v} = \frac{kB}{\sqrt{4RFKTB}} = \sqrt{\frac{Bk^2}{4RFKT}} \tag{3-1}$$

where:

B = bandwidth in Hertz (Hz)
N_v = internal noise voltage
R = resistive component of the equivalent input impedance in ohms
F = equivalent noise figure of receptor
$KT = 4 \times 10^{-21}$ watts/Hz bandwidth
K = Boltzmann constant (watts/Hz-°K)
T = absolute temperature (°K)

and:

$$k = \frac{C_{TR}\tau\,\Delta e}{\sqrt{ZFKT}}$$

where:

C_{TR} = emission source to receptor coupling coefficient
τ = pulse duration producing broadband coherent emissions
Δe = amplitude of pulse producing broadband coherent emissions
Z = receptor input impedance

If the EMI source is incoherent (e.g., bandwidth-limited white noise, such as an unmodulated arc discharge), the susceptibility voltage-to-noise ratio,

RS_i, is proportional to the square root of bandwidth:

$$RS_i = \frac{k\sqrt{B}}{N_v} = \frac{k\sqrt{B}}{\sqrt{4RFKTB}} = \frac{k}{\sqrt{4RFKT}} \qquad (3\text{-}2)$$

where the subscript i in RS_i stands for incoherent emissions.

Higher values of RS correspond to greater tendencies for EMI susceptibility. Since potential EMI sources may be either coherent or incoherent, equation (3-1) for coherent sources represents the more damaging of the two situations; it becomes the worst case basis for latent receptor susceptibility.

Relative susceptibility levels are determined below for a few classes of systems. Without loss of generality when determining relative susceptibilities, we can set $k = 1$. Determining susceptibility in terms of voltage ratio (as indicated by the subscript v), equation (3-1) becomes:

$$RS_v = \sqrt{\frac{B}{4R \times 4 \times 10^{-21}F}} = 0.8 \times 10^{10}\sqrt{B/RF} \qquad (3\text{-}3a)$$

$$RS_v = 198 + 10 \log{(B/RF)} \text{ dB} \qquad (3\text{-}3b)$$

Equation (3-3b) is obtained by taking 20 times the common logarithm of equation (3-3a).

Sometimes it is useful to calculate RS in terms of power sensitivity, N_p, whereupon equation (3-3a) becomes (5):

$$RS_p = RS_v{}^2 = \frac{B^2}{N_v{}^2} = \frac{B^2}{4RN_p} = 20 \log{(B)} - N_{\text{dBW}} - 10 \log{(4R)}$$

$$= 20 \log{(B)} - (N_{\text{dBm}} - 30 \text{ dB}) - 10 \log{(4R)}$$

$$= 24 \text{ dB} + 20 \log{(B)} - 10 \log{(R)} - N_{\text{dBm}} \qquad (3\text{-}4)$$

where:

N_{dBW} = sensitivity in dB referenced to 1 watt
N_{dBm} = sensitivity in dB referenced to 1 milliwatt = $N_{\text{dBW}} + 30$ dB
$\log$ = common logarithm (base 10)

Illustrative Example 3-1

Determine the RS (receptor susceptibility) rating of a receiver having a sensitivity $N = -104$ dBm, an input impedance of 50 ohms, and a bandwidth of 1 MHz. For this situation, equation (3-4) is used as follows:

$$RS_p = 24 \text{ dB} + 20 \log (10^6) \text{ Hz} - 10 \log (50) - (-104 \text{ dBm})$$
$$= 24 \text{ dB} + 120 \text{ dB} - 17 \text{ dB} + 104 \text{ dBm}$$
$$= 231 \text{ dB}$$

This same receiver exhibits a noise figure of 10 dB or a noise factor of 10. For this situation, equation (3-3) yields (5):

$$RS_v = 198 + 10 \log (10^6/(50 \times 10)) \text{ dB}$$
$$= 198 + 33$$
$$= 231 \text{ dB}$$

To obtain a feel for typical ranges of RS, a few values are computed and tabulated in Table 3-2 for different amplifier types.

From Table 3-2, an RS level ranking can be created in terms of susceptibility value judgments corresponding to all types of receptors. One such relative rating appears in Fig. 3-5 for typical receivers and amplifiers. The RS rating scores shown range from > 230 dB (extremely susceptible) to < 80 dB (rarely susceptible). A number of interesting observations can now be made:

- Amplifiers that have a voltage sensitivity independent of bandwidth (e.g., base-band video amplifiers driven from a second detector with about 0.2 volt threshold) have an RS slope of 20 dB/decade of receptor bandwidth [N_{dBm} is independent of bandwidth in equation (3-4)].
- Amplifiers and receivers that have noise-limited front ends have an RS slope of about 10 dB/decade of receptor bandwidth [N_{dBm} decreases at

Table 3-2. Relative susceptibilities of selected amplifier types.

Amplifier type	N_{dBV}	Z-Level (ohms)	N_{dBm}	Bandwidth	RS (dB)
TWT amplifier	+37	50	−70	26 Hz	263 dB
I-F amplifier	+37	50	−70	1 MHz	197 dB
Crystal-video RX	+75	1k	−45	3 MHz	169 dB
Video amplifier	+106	50	−1	1 MHz	128 dB
Audio amplifier	+100	600	−18	10 kHz	94 dB
Sensor amplifier	+60	1k	−60	100 Hz	94 dB
Digital data RX	+126	150	+14	1 MHz	108 dB

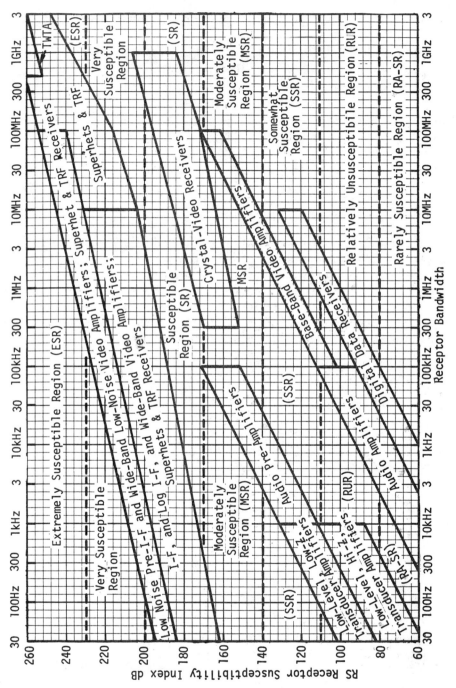

Fig. 3-5. Typical receptor susceptibility scores of receivers and amplifiers. (*Courtesy of Interference Control Technologies*)

10 dB/decade, and $20 \log (B)$ increases at 20 dB/decade in equation (3-4)].

- The spread in RS values for receivers and amplifiers is in excess of 200 dB—an enormous range!
- Broadband amplifiers and receivers are more susceptible than their narrowband counterparts.
- Low-noise-figure (2–20 dB) receivers, pre-I-F, and wide-band video amplifiers are very susceptible receptors.
- Digital circuits are less susceptible than low-noise video amplifiers having the same bandwidth because the noise immunity level of digital circuits is higher.
- Low-level (0.1 mV), narrowband transducer amplifiers are relatively unsusceptible. However, low-impedance, base-band transducer amplifiers are more susceptible than high-impedance amplifiers at low frequencies.
- Assuming all amplifiers and receivers exhibit a 60-dB rejection to out-of-band EMI emission, wide-band, low-noise devices still evidence moderately susceptible characteristics.

Communications-Electronics (C-E) Receivers

Figure 3-1 shows that C-E receivers comprise one of the categories of man-made receptors. In Chapter 2, it was shown that these systems are also potential emitters. This class is generally the most susceptible to EMI as indicated by their RS rating, which varies from 160 dB to over 260 dB (see Fig. 3-5). Wherever these systems are employed, their EMI susceptibility must be considered a significant factor.

Broadcast C-E Receivers. A tabulation of broadcast band receivers, their RS ratings, and (somewhat subjective) susceptibility ratings is provided for comparison.

Broadcast band	Frequency range (MHz)	RS rating (dB)	Relative susceptibility
AM	0.530–1.650	195	Susceptible
VHF FM	88–108	215	Very susceptible
VHF TV:			
Lower band	54–88	230	Extremely susceptible
Upper band	174–216	230	Extremely susceptible
UHF TV	470–890	225	Extremely susceptible

In addition to man-made sources of EMI, AM broadcast receivers are also susceptible to broadband atmospheric noise. FM and TV receivers, while immune to atmospheric noise, are quite susceptible to automobile ignition noise if their associated antennas are located near roads and are not located well above the ground (e.g., are less than 10 meters in height).

C-E Communication Receivers. Communication equipments are the greatest in number and most varied of all C-E types. They occupy portions of the frequency spectrum interlaced between other activities from about 10 kHz to about the 2-GHz range. Many of the C-E receivers are of the land-mobile type. Above 2 GHz, point-to-point communications are generally of the relay type. *RS* ratings for communications receivers vary from 150 dB (moderately susceptible) at VLF to about 235 dB (extremely susceptible) at UHF.

Since many C-E receivers are of the land-mobile type, they are susceptible to automobile ignition and nearby industrial-area noise. This is a lesser problem for military land-mobile receivers than for other types, since military vehicles employ more ignition suppression devices than other vehicles. There, co-channel and adjacent channel interference can be particularly troublesome because of the rapid change of combat frequency assignments.

C-E Relay Communications Receivers. Point-to-point relay communications generally consist of one or more of four types (see tabulation).

Relay type and frequency range	*RS* rating	Relative susceptibility
Common carrier, microwave relay (2.1–11.7 GHz interspersed)	245 dB	Extremely susceptible
Satellite relay (2.4–16 GHz interspersed)	225 dB	Very susceptible
Ionospheric scatter (400–500 MHz)	220 dB	Very susceptible
Tropospheric scatter (1.8–5.6 GHz interspersed)	230 dB	Extremely susceptible

It is interesting to note that microwave relay receivers, per se, are the most susceptible to EMI (*RS* rating of 245 dB). Yet, pragmatically, microwave relay links are relatively immune to EMI. This is due to both the enormous off-axis interference rejection of their antennas (typically 60–70 dB) and high *S/N* (signal-to-noise) ratios (typically about 50 dB) of the links. This illustrates what can be done to obtain good EMI control in system equipment design.

C-E Navigation Receivers. Navigation receivers in this classification exclude radars, since radars are covered in the next section. The *RS* rating of navigation receivers varies from about 170 dB (moderately susceptible for certain VLF receivers) to about 230 dB (extremely susceptible) for certain UHF types.

C-E Radar Receivers. Radars are used in intermittent portions of the spectrum from about 225 MHz to 35 GHz. They are employed in many civilian and military capacities including: air traffic control, air and surface search, harbor surveillance, mapping, tracking and fire control, police speed-monitoring, and weather.

Except for narrowband Doppler police and CW radars, radar receivers are usually broadband and operate at higher frequencies. Hence, their *RS* rating typically is about 230 to 240 dB (extremely susceptible). This is a bit ironic, since radar transmitters are significant offending sources of EMI because of their high effective radiated powers (+95 dBm to +135 dBm), and high operating frequencies.

Amplifier Susceptibility

The preceding section summarized C-E receivers and stated that they are generally the most susceptible of the four man-made receptor categories shown in Fig. 3-1. There exist a few exceptions such as when low-noise, wide-band video amplifiers exhibit higher *RS* ratings than receivers. Many applications of such amplifiers, however, improve the sensitivity of receiver front-ends, and therefore become a part of the C-E receiver per se. On the other hand, low-noise, wide-band video amplifiers are used in other applications.

Intermediate-frequency (I-F) amplifiers are part of C-E receivers by definition because they are an integral part of the superheterodyning process. Video (not low-noise, wide-band types) and/or audio amplifiers are also always found in superheterodyne receivers. In contrast to I-F amplifiers, however, video and audio amplifiers are used in many applications other than receivers. Since all these amplifiers can constitute a separate class of potential EMI victims, whether or not they appear in C-E receivers, they are considered as a distinct category of man-made receptors in Fig. 3-1.

I-F Amplifiers. I-F amplifiers constitute another input terminal pair to which a receiver may be susceptible. Culprit emissions ordinarily enter a receiver input terminal by the antenna and sometimes as pickup (leakage) by the lead-in RF transmission line. If the input interference signals exist at intermediate frequencies, the receiver front-end rejection may or may not be adequate to prevent a susceptible situation. Assuming such rejection to I-F emissions to be adequate, it may still be possible for these emissions to bypass the receiver front-end and be picked up by the I-F input cable. One possible example occurs when the receiver front-end is located upon a ship's mast with an antenna, and the I-F signal is piped down to the deck where the remainder of the receiver is located. Here, the long I-F coaxial cable acts as a pickup antenna, rendering the I-F amplifier (and remainder of the receiver) susceptible.

An I-F amplifier may have more or less sensitivity than its associated overall receiver sensitivity. This depends upon the loss of the R-F input cable; the loss or gain, noise figure, and bandwidth of the pre-mixer circuitry; the conversion loss or gain of the converter; and the noise figure and bandwidth of the I-F amplifier. This topic is discussed in further detail in Chapter 5.

Figure 3-5 shows that low-noise, pre-I-F amplifiers may have *RS* ratings comparable to receivers. These ratings typically vary from about 190 dB (susceptible) for small bandwidths to about 240 dB (extremely susceptible) for very large bandwidths.

Figure 3-5 also shows less sensitive I-F amplifiers which typically accommodate receivers having pre-mixer R-F gain, first-converter gain, and/or pre-I-F amplifiers. Such I-F amplifiers may have *RS* ratings that are 10 to 30 dB less than those of low-noise, pre-I-F amplifiers. Their susceptibilities, however, are still significant, especially when operated under conditions in which they are physically removed from their receiver front-ends by a long interconnecting cable as discussed above.

Video Amplifiers.   There are many types of video amplifiers. One simple classification of such amplifiers is: (1) low-level video amplifiers and (2) baseband, high-level video amplifiers. A few examples of each classification are:

1. Low-level video amplifers:
   - Receiver front-end noise figure improvement
   - Crystal-video receiver pre-amplifiers
   - Nontunable, low-noise receivers
   - Low-noise oscilloscope amplifiers

2. High-level video amplifiers:
   - Receiver post-detector amplifier
   - Digital-data receivers
   - Telemetry data amplifiers
   - A/D and D/A converters

The distinguishing features of the above video amplifier classification are that low-level types are generally noise-limited by either typical 50-ohm (sometimes 72-ohm, or other) input impedances, or high-level impedances (100-KΩ, 1-mΩ, or other). On the other hand, high-level video amplifiers are generally characterized by requiring a minimum voltage to perform. For example, receiver post-detector video amplifiers typically need a level of about 0.2 volt because of the second detector characteristics. A second example is digital-data receivers which may require a minimum level of about 2 volts to sense a mark or "1-bit."

Figure 3-5 shows that low-noise video amplifiers have *RS* ratings comparable to receivers and pre-I-F amplifiers (e.g., from about 190 dB to 240 dB). Unless protected by a low- and/or a high-pass filter, the wide-band, low-noise amplifier is highly susceptible to EMI emissions and intermodulation (see Chapter 5). This happens because these amplifiers generally operate over a broad-frequency base-band, having a low-frequency cutoff of perhaps 100 Hz to 1 kHz, where the amplitudes of many broadband EMI emissions are the highest.

Low-noise video amplifiers can even have pass bands of 1 GHz or more due to recent advances in solid-state circuitry, where gain–bandwidth products of active devices may be of the order of 3 GHz or more. Here, Fig. 3-5 can be extrapolated to yield *RS* ratings of 250 to 260 dB—an extremely susceptible range.

Finally, when the low-noise, wide-band video amplifiers have a high low-frequency cutoff, they take on other names such as traveling-wave tube (TWT) amplifiers, as shown in Fig. 3-5. Low-noise TWTAs have the highest RS ratings of all (250 to 270 dB). Fortunately, broadband EMI noise and/or C-E transmitter spectrum activity in the 10-GHz and higher-frequency regions renders the TWTAs less susceptible than their latent *RS* ratings would indicate. From 500 MHz to 10 GHz, however, they are extremely susceptible to EMI.

The high-level video amplifiers pose an altogether different susceptibility situation. Figure 3-5 indicates that *RS* ratings may vary from about 90 dB (relatively unsusceptible) for 100-KHz, base-band bandwidths, to about 160 dB (moderately susceptible) for 30-MHz bandwidths (high data rates and/or short pulse widths). For similar bandwidths, this corresponds to 80- to 130-dB smaller susceptibilities than those of their low-level video amplifier counterparts. Nevertheless, the high-level, wide-band video amplifiers such as those used in high-clock-rate digital receivers and computers can be susceptible to EMI.

Audio Amplifiers.   There are also many types of audio amplifiers. One simple classification of such amplifiers somewhat parallel that for video amplifiers, except that 100 kHz is arbitrarily defined here as the upper frequency limit for audio amplifiers. The classification suggested as: (1) low-level audio amplifiers and (2) high-level audio amplifiers. A few examples of each classification are:

1. Low-level audio amplifiers:
   • Audio pre-amplifiers
   • Low-noise oscilloscope amplifiers

- Low-level sensor amplifiers
- Biomedical instruments

2. High-level audio amplifiers:
   - Telephone-line repeaters
   - Public address amplifiers
   - High-fi stereo power amplifiers
   - Chart recorder drivers

The distinguishing feature of the above audio amplifier classification is that low-level types are often noise-limited with either low input impedances (approximately 10 ohms) or high input impedances (10 K, 100 K, or other). If not noise-limited, they have input sensitivities of less than about 10 mV. On the other hand, high-level audio amplifiers are generally characterized by requiring a minimum voltage to perform, such as about 100 mV, as in the case of many hi-fi audio power amplifiers.

Figure 3-5 shows that audio pre-amplifiers have RS ratings ranging from about 100 dB (somewhat susceptible) to 170 dB (susceptible). If the pre-amplifier is used for a low-level, high-impedance transducer or oscilloscope application, Fig. 3-5 shows that it may be 30 to 40 dB more susceptible than its lower-impedance counterpart.

Another EMI problem associated with low-level video amplifiers is their out-of-band rejection to AM and FM broadcast, TV, radar, and other high-level emissions. This is the audio rectification problem discussed in connection with Fig. 3-3. Typical rejections vary from about 60 dB to 100 dB. Thus it is not uncommon for EKG biomedical instruments having a 100-volt sensitivity and 100-Hz bandwidth, for example, to be susceptible to a 100-watt, 30-MHz physician's paging system in a hospital. Other, better-known examples are hi-fi stereo amplifiers and public address systems, which are susceptible to nearby FAA or military radar, and land-mobile transmitters.

The high-level video amplifiers pose an altogether different susceptibility situation. Figure 3-5 indicates that RS ratings may vary from about 60 dB (rarely susceptible) to 110 dB (somewhat susceptible). Thus, audio power amplifiers and audio signal boosters are not often susceptible to EMI.

Industrial and Consumer Receptors

The preceding two sections discussed communications-electronics receivers and amplifiers as two of the four classes of man-made receptors shown in Fig. 3-1. This section presents a third class: industrial and consumer receptors. The concept of an RS (receptor susceptibility) rating, introduced previously, will be continued here to establish some measure of relative susceptibility to EMI.

Industrial and consumer receptors are electrical, electromechanical, and electronic systems, equipments, and/or products that may malfunction totally or degrade in performance in the presence of EMI ambients indigenous to these items. Most of them require amplifiers of one form or another in order to accomplish their intended function. As such, they sense lower-level signals and deliver higher-level outputs, which accounts for their susceptibility with respect to amplifiers. Some examples of such systems, equipments, and products are:

1. Industrial receptors:
   - Computers
   - Industrial process controls
   - Electronic test instruments
   - Biomedical instruments
   - Public address systems
   - Telephone and teletype
   - Electronic security alarms
   - Electric recorders
   - Land-mobile receivers

2. Consumer receptors:
   - Radio and TV receivers
   - Hi-fi stereo equipment
   - Electronic musical instruments
   - Climate control systems
   - Heart pacemakers
   - Automobile, boat, and aircraft microprocessor systems

The following sections outline a few of the industrial and consumer receptors. It is noteworthy that relatively little is reported in the technical literature about the susceptibility of many of these receptors. Consequently, relatively few quantitative results are reported here.

Digital Computers.   Chapter 2 indicated that computers and peripherals both conduct and radiate electromagnetic energy. Most such emissions emanate from the higher-level peripherals, including disk drives, character and line printers, and the like. Computers, on the other hand, may be susceptible to EMI because they generally operate at lower levels than their peripherals. For example, the computer areas most sensitive to pulse and transient emissions are the amplifier circuits whose direct inputs are from low-level output storage devices, such as magnetic tapes and disks (most susceptible) and drum and ferrite cores (less susceptible).

Other aspects of computers that make them susceptible to EMI include power-line conducted transients, ground shifts, and electrostatic discharge.

All are characterized by short-duration emissions that are sensed as a mark (1-bit) or space (0-bit), resulting in character or word-error readouts. Line transients are superimposed on the AC power system and may couple to the computer logic bus, sense amplifiers, or logic circuits. Ground shifts are transient potential differences between two or more portions of the computer ground reference. Thus, common-mode impedance drops result in sensed logic errors. Personnel-induced electrostatic discharge where carpets are located near computers or static discharge associated with moving belts, paper, and tape can also couple common-mode and differential-mode noise into logic circuits.

Industrial Process Controls. Many manufacturing and other industrial processes are becoming increasingly automated. A myriad of operations are affected, including: automatic machinery operations, in-plant climate control, oil refining and processing, item weighing and counting, steel fabrication parameter control, and so on. Nearly all are characterized by a closed-loop servo system in which intended controls are injected (often from minicomputers and microprocessors), results are sampled, departures or errors are sensed, and an adjusted control signal is generated. Thus, these industrial devices contain analog amplifiers and digital control circuitry. Consequently, they are susceptible to the same type of EMI sources as amplifiers and computers.

Because servo amplifiers may have a frequency response to DC of perhaps only a few Hertz, it does not follow that they are not responsive to higher-level AM broadcast, FM, TV, radar, and other high-frequency ambients. While offering 60- to 120-dB rejection to such radiated out-of-band EMI emissions, servo amplifiers and associated control processes will often malfunction when the electric-field intensity exceeds about 5 V/m. Not infrequently, this results from EMI pickup in the sense amplifier harness or cable. Thus, cable shielding and shield grounding becomes an EMI-control requirement.

Electronic Test Instruments. This class of devices is also very numerous in type and variety. A few examples include: oscilloscopes, frequency and time-event counters, wave and spectrum analyzers, recorders and $X-Y$ plotters, audio frequency (AF) and radio frequency (RF) millivoltmeters, impedance bridges, and so forth. Again, nearly all of these instruments contain amplifiers of one form or another. Thus, they may be susceptible in the same sense as the amplifiers discussed in a previous section. The mode of susceptibility (EMI entry) may be either power-line conduction or cable pickup of radiation, as well as case leakage.

Biomedical Instruments.   This class of devices is similar to electronic test instruments except that upper 3-dB frequency responses are often limited to about 100 Hz or less. Consequently, their susceptibility, on the average, is less than that of electronic test instruments. A few typical biomedical instruments are EKG and EEG recorders, arterial and venous pressure monitors, and respiration indicators. Because they all use amplifiers to boost lower-level transducer sensors for display purposes, they may be susceptible to both 60-Hz power-line emissions and RF ambients. Regarding the latter, it is not uncommon in a hospital for an EKG monitor to pick up a 30-MHz physician's paging signal through its sensor leads via audio rectification, as discussed previously.

Public Address Systems and Intercoms.   Public address systems are occasionally susceptible to EMI emissions because their microphone lead-in cables to the amplifiers may be several feet long. The cable thus can act as a pickup antenna to electromagnetic ambients, and the latter may be demodulated and amplified along with a voice message. (See discussion of audio rectification.) Intercoms (intercommunication system between different rooms of an office or private residence) may pose an even greater susceptibility problem, due to the substantial amount of multi-conductor wiring that is routed back and forth between station speakers and a central control unit and amplifier. Here, the cable to the master station can run 100 feet or more and act as a pickup antenna.

Radiation Hazards to Ordnance.   The fourth class of man-made receptors shown in Fig. 3-1 involves radiation hazards to electro-explosive devices (EEDs) and fuels. Electro-explosive devices are used to ignite explosives electrically by the application of a specified electrical current; input cables may act as pickup antennas (by induction or radiation coupling), thereby delivering unwanted electromagnetic energy to the susceptible igniter. The ignition of fuels, on the other hand, generally is due to sparks from structural members immersed in a high-RF field, motor brush–commutator interfaces, inductive switching transients, or the electrostatic discharge that results from the accumulation of charge on a nonconductive or electrically isolated metal body.

  The U.S. Department of Defense is actively engaged in determining the extent of radiation hazards and methods for controlling the same under the program name RADHAZ (for RADiation HAZards). The Navy has been involved in the RADHAZ problem in connection with ordnance programs known as Hazards of Electromagnetic Radiation to Ordnance (HERO). Initiated in 1958, HERO covers research and development with special attention focused on EEDS. Primary research by the Air Force has been related

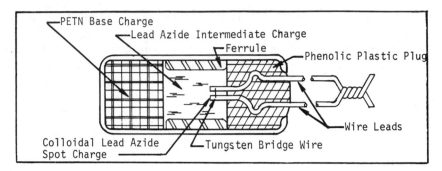

Fig. 3-6. Wire-bridge-type electric initiator. (*Courtesy of Interference Control Technologies*)

to the biological effects of radiation, and some studies have covered the effects of RF radiation on volatile fuels (SPARKS).

The RF power dissipated in an EED depends upon the characteristic of an EMI signal appearing across the EED leads, the impedance characteristics of both the EED and the leads, and other factors including circuit geometry. Electric initiators are classified under the following types: high-resistance wire, low-resistance wire, carbon (graphite) bridge, conductive film, semiconductor, and spark gap. The resistance wire and the carbon bridge are the most commonly used EEDs. Combinations of more than one of these types are used to obtain special characteristics for a particular application.

The wire-bridge detonator uses a fine wire of tungsten or some noble metal between two electrodes to form the bridge. Electricity flowing in the bridge heats the wire and ignites a spot charge that in turn, sets off the detonator base charge. A typical low-energy bridge detonator is shown in Fig. 3-6. The bridge resistance ranges from 2 to 5 ohms and is made of tungsten wire. The detonator is rated at a nominal energy level of 5000 ergs at less than 10 microseconds, with a capacitor charged to 50 volts. The bridge wire is coated with an explosive spot charge.

The carbon-bridge initiator shown in Fig. 3-7 uses a colloidal-graphite charge to form the electrical detonation circuit path. It functions like the wire bridge, but is more sensitive than the wire. The lead wires inside the plug are coated with insulating varnish and twisted to attain the small separation at the face of the plug needed for graphite-bridge detonators. This detonator will function in 10 microseconds, with 300 volts applied from a 2200-pf capacitor.

The explosive mixture contains conductive material that forms the electric circuit. The heating effect of current in this mixture causes detonation. The conductive material is a mixture of metals and graphite, and is mixed with explosives. Conductive mixtures vary in resistance from approximately 1 ohm to 800 ohms.

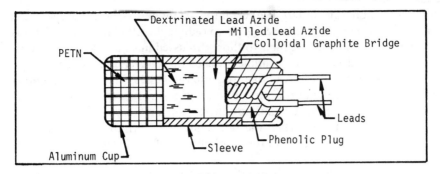

Fig. 3-7. A carbon bridge electric initiator. (*Courtesy of Interference Control Technologies*)

Because of manufacturing difficulties with the wire-type bridge and its relatively low sensitivity, a deposited-metal film initiator is sometimes used. Titanium smears on a glass base have detonated the explosive with an average energy of 50 ergs at 30 to 40 ohms resistance.

Three principal modes of electrical initiation of EEDs are: arcing, heating, and shock wave. Arcing exists at levels greater than 25 volts and usually occurs in carbon-bridge and thin-film initiators. Hot-wire initiators, as well as conductive film and conductive-mix initiators, operate by the heating mode. The third mode of initiation occurs in the exploding-bridge, wire-type initiator. The shock-wave mode requires more than 300 volts to cause the formation of an intense shock wave that directly initiates a secondary high-explosive. Most initiators can operate under more than one mode, depending on the magnitude of the electrical stimulus.

EEDs also may be activated by unintentional sources. The conventional sources are those designed into a firing circuit to supply a controlled amount of energy to initiate the EED at the time of firing. Some circuits use a capacitor bank that is slowly charged some time prior to firing. Unintentional activating sources are those that couple sufficient EMI energy to the EED to result in inadvertent ignition.

When RF radiated energy may be a potential cause for unintentional firing of EEDs, the transmission path of a firing circuit becomes an important factor. Parameters that define the RF source and thereby provide information for the design of protective measures are field intensity or power density, type of modulation and duty cycle, radio frequency, and polarization. Other important factors include electromagnetic coupling and thermal parameters. Here RF waves propagating in free space become guided waves when coupled into a firing circuit. The thermal parameter represents the mechanism by which RF initiation of EEDs takes place by heating.

A number of mathematical models have been used to predict the probability of firing an EED when the EMI sources are radars and/or nearby communication transmitters. Rather simplified models assume that the source transmitter antennas were boresighted to radiate directly into the EED leads. Using prediction models similar to those discussed in Chapter 5, the effective area of the lead circuit is an indication of the conversion of the arriving power densities to power available to ignite the EEDs. The most susceptible path is usually not the pin-to-pin situation, but involves the common-mode pin-to-case circuit geometry and parameters such as the pin-to-case capacitance and conductance. When the prediction is superimposed on the EED known firing characteristics, the probability of EED initiation can be determined.

Nearly all the above mathematical models yield pessimistic results for radar EMI sources and optimistic results for some communication transmitter sources. This is so for radar models because the simultaneous superposition of a boresight condition and matched polarization, ignoring the effect of EED lead twist and/or lossy material and underestimating the shielding effect of an intervening barrier (e.g., missile skin), necessarily results in a dangerous cumulative probability that, in effect, may exist arbitrarily close to 0% of the time. Thus, the recommended measures are to shut down radars, to the chagrin of a ship's captain or a range safety officer. Some communication models, on the other hand, fail to recognize that an entire missile skin or aircraft can act as a pickup antenna and capacitively couple intercepted energy to the EED leads.

Among the additional parameters complementing the prediction process are personnel proximity to or contact with the EED or its housing, structure of the EED container and ground plane, openings in the EED container, configuration of lead wires from control stations to EED or container, and RF impedance of the firing circuit. As a result, some attempt has been made to establish safe distances of separation between explosives and EMI emitter sources. Although not particularly useful, they have been used as one measure of RADHAZ protection (see References).

SUSCEPTIBILITY CRITERIA

The preceding sections discussed man-made receptors of EMI. The receptor susceptibility (RS) rating concept was based upon the notion of signal (S) equal to noise (N) as part of the basis for developing relative susceptibility scores. The $S = N$ basis may not always be an acceptable criterion for determining susceptibility because:

- Different detection systems are affected differently by the same interference or noise spectral intensity levels.

- Different readout and display systems may respond differently to the same detection systems.
- Different user requirements impose different interpretations on any given level of receptor performance.

Thus, it is necessary to introduce the concept of receptor susceptibility criteria in order to relate any given interference level referenced to the input of a receptor to the impact upon the user's satisfaction with the degree of performance.

Another way to introduce the receptor susceptibility criteria is to recognize that some receptors either perform correctly or malfunction entirely. A few examples of these black and white (binary) situations are:

- An electroexplosive device (EED) either detonates, or it does not (e.g., for explosive bolts, an associated fuel wing tank is dropped, a canopy and pilot are jettisoned, a missile is launched, etc.).
- A carrier-operated relay is closed, or it is not.
- Any two-state output device from the receptor is either in state "0" or state "1."

The problem, however, usually is more complex than this because most receptor output devices are not two-state output devices. In fact, they are not three-state or n-state, but rather they represent a continuum of degradation (a very large or infinite state) from white to black, that is, from perfect performance to complete malfunction. Consequently, it is necessary to relate the status of performance versus $S/(N + I)$ referred to either the input of the receptor or to its output, where I is the interference signal level. The next sections summarize this for voice intelligibility, digital-error acceptance, TV-picture displays, radar displays, and other outputs.

Voice Intelligibility

The problem of specifying an operational performance measure for voice communication systems is complicated by the random nature of transmitted and received voice signals, variations in message content, and differences in hearing and understanding abilities from one receiver operator to another.

One performance measure used for voice systems is an articulation score, obtained by using trained talkers and listeners to determine the percentage of words scored correctly by the listener out of the total number of words contained in the test.

The procedure used in an articulation test consists of a talker (or a standardized voice generator such as a tape-recorded voice) reading a set of

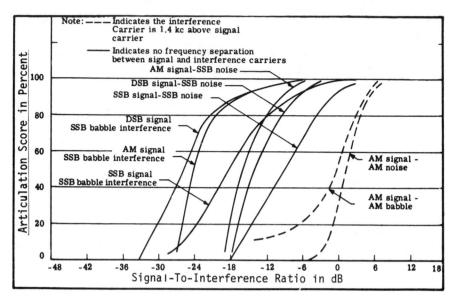

Fig. 3-8. Voice system performance. (*Courtesy of Interference Control Technologies*)

selected words or syllables over a communication system (which may be subjected to interference). The listener panel interprets what it hears. Various levels of interference may be introduced into the receiver system along with the selected words. The percentage of words interpreted correctly by the listener indicates the intelligibility level or articulation score for the particular set of conditions tested. Resulting empirical data can then be translated into suitable electrical characteristics (such as signal-to-interference ratio) which in turn can be used in an EMI prediction process to determine voice system performance.

Figure 3-8 shows the relationship between signal-to-interference ratio and articulation score for different combinations of desired and interfering signal conditions. All of the cases illustrated are for co-channel interference conditions. One very significant factor, evident in the figure, is that there is a fairly rapid transition from good to poor performance versus S/I ratio.

Another method for specifying performance of voice communications systems is the articulation index. This method and the above articulation score are both treated at length in Volume 5 of the DWCI handbook series (6).

Digital Communication Systems

The evaluation of performance for a digital system consists of calculating the probability of error. Two basic types of errors are false acceptance (i.e., mistaking interference or noise for the signal) and false dismissal (i.e., not

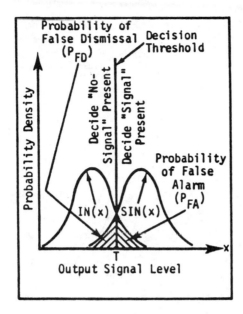

Fig. 3-9. Binary decision process. (*Courtesy of Interference Control Technologies*)

recognizing the presence of the signal). For on–off binary transmission, false acceptance is equal to the probability of false alarm, and false dismissal is equal to one minus the probability of detection. Probability of false alarm is the conditional probability of deciding that a signal is present when no signal was transmitted. Probability of detection is the conditional probability of deciding that a signal is present, given that a signal was transmitted. The relative occurrence of false acceptance and/or dismissal can be determined from the probability densities for signal, interference, and noise at the receiver output.

The relationship between the basic decision process and the two types of errors is illustrated in Fig. 3-9. The density function, designated $IN(x)$, refers to the output probability distribution density when interference and noise are present, while $SIN(x)$ is the output distribution density when a signal, interference, and noise, $S + I + N$, are present. Decision regions are defined such that when the output exceeds a certain threshold, T, the decision is "signal present," whereas if the output is less than T the decision is "no signal present" (3).

Figure 3-10 illustrates the relationship between signal-to-noise ratio and error rate. Many types of interference result in noiselike signals at the receiver output, and for these situations Fig. 3-10 provides a good approximation of the error rate. One very significant factor, evident in the figure, is that there is a rapid transition from good to poor performance.

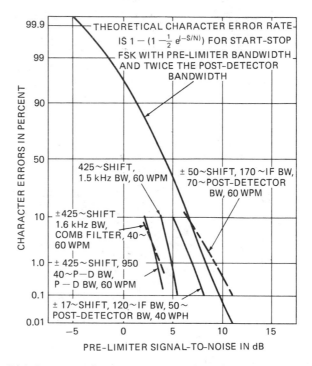

Fig. 3-10. Digital system performance. (*Courtesy of Interference Control Technologies*)

Picture Communication Systems

Television and facsimile systems transmit information that is eventually displayed in the form of a picture, an important form of communication. Aside from the many TV sets currently in use, this form of communication is experiencing increasing use by law-enforcement and criminal justice agencies for transmitting mug-shots and line-ups to neighboring agencies. The picture phone, another example of picture communication, is expected to be widely used in the future.

Interference can degrade picture transmission by introducing dots, lines, or bars, causing the picture to be blurred, or causing the receiver to lose synchronization and roll. Figure 3-11 shows typical effects of different types of interference on TV. Television receivers (particularly color TV receivers) are relatively sensitive to interference. For example, with pulse interference such as would be produced by a radar, a 15-dB ratio of peak signal to peak

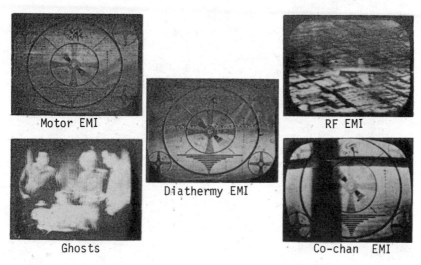

Motor EMI

Diathermy EMI

RF EMI

Ghosts

Co-chan EMI

Fig. 3-11. Interference to TV pictures. (*Courtesy of Interference Control Technologies*)

interference is required to avoid picture degradation in the form of snow in the picture.

The effects of interference on the performance of picture communication systems are somewhat subjective. One technique for rating television performance in the presence of interference is based on establishing six rating grades, as follows: (1) excellent, (2) fine, (3) passable, (4) marginal, (5) inferior, and (6) unusable. An extensive measurement program was conducted using the above rating scheme, and approximately 38,000 ratings were obtained on color and monochrome TV pictures having different injected interference. Nearly 200 observers participated in these experiments (7).

Representative results are shown in Figs. 3-12 and 3-13 for co-channel interference from another TV station and for random noise interference, respectively. For co-channel interference, the signal-to-interference ratios required for 50% or more of the observers to give the viewed picture a passable score (or better) show a decrease of S/I with an increase in separation of co-channel interference frequency. For random noise interference, the S/I requirement for at least a passable rating by 50% of the observers was +27 dB on the basis of root-mean-square (RMS) sync amplitude to RMS noise over the 6-MHz TV channel.

Radar Systems

It has been estimated that over 12,000 high-power radars are currently operating in the continental United States. Many of these radars are situated

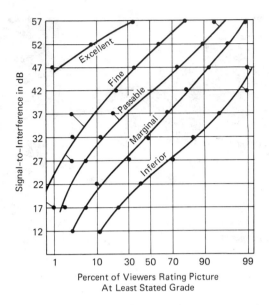

Fig. 3-12. Co-channel interference 604 Hz carrier separation. (*Courtesy of Interference Control Technologies*)

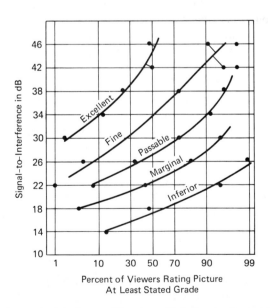

Fig. 3-13. Random noise interference. (*Courtesy of Interference Control Technologies*)

in congested areas, and, because of real estate considerations, they are fre quently co-located with other equipment at specific sites such as airports, military bases, and missile-launching sites. In addition, there has been, and will continue to be, increasing use of radar in airborne systems, navigational aids, weather observation, and satellites and space probes. This trend toward greater spectrum usage and congestion has resulted in a mutual interference problem that is becoming increasingly complex.

Surveillance radar systems are used in a wide variety of applications where it is necessary to monitor a relatively large area. The primary use of surveil lance radars is to monitor marine and air traffic for the purpose of national defense and traffic control. A typical surveillance radar provides an operator with a scope display of both angular position and range of objects within the radar coverage area. The scope used for this type of display is referred to as a plan position indicator (PPI).

The most common form of interference is the appearance of interfering dots or spirals on the radar scope presentation caused by pulse interference from other radars. This type of interference (called rabbits) is usually moving continuously and may cover a large portion of the scope face, making targets difficult to detect. Interference of this type is annoying to the operator, and over a period of time it causes fatigue, thus reducing effectiveness. If the interference sector contains a target, delayed detection is likely to result. If the interference is extreme, false target reports are likely.

One measure of operational performance that is used for this type of inter ference to surveillance radars is scope condition. Interference effects on radar PPI scopes have been classified into five scope conditions, which are illus trated in Fig. 3-14.

Other Receptor Outputs

The foregoing discussion emphasized communications-electronics (C-E) equipment susceptibility criteria. However, there are many other, non-C-E receptors. For example, the following systems and outputs are likely to be impacted by different EMI signals:

- A digital computer intervening between monitoring and displaying the status of many patients in a hospital intensive care ward.
- A digital computer time-sharing the debiting and crediting of bank cus tomers' accounts.
- The control systems of an urban rapid transit rail system.
- The control system of a computer-fed automatic milling machine.
- The monitoring and control system of an electrical utility substation.
- The transmission and status display of stock-market transactions.
- The recorder output of a patient's EKG.
- The output of a polygraphic recorder.

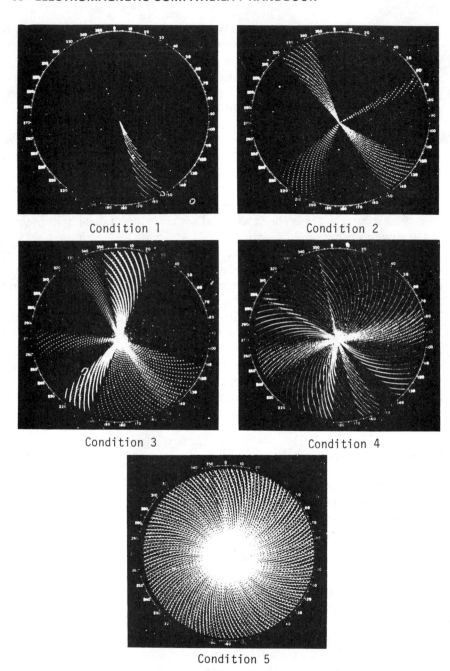

Condition 1

Condition 2

Condition 3

Condition 4

Condition 5

Fig. 3-14. Typical scope conditions.

An analysis of the impact of EMI on the situations described above indicates a wide range in the degree of EMI and variation in the potential consequences. Depending upon the degree of manifested EMI, some of the above situations might only flag the operator that the data are not reliable or repeatable; some might cause medical doctors to improperly diagnose a patient's condition and result in incorrect treatment; some might cause a customer to lose many dollars; but some could result in death to a number of people or cause a power blackout over wide areas affecting millions.

The topic of susceptibility criteria is often dependent upon specific scenarios. Since each of the above EMI situations has a precedent, this discussion is intended to alert the reader to the fact that susceptibility criteria are necessary and should be determined in quantitative terms if possible. Sometimes the criteria are established by negotiation and mutual agreement whereby unacceptable failure or system degradation becomes defined for specific situations.

ANALOG NOISE THRESHOLD

Analog and digital circuits are essentially subjected to conducted and radiated EMI from the same sources. However, analog and digital circuit response to EMI differs widely between the two classes of circuits, with the analog circuits generally being the more sensitive to EMI and therefore the more susceptible. The undesired response, or degree of system degradation due to EMI, depends upon the susceptibility level of the most sensitive system circuits and the level of noise coupling. For analog systems, the susceptibility can be expressed in terms of the noise level (N), with the system performance predicated upon the signal-to-noise ratio, S/N. The thermal noise level is the principal factor that limits the sensitivity of a receiving system. It is expressed as a function of the system absolute temperature (T) [degrees Kelvin (°K) above absolute zero] and the receiver noise bandwidth (B) in Hertz (8). For an analog system (such as a communication receiver) operating at an absolute temperature T with a specified bandwidth B, the noise level N is given by:

$$N = KTB \text{ watts} \qquad (2\text{-}5)$$

where K is Boltzmann's constant, equal to 1.38×10^{-23} joule/°K.

The sensitivity of an analog system is determined by the level of N, and, although this quantity is specified for certain systems such as a receiver, an absolute "threshold" for signal detection is difficult to define deterministically. Any signal that exceeds the level of N can be said to have a "high probability" of being detected. Values of N for communication receivers can be as low as -100 dBm to -110 dBm. For a receiver with a 50-ohm input impedance, an input power level of -107 dBm corresponds to an input voltage of 1.0 microvolt, a level that can be exceeded easily by EMI signals coupled into the input circuit, such as input signal cables.

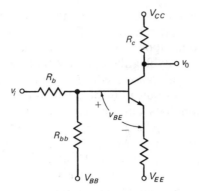

Fig. 3-15. NPN transistor inverter gate.

DIGITAL DC NOISE MARGIN

For digital systems, a threshold of susceptibility to noise can be defined in terms of a noise immunity, or noise margin, based upon the binary or two-state nature of digital circuits (8–11). Figure 3-15 shows a simplified circuit for a logic inverter gate whose transfer characteristic is essentially similar to the illustration of Fig. 3-16, which shows the output voltage, v_o, as a function of the input voltage, v_i (9, 10). In Fig. 3-15, the values of the resistors R_b and R_{bb} are chosen such that when $v_i = V_{iH\,min}$, v_{BE} will be positive, the transistor will be conducting, and the output voltage will be in a low state with $v_o \sim V_{oL\,max}$, neglecting the small drop across the transistor in saturation. When $v_i = V_{iL\,max}$, v_{BE} will be negative, the transistor will be at cutoff, and $v_o = V_{oH\,min}$ (9). With a low-level input voltage v_i, the output voltage v_o will be at a high level, and vice versa.

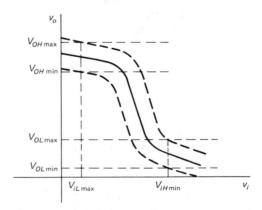

Fig. 3-16. Logic gate inverter transfer characteristics.

It is noted that the inverter shown in Fig. 3-15 is representative only for some technologies. Transistor-transistor logic (TTL), on the other hand, uses multiple emitter transistors as inputs. In this case, the DC/AC performance will depend on the RC time constant of the input gate.

The corresponding values of v_i and v_o will vary from one gate circuit to another because of tolerances in component values, manufacturing processes, and so on. To place bounds upon the variations in v_i and v_o, manufacturers specify the following voltage levels for logic gates (9, 10):

- $V_{oH\,max}$: maximum value of v_o, with the output in the high state, for a given low-level input voltage v_i causing the transistor to be at cutoff.
- $V_{oH\,min}$: minimum value of v_o, with the output in the high state, for a given low-level input voltage v_i. The manufacturer guarantees that with the output in the high state, v_o will never be less than this value.
- $V_{oL\,max}$: maximum value of v_o, with the output in the low state, for a given high-level input voltage v_i. The manufacturer guarantees that with the output in the low state, v_o will never exceed this value.
- $V_{iL\,max}$: Circuit is guaranteed to interpret v_i as "low-level," and produces a corresponding high output v_o, if input voltage does not exceed this value.
- $V_{iH\,min}$: Circuit is guaranteed to interpret v_i as "high-level," and produce a corresponding low output v_o, if v_i exceeds this value.

The above input and output voltages are used to determine the DC noise margins for various logic families. Figure 3-17 shows one logic gate, $G1$, driving another gate of identical logic, $G2$. It is assumed that the inputs to $G1$, designated collectively as v_1, are such that the output v_o is at a low level. According to specifications, v_o is guaranteed not to exceed $V_{oL\,max}$. Since v_o is also an input to $G2$, the latter will interpret v_o as a low-level input, provided that v_o does not exceed $V_{iL\,max}$. The amount of "noise" that can appear superimposed on v_o without causing $G2$ to respond erroneously is defined as the *DC low noise margin* (NM_L), given by reference 9:

$$NM_L = V_{iL\,max} - V_{oL\,max} \tag{3-6}$$

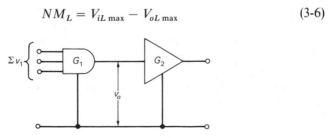

Fig. 3-17. Two digital logic gates coupled together.

Conversely, if the output of $G1$ is at the high level, then v_o will be no less than $V_{oH\,min}$. Gate $G2$ will interpret v_o as a high-level input provided that v_o equals or exceeds $V_{iH\,min}$. In this instance, the amount of "noise" that can reduce v_o from V_{oHmin} to $V_{iH\,min}$ is defined as the *DC high noise margin* (NM_H), given by reference 9:

$$NM_H = V_{oH\,min} - V_{iH\,min} \qquad (3\text{-}7)$$

In general, NM_L and NM_H will not be equal for a given logic family; therefore, the specified overall DC noise margin is the smaller of NM_L or NM_H (9).

Example

Figure 3-18 shows an inverter gate with input and output voltages designated v_i and v_o, respectively. It is desirable to design the amplitude of the waveform of v_i such that when $v_i = V_{i1}$, the gate output v_o is in a low state, and when $v_i = V_{i2}$, the gate output v_o switches to a high state. Based upon the preceding discussion, v_i must be designed such that $V_{i1} = V_{iH\,min}$ and $V_{i2} = V_{iL\,max}$. Corresponding values of v_o are indicated, with the low-state output $V_{o1} = V_{oL\,max}$, and the high state $V_{o2} = V_{oH\,min}$.

The right-most column of Table 3-3 shows the DC noise margins of the most popular logic families.

DIGITAL SYSTEM AC NOISE MARGIN

The DC noise margin is an indication of the susceptibility of a logic family to EMI based upon waveform amplitude considerations. However, during

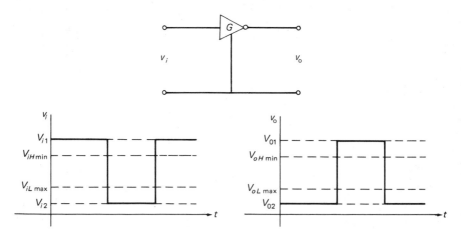

Fig. 3-18. Illustration of inverter gate output voltage waveform (v_o) for given input voltage (v_i).

Table 3-3. Typical characteristics of various logic families. (Courtesy of Interference Control Technologies.)

Logic families	Output voltage swing (V)	Rise/fall time (ns)	Bandwidth (MHz) $1/\pi\tau_r$	DC noise margin (mV)
Emitter Coupled Logic (ECL-10K)	0.8	2/2	160	100
Emitter Coupled Logic (ECL-100K)	0.8	0.75	420	100
FAST (Fairchild)	3	1.75	182	
Transistor-Transistor Logic (TTL)	3	10	32	400
Low Power TTL (LP-TTL)	3	20/10	21	400
Schottky TTL Logic (STTL)	3	3/2.5	120	300
Low-Power Schottky (LS-TTL)	3	10/6	40	300
Complementary Metal Oxide Logic 5V or (15V)	5 (15)	90/100 (50)	3 (6)	1 V (4.5)
High Speed CMOS (5V)	5	10	32	1 V

the dynamic operation of a digital system, switching time delays affect the processing of both desired and EMI signals by logic gates. In a sense, the DC noise margin may be considered as a "static" measure of system logic susceptibility.

Figure 3-19 illustrates a simplified inverter gate with base-emitter and collector-emitter parasitic capacitances, C_{BE} and C_{CE}, indicated. Figure 3-20

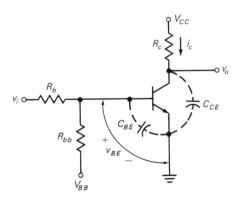

Fig. 3-19. Transistor NPN inverter illustrating parasitic capacitances between base emitter and collector emitter.

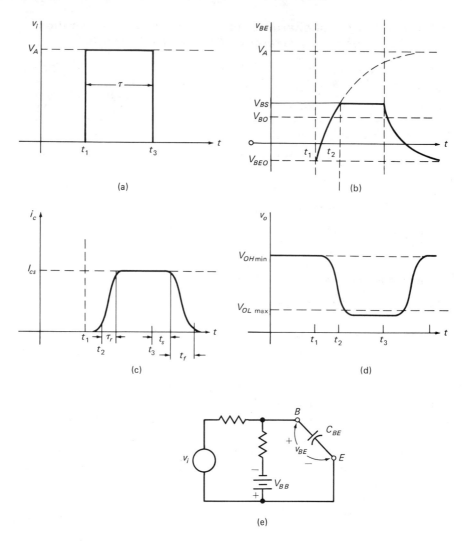

Fig. 3-20. Response of logic inverter gate to a rectangular input pulse.

illustrates the response of the circuit in Fig. 3-19 to a rectangular input voltage pulse, v_i, shown in Fig. 3-20a. The amplitude of v_i is V_A, and the pulse duration is given by τ. It is assumed that when $v_i = 0$, the transistor is biased so that it is either at or heading toward cutoff. A sketch of the base-emitter voltage, v_{BE}, is shown in Fig. 3-20b. At $t = t_1$, v_{BE} departs from

V_{BEO}, the quiescent voltage, and tends toward V_A, following the dotted exponential curve. The time constant of the curve is determined by C_{RF} and the parallel combination of R_b and R_{bb}. The waveform of v_{BE} follows the dotted curve, which is essentially the voltage buildup across C_{BE} with the transistor at cutoff, until it reaches V_{BO}, the voltage at which the transistor starts to conduct (at $t = t_2$), and collector current i_c starts to flow, as shown in Fig. 3-20c. The capacitance C_{BE} accounts for the time delay from t_1 to t_2. The collector current i_c rises toward saturation I_{cs} with a rise-time τ_r, determined by C_{CE}. On the pulse trailing edge, after v_i goes to zero, there is a time delay t_s before the current starts to drop due to C_{BE} and also due to the charge storage in the base region during saturation (9). The fall-time, t_f, is attributable to C_{CE}. The resulting output voltage, v_0, is indicated in Fig. 3-20d.

The sequence of Fig. 3-20 is repeated in Fig. 3-21, except that in the latter the input pulse v_i has the same amplitude V_A but a much shorter time duration (less than $t_2 - t_1$). Under these conditions, the base-emitter voltage will not reach V_{BO}, the voltage at which the transistor conducts. Therefore, the gate will not respond to this pulse of short duration, and the transistor remains at cutoff.

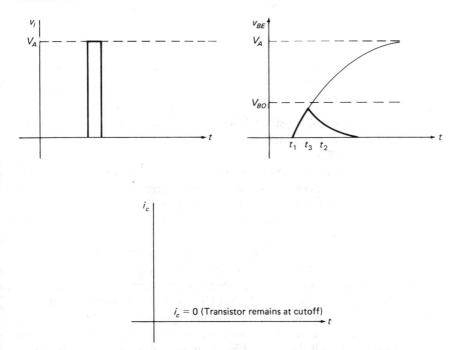

Fig. 3-21. Response of logic inverter gate to a rectangular pulse of insufficient duration to cause gate to respond.

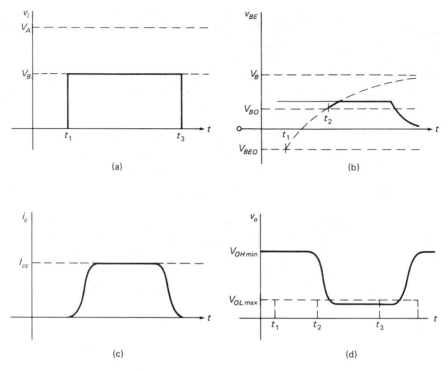

Fig. 3-22. Response of logic inverter to a rectangular input pulse of sufficient amplitude and time duration to cause gate to respond.

A similar situation is illustrated in Fig. 3-22. In this case, the gate input voltage v_i is of lower amplitude than the value V_A in Figs. 3-20 and 3-21, but the amplitude is sufficient to cause the transistor to conduct, that is, $V_B > V_{BO}$. If the pulse duration of v_i is sufficient, v_{BE} will eventually reach V_{BO}, as shown in Fig. 3-22b, with the corresponding collector current i_c and output voltage v_o waveforms illustrated by Fig. 3-22c and Fig. 3-22d.

The foregoing discussion illustrates the response of logic gates to signals of sufficient amplitude and time duration to cause the circuits to react. This suggests a dynamic, or AC noise margin, that is a function of both signal amplitude and time duration. A good discussion on this subject is presented in reference 10, which includes suggested test procedures for evaluating AC noise immunity of Motorola ECL (MECL) 10,000 integrated circuits. This reference suggests that the AC noise margin parameter can be obtained by cascading worst-case gates and measuring the minimum "noise" input that will propagate through the gates. This technique may be more indicative of actual system operation than DC noise margin. However, it is a difficult parameter to measure and is not specified on data sheets. Since noise can enter a circuit at more than one point, the AC noise margin should be determined at all possible points of entry, as the response of the circuit depends upon where the noise enters (10).

REFERENCES

1. "Safety Levels with Respect to Human Exposure to Radio Frequency Electromagnetic Fields, 300 kHz to 100 GHz." *ANSI C95.1—1982*, American National Standards Institute. New York: IEEE, September 1, 1982.
2. White, Donald R. J. "Audio Rectification, the Nemesis of Many EMI Problems." *EMC Technology*, Volume 1, Number 1, January 1982.
3. White, Donald R. J. *EMI Control Methodology and Procedures*, Third Edition, Second Printing. Gainesville, VA: Don White Consultants, Inc., 1982.
4. "Bead, Balun, and Broadband Kit." Wallkill, NY: Fair Rite Products Corporation, March 1981.
5. White, Donald R. J. *EMI Control Methods and Techniques*, A Handbook Series on Electromagnetic Interference and Compatibility, Volume 3. Germantown, MD: Don White Consultants, Inc., 1973.
6. Duff, William G., and White, Donald R. J. *EMI Prediction and Analysis Techniques*, A Handbook Series on Electromagnetic Interference and Compatibility, Volume 5. Germantown, MD: Don White Consultants, Inc., 1972.
7. Dean, Charles E. "Measurements of the Subjective Effects of Interference in Television Receivers." *Proceedings of the IRE*, pp. 1035–1049, 1960.
8. Freeman, Roger L. *Telecommunication System Engineering, Analog and Digital Network Design*. New York: John Wiley and Sons, 1980.
9. Hill, Frederick J., and Peterson, Gerald R. *Introduction to Switching Theory and Logical Design*, Third Edition. New York: John Wiley and Sons, 1981.
10. Blood, William R., Jr. *MECL System Design Handbook*, Third Edition, Second Printing. Phoenix: Motorola Semiconductor Products, Inc., Compiled by the Motorola Computer Applications Engineering Department, Motorola Inc., 1982.
11. White, Donald R. J. *EMI Control in the Design of Printed Circuit Boards and Backplanes*, Third Edition. Gainesville, VA: Don White Consultants, Inc., 1982.

Chapter 4
INTERSYSTEM EMI PREDICTION
AND CONTROL

This chapter summarizes interference prediction and control between one or more potential culprit transmitters and one or more victim receivers immersed in a common electromagnetic ambient environment. The EMI prediction emphasizes the antenna-to-antenna mode of coupling. Reference 1 is suggested for the reader who desires a comprehensive treatment of intersystem EMI.

The basic EMI prediction equation is developed, with a summary of its comprehensive version involving many parameters. Separate mathematical models for transmitters, receivers, antennas, propagation, and signal acceptability criteria are summarized. Illustrative examples show how to make EMI predictions and how to effect EMI control in the presence of the signal of interest.

Most EMI intersystem predictions involve two or more potential culprit emitters, each having several spurious emissions in addition to the desired (fundamental) signal. Additionally, each receiver exhibits several spurious responses, in addition to the fundamental response, one or more of which may result in interference susceptibility. Thus, to facilitate selection and computation of EMI, the multi-transmitter/receiver prediction problem is based upon developing an effective culling process. This eliminates non-EMI (i.e., low probability) situations early in the prediction process to provide an efficient procedure.

BASIC PROPAGATION EQUATION

Figure 4-1 illustrates the geometry used to derive basic relationships between power delivered by a transmitter (P_t), transmitter antenna gain (G_t), power flow into a receiving antenna (P_r), receiver antenna gain (G_r), and the distance between the transmitter and receiver in meters (R_m). In Figure 4-1a, a transmitter Tx is located at the origin of a spherical coordinate system with a segment (octant) of a sphere of radius R_m shown. If the transmitter delivers P_t watts to an ideal isotropic radiating antenna, the power will be radiated homo-

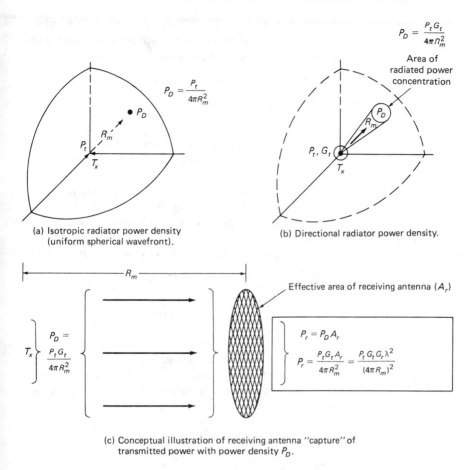

(a) Isotropic radiator power density
(uniform spherical wavefront).

(b) Directional radiator power density.

(c) Conceptual illustration of receiving antenna "capture" of
transmitted power with power density P_D.

Fig. 4-1. Geometry illustrating antenna gain, transmitted power density, and power received.

geneously in all directions in the form of a spherical wavefront. The power density (power per unit area, P_D) will be uniform throughout the surface of the sphere and equal to: (2, 3)

$$P_D = \frac{P_t}{4\pi R_m^{\ 2}} \qquad \text{watts/m}^2 \qquad (4\text{-}1)$$

In practice, radiating elements (antennas) do not radiate electromagnetic energy uniformly in all directions but tend to produce radiation patterns with some associated directivity; that is, a greater portion of the energy is transmitted along some propagation paths than others. This is referred to as antenna gain, G, or directivity, usually along a defined specific direction (axis) for a given antenna design. For the ideal isotropic radiator, $G = 1$, which corresponds to zero dB. Assuming that the radiating element has a given gain, G_t, along a specified direction or axis, then equation (4-1) becomes:

$$P_D = \frac{P_t G_t}{4\pi R_m^{\,2}} \qquad \text{watts/m}^2 \qquad (4\text{-}2)$$

This is illustrated in Fig. 4-1b where the radiated power is concentrated over a segment of the spherical surface area as a result of the transmitting antenna directivity or gain, G_t.

A receiving or intercepting antenna exhibits an effective "capture area," A_r, relative to the amount of the incoming power that it "captures," or receives. If the normal to the receiving antenna effective area (or aperture, as it is often called) is coincident with the main directivity axis (often referred to as the boresight axis) of the transmitting antenna, then the net power flow, P_r, into the receiving antenna becomes:

$$P_r = P_D A_r = \frac{P_t G_t A_r}{4\pi R_m^{\,2}} \qquad \text{watts} \qquad (4\text{-}3)$$

The concept of received power is illustrated in Fig. 4-1c.

Equation (4-3) is not particularly useful in the format presented. The relationship between receiving antenna capture area (effective aperture), Ae and its gain, G_r, is given by:

$$G_r = \frac{4\pi Ae}{\lambda^2} = \frac{4\pi A_r}{\lambda^2} \qquad (4\text{-}4)$$

where:

A = actual (physical) area of antenna
e = antenna efficiency
λ = free-space signal wavelength in meters
$Ae = A_r$ = effective area of antenna in square meters

Table 4-1 lists the efficiencies and effective areas of some antennas that are used frequently.

Substituting equation (4-4) into equation (4-3) for A_r:

$$P_r = \frac{P_t G_t G_r \lambda^2}{(4\pi R_m)^2} \qquad \text{watts} \qquad (4\text{-}5)$$

Table 4-1. Typical antenna efficiency and effective areas. (Courtesy of Interference Control Technologies.)

Antenna type	Antenna efficiency	Effective area
Isotropic radiator	1	$\lambda^2/4\pi$
Small dipole or loop ($l < \lambda/10$)	1	$1.5\lambda^2/4\pi$
Half-wave dipole	1	$1.64\lambda^2/4\pi$
Horn	0.45	0.45A
Parabola	0.5–0.6	0.5A–0.6A

Equation (4-5) is one form of the basic one-way, free-space propagation (transmission–reception) equation. The equation is more useful if λ is expressed in terms of signal frequency and free-space electromagnetic wave propagation velocity, $c = 3 \times 10^8$ meters/second:

$$\lambda = \frac{c}{f_{\text{Hz}}} = \frac{300}{f_{\text{MHz}}} \quad \text{meters} \tag{4-6}$$

Substituting equation (4-6) into equation (4-5):

$$P_r = \frac{P_t G_t G_r c^2}{(4\pi R_m f_{\text{Hz}})^2} \quad \text{watts} \tag{4-7}$$

or:

$$P_r = \frac{P_t G_t G_r (9 \times 10^4)}{(4\pi R_m f_{\text{MHz}})^2} \quad \text{watts} \tag{4-8}$$

The constants of equation (4-8) can be combined to obtain:

$$P_r = \frac{(570) P_t G_t G_r}{(R_m f_{\text{MHz}})^2} \quad \text{watts} \tag{4-9}$$

If the transmitter-to-receiver distance is expressed in kilometers (R_{km}), then equation (4-9) becomes:

$$P_r = \frac{(0.00057) P_t G_t G_r}{(R_{km} f_{\text{MHz}})^2} \quad \text{watts} \tag{4-10}$$

It is common practice to express equation (4-10) in terms of decibels referenced to 1 milliwatt. This is accomplished by expressing P_r and P_t in milli-

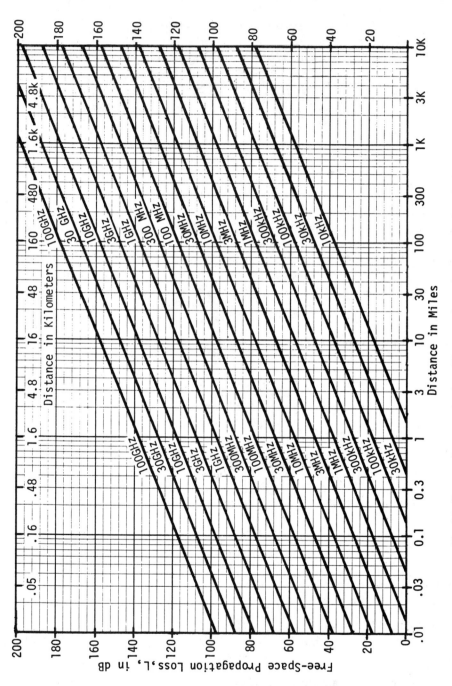

Fig. 4-2. Free-space propagation loss. (*Courtesy of Interference Control Technologies*)

watts and by taking ten times the (common) logarithm of both sides of equations (4-10) to obtain:

$$P_r(\text{dB}_m) = -(32 + 20 \log R_{km} + 20 \log f_{MHz}) + P_t(\text{dB}_m) + G_t(\text{dB}) + G_r(\text{dB})$$
$$(4\text{-}11)$$

The term in parentheses in equation (4-11) constitutes the free-space propagation loss (L) between a transmitter and receiver. Expressed in terms of frequency in MHz and the transmitter–receiver distance in kilometers, L becomes:

$$L = 32 + 20 \log R_{km} + 20 \log f_{MHz} \qquad (4\text{-}12)$$

Equation (4-12) is plotted in Fig. 4-2.

INTERSYSTEM EMI PREDICTION FORM

Equation (4-12) is often useful for predicting the intercepted power between an *intentional* transmitter—receiver (Tx-Rx) pair. However, it is seldom useful for predicting EMI between an *unintentional* Tx-Rx pair because many conditions are assumed, including:

- Free-space (far-field) propagation applies.
- Radio line-of-sight (RLOS) conditions exist.
- The intervening atmosphere has no effect.
- Antenna polarizations are matched.
- The transmitter antenna is looking at the receiver antenna.
- The receiver antenna is looking at the transmitter antenna.
- Antenna transmission line loss is negligible.
- Antenna receiver line loss is negligible.
- Frequency alignment exists between transmitter and receiver.
- Receiver bandwidth equals or exceeds transmitter bandwidth.

More comprehensive radio propagation models have been developed and are recommended if the requirement exists to account for the effects on propagation of frequency, distance, polarization, antenna heights, curvature of the earth, atmospheric conditions, reflections, and the presence of obstacles along the propagation path. These models are generally applicable to the determination of propagation factors of desired signals, which, on occasion, constitute the EMI signal. The models can also be utilized to obtain approximations for the propagation of undesired signals, such as transmitter harmonics.

ELECTROMAGNETIC INTERFERENCE PREDICTION FORM

Signal-Source TX:_____MHz; _____Watts; Other:_____

Potential EMI TX:_____MHz; _____Watts; Other:_____

Victim Receiver:_____MHz; _____kHz Predetection BW; _____ kHz Baseband BW

	SIGNAL +dB	SIGNAL −dB	STDV dB	INTERFERENCE +dB	INTERFERENCE −dB
TRANSMITTERS (Signal and Interfering Sources)					
1. Transmitter Power Available in dBm	30			93	
2. Power Reduction for Out-of-Band Emission					64
3. Antenna Transmission Line Loss		2			O
4. Antenna Gain	35			30	
5. Gain Reduction for Out-of-Band Frequency					O
6. Gain Reduction in Direction of Receiver					O
7. TOTAL MODIFIED ERP (Sum Lines 1 thru 6)	63 dBm			59 dBm	
PROPAGATION PATH (Signal and Interfering Sources)					
8. Free-Space Loss (Sig:____mi; EMI:____mi.)		140			110
9. Non-RLOS Correction Loss		O			O
10. Fade Margin Required (____% Reliability)		30			
11. Rain, Water Vapor, and Oxygen Loss		O			O
12. TOTAL PROPAGATION LOSS (8+9+10+11)		170			110
RECEIVER (Signal and Interfering Victim)					
13. Receiver Antenna Gain	35			35	
14. Gain Reduction in Direction of EMI TX)					45
15. Gain Reduction due to Polarization					—
16. Antenna Transmission Line Loss		2			2
17. TOTAL ANTENNA GAIN OR LOSS (13+14+15+16)	33			12	
18. RECEIVER-ANTENNA INPUT POWER (7+12+17)	−74 dBm			−63 dBm	
19. Frequency Mis-Alignment Correction				12	
20. Bandwidth Correction (Channel Baseband)				O	
21. Receiver Sensitivity (RMS Noise Level)	−98 dBm			−98 dBm	
22. S/N & I/N Before Detection (18+19+20-21)	26			35	
23. Modulation Noise Improvement					
24. Video/Audio Discrimination and A-J Rejection					
25. S/N and I/N at Output (22+23+24)	26			35	
26. Combined S/(N+I) (Line 25: S/N-I/N if I/N>3)	−9				
27. Interference Margin: S/(N+I) −____Std. Dev.					

CONCLUSIONS:_____

Fig. 4-3. Electromagnetic interference prediction form. (*Courtesy of Interference Control Technologies*)

As a result of the above, it is necessary to provide correction terms to equation (4-12) so that the basic equation may be used for either intentional transmission–reception or EMI prediction and analysis. To introduce these corrections, a composite signal and EMI prediction form has been developed, as shown in Fig. 4-3. Details of how to determine or compute the entries are discussed in the next section (2).

For convenience of groupings, the EMI prediction form is divided into topics involving: (1) both the intentional signal and potential EMI transmitters (lines 1 through 7), (2) the intervening propagation path loss to both signal and EMI paths (lines 8 through 12), (3) the victim receiver input (lines 13 through 18), and (4) the processed signal and potential EMI through the victim receiver in terms of output performance (lines 19 through 26). The resultant performance (line 26) is $S/N + I)$, the signal-to-noise-plus-interference ratio.

The headings identified on the top right columns of the form provide for $+\mathrm{dB}$ and $-\mathrm{dB}$ entries for both the signal and potential EMI source. The central column, marked STDV, provides for entering the standard deviation associated with each parameter value in dB. This permits calculating either (1) the overall standard deviation of the output $S/(N + I)$ or (2) the probability that malfunction degradation due to specific EMI may result. All parameter entries in dB are usually rounded off to the nearest integer.

Classically, the communications-electronics system designer will design the system to give a desired S/N performance under stipulated conditions. This corresponds to entries in the column labeled SIGNAL in Fig. 4-3, which ends up with the S/N determination on line 25. The column labeled INTERFERENCE includes entries that end up with the interference-to-noise (I/N) ratio on line 25.

If both the numerator and the denominator of $S/(N + I)$ are divided by the internal receiver noise, N, converting to dB, the following is obtained:

$$\frac{S}{N + I}\,(\mathrm{dB}) = 10 \log\left(\frac{S}{N + I}\right) = 10 \log\left(\frac{S/N}{1 + I/N}\right)$$

$$\cong 10 \log\left[S/N\right] - 10 \log\left[I/N\right] \qquad (4\text{-}13)$$

Again, S/N is the controlled design parameter while the I/N ratio is what develops with the presence of EMI. The I/N computation results from processing the top right INTERFERENCE column in Fig. 4-3, which leads to I/N on line 25. Then line 26 is computed to yield $S/(N + I)$ from equation (4-13). The approximation in equation (4-13) applies within 1 dB when $I/N > 4$.

SIGNAL-TO-NOISE (S/N) PREDICTION

Transmitter-Antenna Effective Radiated Power (ERP)

The first entry in the prediction form is the measured, rated, or nominal transmitter power output, P_t, expressed in units of dBm. The reader may choose to use Table 4-2 to determine the nearest dBm equivalent of the power output in watts. For example, a 10-watt microwave relay transmitter power output corresponds to 40 dBm.

The entry on line 3 in the EMI prediction form involves the antenna transmission line loss, which is rated in dB attenuation at the prediction frequency. Figure 4-4 shows the attenuation in dB per 30 meters (100 feet)

Table 4-2. Transmitter power in watts vs. equivalent dBm. (Courtesy of Interference Control Technologies.)

dBm	Watts	dBm	Watts	dBm	Watts	dBm	Watts
110	100 MW	80	100 kW	50	100 W	20	100 mW
109	80 MW	79	80 kW	49	80 W	19	80 mW
108	63 MW	78	63 kW	48	63 W	18	63 mW
107	50 MW	77	50 kW	47	50 W	17	50 mW
106	40 MW	76	40 kW	46	40 W	16	40 mW
105	32 MW	75	32 kW	45	32 W	15	32 mW
104	25 MW	74	25 kW	44	25 W	14	25 mW
103	20 MW	73	20 kW	43	20 W	13	20 mW
102	16 MW	72	16 kW	42	16 W	12	16 mW
101	13 MW	71	13 kW	41	13 W	11	13 mW
100	10 MW	70	10 kW	40	10 W	10	10 mW
99	8 MW	69	8 kW	39	8 W	9	8 mW
98	6.3 MW	68	6.3 kW	38	6.3 W	8	6.3 mW
97	5 MW	67	5 kW	37	5 W	7	5 mW
96	4 MW	66	4 kW	36	4 W	6	4 mW
95	3.2 MW	65	3.2 kW	35	3.2 W	5	3.2 mW
94	2.5 MW	64	2.5 kW	34	2.5 W	4	2.5 mW
93	2 MW	63	2 kW	33	2 W	3	2 mW
92	1.6 MW	62	1.6 kW	32	1.6 W	2	1.6 mW
91	1.3 MW	61	1.3 kW	31	1.3 W	1	1.3 mW
90	1 MW	60	1 kW	30	1 W	0	1 mW
89	800 kW	59	800 W	29	800 mW	-1	800 μW
88	630 kW	58	630 W	28	630 mW	-2	630 μW
87	500 kW	57	500 W	27	500 mW	-3	500 μW
86	400 kW	56	400 W	26	400 mW	-4	400 μW
85	316 kW	55	316 W	25	316 mW	-5	316 μW
84	252 kW	54	252 W	24	252 mW	-6	252 μW
83	200 kW	53	200 W	23	200 mW	-7	200 μW
82	159 kW	52	159 W	22	159 mW	-8	159 μW
81	126 kW	51	126 W	21	126 mW	-9	126 μW
80	100 kW	50	100 W	20	100 mW	-10	100 μW

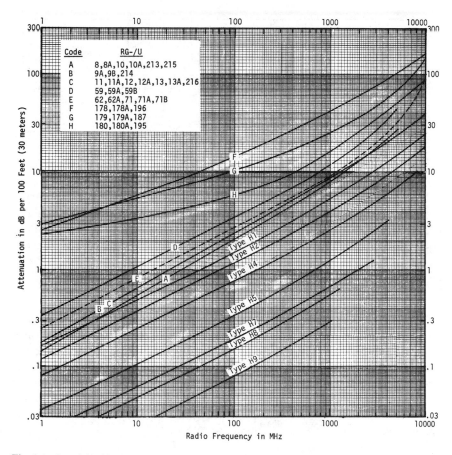

Fig. 4-4. Coaxial cable attenuation vs. frequency. (*Courtesy of Interference Control Technologies*)

for several coaxial lines, and Fig. 4-5 shows the attenuation for several common waveguides.

The transmitter antenna gain (line 4, Fig. 4-3) corresponds to the nominal rated gain for the intentional signal.

Line 7 corresponds to the total effective radiated power (ERP) output (transmitter power output in dBm plus antenna gain in dB minus transmission line loss).

Propagation Path Correction

Line 8 of the prediction form corresponds to the free-space propagation loss in dB obtained from either equation (4-12) or Fig. 4-2.

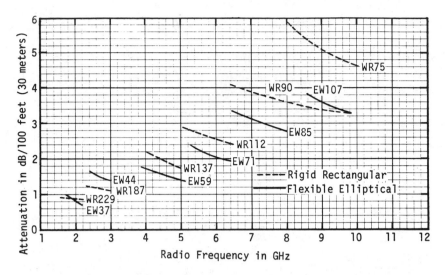

Fig. 4-5. Waveguide attenuation vs. frequency. (*Courtesy of Interference Control Technologies*)

When non-RLOS conditions exist (line 9), such as when the view is blocked by an intervening hill or mountain, an additional propagation correction is required. This correction information is beyond the scope of this book but may be obtained from references 1 and 4. Similarly, for transmission above about 5 GHz, rain, water vapor, and/or oxygen losses (line 11) should be computed from the references (4, 6). When transmission reliability is important and fade margins must be provided (line 10), the references (4, 5) are again used.

Receiver-Antenna Input Power

Line 13 of the prediction form corresponds to the measured or nominal receiver antenna gain for conditions of boresight and polarization alignment.

Line 16, like line 3, corresponds to the receiver antenna transmission line loss.

The modified antenna gain on line 17 is obtained by subtracting from line 13 the antenna transmission line or waveguide attenuation.

Pre-detection *S/N* Ratio

Line 21 is the receiver sensitivity corresponding to $S + N = 2N$ or $S = N$. It is defined in terms of receiver internal noise, N:

$$N = FKTB \qquad (4\text{-}14)$$

where:

F = receiver noise figure
K = Boltzmann's constant = 1.38×10^{-23} joules/$^\circ$K
T = absolute temperature in $^\circ$K
B = receiver bandwidth

Equation (4-14) may be expressed as follows:

$$N_{dBm} = -144 + F_{dB} + 10 \log B_{kHz} \qquad (4\text{-}15)$$

where:

F_{dB} = receiver noise figure in dB
B_{kHz} = receiver bandwidth in kilohertz

Equation (4-15) is plotted in Fig. 4-6 on the right ordinate (Y-axis) for bandwidths (X-axis) ranging from 100 Hz to 10 MHz. The left ordinate is in units of dBμV and corresponds to the right axis *only when the input impedance is 50 ohms*. Receiver sensitivity is entered on line 21 of Fig. 4-3. The parameter for each curve in Fig. 4-6 is the noise figure (F_{dB}). For example, for a microwave link receiver noise figure of 6 dB and a bandwidth of 10 MHz (10,000 kHz), either equation (4-15) or Fig. 4-6 indicates that the sensitivity is -98 dBm.

The S/N ratio on line 22 is obtained by subtracting the receiver sensitivity (line 21) from the receiver antenna input signal on line 18.

Post-detection S/N Ratio

After final detection, this permits further S/N improvement in demultiplexed channels, or by the application of discriminators and other anti-jamming techniques.

INTERFERENCE-TO-NOISE (I/N) PREDICTION

The preceding section developed the computation of the S/N ratio for the desired signal. Basically, the part of the prediction form in Fig. 4-3 used so far is a path-loss analysis. Computing the S/N ratio comprises only half of the procedure to determine the effects of EMI. This section treats the other half of the problem, namely, computation of the I/N ratio. Essentially, computing the I/N ratio may be much more than half the problem because it involves many more parameters than the S/N ratio. As previously discussed, S/N involves many practical simplifications, since the transmitter and receiver are cooperatively designed as a system. Such is not the case for I/N.

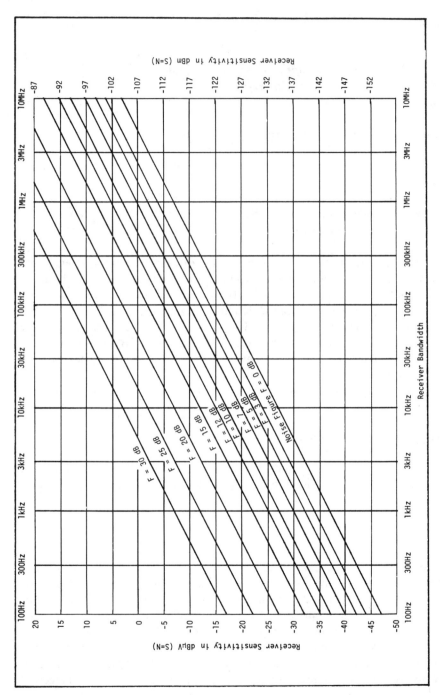

Fig. 4-6. Receiver narrowband sensitivity vs. bandwidth and noise figure. (*Courtesy of Interference Control Technologies*)

Illustrative Problem

This example shows EMI intersystem prediction techniques and exercises the structured prediction form of Fig. 4-3 in order to compute the S/N and I/N ratios. The example is illustrated by Fig. 4-7, where both the intentional microwave relay link transmitter and an unintentional third harmonic from the air traffic control (ATC) radar transmitter appear as radiated energy to the victim microwave receiver. The objective is to compute the S/N and I/N ratios and then the resulting $S/(N + I)$ ratio. This will provide an application of the prediction form and set up the problem for subsequent solution and critique.

The top half of Fig. 4-7 shows a plan view of the EMI geometry. While the intentional transmitter located at the left is 48 kilometers away, the potential EMI source, the L-band ATC radar, is only 3 kilometers away. Once during each complete revolution, the ATC radar antenna is momentarily aimed directly at the victim relay receiver antenna. On the other hand, the radar emissions are arriving 20° off-axis to the direction in which the receiving antenna is looking. The prediction procedure accounts for this.

The lower half of Fig. 4-7 shows what is happening in the frequency domain. The (sin x)/x envelope of the ATC radar is shown at the left with

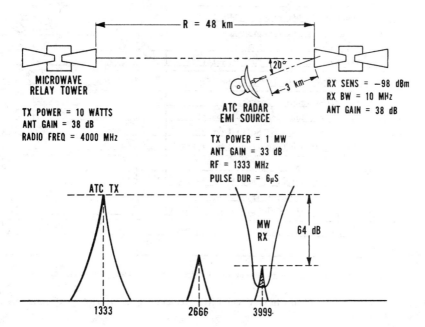

Fig. 4-7. Illustrative example: EMI between air traffic control (ATC) radar and microwave relay link. (*Courtesy of Interference Control Technologies*)

ELECTROMAGNETIC INTERFERENCE PREDICTION FORM

Signal-Source TX: **4000** MHz; **10** Watts; Other: _____

Potential EMI TX: **1333** MHz; **1M** Watts; Other: _____

Victim Receiver: **4000** MHz; _____ kHz Predetection BW; _____ kHz Baseband BW

	SIGNAL		STDV	INTERFERENCE	
TRANSMITTERS (Signal and Interfering Sources)	+dB	-dB	dB	+dB	-dB
1. Transmitter Power Available in dBm	40			90	
2. Power Reduction for Out-of-Band Emission					-64
3. Antenna Transmission Line Loss		2			0
4. Antenna Gain	38			33	
5. Gain Reduction for Out-of-Band Frequency					0
6. Gain Reduction in Direction of Receiver					0
7. TOTAL MODIFIED ERP (Sum Lines 1 thru 6)	76 dBm			59 dBm	
PROPAGATION PATH (Signal and Interfering Sources)					
8. Free-Space Loss (Sig: **30** mi; EMI: **1.9** mi.)		138			116
9. Non-RLOS Correction Loss		0			0
10. Fade Margin Required (**99.9** % Reliability)		40			
11. Rain, Water Vapor, and Oxygen Loss		0			0
12. TOTAL PROPAGATION LOSS (8+9+10+11)		178.			116
RECEIVER (Signal and Interfering Victim)					
13. Receiver Antenna Gain	38			38	
14. Gain Reduction in Direction of EMI TX)					48
15. Gain Reduction due to Polarization					0
16. Antenna Transmission Line Loss		2			2
17. TOTAL ANTENNA GAIN OR LOSS (13+14+15+16)	36			-12	
18. RECEIVER-ANTENNA INPUT POWER (7+12+17)	-66 dBm			-69 dBm	
19. Frequency Mis-Alignment Correction					0
20. Bandwidth Correction (Channel Baseband)					0
21. Receiver Sensitivity (RMS Noise Level)	-98 dBm			-98 dBm	
22. S/N & I/N Before Detection (18+19+20-21)	32			29	
23. Modulation Noise Improvement					
24. Video/Audio Discrimination and A-J Rejection					
25. S/N and I/N at Output (22+23+24)	32			29	
26. Combined S/(N+I) (Line 25: S/N-I/N if I/N>3)	-3				
27. Interference Margin: S/(N+I) - Std. Dev.					

CONCLUSIONS: _____

Fig. 4-8. Illustrative problem—Intersystem EMI prediction form. (*Courtesy of Interference Control Technologies*)

a fundamental frequency of 1333 MHz. The 3999-MHz third harmonic from the radar can potentially interfere with the 4000-MHz microwave link receiver. The picture qualitatively shows the amplitude of the third harmonic above the sensitivity in the receiver selectivity curve. The objective, then, is to quantitatively compute both the S/N and I/N ratios.

Signal and Interference Effective Radiated Power (ERP) Output

Figure 4-8 provides a prediction form for the illustrative problem. Figure 4-7 indicates the microwave signal power to be equal to 10 watts, which corresponds to 40 dBm entered in the SIGNAL column of Fig. 4-8.

The ATC radar power output at the fundamental frequency shown in Fig. 4-7 is 1 megawatt (MW). Thus, in line 1 of the EMI prediction form shown in Fig. 4-8, 90 dBm is entered. Since the third harmonic is the potential interference, the power reduction of the third harmonic relative to the fundamental is needed. This may be obtained either by measurement or from the manufacturer's literature if either is available. On the other hand, if neither is known, default data may be used to provide typical values for heretofore unknown data to allow analysis of a specific problem on a temporary basis pending the availability of more specific data. Typically, default values may be selected conservatively, that is, on the pessimistic side of a possible range of values.

Under the Department of Defense Spectrum Signature Collection Program (see MIL-STD-449D) transmitter power output at harmonic frequencies, receiver spurious responses, antenna out-of-band performance, and other data are provided and may be used as default data. Table 4-3 lists such default data in summary form for transmitter average harmonic frequency emission levels relative to the fundamental. It is observed that the second through tenth harmonics are listed and grouped for all transmitters operating with fundamentals below 30 MHz, between 30 MHz and 300 MHz, and above 300 MHz. Since the illustrative problem shown in Fig. 4-7 involves the third harmonic, Table 4-3 shows that the mean value of this harmonic is 64 dB relative to the fundamental. Thus, 64 dB is entered on line 2 of the prediction form shown in Fig. 4-8 under INTERFERENCE. Since the signal is transmitted at the fundamental frequency, no entry is required under the SIGNAL column.

Line 3 provides for the insertion of the antenna line loss. The microwave relay transmitter is assumed to be at the bottom of a 30-meter tower. From Fig. 4-5, the waveguide attenuation is 2 dB, which is entered on line 3 of Fig. 4-8. However, since, for waveguide transmission lines, the harmonic levels measured in the far-field to develop the data in Table 4-3 include the

Table 4-3. Summary of transmitter (TXMR) harmonic average emission levels in dB above fundamental.* (Courtesy of Interference Control Technologies.)

TXMRS harmonic number	All TXMRS combined ($\sigma = 15$ dB)	Transmitters categorized according to radio frequency		
		Below 30 MHz ($\sigma = 10$ dB)	30 MHz–300 MHz ($\sigma = 15$ dB)	Above 300 MHz ($\sigma = 20$ dB)
2	-51	-41	-54	-55
3	-64	-53	-68	-64
4	-72	-62	-78	-70
5	-79	-69	-86	-75
6	-85	-74	-92	-79
7	-90	-79	-97	-82
8	-94	-83	-102	-85
9	-97	-87	-106	-88
10	-100	-90	-110	-90

* This includes the contributions from high-power emitters having integral transmitter antennas such as radar, troposcatter communications, and TV broadcast.

effect of the power output transmission line loss and antenna gain, no further reduction in the antenna transmission line loss is required in the INTER-FERENCE column.

For coaxial systems, the data compiled and presented in Table 4-3 correspond to the transmitter output as measured into a dummy load. The antenna characteristics are then measured separately. The effect of radiated power for coaxial systems is obtained by combining the transmitter power output, its reduction at harmonics, and the antenna transmission line loss for the harmonic components.

For waveguide systems, however, it is impossible to measure the above in the same manner because of the large number of multi-modes that are developed and propagated. As a result, all waveguide-operated transmitter-antenna systems are measured in the far-field as a single unit. Table 4-3 corresponds to transmitter power outputs only for coaxial systems but represent combined transmitter output, antenna gain reduction, and transmission line loss for waveguide systems.

The rated antenna gains of the microwave transmitter and potentially offending ATC radar are shown in Fig. 4-7 as 38 and 33 dB, respectively, at the fundamental frequency, and these values are entered on line 4 of the EMI prediction form. For the interfering radar, the desired antenna gain is that at the third harmonic interfering frequency. The gain reduction for out-of-band frequency operation (viz., at the third harmonic) is desired, as indicated on line 5. Again, for waveguide systems, the antenna gain reduction for out-of-band frequency measurements is included in the power reduction measured in the far-field; that is, it is included in the -64 dB entered on line 2 of the form. Therefore, enter 0 on line 5.

The final entry to be made under the transmitter ERP in the prediction form involves the gain reduction when the potentially offending transmitter antenna (in this case, the ATC radar) is not looking at the potential victim microwave relay antenna. However, as shown in Fig. 4-7, there exists a point in the 360° horizon scan of the radar antenna when it is momentarily boresighting the victim. For this condition, there is no antenna gain reduction in the direction of the receiver. Consequently, 0 is entered on line 6 of the EMI prediction form in Fig. 4-8.

To obtain the modified effect of radiated power for the signal and potentially offending ATC radar, lines 1 through 6 of the prediction form are now added, with 76 and 59 dB computed and entered on line 7. It is observed that the level of the radiated power from the ATC radar is 17 dB less than that from the intentional signal. However, no conclusion can be drawn yet because there are several other factors to be taken into account before the interference effects can be evaluated at the victim receiver.

Propagation Path Loss for Microwave Relay and ATC Radar Signals

As shown in Fig. 4-7, the microwave transmitter and the ATC radar are located 48 km and 3 km, respectively, from the microwave link receiver. From Fig. 4-2, it is seen that the respective free-space propagation losses are 138 dB and 116 dB and 4 GHz, the signal and third harmonic frequency. These figures are entered on line 8 of the EMI prediction form in Fig. 4-8.

It is assumed that the ATC radar is within line-of-sight of the victim receiving antenna. Accordingly, there is no RLOS correction due to blockage by hills or other obstacles, and, as a result, 0 is entered on line 9 of the prediction form for both SIGNAL and INTERFERENCE blocks.

The fade margin accounts for signal fades along the propagation path that result in a decreased signal-to-noise ratio at the receiver. The required fade margin can be determined as a function of propagation path length and desired system reliability (4, 5). For a reliability of 99.99%, a fade margin of 40 dB is required (5). This is entered on line 10 in Fig. 4-8.

It is assumed that rain, water vapor, and oxygen losses are negligible below 5 GHz. Consequently, 0 is entered for SIGNAL and INTERFERENCE on line 11. The total propagation losses for the signal and interference are obtained by adding lines 8 through 11, with 178 and 116 dB, respectively, entered on line 12.

Receiver Antenna Input Power

Since the same receiver antenna is processing both the intentional signal and the interference at nearly the same frequency (4000 vs. 3999 MHz), a gain of 38 dB is entered on line 13 for both SIGNAL and INTERFERENCE.

Although the radar antenna is shown looking at the victim receiving antenna in Fig. 4-7, the reverse does not apply. The figure shows that the microwave relay link antenna is boresighting its intentional transmitter but is off the boresight axis of the ATC radar antenna by 20°. Thus, it is expected that there will be an "out-of-beam" reduction in the receiving antenna gain. This reduction is generally obtained from data supplied by the manufacturer or measured antenna pattern data corresponding to a 20-dB off-axis situation.

Should the antenna pattern data not be known or at least available to the user of the prediction form, temporary default data may be used pending the availability of better data. Accordingly, Fig. 4-9 shows quantized default antenna pattern data for both a two-step and a four-step quantization. For the present problem the latter will be used. The figure shows that within ±0.5 beam width, the fully specified antenna gain (G_0) is used. For an off-axis condition between 0.5 and 1 beam width, a gain reduction of 5 dB is used. For off-axis conditions ranging from 1 to 3 beam widths, an isotropic antenna gain level of 0 dB is used, corresponding to a reduction of the full rated gain. Finally, for off-axis conditions greater than ± 3 beam widths, a

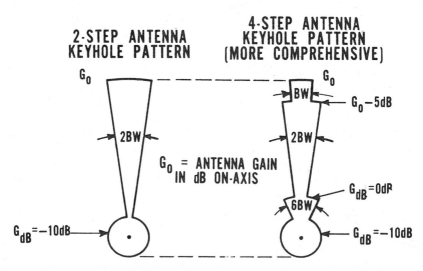

Fig. 4-9. Two- and four-step quantized antenna default patterns. (*Courtesy of Interference Control Technologies*)

scatter level gain of -10 dB is used. This corresponds to a rated gain reduction of the full gain of $+10$ dB.

The data presented in Fig. 4-9 are in universal units of beam width. Thus, it is necessary first to compute the beam width corresponding to a microwave relay link antenna. The beam width, BW, for any pencil-type antenna is:

$$BW° = \sqrt{\frac{30,000}{G}} = 173(1/G)^{1/2} \qquad (4\text{-}16)$$

or:

$$BW° \cong 70\lambda/D \qquad (4\text{-}17)$$

where:

$G =$ anti-log $(G_{dB}/10)$
$\lambda =$ wavelength in meters
$D =$ diameter of dish or diagonal of horn in meters

The power gain corresponding to 38 dB is 6310. When this is substituted into equation (4-16), a beam width of 2.2° results. When this beam width is divided into the off-axis angle of 20°, there results an off-axis angle of arrival of 9.2 beam widths. Consequently, since the off-axis angle exceeds 3 beam widths, the scatter level gain of -10 dB is used, according to Fig. 4-9. This corresponds to a $+38$ dB $-(-10$ dB) or a net gain reduction of 48 dB. This value is entered on line 14 of the EMI prediction form in Fig. 4-8.

Line 15 requires the entry of gain reduction due to mismatched polarization between the transmitting antenna and the receiver as shown in Table 4-4. As it develops, however, the polarization appearing at the fundamental and that of its harmonics from a transmitter are no longer related. This means that at harmonic frequencies, both vertical and horizontal polarization have an equal probability of existing, and the polarization purity

Table 4-4. Antenna polarization correction in dB. (Courtesy of Interference Control Technologies.)

Tx		Horizontal		Vertical		Cir
Rx		$G < 10$ dB	$G \geq 10$ dB	$G < 10$ dB	$G \geq 10$ dB	
Hor $G < 10$ dB		0	0	-16	-16	-3
$G \geq 10$ dB		0	0	-16	-20	-3
Ver $G < 10$ dB		-16	-16	0	0	-3
$G \geq 10$ dB		-16	-20	0	0	-3
Circular		-3	-3	-3	-3	0

associated with the fundamental is lost. It is assumed that the signal transmitting and receiving antennas are compatibly polarized. Accordingly, 0 is entered on line 15 for both SIGNAL and INTERFERENCE.

Since the victim receiver is processing both the intentional and interfering emissions at nearly the same frequency, the previous 2-dB antenna transmission line loss is assumed to apply at the receiver, and 2 is entered in both blocks on line 16. The total antenna gains or losses are then obtained by adding lines 13 through 16, resulting in $+36$ dB for the SIGNAL, and -12 dB for the INTERFERENCE on line 17.

Receiver Input Power and S/N and I/N Ratio Prediction

The receiving antenna input power for line 18 is now computed by adding the transmitter ERP of line 7, the total propagation loss of line 12, and the receiver antenna gain of line 17. When this is done, values of -66 dB and -69 dB are entered on line 18.

The frequency separation between the third harmonic frequency from the ATC radar (3999 MHz) and the fundamental frequency of the microwave receiver (4000 MHz) is 1 MHz. A correction must be determined for this misalignment, and Fig. 4-10 provides default data if more exact information

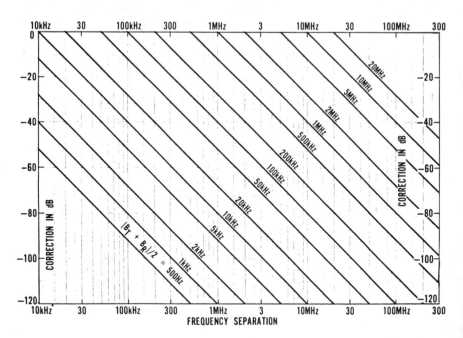

Fig. 4-10. Transmitter-receiver frequency separation correction. (*Courtesy of Interference Control Technologies*)

is not available. The 3-dB bandwidths, B_T and B_R, for the transmitter and receiver must be known. The receiver bandwidth B_R is provided as 10 MHz in Fig. 4-7. The ATC radar transmitter bandwidth is related to the pulse duration of 6 microseconds. The pulse–bandwidth relationship for pulsed-carrier signals is defined as (1):

$$BW = 2/(\pi\tau) \text{ Hz} \tag{4-18}$$

where τ = pulse duration in seconds.

For a pulse duration of 6 microseconds, the ATC radar transmitter bandwidth B_T is 106 kHz. For the values of B_R and B_T, Fig. 4-10 shows zero correction required for frequency misalignment, and 0 is entered in line 19.

In bandwidth correction, the receiver will only intercept part of the transmitter emissions when the transmitter bandwidth exceeds that of the receiver. When this condition exists, the bandwidth correction shown in Table 4-5 may be used.

To apply Table 4-5, both the transmitter and the receiver bandwidth must be known or computed. Figure 4-7 indicates that the receiver bandwidth is 10 MHz, and the ATC radar transmitter bandwidth was previously computed as 106 kHz. Since the transmitter bandwidth is very much less than the receiver bandwidth, Table 4-5 indicates that no correction is required. Consequently, 0 is entered on line 20 of the prediction form.

Since the same receiver is processing both the signal and interference emissions, its sensitivity will be the same for both. Thus, -98 dB is inserted in both blocks in line 21.

To obtain the resulting signal-to-noise and interference-to-noise ratios, lines 18 through 21 are used to produce ratios of 32 dB and 29 dB, respectively, and these values are entered on line 22 of the prediction form.

Table 4-5. Bandwidth corrections in dB.
(Courtesy of Interference Control Technologies.)

Modulation type	Bandwidth conditions	On-Tune $\Delta f \leq (B_T + B_R)/2$	Off-Tune $\Delta f > (B_T + B_R)/2$	Remarks
Noise Like Continuous	$B_R \geq B_T$	No Correction	$10 \log_{10}\left(\dfrac{B_R}{B_T}\right)$	RMS Power Proportional to Bandwidth
	$B_R < B_T$	$10 \log_{10}\left(\dfrac{B_R}{B_T}\right)$		
Pulse	$B_R \geq B_T$	No Correction	$20 \log_{10}\left(\dfrac{B_R}{B_T}\right)$	Peak Voltage Proportional to Bandwidth
	$PRF < B_R < B_T$	$20 \log_{10}\left(\dfrac{B_R}{B_T}\right)$		
	$B_R < PRF$	$20 \log_{10}\left(\dfrac{PRF}{B_T}\right)$	$20 \log_{10}\left(\dfrac{PRF}{B_T}\right)$	Power in $B_R < PRF$

Post-detection S/N, I/N, and $S/(N + I)$ Ratios

Post-detection processing allows for the reduction of I/N by the application of discriminators and other anti-jamming (noise-improvement) techniques. These techniques will not be considered here, but they are addressed in reference 1. Therefore, lines 23 and 24 are left blank, and the values of line 22 are entered on line 25 to yield the final S/N and I/N ratios.

In order to obtain the desired $S/(N + I)$ ratio, the difference between the two performance ratios is computed. The approximation expression of equation (4-16) may be used, since an I/N ratio of 29 dB yields a value of $I/N > 4$ in basic power units. Therefore, an $S/(N + I)$ ratio of 3 dB results and is entered on line 26.

In surveying the results of using the EMI prediction form in Fig. 4-8, it is seen that a 32 dB S/N ratio was obtained in the absence of interference. On the strength of this alone, performance would appear to be good. However, since an I/N ratio of 29 dB resulted from the undesired ATC radar emissions, it is now seen that the combined performance of both signal and interference yields an undesirable situation, namely, $S/(N + I) = 3$ dB. Some measures must be taken to substantially reduce the I/N ratio so that the resulting performance will represent little degradation, if any, from the original S/N ratio of 32 dB.

The question arises of just how much to reduce the I/N ratio from the present 29 dB. If it were reduced to 0 dB (i.e., a net reduction of 29 dB), there would still result a 3-dB degradation in the S/N ratio as seen from equation (4-16). This means that when both noise and interference are equal, a net reduction of 3 dB develops. Because of the desire not to permit a 3-dB reduction and because of the uncertainty in the prediction process, an I/N objective less than 0 dB is needed. If it is made too negative, then an overdesign or "overkill" EMC condition could result that might not be justified or cost-effective.

In Chapter 5, the criterion for the selection of the I/N objective is established. For the present, an objective I/N ratio of -10 dB is sought. This results in a requirement to reduce the I/N ratio from 29 dB to -10 dB for a net reduction requirement of 39 dB. The next section discusses a number of alternatives that can be used to achieve these objectives, showing that several EMC actions may be possible in theory but that many of these are not practically achievable because of logistics, cost, or other considerations.

INTERSYSTEM EMI CONTROL OPTIONS AND CRITIQUE

This section discusses a number of EMI control components and techniques. Some techniques pertain to the design stage, whereas others pertain only to field situations where an EMC retrofit is needed. A few techniques will apply to both situations. To facilitate discussion, the ongoing example presented in the previous sections will be continued. Among the various total or partial EMC solutions to be considered and critiqued are the following:

- Change assigned frequency.
- Add a harmonic filter.
- Use directional or time blanker.
- Erect obstacles in the propagation path.
- Increase antenna off-axis rejection.
- Increase antenna gain.
- Increase link transmitter power output.

Change Assigned Frequency

If the preceding EMI problem corresponds to an existing situation, one of the first potential solutions, in whole or in part, would be to consider changing one of the transmitter frequencies. This immediately raises the question, "Should the frequency of the microwave relay transmitter or the offending ATC radar transmitter be changed?" The purpose behind this question is to alter a third-harmonic, co-channel interference situation, as shown in Fig. 4-11.

For the purpose of discussion it is assumed that the victim microwave relay link is under control of the reader (in contrast to the ATC radar) so that changing the microwave link transmitter is considered first. In reality, the frequencies of a single microwave link path cannot be changed without

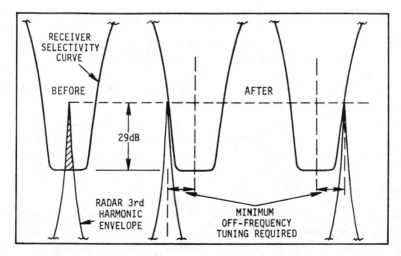

Fig. 4-11. Change in frequency of either the microwave relay receiver or the ATC radar to eleminate EMI. (*Courtesy of Interference Control Technologies*)

affecting other frequencies in the link. For example, the frequencies of the microwave link paths are likely already assigned to respective users, and this option is usually not available to the field engineer as an EMC solution. There could be exceptions in more sparsely populated areas where unassigned frequencies may be available.

Next, consider the option of changing the frequency of the ATC radar. Since this radar is not under the control of the organization responsible for the microwave relay link, it will be necessary to contact Federal Aviation Administration (FAA) and possibly Federal Communication Commission (FCC) authorities to request and negotiate a radar frequency change. One should be prepared to answer all possible questions posed by these authorities concerning what frequency bands may be desirable (or available) to avoid EMI. The nature of the required action dictates that a long period of time (at least months) will likely pass before this action is fully coordinated, approved, and accomplished, if it is accomplished at all.

If the shape factor (60 dB to 6 dB bandwidth ratio) of the microwave receiver is known to the reader, a detailed computation can be made to compute the required frequency shift suggested in Fig. 4-11. If the shape factor is not known, an estimation of frequency separation can be made by referring to the default relations in Fig. 4-10. The latter situation will now be assumed. From the EMI prediction form shown in Fig. 4-8, it was observed that a net reduction in the I/N ratio of 39 dB is necessary. Before this number is slected on the X-axis of Fig. 4-10, the $(B_T + B_R)/2$ parameter factor must be calculated:

$$(B_T + B_R)/2 = (0.106 + 10)/2 = 5.05 \text{ MHz}$$

where:

B_T = transmitter bandwidth = 106 kHz [see equation (4-21)]
B_R = microwave receiver bandwidth = 10 MHz (see Fig. 4-7)

In Fig. 4-10, it is seen that the intersection of the 5-MHz parameter and the -39 dB correction exists at about 48 MHz. This off-frequency correction is required for the third harmonic of the ATC radar. Thus, the fundamental radar frequency at 1333 MHz needs to be changed by only one-third of the 48 MHz, or 16 MHz.

Add a Harmonic Filter

Reference to Fig. 4-8, line 2, shows that the out-of-band reduction for the third harmonic is 64 dB down relative to the fundamental. To achieve an additional 39 dB reduction in I/N ratio, another possibility would be to add a low-pass filter into the radar such that the third harmonic would then become 103 dB down.

The requirements for the filter would be to pass the fundamental with very small attenuation, such as 0.1 dB, to process the 1-MW peak-pulse power of the radar. Even a 0.1-dB insertion loss corresponds to 23 kW of peak-pulse power. Therefore, forced cooling may be required, as the insertion loss is still significant.

The harmonic filter would have to have its cutoff frequency above 1333 MHz in order to decrease the insertion loss to the fundamental. Thus, the cutoff frequency might be about 1500 MHz. Since it must offer at least 39 dB transmission loss at 4 GHz, this implies that a four- or five-stage, low-pass filter is required.

The above filter would normally not be stocked on the shelf by any manufacturer because of its complexity and low-use requirement. Thus, its cost would be high, and several months might be needed to obtain it.

In conclusion, the filter is a good long-term solution to the ongoing EMI problem, but would give no immediate relief for the problem in the field. It also raises the question of who is going to pay for its purchase and installation. Furthermore, arrangements would have to be made with the authorities to have the filter installed at a time when the radar could be taken off the air.

Use Directional or Time Blanker

Another possible solution to the EMI problem is to consider the use of a blanker during the portions of the scan when the radar is illuminating the victim microwave antenna. Since most radars have built-in sector blankers, this would appear to be a simple solution, namely, to blank the radar for a degree or two during rotation. Although the solution seems easy to implement, it would not be permissible because the ATC radar would develop a blind sector in which air traffic could be hidden. Thus, this dangerous situation would be forbidden, and the blanker could not be used.

Another possibility would appear to be a time blanker at the victim relay link receiver. This blanker, in essence, would position time holes in the transmission during periods when the radar is boresighting the victim. This solution would also be impractical because information would be lost. Voice channels would suffer very little degradation from such time holes. However, were digital data being multiplexed on one or more voice channels, this would be unacceptable because the bit error rate could become prohibitive.

Erect Obstacles in the Propagation Path

Another approach to lowering the I/N ratio would be to block the direct line of sight between the radar transmitter and the victim receiver (7). This raises the question of whether the erected obstacle should be located closer to the radar or to the victim receiver. If something like a radar fence were

TWO BORESIGHTED ANTENNAS WITH NO TOWER WIND

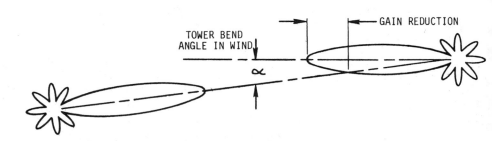

OFF-AXIS ANTENNA POINTING IN PRESENCE OF STRONG WIND AND GUSTS

Fig. 4-12. Effects of antenna axis shift due to wind. (*Courtesy of Interference Control Technologies*)

to be used, this would be unacceptable because it would put blind sectors in the air traffic control coverage. If the obstacle were a billboard and to be mounted near the victim antenna and supported by its tower, another kind of problem could exist. The tower with billboard would then have an increased susceptibility to wind. It would result in bending with high-velocity winds and/or gusts. This has the effect of increasing the fades due to off-boresight antenna modulation, as shown in Fig. 4-12. The effect would be to further aggravate the EMI problems.

One possible solution for the increased wind resistance would be to replace the billboard with a wire mesh screen. In order to provide about 40 dB of shielding effectiveness, the center-to-center spacing of the screen elements would have to be $\lambda/200$ (see Chapter 11). Since 4 GHz corresponds to a wavelength of 7.5 cm, the center-to-center screen spacing would be 0.04 mm. This is so close that it would correspond to a continuous sheet. In a freezing rain, the screen would accumulate ice and act again as a windsock, and the performance would degrade because of antenna modulation. Hence this approach is impractical.

Increase Antenna Off-Axis Rejection

Another approach to erecting an obstacle in the propagation path mentioned in the previous section is to use hooded antennas. These antennas are essentially a five-sided box lined with absorber material. Here, the mission is to reduce the off-axis rejection of the microwave relay antenna. The problem with hooded antennas is that they also increase the wind resistance and degrade performance by antenna modulation.

A better way to increase the antenna off-axis rejection is simply to use a higher-quality antenna. Referring to line 14 in Fig. 4-8, it was noted that the antenna gain reduction was only 48 dB. However, that is so because a default number was used. Microwave relay link antennas have a much higher performance, typically corresponding to a gain reduction in excess of 70 dB for a 20° off-axis emission arrival. The reader is reminded that default data are only used when no other data are available. Default data should always be replaced with real data at the first opportunity. Thus, the true antenna pattern data would be used instead of the default data. Here, it is likely that an improvement in off-axis rejection over the 48 dB might be about 30 dB or so. This also demonstrates that the problem is not as severe as originally predicted.

Increase the Antenna Gain

One way to apparently increase the S/N ratio (i.e., to reduce the I/N ratio) is to increase the gain of the microwave transmitting antenna link from the present 38 dB. This approach, however, is of no practical merit because any antenna gain increase is not only expensive, but corresponds to an increase in its effective area. Once again, as discussed in connection with Fig. 4-12, the antenna tower would bend more in winds because the antenna acts like a larger windsock.

Increase Link Transmitter Power Output

One classical solution was simply to increase the power of the intentional transmitter. To a marked extent this practice has contributed much to the growth of EMI problems because it tries to brute-force the solution. In general, the net effect has been to improve the performance of the particular intentional transmitter–receiver pair and to create an increased probability of interference to other victims via either co-channel, adjacent channel, or audio rectification mechanics.

Perhaps an even more important reason for not increasing the transmitter power output involves the subject of reliability. One of the reasons why microwave relay link transmitters have a very low power output (from 1 to 10 watts) is that their continuous service is of the order of tens of thousands of hours. If one were to increase the 10-watt transmitter by 39 dB, then 100 kW output power would be required. Not only would other kinds of problems develop from this, but the reliability would now represent a MTBF of less than 1000 hours. Thus, increasing the transmitter power output is impractical.

MULTI-TRANSMITTER/RECEIVER
PREDICTION TECHNIQUES

In executing the EMI prediction form introduced in the previous section, a prediction may be carried out in which it is concluded that no EMI exists. Although this is good in the sense that EMI is not wanted, it can be a waste of time and resources to execute many such EMI predictions only to find that no EMI problems exist. These situations may result when there are requirements for determining where EMI exists in a multi-transmitter and/or multi-receiver population.

For example, a transmitter can radiate 10 harmonics (including its fundamental, f_{0T}) over a decade, and a superheterodyne receiver may exhibit 40 spurious responses (including its fundamental, f_{0R}) over ± 1 decade of frequency. Thus, 400 possible transmitter–receiver (Tx-Rx) combinations exist for a single Tx-Rx pair over $f_{0R} \pm 1$ decade. Further, if there exist 10 transmitters and 10 potentially victim receivers, 100 fundamental emission-reception combinations exist, and about 40,000 spurious situations may be possible. Accordingly, a technique is required to permit prediction efficiency by rapidly eliminating at the onset most situations for which EMI is not possible. A technique for doing this is the *multi-level culling process.*

Multi-level Prediction Process

For large transmitter and/or receiver populations, it is possible to take a quick look at the total EMI prediction problem using simple prediction models, and eliminating from further consideration many output–response pairs that probably would not represent EMI situations. This culling process can be extended to several levels of prediction detail. At each subsequent level, additional factors are introduced, and more noninterfering cases are eliminated from consideration.

When the prediction is performed in the above manner, it is necessary to carry out a complete prediction only on those surviving cases that exhibit a significant potential for creating EMI. The number of culling levels that gainfully should be used and the specific assumptions and calculations that should be performed at each level depend on the particular EMI prediction problem. As a general objective, each culling level of prediction should remove about 90% of the noninterfering situations.

Four culling prediction levels are considered. They are based on amplitude, frequency, parameter detail, and performance. These prediction levels, the factors that are considered in each, and the results that are obtained are illustrated in Fig. 4-13.

The most fundamental level of EMI prediction is based on emission-response amplitude. Frequency, time, distance, and direction effects are considered only in a relatively gross sense. The amplitude cull uses simple,

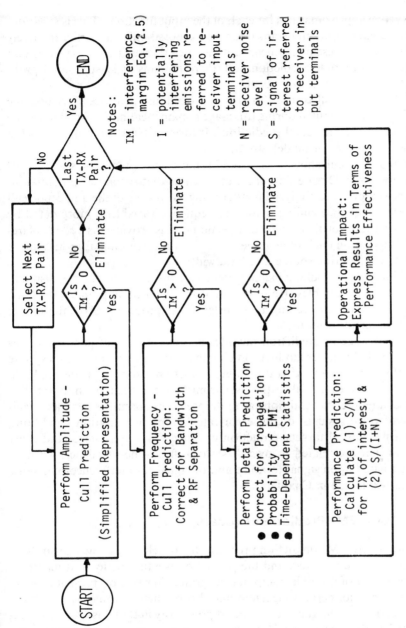

Fig. 4-13. The EMI prediction process for multiple transmitter/receiver pairs showing four levels of prediction culling. (*Courtesy of Interference Control Technologies*)

conservative approximations for each of the input functions. This is separates a large number of unimportant interference possibilities from the relatively few cases that represent significant interference threats. For those cases that survive the amplitude prediction cull, there remains at least a small probability that interference will occur.

The amplitude cull is more than just a weeding-out process because the basic functions computed during this stage remain meaningful throughout the remainder of the finer-level predictions. It forms a basic nucleus to which a number of adjustment models are applied.

The frequency cull shown in Fig. 4-13 is the second prediction sorting level to be considered. During this stage, those cases that remain as potential interference culprits are analyzed further using the results of the amplitude cull as a base. Frequency culling treats the frequency variable in more detail by considering additional interference rejection that is provided as a result of frequency separation (lack of alignment) between a susceptible response and a potentially interfering source and bandwidths.

The third stage of culling includes detailed consideration of time, distance, and directional variables. Here, potential interference amplitudes resulting from previous stages of culling are translated into probabilities of interference and time-dependent terms.

The final stage of prediction includes a consideration of factors such as transmitter and receiver modulation characteristics and operational response and performance analysis. At this stage, prediction results are translated into terms that are meaningful to the user from an operational standpoint. For example, interference conditions are interpreted in quantities such as intelligibility of voice systems, digital error-rates, scope presentation conditions, false alarms, missed targets, reduction in range of radar systems, and other measures that are related to overall system effectiveness and performance. They are based upon signal-to-noise and/or signal-to-interference-plus-noise ratios, as explained in Chapter 3.

Short Form EMI Prediction Process

There are many EMI prediction problems for which only a few transmitter–receiver pairs are involved, and the prediction is either performed manually or with the aid of a simple computer program. This section presents a step-by-step process for performing a manual EMI prediction using a special form. This form differs from the one presented previously in that it is more comprehensive than the earlier form. However, it does not contain the intended signal level prediction because the I/N ratios (interference margins) are emphasized here rather than $S/(N + I)$ ratios. The user simply fills out the prediction form in the same manner as an income tax form. References are provided to specific parts of this chapter that may be used to obtain further description of EMI prediction models or techniques.

A brief description follows of some of the major EMI prediction considerations that apply to a mannual prediction process. There are four possible combinations of Tx-Rx emissions and responses for each pair accommodated on the form (see Fig. 4-14):

- Class FIM, Fundamental Interference Margin: Tx fundamental emission and Rx fundamental response.
- Class TIM, Transmitter Interference Margin: Tx fundamental emission and Rx spurious response.
- Class RIM, Receiver Interference Margin: Tx spurious emission and Rx fundamental response (this was the ongoing problem in the previous section).
- Class SIM, Spurious Interference Margin: Tx spurious emission and Rx spurious response.

Factors Pertinent to the Use of the EMI Prediction Form

The considerations that apply and the calculation procedures used are different for each of the four categories. The first step in a prediction process is to determine the frequency limits associated with each Tx-Rx pair, as provided for in entries 1 through 7 of the short form. To determine which emission–response categories apply, the frequency limits are processed under

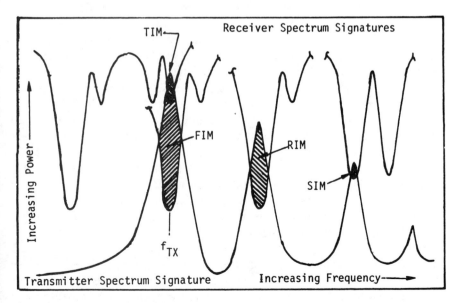

Fig. 4-14. The four classes of transmitter-to-receiver interference margins. (*Courtesy of Interference Control Technologies*)

"APPLICABILITY OF FOUR EMI PREDICTION CASES" in the short form shown in Fig. 4-15.

After it is determined which of the four cases apply, the next step is to calculate a preliminary level based on emission and response amplitudes, and free-space propagation loss from Fig. 4-2. This is accomplished in lines 8 through 16 on the short form. Line 16 then shows which, if any, interference margin (IM) cases survive the amplitude cull. A correction or adjustment is then made as the first step of the frequency-cull process to account for differences between the transmitter and receiver bandwidths. Entries are made on lines 17 through 21 of the short form. In the next step, a correction is made to account for frequency separation between the particular transmitter output and receiver response. Lines 22 through 38 are used for this purpose. In this step, if specific transmitter and receiver frequencies are not known, calculations may be repeated for different frequencies, if desired. Alternately, the relationship plotted in Fig. 4-10 may be used to determine the frequency separation required for compatibility.

If none of the four EMI prediction cases survives after totaling entries on line 18, then the prediction process ends, and there exist no intersystem EMI problems for the particular Tx-Rx pair. However, should one or more cases survive (entry on line 38 algebraically greater than -10 dB), then the prediction continues into the third level of culling, detail cull. These entries are listed as lines 39 through 46 on the second page of the short form. The supporting subforms, forms A–E, for detail culling are included on the third and fourth pages of the short form. Details regarding the use of these subforms are discussed in reference 1.

One problem in performing an EMI prediction of the type described is obtaining all input information required for the prediction. In the absence of certain specific data, values that may be used (i.e., default data) are provided.

For the EMI prediction process, the specific values and techniques provided in the short form are intended to be as realistic as possible. Users may want to regard situations that result in an interference margin between -10 and $+10$ dB as marginal. They may determine whether specific values and relationships used in the prediction process are realistic for the specific problem. In this case, users should refer to Reference (1) to obtain more detailed descriptions and models for the relationships. If significant differences exist, it may be necessary to conduct a detailed EMI prediction.

The major assumptions used or implied by the short-form EMI prediction process and the suggested values provided are:

- Frequency limits for transmitter spurious emissions and receiver spurious responses are from 0.1 to 10 times fundamental frequency. This assumes that there are no significant emissions or responses outside these limits.

SHORT FORM EMI PREDICTION

TRANSMITTER AND RECEIVER FREQUENCY LIMITS

Frequency: ☐ kHz; ☒ MHz

1. TX Fundamental Frequency, (f_{OT}) — **220**
2. TX Minimum Spurious Frequency, $(f_{ST})_{min}$ or $.1f_{OT}$ — **22**
3. TX Maximum Spurious Frequency, $(f_{ST})_{max}$ or $10f_{OT}$ — **2200**
4. RX Fundamental Frequency (f_{OR}) — **360**
5. RX Minimum Spurious Frequency, $(f_{SR})_{min}$ or $.1f_{OR}$ — **36**
6. RX Maximum Spurious Frequency, $(f_{SR})_{max}$ or $10f_{OR}$ — **3600**
7. TX-RX Maximum Allowable Frequency Separation for Fundamental EMI, Δf_{max} or $.2f_{OR}$ — **72**

APPLICABILITY OF FOUR EMI PREDICTION CASES:

SIM = TX Harmonic & RX Spurious
Is (2)____**22**____ < (6)**3600** ?   ☒ Yes, ☐ No
Is (3)**2200**____ > (5)____**36** ?   ☒ Yes, ☐ No
If either is No, there is no EMI Problem-STOP

RIM = TX Harmonic & RX Fundamental
Is (2)____**22**____ < (4)__**360**__ ?   ☒ Yes, ☐ No
Is (3)**2200**____ > (4)__**360**__ ?   ☒ Yes, ☐ No
If either is No, skip RIM and enter N/A on line 38.

TIM = Tx Fundamental & RX Spurious
Is (1)____**220**____ < (6)**3600** ?   ☒ Yes, ☐ No
Is (1)____**220**____ > (5)__**36**__ ?   ☒ Yes, ☐ No
If either is No, skip TIM and enter N/A on line 38.
If both RIM and TIM were N/A, skip FIM and enter N/A on line 38.

FIM = TX Fundamental & RX Fundamental
Is |(1)__**220**__ -(4)__**360**__ |<(7)__**72**__ ?   ☐ Yes, ☒ No
If No, skip FIM and enter N/A on line 38.

Surviving cases ☒ SIM, ☒ RIM, ☒ TIM, ☐ FIM
☐ No cases survived - No EMI problem.

AMPLITUDE CULLING (See Sec. 5.3.1.2,5.3.2.2,5.3.3.2)

	FIM	TIM	RIM	SIM	
8. TX Power, $P_T(f_{OT})$, (peak power if pulsed)	NA	80	░░░	░░░	dBm
9. TX Spurious Power Output:$P_T(f_{ST})$ or $P_T(f_{OT})$-60dB	░░░	░░░	20	20	dBm
10. TX Antenna Gain in RX Direction: $G_{TR}(f)$ or 0dB	NA	0	0	0	dB
11. RX Antenna Gain in TX Direction: $G_{RT}(f)$ or 0dB	NA	0	0	0	dB
12. Propagation Loss, L Using Frequency No.	(1)	(1)	(4)	(2)	
Loss in dB from Fig. 4.2	-NA	-49	-55	-30	dB
13. Unintentional Power Available, $P_A(f)$ Add 8 to 12	NA	31	-35	-10	dBm
14. RX Fundamental Susceptibility, $P_R(f_{OR})$	NA	░░░	-100	░░░	dBm
15. RX Spurious Suscept.:$P_R(f_{SR})$ or $P_R(f_{OR})$ + 80dB	░░░	-20	░░░	-20	dBm
16. Preliminary EMI Prediction: line 13-14 or 13-15	NA	51	66	10	dB

If EMI margin < -10 dB, EMI Highly Improbable - STOP
If EMI margin > -10 dB, Start Frequency Culling

Fig. 4-15. Short form EMI prediction. (*Courtesy of Interference Control Technologies*)

FREQUENCY CULLING (See Sec. 5.3.1.3 and 5.3.2.3)

BANDWIDTH CORRECTION

Frequency:
☒ kHz; ☐ MHz

17. TX PRF (if pulse) — `100` pps
18. TX Bandwidth, ($B_T = 2/\pi\tau$ if pulse; τ = width) — `64`
19. RX Bandwidth, B_R — `1.0`
20. Adjustment(from lines 17 to 19; Fig.5.11&5.12) — `NA` `-15` `-15` `-15` dB
21. Bandwidth Corrected, EMI Margin = lines 16 + 20 — `NA` `36` `56` `-5` dB

IF EMI MARGIN ≤ -10 dB, EMI HIGHLY IMPROBABLE - STOP

FREQUENCY CORRECTION

Frequency:
☐ kHz; ☐ MHz

22. RX Local Oscillator Frequency, f_{LO} — `400`
23. RX Intermediate Frequency, f_{IF} — `40`

	FIM	TIM	RIM	SIM
24. TX-RX Frequency Separation: $\Delta f = \lvert(1)-(4)\rvert$	NA	☒	☒	☒
25. $\Delta f > (B_T + B_R)/2$ (from line 24, use Fig.5.10)	NA	☒	☒	☒
26. Calculate $(f_{OT} \pm f_{IF})/f_{LO}$ to nearest integer	☒	1	☒	☒
27. Multiply lines 22 x 26	☒	400	☒	☒	MHz
28. $\Delta f = \lvert1-23-27\rvert$ = _____ : $\lvert1+23-27\rvert$ = _____				
29. Select smaller Δf from line 28	☒	140	☒	☒	MHz
30. $\Delta f > (B_T + B_R)/2$(from line 29, use Fig.5.10)	☒	-100	☒	☒	dB
31. Calculate f_{OR}/f_{OT} to nearest integer	☒	☒	2	☒
32. Multiply lines 1 x 31	☒	☒	440	☒	MHz
33. $\Delta f = \lvert(4) - (32)\rvert$	☒	☒	120	☒	MHz
34. $\Delta f > (B_T + B_R)/2$ (from line 33 use Fig.5.10)	☒	☒	-100	☒	dB
35. Calculate Minimum Δf (see Form A)	☒	☒	☒	0	MHz
36. If $\Delta f > (B_T + B_R)/2$ (from line 35 use Fig.5.10)	☒	☒	☒	0	dB

EMI FREQUENCY CORRECTED SUMMARY

37. Add line 21 to line — `25` `30` `34` `36`
38. Total Here — `NA` `-49` `-29` `-5` dB

IF EMI MARGIN < -10 dB, EMI HIGHLY IMPROBABLE

DETAILED CULL (see Sec. 5.3.2.4, 5.3.3.3 and 5.3.4.4)

HARMONIC OR SPURIOUS RESPONSE CORRECTION*

	FIM	TIM	RIM	SIM
39. Correction from Form B	☒	NA	NA	26	dB
40. Corrected EMI add line 38 to line 39*	NA	NA	NA	21	dB

MODULATION CORRECTION

| 41. Correction from Form C | NA | NA | NA | 0 | dB |
| 42. Corrected EMI Add line 40 to line 41 | NA | NA | NA | 21 | dB |

POLARIZATION CORRECTION

| 43. Correction from Form D | NA | ☒ | ☒ | ☒ | dB |
| 44. Corrected EMI Add line 42 to line 44 | NA | NA | NA | 21 | dB |

PROPAGATION LOSS CORRECTION

| 45. Correction from Form E | NA | NA | NA | 0 | dB |
| 46. Corrected EMI Add line 44 to line 46 | NA | NA | NA | 21 | dB |

IF EMI < -10 dB EMI HIGHLY IMPROBABLE
-10 dB ≤ EMI ≤ 10 dB EMI MARGINAL
EMI > 10 dB EMI PROBABLE

*Apply corrections only if nominal -60dB TX (line 9) and/or +80dB RX (line 15) corrections were used.

© Copyright 1972, Don White Consultants

Fig. 4-15. (Continued)

TX Fund(f_{OT}): __220__ ; RX LO(f_{LO}). __400__ ; RX IF (f_{IF}): __40__

N	$(Nf_{OT} + f_{IF})/f_{LO}$	Δp*	$(Nf_{OT} - f_{IF})/f_{LO}$	Δp *
2	1.2	.2	1.0	0
3	1.75	-.25	1.55	-.45
4	2.30	.30	2.10	.10
5	2.85	-.15	2.65	-.35
6	3.40	.40	3.20	.20
7	3.95	-.05	3.75	-.25
8	4.50	.50	4.30	.30
9	5.05	.05	4.85	-.15
10	5.60	-.40	5.40	.40

Select Minimum Value For Δp

Minimum Spurious Frequency Separation = $(\Delta p)_{min} f_{LO}$

---

*Δp is the magnitude of difference between value obtained for $p \pm \Delta p$ = $N(f_{OT} \pm f_{IF})/f_{LO}$ and the nearest integer.

FORM B - HARMONIC AND/OR SPURIOUS RESPONSE CORRECTION IN dB

FIM - No Correction
TIM - Spurious Response Correction use p from line 26
RIM - Harmonic Correction use N from line 31
SIM - Add Spurious and Harmonic Correction use p and N from Form A above.

Category		N or p*								
		2	3	4	5	6	7	8	9	10
HF & Below	Harmonic	+19	+ 7	- 2	- 9	-14	-19	-23	-27	-30
	Response	-13	-17	-20	-22	-24	-26	-27	-29	-30
VHF	Harmonic	+ 6	- 8	-18	-26	-32	-37	-42	-46	-50
	Response	-15	-22	-26	-29	-32	-35	-37	-38	-40
UHF & Above	Harmonic	+ 5	- 4	-10	-15	-19	-22	-25	-28	-30
	Response	+ 8	+ 1	- 4	- 8	-11	-14	-16	-18	-20

*For the receiver image response use a +20 dB correction.

FORM C - MODULATION CORRECTION IN dB*

(Use Δf's from lines 24,29,33, or 35)

Modul Type	B_T line 18	B_R line 19	Modulation Correction in dB
Pulse			0
AM			$20 log (B_T+B_R)/2\Delta f$
FM			$40 log (B_T+B_R)/2\Delta f$

The combined Frequency Separation Correction from lines 25,30,34,or 36 plus Modulation Correction must not exceed 100 dB

FORM D - ANTENNA POLARIZATION CORRECTION IN dB *

TX Pol: ☐ H;☐ V;☐ C.
RX Pol: ☐ H;☐ V;☐ C.

RX \ TX	Horizontal		Vertical		Cir
	G<10dB	G≥10dB	G<10dB	G≥10dB	
Hor G<10dB	0	0	-16	-16	-3
Hor G≥10db	0	0	-16	-20	-3
Ver G<10dB	-16	-16	0	0	-3
Ver G≥10dB	-16	-20	0	0	-3
Circular	-3	-3	-3	-3	0

*Applies only to intentional radiation region and design frequency

Fig. 4-15. (Continued)

Form E - Propagation Loss Correction in dB

- Sketch Path Profile with TX and RX Antennas on Chart Below

- Check for Line of Sight Between TX and RX

- If Line-of-Sight Exists, Propagation Loss Correction = 0

- If Line-of-Sight is Blocked Calculate Line of Sight Distance, d_{LOS}, from TX to Terrain

- Propagation Loss Correction (dB) = $40 \log_{10} d_{LOS}/d$ where d = Distance From TX to RX.

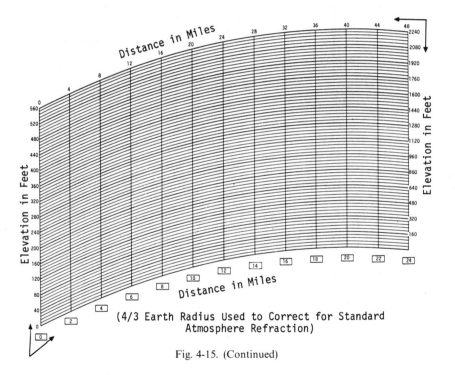

Fig. 4-15. (Continued)

- Maximum Tx-Rx frequency separation for fundamental interference is 0.2 times receiver fundamental. This assumes fundamental interference is not significant for larger frequency separations.
- Free-space propagation loss is assumed in amplitude culling.
- Levels for transmitter spurious emissions are 60 dB below fundamental emission.
- Levels for receiver spurious susceptibility are 80 dB above fundamental susceptibility.

- An additional 20-dB rejection is assumed for each transmitter and receiver minor emission and response.
- Values for antenna gains in unintentional radiation directions and at unintentional frequencies are 0 dB.
- Differences in transmitter and receiver bandwidth are assumed to modify the power available in the manner specified in Table 4-5.
- Frequency separation, Δf, between transmitter emission and receiver response is assumed to reduce the effective power available by an amount given by $40 \log [(B_T + B_R)/2\Delta f]$.
- A go/no-go cull level of -10 dB is used. Thus, potentially interfering situations are culled only if the mean signal level is less than -10 dB relative to the receiver susceptibility threshold.

A relatively detailed application of the short-form EMI prediction process is presented below. Additional detailed descriptions of the procedures and techniques are provided in reference 1.

Illustrative Example—EMI Prediction

Consider the potential EMI that may exist between a P-band radar transmitting at 200 MHz and a UHF AM voice receiver tuned to 360 MHz used for communicating with aircraft. Assume that both the P-band radar and the UHF receiver are located on board the same ship, and their antennas are only 30 meters apart but are located with sufficient vertical separation so that the radar antenna main beam does not illuminate the UHF antenna. Assume that the nominal Tx-Rx characteristics are as listed.

Parameter	P-band radar	UHF receiver
Operating frequency (f_0)	220 MHz	360 MHz
Peak power output	100 kW	NA
Pulse width (τ)	10 μsec	NA
Pulse repetition rate	100 pps	NA
Transmitter bandwidth ($2/\pi\tau$)	64 kHz	NA
Antenna gain	23 dB	0 dB
Receiver bandwidth	NA	10 kHz
Receiver sensitivity	NA	-100 dBm
Receiver IF	NA	40 MHz
Receiver local oscillator (LO)	NA	$(f_{0R} \rightarrow f_{IF})$

The EMI prediction short form is used for this problem, and the results are shown in Fig. 14-15. Referring to the short form, the spurious (SIM), receiver (RIM), and transmitter (TIM) interfering cases are applicable to this problem. The fundamental (FIM) case is not applicable because the fundamental frequency separation is greater than $0.2 f_{OR} = 0.2$ (360 MHz) = 72 MHz.

As a result of performing the amplitude cull, a positive EMI margin is obtained for the TIM, RIM, and SIM cases, as shown on line 16. Hence, it is necessary to continue with the frequency cull. When the frequency cull is applied, the corrected TIM and RIM are both below the -10 dB cull level, as shown on line 38. Hence, EMI is highly improbable for these cases, and it is not necessary to examine them further in the detailed cull. On the other hand, the frequency-cull corrected SIM case results in an EMI margin that is above the -10 dB cull level (-5 dB on line 38). Hence, the SIM case must be examined further during the detailed cull.

When the detailed cull is applied to the SIM case, it is seen that a $+26$ dB correction is obtained for the harmonic or spurious response correction, as shown on line 39. This positive correction results because the SIM case in this particular example involves the transmitter second harmonic ($N = 2$; subform A) interfering with the receiver image response. The transmitter second harmonic results in a $+6$ dB harmonic correction for VHF equipments. This positive correction results because second harmonics tend to exceed the nominal level [i.e., $P_T(f_{0T}) - 60$ dB] used in line 9. Receiver image responses result in a $+20$ dB correction because receivers tend to be more susceptible to the image than the nominal level [i.e., $P_R(f_{0R}) + 80$ dB] used in line 15. Application of the detailed cull results in a $+21$ dB EMI margin for the SIM case, as shown in lines 40 and 46.

To summarize results of this example, EMI is probable for the particular conditions considered. Potential EMI results from the transmitter second harmonic existing at the receiver image response frequency of 440 MHz. This potential EMI problem may be avoided by changing the receiver operating frequency so that there is sufficient frequency separation between the transmitter second harmonic and the image response to provide a -31 dB rejection, required to reduce the EMI margin below the -10 dB cull level. Referring to Fig. 4-10, if the output–response frequency separation for $(B_T + B_R)/2 = 37$ kHz is greater than 200 kHz, more than 31 dB of rejection will be provided, and the potential EMI problem will be eliminated. For example, if the receiver is tuned to a frequency of 360.5 MHz, there should be no EMI problem resulting from the transmitter second harmonic and the receiver image response.

REFERENCES

1. Duff, William G., and White, Donald R. J. *EMI Prediction and Analysis Techniques*, A Handbook Series on Electromagnetic Interference and Compatibility, Volume 5. Germantown, MD: Don White Consultants, Inc., 1972.

2. White, Donald R. J. *EMI Control Methods and Techniques*, A Handbook Series on Electromagnetic Interference and Compatibility, Volume 3. Germantown, MD: Don White Consultants, Inc., 1973.

3. Stutzman, Warren L., and Thiele, Gary A. *Antenna Theory and Design.*, New York: John Wiley and Sons, 1981.

4. Bullington, Kenneth. "Radio Propagation at Frequencies above 30 Megacycles." *Proceedings of the IRE*, October, 1947.

5. Freeman, Roger L. *Telecommunication System Engineering Analog and Digital Network Design*. New York: John Wiley and Sons, 1980.

6. *Reference Data for Radio Engineers*, Sixth Edition, Third Printing. Indianapolis, IN: Howard W. Sams, 1979.

7. Freeman, Ernest R. *Interference Suppression Techniques for Microwave Antennas and Transmitters*. Dedham MA: Artech House, Inc., 1982.

Chapter 5
INTRASYSTEM EMI PREDICTION
AND CONTROL

This is a sequel to Chapter 4, in which the emphasis was on *inter*system (antenna-to-antenna) EMI prediction and control. In this chapter, the emphasis is on *intra*system problems. Antennas generally are not involved except when the EMI source is an offending transmitter. If the distinction between *inter*- and *intra*system is not clear, it is suggested that the reader review Chapter 1 of this *Handbook*. EMI ranges from problems at the chip level to ensembles of systems and subsystems. Nine identifiable levels of EMI complexity are shown in Fig. 5-1 with a middle level defined at the subsystem or collection of equipments. It is at this level that this chapter develops concepts and quantitative models for determining EMI coupling based upon a "two-box" situation. If quantitative solutions to EMI problems can be achieved at this middle level, then the user can either proceed toward more complex levels at higher echelons or work down toward the motherboard, printed-circuit board, or individual component levels.

This chapter defines and illustrates many different EMI coupling paths. There is common impedance coupling, or coupling due to a conducted current flowing through a common impedance, across which two or more circuits are connected, thereby inducing a common-mode (CM) voltage into a victim circuit. Also defined is radiated electromagnetic field coupling into circuits and/or cables producing common-mode and/or differential-mode (DM) voltages, as well as cable-to-cable (crosstalk) capacitive and inductive DM coupling. In addition, coupling via power lines and power supply circuits, a mix of the previously mentioned topics, is covered. Basically, these five coupling paths (illustrated by Fig. 5-5, below) predominate over many others, and perhaps account for up to 95% of EMI situations.

Effective EMI control is often difficult because the large number of possible combinations of EMI fixes and coupling paths involve many variables. Because of the inherent complexity of EMI situations, an organized approach (i.e., a methodology and procedure) is paramount if EMI problems are to be solved in a timely, efficient, and cost-effective manner (12).

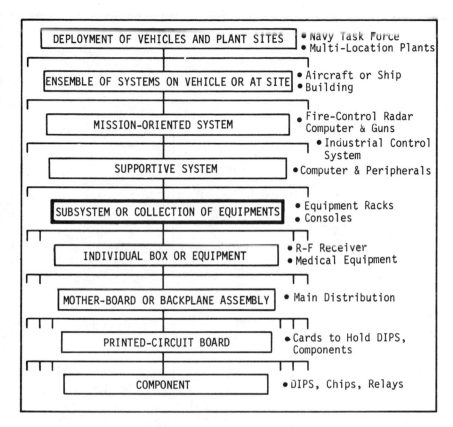

Fig. 5-1. The many levels of EMI manifestations. (*Courtesy of Interference Control Technologies*)

The situation shown in Fig. 5-2 corresponds to a two-box equipment-level EMI situation. (See middle level, Fig. 5-1.) Here, the word "box" means equipment case, console, printed circuit board, and so on. In viewing the picture, it is seen that box case number one at the left is interconnected through power and/or data and control leads to box case number two on the right.

In examining Fig. 5-2 more thoroughly, it is seen that there exists a total of 29 question marks. A question mark is shown to signify a binary situation, illustrated as a toggle switch, meaning a yes or no answer, or around a component, indicating that the component may or may not be used. In the figure, FR stands for ferrites, IT for isolation transformers, IS for isolators (optical or transformers), F for filters, C for connectors, and PS for power supply.

The question marks under the local grounds mean, "Is that point to be grounded: yes or no?" For the question marks next to the toggle switches, the question is, "Is that point connected (grounded): yes or no?" Where grounding of cable shields is shown, in addition to asking the first question

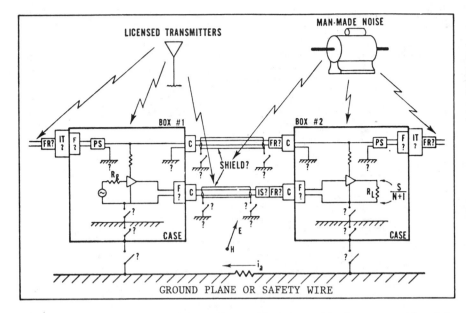

Fig. 5-2. A two-box equipment EMC situation—500,000,000 combinations, many with an infinite range of variables. (*Courtesy of Interference Control Technologies*)

("Is it grounded: yes or no?"), the next question mark at the local grounds means, "Where is it to be grounded?" Because each of the questions requires one of two choices, the 29 questions correspond to 2^{29} or over 500,000,000 possible combinations. This shows why EMI control can be difficult! A relatively simple situation such as that of Fig. 5-2 results in an enormous number of possible combinations.

In pursuing the topic still further in Fig. 5-2, it is recognized that the number of combinations far exceeds 500,000,000. For example, the question is not simply asked, "Is a filter to be used or not?" Rather, it is recognized that a filter can assume an arbitrary large number of parameter values corresponding to different attenuation slopes in the stop band. Consequently, there do not exist just 500,000,000 combinations in Fig. 5-2, but possibly an infinite number.

Therefore, a disciplined methodology and technique are required to quantitatively solve the EMI problem suggested in Fig. 5-2. That is the motivation for this chapter.

Coupling Paths

Electromagnetic interference may enter a collection of circuits or a subsystem by a number of different paths from the offending emission source(s) to the

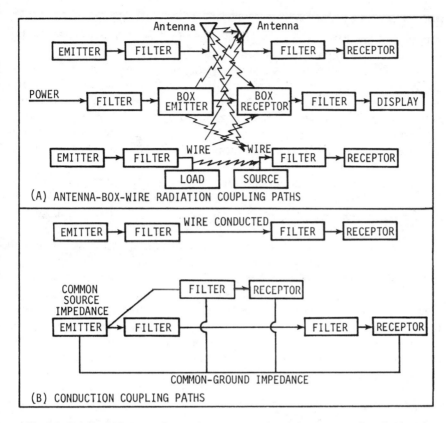

Fig. 5-3. Principal EMI coupling paths. (*Courtesy of Interference Control Technologies*)

victim receptor(s). Figure 5-3 illustrates such a situation for both radiated and conducted coupling paths.

For radiated coupling paths, three different classifications of both the offending emission and the victim reception path are shown in the top part of the figure. One path involves an antenna used for radiation and/or reception; the second is emission from a box (enclosure) containing noise-making parts or circuits to a victim box; and the third is wire-to-wire or cable-to-cable coupling. Thus, the combinations correspond to nine different possible paths; namely, antenna-to-antenna, antenna-to-box, antenna-to-wire, box-to-antenna, box-to-box, box-to-wire, and wire-to-antenna, wire-to-box, and wire-to-wire (cable-to-cable).

Not all nine paths contribute equally to electromagnetic interference. In cases where systems are at different sites and baseband information modulates a carrier, the antenna-to-antenna path usually predominates. Where a transmitting antenna is involved and the victim does not have an antenna, the antenna-to-wire (field-to-wire) path usually predominates. Finally, in situation where cables are routed relatively close to each other, the cable-to-cable (crosstalk) path may predominate.

It is rare that the box-to-box coupling path predominates. Other situations involving the box as either an emission source or a victim result in relatively infrequent EMI problems. Consequently, in many cases the nine possible radiated coupling paths shown in Fig. 5-3 reduce to antenna-to-antenna, antenna-to-cable, cable-to-antenna, and cable-to-cable. These account for a high percentage of the radiated EMI situations found in widespread sampling of practical situations.

Where conducted emission over hard wire is involved, the lower part of Fig. 5-3 summarizes the existing situations. One path corresponds to direct interference propagated over a wire line and coupled via the cable to the victim receptor. More often, the EMI problem involves the power mains to which an interfering source is coupled. For example, switching-regulator-power-supply emissions or transients from other user switching actions are developed on power busses and coupled to the victim via the power lines.

Occasionally, common-impedance coupling exists whereby two circuits, networks, or systems share a common section of a ground plane or safety-wire bus due to multi-point grounding or sharing the same power supply. This problem becomes quite prevalent in complex installations or in situations wherein many equipments are interconnected and attached to a common ground reference.

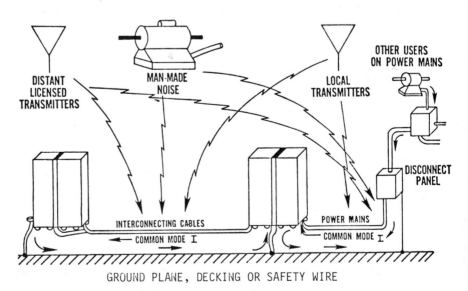

Fig. 5-4. Principal EMI coupling paths. (*Courtesy of Interference Control Technologies*)

Figure 5-4 illustrates "real life" noise-making sources coupling into victim cabinets and racks by both radiation and conduction. Radiation coupling involves coupling into loop areas formed by cables, cabinets, and ground planes. Conduction coupling includes the power mains servicing the victim equipment and voltages induced by currents flowing through ground plane impedances. Figure 5-4 is simply another way of looking at the EMI sources, victims, and coupling paths that often appear in practice.

The antenna-to-antenna coupling path is excluded here because it is an intersystem EMI problem and covered in Chapter 4. The antenna-to-antenna situation, however, may be included in the overall EMC methodology.

OVERVIEW OF EMC METHODOLOGY AND PROCEDURES

Figure 5-5 is a general flow diagram used for prediction and for solving EMI problems. This summary diagram presents a qualitative picture of the EMC analysis methodology. The introduction and general discussion of this diagram provide an overview of the techniques used in this chapter.

The methodology begins at the "start" circle (Fig. 5-5), where the user identifies the EMI problem data and EMI application constraints (block 1). The problem data block pertains to performing actions to identify intentional

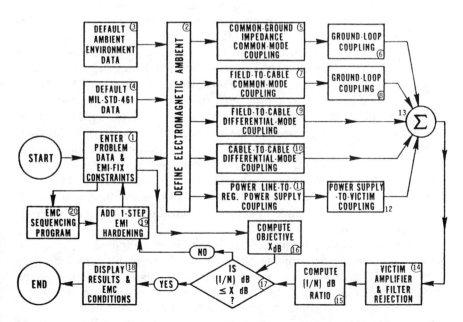

Fig. 5-5. General flow diagram for predicting and solving EMI problems. (*Courtesy of Interference Control Technologies*)

signal levels, receptor or victim frequency response characteristics, and one or more expected coupling paths. Examples of intentional signals include signal input voltages, power amplitude, operational signal frequency, digital pulse duration, and rise times. Examples of the potential EMI victim characteristics are in-band sensitivities, as a function of frequency, corner or cutoff frequencies of the pass band, roll-off slope in the rejection band, and out-of-band susceptibilities. Default data are available to the user whenever required actual data are unknown or not immediately available.

In anticipation of an interference problem, the user supplies a list of different EMI fix constraints that should be employed (also in block 1) (i.e, whether particular EMI components or techniques can indeed be used). Constraints are required because some applications of the EMI analysis and/or synthesis method may emphasize conditions in a laboratory design, whereas others may apply only for field installations. Furthermore, some applications may involve weight or size limitations, and others may not. The ultimate EMC constraint is usually cost. In many actual situations, limited resources are allocated to EMC until a critical situation develops. Therefore, only the user, in his or her particular applications, knows what constraints exist.

The next step in the user EMI problem entry is to define the electromagnetic ambient environment (block 2). This involves the electric-field strength or magnetic-flux density, voltage or current, or other ambient parameters corresponding to a radiated and/or conducted environment.

Should the user not know the electromagnetic ambient, a number of default conditions are supplied (block 3). This covers typical industrial and commercial environments, often specified by regulatory agencies such as the FCC. Another of the family of default data corresponds to specification limits such as those that appear in MIL-STD-461 (block 4).

Once input data, EMI fix constraints, the electromagnetic ambient, and the allowed influence on the ambient have been defined, the next step in the methodology involves processing one or more of the applicable coupling paths (blocks 5, 7, 9, 10, and 11), which cover most of the intrasystem problems found in government, industrial, and commercial situations. The first coupling path involves computing the common-mode coupling by a common-ground impedance shared between an undesired emission source and the victim (block 5). (Common-mode in a cable or harness derives its name from the induced currents flowing in phase.)

The common-mode coupling voltage having been computed, the coupling path is completed by determining how much of the available common-mode voltage manifests itself at the victim input terminals as differential mode. This is called ground-loop coupling (block 6). Thus, the common-impedance coupling path is divided into two parts to facilitate computations. Another common term for ground-loop coupling is common- to differential-mode conversion as applied to a system containing a ground loop.

When the potentially interfering source contains a radiating antenna, or when an ambient exists in the form of a time-changing electric and/or magnetic field-strength at the location of the victim, a second coupling path is identified as field-to-cable common-mode coupling (block 7). Here, the cable is defined as the signal and/or power cables between the pieces of equipment or boxes, as previously shown in Figs. 5-2 and 5-4. The time-changing electromagnetic field induces a voltage around any closed loop through which the field propagates. This voltage appears as a common-mode voltage that is partially coupled to the victim (amplifier, digital circuit, etc.) via the ground-loop coupling path (block 8). This ground-loop coupling is the same as that mentioned in the preceding paragraph (block 6).

Differential-mode coupling (block 9) is another identified coupling path, involving field-to-cable coupling. This is a direct radiation path from the electromagnetic ambient field into a signal cable or harness. This coupling directly develops a potentially interfering differential mode signal at the input terminals of the victim amplifier, digital circuit, and so on. A fourth coupling path, which also develops a direct differential-mode interference across the victim input terminals, is identified as cable-to-cable coupling (block 10). This coupling involves near-field electric field (capacitive) or magnetic field (inductive) coupling conditions only, and corresponds to one "culprit" cable (or harness) carrying a signal that is coupled into a neighboring "victim" cable (or harness) feeding the victim. Actually, both source and victim wires may be located in the same harness, but they are still separated by a definable distance. The cables may also be located in a common cable tray, raceway, or cable hanger.

The fifth and final coupling path shown in Fig. 5-5 corresponds to conducted EMI coupling on AC and/or DC power mains via regulated power supplies (block 11) to the signal circuit. A form of undesired emission may be coupled onto power lines by other users also connected to the same power lines. Common-impedance coupling results whenever two or more loads derive power from a common power supply and distribution system. Whenever one of the loads draws power from the distribution system, a steady-state or transient differential-mode (DM) voltage drop occurs across the power system and conductors due to the power distribution system internal impedance. This DM voltage drop is seen by all the loads on the power system, and it can cause system malfunction should it exceed equipment susceptibility levels. This is especially critical in power distribution systems serving rapid-switching digital systems with fast rise times, and high-frequency analog systems (12, 13). Should the power line, however, be exposed to a radiated electromagnetic ambient at another location, the undesired emissions thereon would appear as common-mode noise.

The second part of the last coupling path involves processing the conducted emissions through the power supply and onto the victim power bus and then into the susceptible victim signal circuits (block 12). Thus, power-supply-to-victim coupling accounts for the need of power circuit isolation offered by power transformers, switching regulator power supplies, and filters.

If more than one coupling path is involved, then the resultant voltage produced at the victim input is determined by summing the voltages produced by all coupling paths (circle 13). It is realized that the voltages produced by the respective coupling paths are generally not in phase. As a "worst case" upper bound, the resultant EMI voltage at the victim input terminals can be determined by assuming that the voltages produced by all applicable coupling paths are in phase.

The resultant victim input voltage from all coupling paths, having been calculated, it is next processed through the victim (block 14). The victim is defined in terms of its analog pass-band sensitivity, the corner or cutoff frequency, the roll-off slope and out-of-band characteristics in the rejection band of the frequency response, or the digital noise margin or immunity level. It also includes the victim out-of-band susceptibility as a function of frequency known as audio rectification (see Chapter 3). Thus, the victim frequency response usually tends to somewhat delimit the effects of the interference unless the EMI exists totally within its pass band.

The intentional signal (S) and interference (I) levels have now been calculated, and the victim inherent sensitivity (N) is identified. Thus, both the interference-to-noise (I/N) ratio (block 15) and the signal-to-noise-plus-interference ratio ($S/(N + I)$) are computed. The I/N ratio is a direct quantitative measure of the amount of EMI hardening that may be required.

To determine how much EMI hardening is needed, an objective I/N ratio (labeled X_{dB}) must be determined (block 16). This objective is indirectly calculated from the user design goal and constraints (block 1), and it involves the allowable probability that EMI will exist (12). Determining the proper I/N ratio is critical to avoid underdesign, whereby the probability of interference is unacceptably high; or overdesign, which can be (and usually is) costly, unless the cost of extra EMI hardening is justified for certain applications, such as for electrically initiated explosive devices, medical or critical military electronic equipment, and so on.

A system designer establishes an objective signal-to-noise ratio, S/N. In reality, an operational system performs in accordance with a signal-to-noise-plus-interference ratio, $S/(N + I)$. If $I/N \ll 1$, then:

$$S/(N + I) = S/N$$

and the system performance limitation will essentially be determined by the inherent analog system noise or digital system noise immunity level or noise margin, represented by N. The objective then becomes to determine a realistic value for I/N based upon a statistical level of confidence developed from system requirements.

The objective is to select $I/N < 1$, but not $I/N \ll 1$, which may result in costly overdesign. If, as is usually the case, statistical data are not available, the following guidelines are suggested for determining an objective I/N ratio, X_{dB} (1):

- $X_{dB} < 10$ dB: usually results in underdesign; $X_{dB} = 10$ dB may be used when system operation is anticipated in relatively benign EMI environments.
- 10 dB $< X_{dB} < 30$ dB: recommended range, nominal 20 dB suggested.
- $X_{dB} > 30$ dB: usually results in overdesign, recommended for special requirements or anticipated system operation in hard-to-estimate EMI environments.

The computed I/N ratio is compared with the design objective (diamond 17) to determine whether EMI or EMC exists. If I/N in dB is equal to or less than X_{dB} ("yes" decision, diamond 17), then EMC is construed to exist. Then, all results including the historical record are displayed as required (block 18), and the program objective has been achieved.

If the I/N ratio is not equal to or less than X_{dB} ("no" decision, diamond 17), then additional EMI hardening is required. One EMI hardening component or technique is added at a time (block 19) with each fix based upon reducing the most significant applicable coupling path, and the entire procedure is repeated to determine a new I/N ratio. While the user can select his own EMI fix (block 1), the selection of usable fixes is prompted by the EMC sequencing program (block 20), based upon the most significant coupling path. This sequences and iterates the procedure until an EMC solution has been achieved.

Following the addition of each EMI fix (component or technique) (block 19), the new I/N ratio is determined (block 15). It is again compared with the objective (block 16), and the process is continued until $I/N < X_{dB}$, whereupon the program ends. This procedure assures a method for determining appropriate EMC fixes, one at a time, thus reducing the number of variables that one must monitor. It also provides an estimate of the benefit derived from each fix with the computation of the new I/N ratio with each iteration. Also, with an appropriately designed form, the procedure illustrated in Fig. 5-5 provides a structured, efficient method for record keeping.

EMI PREDICTION AND PERFORMANCE DISPLAYS

This section presents the executive program for using and controlling the input from previous sections and processing it on a summary basis to determine the EMI/EMC performances in accordance with the procedure in Fig. 5-5. As indicated previously, the resulting output will determine whether more EMI hardening is required, or the objective has been achieved. In either case, the accounting of performance is systematically recorded. Subsequent sections present details of individual mathematical modules with methods to determine coupling and EMI hardening fixes, and the EMC overall sequencing process.

$S/(N + I)$—Overall System Performance

This section is an overview of signal-to-noise (S/N), interference-to-noise (I/N), and signal-to-noise-plus-interference $S/(N + I)$ ratios. All three are important in design and real-life situations because:

- S/N corresponds to a functional design engineer's objective if the environment were free of EMI.
- I/N corresponds to an EMC engineer's emphasis, since interference must be suppressed sufficiently below internal noise, or the noise immunity level.
- $S/(N + I)$ corresponds to the overall performance measure of a design when it is immersed in the real world containing EMI emissions from many sources.

$S/(N + I)$ may be expressed in an operationally useful form:

$$\frac{S}{N + I} = \frac{S/N}{1 + I/N} \tag{5-1}$$

$$= S/N \text{ for } I/N \ll 1 \tag{5-2}$$

$$= S/I \text{ for } I/N \gg 1 \tag{5-3}$$

The approximation in equation (5-2) is adequate (i.e., the error is less than 0.5 dB) when (12):

$$I/N < 0.122 = -9 \text{ dB} \tag{5-4}$$

Similarly, equation (5-3) applies when $I/N > 9$ dB.

Equation (5-1) is especially useful because it is broken into two separate parts, thereby illustrating the way a system designer usually thinks, in terms of S/N, versus the way that an EMC engineer usually thinks, in terms of I/N. $S/(N + I)$ and I/N ratios are also separately used and displayed in performance summaries below. Usually, the EMC engineer does not challenge S/N because this is the design engineer's perogative.

Performance Reporting and Organization

This section describes a novel method for organizing the results of the EMI control methodology and procedures illustrated in Fig. 5-5. A good bookkeeping system is essential to the success of the technique. Many EMC methods fail because the user is unable to keep track of procedures taken, and is often unable to note which ones are successful and which are not.

One display that offers high visibility is the organization chart shown in Fig. 5-6. This can be visualized as either a CRT display or an LED display board showing the EMI/EMC performance. The I/N ratio is displayed at the top center in dB. Supporting information is at either side: the overall $S/(N + I)$ ratio and the objective I/N ratio.

The victim amplifier (or digital circuit) is shown next in line on the organization chart because all potentially interfering energy must be processed through it. The total amplifier rejection (shown in the center box) is composed of in-band/out-of-band frequency response and susceptibility to audio rectification (left center box) and that due to a supporting EMI filter, if applicable (right center box).

The third tier in the I/N organization chart is composed of the five principal EMI coupling paths, previously shown in the flow diagram of Fig. 5-5. The

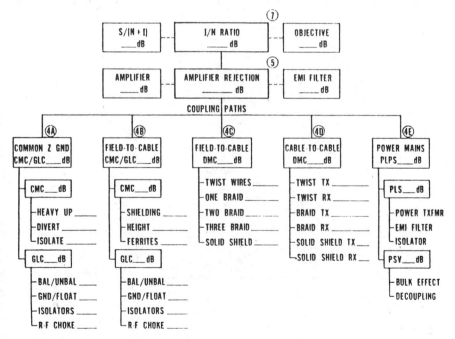

Fig. 5-6. I/N ratio and system performance organization chart (Courtesy of Interference Control Technologies)

EMI contribution is shown in Fig. 5-6 for each of the five coupling paths involved in any one problem. Some of the coupling paths contain two step parts, with each step depicted separately. For example, the two coupling paths on the left, designated common-ground impedance and field-to- cable coupling, include both common-mode coupling (CMC) and ground-loop coupling (GLC), the latter being the conversion of CM to DM. Finally, Fig. 5-6 illustrates the contribution from each of the potential EMI fixes, as applicable. These entries are indicated on the lowest level on the organization chart.

The display shown in Fig. 5-6 is intended to portray the entire EMI/EMC performance resulting from a selected problem and its corresponding input data. However, the display provides no record of what previous actions have accomplished; that is, it provides no indication of benefits (or lack of same) derived from previous EMC fixes. A display that accomplishes this objective, in part, is explained in the next section.

The EMI/EMC Performance Display

Figure 5-7 shows that the captions and columnar headings for the EMI/EMC performance display were selected in accordance with the considerations presented in the previous discussion and related to appropriate blocks in Fig. 5-6. The numerical headings in circles 4-7 correspond to the same numerical headings in both Fig. 5-7 and the flow diagram of Fig. 5-6.

Column 1 corresponds to the electromagnetic ambient in appropriate radiated or conducted units: $dB\mu V/m$, $dBpT$, $dB\mu V$, or other units, as applicable. Column 2 is the sensitivity (N) of the victim amplifier or digital logic device noise immunity level (NIL). While columns 1 and 2 are not necessarily of the

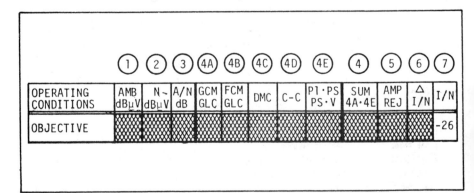

Fig. 5-7. Worksheet for EMI prediction and control. (*Courtesy of Interference Control Technologies*)

same units, column 3 is their ratio, corresponding to their difference in dB, and is one measure of the potential severity of the latent EMI problem should one exist.

Columns 4A through 4E correspond to the five principal EMI coupling paths, as applicable. Column 4 is their sum, as shown in Fig. 5-7. Essentially, the column 4 entry is the "worst case" situation of the determined entries appearing in columns 4A through 4E. The "summation" of column 4 essentially involves the determination of this worst case. The amplifier rejection appears in column 5. The objective I/N ratio is listed under column 7, and the -26 dB corresponds to a 95% probability of EMI. The difference in I/N ratios (improvement of I/N) in column 6 corresponds to each succeeding design/retrofit trial run.

The Initial Performance Trial Run

To illustrate the usefulness of the performance display of Fig. 5-7, an ongoing illustrative example is shown in Fig. 5-8. A potential interfering radiated emission from an AM broadcast transmitter is shown at 1 MHz. The field strength is 10 V/m.

Figure 5-9 shows the result of the first trial EMC analysis run. Column 1 contains the ambient E-field strength of 10 V/m converted to 140 dBμV/m. Column 2 is receiver sensitivity of 1 μV (see Fig. 5-8) converted to dBμV. Column 3 is the ratio of columns 1 and 2 and is obtained by subtracting column 2 from column 1 in dB. Since this is a field-to-cable coupling problem, only columns 4B and 4C are used. The larger of the two coupling paths (4B = FMC and GLC yield -47 dB) also appears in column 4. The amplifier rejection (column 5) to the out-of-band AM broadcast station corresponds to -14 dB. Finally, the resultant EMI problem corresponds to an I/N ratio of $+79$ dB (obtained by adding columns 4 and 5)—a very formidable situation.

Performance of Subsequent EMC Fixes

It is now observed that 103 dB [79 dB $-(-24)$ dB] of EMI hardening (EMI fixes) is required. Figure 5-10 shows the historical performance record of what the user tried and the corresponding results. (This might also be automatically synthesized if a computer is used, as discussed in a later section.) After the first trial run, the user decided that to reduce the effect of the out-of-band interference ($f_{int} = 1$ MHz) relative to the amplifier cutoff frequency ($f_{co} = 200$ kHz in Fig. 5-8), the equivalent of a five-stage, low-pass filter roll-off (100 dB/ decade) was needed. This increased the amplifier rejection from -14 dB (in column 5) to -60 dB, for a net improvement of 46 dB (column 6). Thus, the new I/N ratio is now 33 dB—still a bad situation, although less formidable than the previous $+79$ dB.

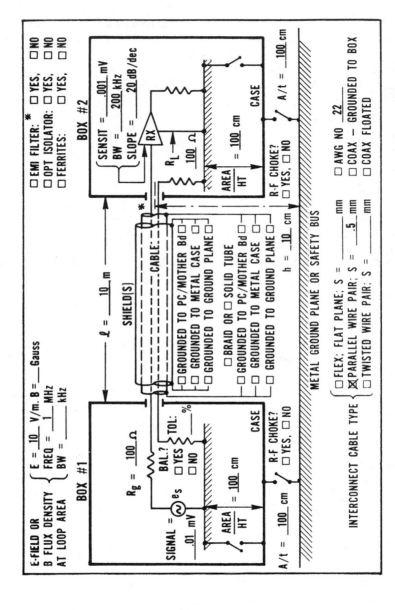

Fig. 5-8. Input data identification for two-box EMI problem. *(Courtesy of Interference Control Technologies)*

OPERATING CONDITIONS	AMB dBµV (1)	N dBµV (2)	A/N dB (3)	GCM GLC (4A)	FCM GLC (4B)	DMC (4C)	C-C (4D)	P1·PS PS·V (4E)	SUM 4A·4E (4)	AMP REJ (5)	Δ I/N (6)	I/N (7)
OBJECTIVE												-26
START: FIRST RUN	140	0	140	NA	-47	-86	NA	NA	-47	-14		+79

Fig. 5-9. The first trial seen corresponding to Fig. 5-8 conditions. (*Courtesy of Interference Control Technologies*)

Figure 5-10 shows that the next user trial EMC fix is to convert to a balanced transmission/reception system from the unbalanced system shown in Fig. 5-8. This improved (reduced) ground-loop coupling from -47 dB to -106 dB (column 4B) although it had no effect on the differential-mode coupling (column 4C). For this reason, the full 59 dB improvement (106 dB $-$ 47 dB) could not be realized, as the -86 dB DMC coupling path is now the larger. Thus, the net improvement is 39 dB (column 6), which corresponds to going from the -47 dB of GLC (column 4B) to -86 dB (column 4C). The resultant I/N ratio of -6 dB represents lowering the previous 33 dB I/N ratio by 39 dB.

OPERATING CONDITIONS	AMB dBµV (1)	N dBµV (2)	A/N dB (3)	GCM GLC (4A)	FCM GLC (4B)	DMC (4C)	C-C (4D)	P1·PS PS·V (4E)	SUM 4A·4E (4)	AMP REJ (5)	Δ I/N (6)	I/N (7)
OBJECTIVE												-26
START: FIRST RUN	140	0	140	NA	-47	-86	NA	NA	-47	-14		+79
AMPL SLOPE = 100 dBD	140	0	140	NA	-47	-86	NA	NA	-47	-60	46	+33
BAL GEN/ AMPL = 5%	140	0	140	NA	-106	-86	NA	NA	-86	-60	39	-6
TWIST WIRES 40TPM	140	0	140	NA	-106	-142	NA	NA	-106	-60	20	-26

$$(I/N)dB: \; (7) = (3) + (4) + (5)$$

Fig. 5-10. Subsequent runs until EMC (-26dB) is achieved. (*Courtesy of Interference Control Technologies*)

The final user-selected EMC fix involves twisting the wires of the cable pair in Fig. 5-8 with 40 twists/meter (TPM). Figure 5-10 shows that only a 20 dB improvement results (column 6) in reducing the DMC by 56 dB (142 dB − 86 dB, column 4C). This improvement truncation resulted in switching from the DMC coupling path as the larger (− 86 dB, column 4C) to the GLC path as the larger (− 106 dB, column 4B). Since the − 24 dB objective has been crossed over (new I/N performance = − 26 dB, column 7), no further EMC design is necessary, and the problem ends.

EMI COUPLING PATHS

This section presents some mathematical models, design data, and details for EMI analysis and prediction based upon the methodology and procedure illustrated in Fig. 5-5.

Common-Ground Impedance Coupling

The first of the five coupling paths in Fig. 5-5 involves the coupling of voltage induced by currents flowing through the common impedance such as a power supply output impedance, common return line, ground plane, or safety wire. This type of coupling results in a common-mode (CM) voltage inducing CM currents flowing into the signal cables of a circuit. The CM voltage is partially converted into a differential mode (DM) voltage at the input terminals of a victim load.

Figure 5-11 illustrates the generation of a CM voltage, V_i, induced by the flow of a ground current, I_g, through a ground impedance, Z_g, where $V_i = I_g Z_g$ (12). If the circuit shown is grounded, that is, the toggle switches are closed thereby connecting the signal reference point C to the case B in box #1, and the load reference point F to the case G in box #2, then the voltage V_i appears between points C and F in the circuit. This voltage, V_i, is then "common" to both signal cable lines, thereby causing currents to flow in both signal lines, designated as I_1 and I_2. Since I_1 and I_2 do not flow through equal impedance paths, a DM voltage V_0 is developed across the load terminals by this impedance unbalance (12). The ratio of V_0/V_i is referred to as ground-loop coupling, described in a later section.

Electromagnetic (EM) Field Coupling into Loops

This section outlines a development for estimating the voltage induced around a closed loop whose path links a sinusoidal, time-varying EM field. Specific relationships are developed for a plane wave EM field configuration.

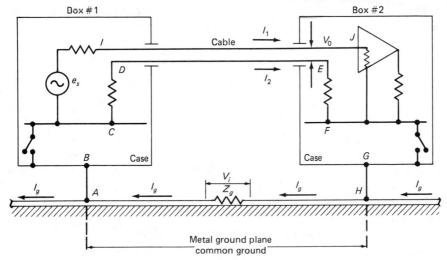

Notes: Z_g = Ground plane impedance between points A and H.

V_i = Voltage drop, $I_g \times Z_g$, between points A and H.

I_g = External ambient current flowing through Z.

V_0 = Differential-mode voltage developed from common-mode voltage, V_i.

Fig. 5-11. Illustration of common-ground impedance coupling.

Figure 5-12 illustrates a rectangular loop, lying in the x-z plane, with dimensions of length l and height h. A radiating source (S) emits an EM wave that propagates from left to right at a velocity $\mathbf{v}$ as shown. The EM wave electric and magnetic field components are designated by vectors $\mathbf{E_i}$ and $\mathbf{H_i}$, respectively, oriented along the x and y axes.

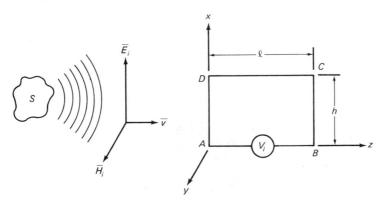

Fig. 5-12. Planewave incident upon rectangular loop.

The induced voltage around the closed loop is designated V_i, and it can be determined from the integral form of Maxwell's equation (5):

$$V_i = \oint \mathbf{E} \cdot d\mathbf{l} = -\frac{\partial \phi}{\partial t} = -\frac{\partial}{\partial t} \iint_A \mathbf{B} \cdot d\mathbf{A} \tag{5-5}$$

where ϕ is the total magnetic flux density passing through the loop area $A = l \times h$. The surface integral of the magnetic flux density (**B**) determines the total magnetic flux, ϕ, when it is evaluated over the surface area, A, of the loop whose perimeter is the path of integration for the **E**-field line integral in equation (5-5). A differential vector path length of the line integral around the loop perimeter is designated by $d\mathbf{l}$.

Evaluating equation (5-5) for V_i requires that analytical expressions for **E** and/or **B** be available around the loop perimeter and/or loop area A, respectively. Equation (5-5) can be solved either in closed-form, or relatively simple approximations developed, whenever the expressions for **E** and **B** are relatively simple, and the geometry of the area A and the path of integration around the loop are easily defined. In the near-field, where the loop is at a distance from the source that is less than $\lambda/2\pi$ (14), analytical expressions for **E** and **B** are quite complex, to the point of usually being undefinable, except for a relatively few cases where the boundary geometry is simple and coincides with the given coordinate system. Under the latter conditions, the field equations can be solved by separation of variables using well-known mathematical techniques (3).

A relatively simple solution can be obtained for V_i for the situation illustrated in Fig. 5-12 if the loop is located far enough from the source [i.e., in the far-field, which corresponds to a distance from the source greater than $\lambda/2\pi$ (see Chapter 10)], and the incident EM field can be approximated by a plane wave. Sinusoidal time-varying plane-wave field components can be expressed in exponential form as follows (5):

$$\mathbf{E_i} = \mathbf{e_x} E_0 e^{j(\omega t - \beta z)} \tag{5-6}$$

$$\mathbf{H_i} = \mathbf{e_y} H_0 e^{j(\omega t - \beta z)} \tag{5-7}$$

where:

$\mathbf{e_x}, \mathbf{e_y}$ = unit vectors in the x, y directions, respectively
ω = radian frequency (radians/second)
E_0 = electric field amplitude (volts/meter)
H_0 = magnetic field amplitude (amps/meter)
$\beta = 2\pi/\lambda$ = plane wave phase constant (meter^{-1})
λ = free-space wavelength (meters)
$j = \sqrt{-1}$

Using equations (5-5) and (5-6), an expression is developed for V_i assuming that the electric field configuration in the vicinity of the loop can be approximated by a plane wave, that is, equation (5-6). To obtain a more rigorous solution, more complex boundary value problem involving incident and scattered fields have to be solved. These are often formulated in the form of integral equations derived by Green's function techniques (7, 9, 10). Generally, it is impractical to solve complex boundary value problems to obtain approximations for EMI quantities, with the exceptional cases fully acknowledged. Evaluating the electric-field line integral in equation (5-5) for a plane wave E-field yields the following expression for the induced voltage amplitude around a closed loop:

$$V_i = E_0 h \sqrt{2(1 - \cos \beta l)} \qquad \text{volts} \qquad (5\text{-}8)$$

where:

E_0 = ambient electric field amplitude
h = loop height in meters (parallel to electric field)
l = loop length in meters

In equation (5-8) and other equations derived therefrom, the $e^{j\omega t}$ sinusoidal time-variation is understood.

Equation (5-8) can be normalized to provide the induced voltage per unit E-field:

$$\frac{V_i}{E_0} = h \sqrt{2(1 - \cos \beta l)} \qquad \text{volts/volts/meter} = \text{meter} \qquad (5\text{-}9)$$

Note that the resulting units of equation (5-9) are "meters."

Expression equation (5-9) in dB:

$$\frac{V_i}{E_0} (\text{dB}) = 20 \log \left[h \sqrt{2(1 - \cos \beta l)} \right] \text{dB}_{\text{meter}} \qquad (5\text{-}10\text{a})$$

$$= 20 \log \left[h \sqrt{2(1 - \cos \beta l)} \right] \text{dB}_{\text{meter}}$$

$$\frac{V_i}{E_0} (\text{dB}) = 20 \log [h] + 10 \log [2(1 - \cos \beta l)] \text{dB}_{\text{meter}} \qquad (5\text{-}10\text{b})$$

Equation (5-10) will be utilized in the following sections to approximate common-mode and differential-mode voltages into circuit loops due to the coupling of a sinusoidal, time-varying electric field.

The second part of equation (5-5) (the surface integral) can be used to derive the induced voltage V_i in terms of an ambient magnetic flux density, **B**.

Using equation (5-6) and the free-space relationship $\mathbf{B} = \mu_0\mathbf{H}$, where μ_0 is the free-space permeability $= 4\pi \times 10^{-7}$ henry/meter, the following expression is obtained by evaluation of the surface integral of equation (5-5):

$$V_i = 3 \times 10^8 \, B_0 h \sqrt{2(1 - \cos \beta l)} \qquad \text{volts} \qquad (5\text{-}11)$$

where:

B_0 = amplitude of ambient magnetic flux density (Tesla)
h = loop height in meters
l = loop length in meters

In Equation (5-11), the unit of B_0 is the Tesla.
For B_0 in gauss, equation (5-11) becomes:

$$V_i = 3 \times 10^4 B_0 h \sqrt{2(1 - \cos \beta l)} \qquad \text{volts} \qquad (5\text{-}12)$$

Equation (5-12) can be normalized to provide the induced voltage per unit **B**-field in volts/gauss:

$$\frac{V_i}{B_0} = 3 \times 10^4 h \sqrt{2(1 - \cos \beta l)} \qquad \text{volts/gauss} \qquad (5\text{-}13)$$

Expressing equation (5-13) in dB:

$$\frac{V_i}{B_0} (\text{dB}) = 20 \log \left[3 \times 10^4 h \sqrt{2(1 - \cos \beta l)}\right] \qquad \text{dB volts/gauss}$$

$$(5\text{-}14a)$$

$$\frac{V_i}{B_0} (\text{dB}) = 89.5 + 20 \log [h] + 10 \log [2(1 - \cos \beta l)] \qquad \text{dB volts/gauss}$$

$$(5\text{-}14b)$$

Equation (5-14) will be utilized in following sections to approximate common-mode and differential-mode voltages into circuit loops due to the coupling of a sinusoidal, time-varying magnetic field.

Equations (5-10) and (5-14) are plotted in Figs. 5-13 and 5-14, respectively. Note that these two equations are functionally identical except for the unit scaling factor of 89.5 dB in equation (5-14). Equations (5-9) and (5-13) indicate that maximum plane wave electric and magnetic field coupling occurs whenever $(\cos \beta l) = -1$, and zero coupling whenever $(\cos \beta l) = +1$. The following equations determine the frequencies at which coupling maxima occur

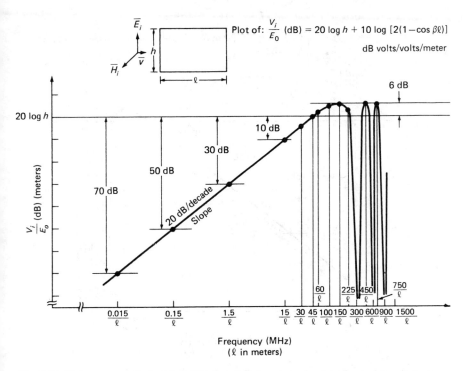

Plot of: $\dfrac{V_i}{E_0}$ (dB) $= 20 \log h + 10 \log [2(1-\cos \beta l)]$

dB volts/volts/meter

Fig. 5-13. Volume per unit electric field induced in a rectangular loop by a plane electromagnetic wave.

when $(\cos \beta l) = -1$:

$$\beta l = \frac{2\pi l}{\lambda} = (2n - 1)\pi, \qquad n = 1, 2, 3, \ldots \tag{5-15}$$

Since the wavelength λ in meters is equal to $300/f_{\text{MHz}}$, then:

$$\frac{2\pi l}{\lambda} = \frac{2\pi l f_{\text{MHz}}{}^{\text{max}}}{300} = \frac{\pi l f_{\text{MHz}}{}^{\text{max}}}{150} = (2n - 1)\pi \tag{5-16}$$

and the frequencies at which the maxima occur are given by:

$$f_{\text{MHz}}{}^{\text{max}} = (2n - 1)\frac{150}{l} \text{ MHz}, \qquad n = 1, 2, 3, \ldots \tag{5-17}$$

where l is the loop length in meters. From equation (5-17), the first maximum occurs when $n = 1$, which corresponds to a frequency of $150/l$ MHz, with

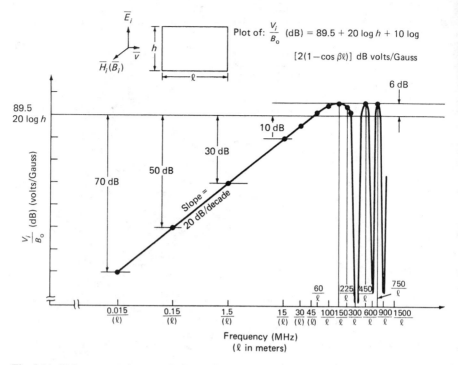

Fig. 5-14. Voltage per unit magnetic flux density induced in a rectangular loop by a plane electromagnetic wave.

other maxima occurring at integral multiples of this frequency, as shown in Figs. 5-13 and 5-14.

The following equations determine the frequencies at which zero (minimum) coupling occurs, when $(\cos \beta l) = +1$:

$$\beta l = \frac{2\pi l}{\lambda} = 2\pi n, \qquad n = 1, 2, 3, \ldots \qquad (5\text{-}18)$$

and:

$$\frac{\pi l f_{\text{MHz}}{}^{\min}}{150} = 2\pi n$$

giving:

$$f_{\text{MHz}}{}^{\min} = (n)\,\frac{300}{l}\ \text{MHz}, \qquad n = 1, 2, 3, \ldots \qquad (5\text{-}19)$$

From equation (5-19), the first zero (minimum) coupling occurs when $n = 1$, which corresponds to a frequency of $300/l$ MHz, with other minima occurring at integral multiples of this frequency, as shown in Figs. 5-13 and 5-14. Note that zero coupling in equations (5-9) and (5-13) corresponds to minus infinite dB in Figs. 5-13 and 5-14 because of the logarithm function used to define the dB.

It must be noted that the "peaks" and "valleys" shown in Figs. 5-13 and 5-14 are determined by using an idealized mathematical model, with the incident plane wave and the loop being oriented spatially for theoretical maximum coupling, or maximum induced voltage V_i, at frequencies where these peaks occur as determined by equation (5-17). The frequencies for theoretical "zero" coupling, determined by equation (5-19), are designated as frequencies for minimum coupling, since, in practice, sharp troughs of absolutely zero coupling do not exist. Some coupling of energy will always occur because the ambient electric field vector is never absolutely oriented with a loop side, actual loops are not rectangular nor absolutely planar, and so on. Other practical approximations/assumptions will be specified as equations (5-10) and (5-14) are applied.

For more detail on the coupling of plane waves into loops, the reader is directed to reference 6.

Radiated Common-Mode and Ground-Loop Coupling

This section describes the second of the five principal coupling paths shown in Fig. 5-5. This coupling path is divided into two parts: (1) field-to-cable common-mode coupling and (2) ground-loop coupling. Common-mode coupling converts an electric or magnetic field to a common-mode, open-circuit voltage in the loop area shown in Fig. 5-15. This voltage then acts as a potential EMI source for common-mode currents, I_1 and I_2, around two loop areas that include the two victim cables. The two loop areas are essentially two loops in parallel, with part of each loop forming a part of the other. Referring to Fig. 5-15, the two loops in question follow the respective paths $EFABCDE$ (first loop) and $EFABC'D'E$ (second loop). It is assumed that the two loops are essentially coincident, and their areas are equal and identically spatially oriented with respect to the ambient EM plane wave, since the separations between the two signal cables and other circuit elements are small compared to the shaded large loop area. Therefore, the induced voltage around each loop is assumed to be equal and indicated by V_i in Fig. 5-15. This voltage, V_i, is referred to as the common-mode voltage induced by the ambient EM field. Ground-loop coupling, which is determined by circuit impedances, converts the common-mode voltage V_i to a differential-mode voltage V_0 across the amplifier or logic circuit input terminals that constitutes the potential EMI threat. This topic is covered in the next few sections.

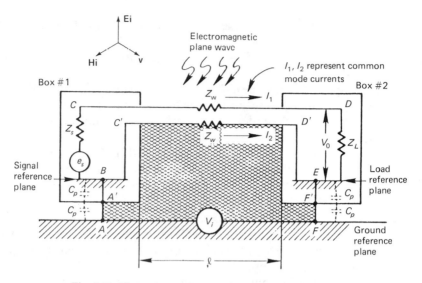

Fig. 5-15. Plane-wave radiation incident upon a closed loop.

Field-to-Cable Common-Mode Coupling

This section presents graphical models for determining estimates of field-to-cable common-mode coupling and methods for reducing this coupling. The objective is to provide means for quantifying common-mode coupling so that induced voltages and/or currents can be quantitatively compared to circuit susceptibility levels to determine whether an EMI situation exists. Equations (5-10) and (5-14), along with their graphical representation in Figs. 5-13 and 5-14, provide a basis for determining induced voltage levels by the coupling of electric and magnetic fields radiated into loop areas. As was previously discussed, the sharp "peaks" and "valleys" in Figs. 5-13 and 5-14 are the result of mathematically "pure" models. For example, should an interfering signal be present at a frequency that coincides closely with a frequency where zero coupling is predicted, such as frequencies determined by equation (5-19), one could assume that no problem exists. However, any slight deviations of the actual circuit from the model used to derive equation (5-19) would result entirely new sets of theoretical frequencies at which maximum and minimum coupling occurs, thereby rendering the precise use of these relationships impractical. However, the above equations and graphs developed from ideal models can still be used to estimate possible "worst case" levels of induced voltages in circuit loops.

Figure 5-16 is a sketch of Figs. 5-13 and 5-14 showing common-mode coupling (CMC) in dB (CMC_{dB}), defined as equation (5-10) for the electric-field ambient, and equation (5-14) for the magnetic-field ambient, as a function

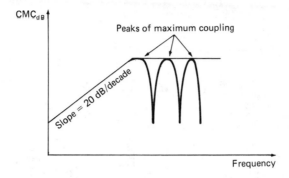

Fig. 5-16. Common mode coupling as a function of frequency.

of frequency. It is noted that the peaks in Figs. (5-13) and (5-14) are all equal in each respective figure. Figure 5-16 suggests that a "worst case" envelope estimate of coupling can be obtained by constructing a graph consisting of two straight-line segments: one consisting of the line segment indicating the 20 dB/decade increase of CMC with frequency until the frequency of the first peak is reached; and the second formed by a line parallel to the frequency axis that passes through the peak values of coupling and also connects with the first line at their point of intersection. This is illustrated in Fig. 5-16, and shown for specific loop dimension combinations in Figs. 5-17 and 5-18. Figure 5-17 provides electric-field CMC in dB(V/E), or in dB volts per unit E-field; and Fig. 5-18 provides magnetic field CMC in dB(V/G), or in dB volts per gauss, assuming that the ambient fields can be approximated by a plane wave.

It is noted that CMC increases with frequency at the rate of 20 dB/decade until the corner frequency is reached at which the loop length is a half-wavelength. Above this frequency, CMC is assumed to be constant at the peak values, as shown in Fig. 5-16. Each set of curves in Figs. 5-17 and 5-18 contains superimposed curves corresponding to specific loop lengths and heights. To assist in determining the proper curve to use, each CMC chart includes a tabular insert listing loop length and height combinations corresponding to the specific curves.

EMI Control Techniques for Reducing Common-Mode Coupling

Common-mode coupling can be reduced by decreasing the area of the loop shown in Fig. 5-15. Either the average height above the ground plane of the interconnecting cable may be lowered (i.e., the cable routed closer to the

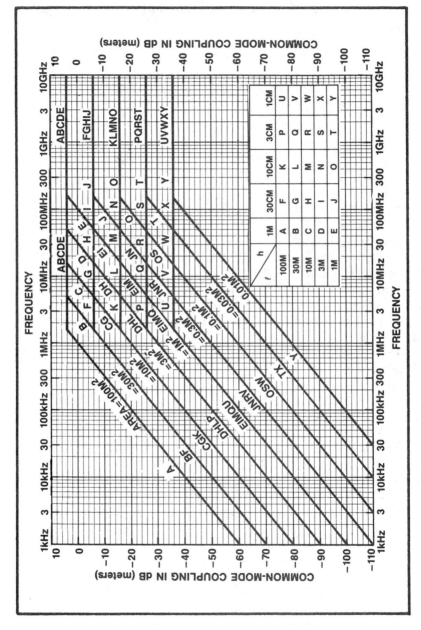

Fig. 5-17. Electric field strength common-mode coupling into box-cable-box-ground loop area. (*Courtesy of Interference Control Technologies*)

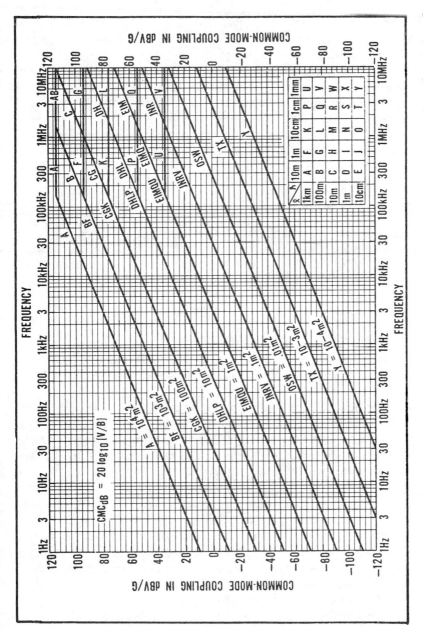

Fig. 5-18. Magnetic flux density, common-mode coupling into box-cable-box-ground loop area. (*Courtesy of Interference Control Technologies*)

ground plane) or the cable length may be reduced by bringing the equipment boxes closer together. These techniques must be considered in light of other practical system design requirements.

Another possible approach to reduce CMC would be to shield the loop area by enclosing all the boxes and interconnecting cables, thereby reducing the amplitude of the ambient fields coupling into the loop. However, this technique may be impractical if the loop area is too large, by having to locate the equipment boxes at a relatively long distance from each other. This could amount to having to construct a shielded room enclosure, and could involve a considerable expense.

Ground-Loop Coupling

The term ground-loop coupling (GLC) is closely related to common-mode rejection and is defined as:

$$\text{GLC}_{dB} = 20 \log (V_0/V_i) \tag{5-20}$$

where:

V_0 = voltage produced at victim input terminals.

V_i = common-mode voltage developed as a result of common ground impedance or electromagnetic radiated field coupling into loop area.

Figure 5-19 illustrates ground-loop coupling for an unbalanced system and Fig. 5-20 for a corresponding balanced system, where equation (5-20) is plotted as a function of frequency (12). The A/t ratio parameter corresponds to an equivalent capacitance, shown as C_p in Fig. 5-15, formed by the area and height of the signal reference plane, such as a printed circuit board, located above a ground plane (such as a metal case). The A/t ratio corresponds to a capacitance formed by two parallel plates of area A, separated by a distance t, with air as a dielectric with a permittivity = 8.85×10^{-12} farad/meter. An A/t ratio of 100 cm, corresponding to a capacitance of 8.85 picofarads (pf), is used as a default value if the actual capacitance value is unknown (12). This capacitance, C_p, becomes part of the ground loop circuit when one of the connections to ground is removed, such as the connection from point B to case, or E to case, in Fig. 5-15.

Figure 5-19 shows that the worst-case GLC is approximately -6 dB at low frequencies when both the signal and load references in Fig. 5-15 are grounded, giving an equivalent capacitance (A/t) of infinity. This demonstrates the unfavorable condition of a low impedance ground loop. Under these conditions, the common-mode voltage V_i is divided between impedances Z_s and Z_L, and if $Z_s = Z_L$, half of V_i appears across both impedances,

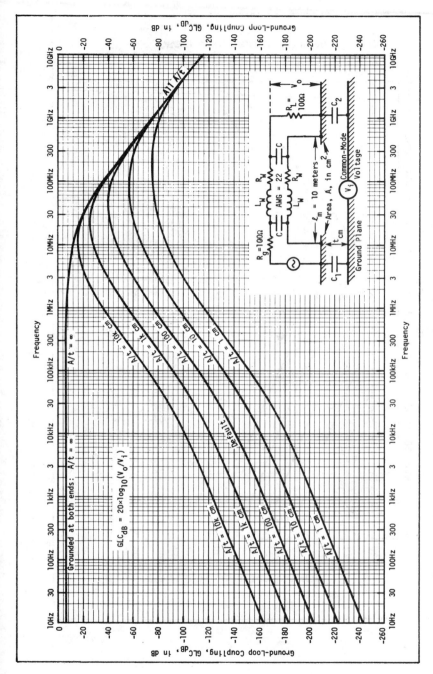

Fig. 5-19. Ground-loop coupling, unbalanced system. (*Courtesy of Interference Control Technologies*)

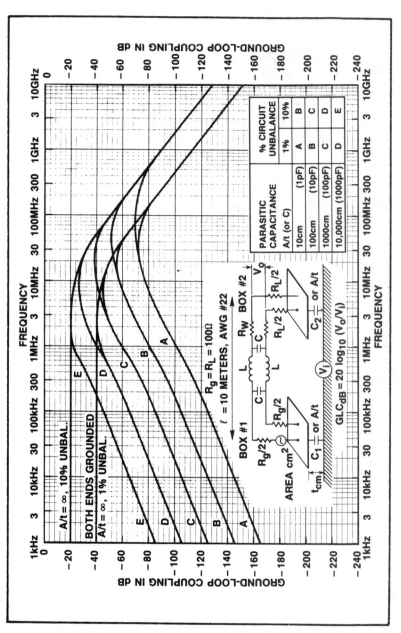

Fig. 5-20. Ground-loop coupling, balanced system. (*Courtesy of Interference Control Technologies*)

and $V_0 = V_i/2$, corresponding to the -6 dB for GLC. Figure 5-19 shows that one way to substantially reduce GLC at low frequencies is to float (unground) either or both the signal and the load reference plane, thereby increasing the ground loop impedance which forces a larger portion of V_i to appear across the floating impedance.

Figure 5-20 shows the overall reduction in GLC that can be realized by balancing circuit impedances at the signal and the load. This is often accomplished by using differential line drivers and balanced receivers, which result in effectively dividing the impedances as shown in the circuit diagram of Fig. 5-20.

Both Figs. 5-19 and Fig. 5-20 indicate that GLC increases with frequency with one or both ends of the circuit floating due to the decreasing impedance of the parasitic capacitance, C_p (Fig. 5-15). As frequency is further increased, the GLC falls off, owing to cable series inductance and parallel capacitance. However, caution must be exercised when the charts are used in this high-frequency region, because the cable length becomes comparable to a significant fraction of a wavelength as frequency is increased. Figure 5-21 illustrates the behavior of GLC at higher frequencies when cable length exceeds 1/4 to 1/2 wavelength. The oscillatory, transmission-line behavior shown in Fig. 5-21 was confirmed by measurements (2). The locations (i.e., exact frequencies of occurrence, amplitudes, etc.) of the maxima and minima shown are dependent upon cable parameters (inductance, capacitance, resistance, and conductance per unit length), cable length, and circuit impedances. These parameters vary with each cable type, source, and load impedances, and must be determined for each application. Unless specific applications require an exact determination of the GLC function, the curves of Figs. 5-19 and 5-20 may be used for high-frequency GLC determination as default values.

Additional data on common-mode coupling and ground-loop coupling are available in reference 12, and additional information on coupling into cables is available in references 6 and 11.

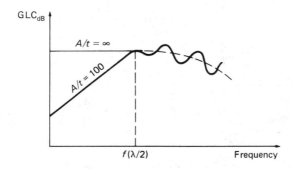

Fig. 5-21. GLC as a function of frequency.

EMI Control Techniques for Reducing Ground-Loop Coupling

Ground-loop coupling may be reduced by an application of one or more of the following:

- Float (unground) either or both signal and load reference points.
- Use a balanced system, such as balanced line drivers and receivers.
- Use an optical isolator in data lines.
- Use an isolation transformer in the signal circuit.
- Install RF chokes in the case-to-ground path, and bond cases together.
- Float box shields inside equipment enclosures.
- Use ferrite beads on signal leads.

These GLC reduction techniques will now be reviewed.

Float Signal Reference Points.   The circuit of Fig. 5-15 is redrawn in Fig. 5-22 to illustrate ground-loop coupling (GLC) at low frequencies with all indicated impedances assumed to be resistive. The interconnecting cable impedance is indicated by Z_w, which is typically a fraction of an ohm (approximately equal to the cable wire DC resistance) at low frequencies (12). From circuit analysis, it can be shown that at low frequencies:

$$\text{GLC} = \frac{V_0}{V_i} = \frac{Z_L}{Z_s + Z_w + Z_L} = \frac{Z_L}{Z_s + Z_L} \qquad (5\text{-}21)$$

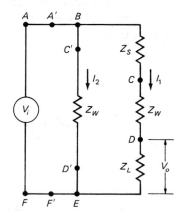

Fig. 5-22. Equivalent circuit to evaluate ground-loop coupling at low frequencies (unbalanced System).

where:

Z_s = signal source impedance (ohms)
Z_L = load (victim) input impedance (ohms)
Z_w = wire (cable) impedance (ohms)
$Z_s, Z_L \gg Z_W$ at low frequencies

If the source and load impedances are equsl ($Z_s = Z_L$), then $V_0/V_i = 0.5$ and:

$$\text{GLC}_{\text{dB}} = 20 \log \frac{V_0}{V_i} = 20 \log [0.5] = -6 \text{ dB} \qquad (5\text{-}22)$$

Equation (5-22) corresponds to the the value of $A/t = $ infinity curve in Fig. 5-19 at low frequencies where GLC is constant until the frequency reaches approximately 10 MHz. Above this frequency, the effects of circuit capacitances and inductances become significant, and the interconnecting cable behaves in a transmission-line manner whenever its length becomes comparable to a wavelength. The high-frequency behavior of GLC will then appear as conceptually illustrated in Fig. 5-21. As previously discussed, the smooth curves in Fig. 5-19 may be used to approximate this behavior, but with an understanding of the difficulty in predicting GLC at high frequencies for specific applications.

If, in Fig. 5-15, the connection between $A'B$ is removed, this corresponds to floating the circuit at the signal end and results in the introduction of the parasitic capacitance, C_p, into the ground loop circuit as shown in Fig. 5-23. This capacitance is the parasitic capacitance between the signal reference

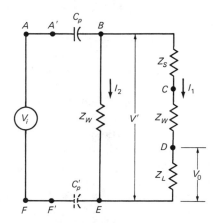

Fig. 5-23. Equivalent circuit to evaluate ground-loop coupling at low frequencies with circuit floated between signal reference plane and box #1 (Fig. 5-15) (unbalanced system).

plane and box #1. Conceptually, the same floating effect would result if the connections were singularly removed between points AA', EF', or FF' in Fig. 5-15, with variations in the capacitance introduced depending upon where the connection is removed.

From Fig. 5-23, it can be shown that at low frequencies:

$$\frac{V_0}{V_i} = \frac{Z_w}{X_c} \cdot \frac{Z_L}{Z_s + Z_L} = \omega C_p Z_w \frac{Z_L}{Z_s + Z_L} \tag{5-23}$$

where Z_s, Z_L, and Z_w are as defined for equation (5-21), and X_c is the capacitive reactance of C_p: $X_c = 1/\omega C_p$ ohms, where $\omega = 2\pi f$. At low frequencies, $X_c \gg Z_w$.

Equation (5-23) shows that the voltage across the victim load terminals, V_0, is directly proportional to frequency ($\omega = 2\pi f$). This direct variation is manifested in Fig. 5-19 by the 20 dB/decade rate of increase in GLC at the low frequencies. Comparing equation (5-23) with equation (5-21), it is observed that by floating the circuit in Fig. 5-15, the GLC has been reduced by the amount of the fraction (Z_w/X_c). This can be a significant reduction when $X_c \gg Z_w$, which is usually the case at low frequencies. This reduction can be seen in Fig. 5-19 when one compares the GLC curve for $A/t = $ infinity for the circuit grounded at both ends, and the other curves that give finite values for A/t depending on the value of the capacitance C_p. At low frequencies, GLC is defined using equations (5-22) and (5-23) for the circuit shown in Fig. 5-15.

As indicated, by floating (ungrounding) one or both of the signal and load reference points in Fig. 5-15 at a specified frequency, the A/t parameter line in Fig. 5-19 changes from infinity (grounded) to either the known parasitic circuit capacitance that is introduced by floating, or, if the actual capacitance is unknown, to a default value of $A/t = 100$ cm (a typical value) that may be used. When only one of the two grounds at either box #1 or box #2 is broken in Fig. 5-15, 6 dB must be added to the corresponding GLC value obtained from Fig. 5-19 or Fig. 5-20 at low frequencies (12). The curves shown are determined for both ends floating, which effectively introduces two capacitances in series in the ground loop circuit as shown in Fig. 5-23, with the second capacitance shown as a dotted symbol and labeled C'_p. The 6 dB correction for floating only one end assumes that in floating both ends of the circuit, two equal capacitances C_p and C'_p are introduced. At low frequencies, the capacitive impedances are much larger than Z_w, and the voltage V_i is essentially divided, as shown in Fig. 5-24.

The parallel combination of $Z_w \| (Z_L + Z_s + Z_w) \simeq Z_w$; therefore:

$$V' = V_i \left[\frac{Z_p}{Z_{C'_p} + Z_p + Z_{C_p}} \right] = V_i \left[\frac{Z_p}{(2Z_{C_p} + Z_p)} \right] \tag{5-24}$$

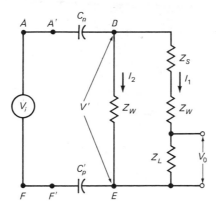

Fig. 5-24. Equivalent circuit to evaluate ground-loop coupling at low frequencies with circuit of Fig. 5-15 floated at signal and load.

with:

$$Z_p = \frac{Z_w(Z_s + Z_w + Z_L)}{Z_w + Z_s + Z_w + Z_L} < Z_w, \qquad Z_p \cong Z_w \text{ (small)}$$

Then:

$$V_0 = V' \left[\frac{Z_L}{Z_s + Z_w + Z_L} \right] \cong V_i \left[\frac{Z_w}{2Z_{C_p}} \right] \left[\frac{Z_L}{Z_s + Z_w + Z_L} \right] \qquad (5\text{-}25)$$

Equation (5-25) gives the voltage V_0 across the victim load Z_L with both ends of the circuit floating. With only one end floating:

$$V_0 = V_i \left[\frac{Z_w}{Z_{C_p}} \right] \left[\frac{Z_L}{Z_s + Z_w + Z_L} \right] \qquad (5\text{-}26)$$

Note that V_0 with only one end floating, equation (5-26), is approximately twice as high as when both ends of the circuit are floating, given by equation (5-25).

Hence, the additional 6 dB must be added to the curves of Fig. 5-19 and 5-20 when only one end of the circuit is floating.

Use of Balanced System. Figure 5-25 illustratés an equivalent circuit with both ends grounded to evaluate ground-loop coupling for a balanced system wherein the source and load impedances are divided equally ($Z_s/2$ and $Z_L/2$) between the two interconnecting cable conductors. The signal source is

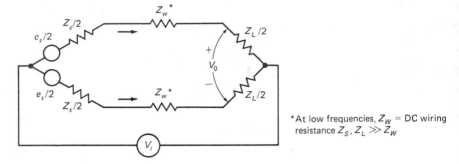

Fig. 5-25. Equivalent circuit to evaluate ground-loop coupling at low frequencies for an ideal balanced system.

also balanced, as indicated by the sources $e_s/2$ in each respective conductors. If the circuit in Fig. 5-25 is perfectly balanced (i.e., the impedances are perfectly divided between the two cable conductors), then V_0, due to V_i, will be zero. However, absolutely perfect balance cannot be attained, and some tolerance of unbalance must be specified.

Figure 5-26 is a modification of Fig. 5-25 and illustrates the impedance distribution with a specified tolerance of unbalance, indicated by X given in percent. The following conditions are assumed to develop an expression of ground-loop coupling as a function of X:

- The total source impedance, Z_s, is divided between the cable conductors as indicated by Z_1 and Z_2 where:

$$Z_1 = (Z_s/2)(1 \pm X), \qquad Z_2 = (Z_s/2)(1 \mp X) \qquad (5\text{-}27a)$$

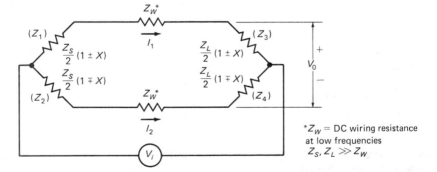

Fig. 5-26. Equivalent circuit to evaluate ground-loop coupling at low frequencies for a balanced system with specified percent tolerance of unbalance (X), with circuit grounded at both ends.

- The total load impedance, Z_L is similarly divided as shown where:

$$Z_3 = (Z_L/2)(1 \pm X), \qquad Z_4 = (Z_L/2)(1 \mp X) \qquad (5\text{-}27b)$$

and, at low frequencies:

$$Z_s, Z_L \gg Z_w \qquad (5\text{-}28)$$

Note that the distribution of unbalance (indicated by the factors $(1 \pm X)$, etc.) is such that:

$$Z_1 + Z_2 = Z_s, \qquad Z_3 + Z_4 = Z_L \qquad (5\text{-}29)$$

From Fig. 5-26:

$$\text{GLC} = \frac{V_0}{V_i} = \left(\frac{Z_3}{Z_1 + Z_3}\right) - \left(\frac{Z_4}{Z_2 + Z_4}\right) \qquad (5\text{-}30)$$

Substituting equations (5-27) and (5-28) into equation (5-30), the latter becomes:

$$\text{GLC} = \frac{V_0}{V_i} = \left(\frac{Z_L(1 \pm X)}{Z_s(1 \pm X) + Z_L(1 \pm X)}\right) - \left(\frac{Z_L(1 \mp X)}{Z_s(1 \mp X) + Z_L(1 \mp X)}\right) \qquad (5\text{-}31)$$

For the case where the signal sources drives a matched load, $Z_L = Z_s = Z$, equation (5-31) then becomes:

$$\text{GLC} = \frac{V_0}{V_i} = \left(\frac{1 \pm X}{(1 \pm X) + (1 \pm X)}\right) - \left(\frac{1 \mp X}{(1 \mp X) + (1 \mp X)}\right) \qquad (5\text{-}32)$$

Equations (5-27), (5-28), (5-29), (5-31) and (5-32) yield four possible combinations of unbalance distribution. The resultant values of GLC are indicated in Table 5-1. This table indicates that if the unbalance X is distributed such that $Z_1 = Z_3$, and $Z_2 = Z_4$, then GLC = 0. If unbalance is distributed such that $Z_1, Z_4 = (Z/2)(1 \pm X)$, and $Z_2, Z_3 = (Z/2)(1 \mp X)$, where $Z_s = Z_L = Z$ (matched source and load), then GLC becomes:

$$\text{GLC} = \frac{V_0}{V_i} \mp X \qquad (5\text{-}33)$$

In the second row of Table 5-1, the negative sign for the ratio of $V_0/V_i = -X$ indicates that the polarity of V_0 is reversed, the GLC_{dB} is defined in

Table 5-1. Ground-loop coupling for a balanced system at low frequencies (Fig. 5-26) for four possible distributions of unbalance toleration (X) in percent.

Z_1	Z_2	Z_3	Z_4	GLC* (V_0/V_i)	GLC$_{dB}$* $(20 \log |X|/100)$
$\dfrac{Z_s}{2}(1 + X)$	$\dfrac{Z_s}{2}(1 - X)$	$\dfrac{Z_L}{2}(1 + X)$	$\dfrac{Z_L}{2}(1 - X)$	0	$-\infty$
$\dfrac{Z_s}{2}(1 + X)$	$\dfrac{Z_s}{2}(1 - X)$	$\dfrac{Z_L}{2}(1 - X)$	$\dfrac{Z_L}{2}(1 + X)$	$-X$	$-40 + 20 \log X$
$\dfrac{Z_s}{2}(1 - X)$	$\dfrac{Z_s}{2}(1 + X)$	$\dfrac{Z_L}{2}(1 + X)$	$\dfrac{Z_L}{2}(1 - X)$	X	$-40 + 20 \log X$
$\dfrac{Z_s}{2}(1 - X)$	$\dfrac{Z_s}{2}(1 + X)$	$\dfrac{Z_L}{2}(1 - X)$	$\dfrac{Z_L}{2}(1 + X)$	0	$-\infty$

* For $Z_L = Z_s = Z$.

terms of the ratio of the amplitudes (absolute values) of V_i and V_0. In terms of dB (12):

$$\text{GLC}_{dB} = 20 \log \frac{|V_0|}{|V_i|} = 20 \log \frac{|X\%|}{100} = -40 + 20 \log |X\%| \quad (5\text{-}34)$$

Figure 5-20 provides a graphical representation (5-34) at low frequencies (up to about 10 MHz) for 10% and 1% unbalance and an A/t ratio of infinity. At frequencies above 10 MHz, circuit capacitances and inductances become significant, and the model becomes more complex (12). Also, the 1-meter cable length becomes a significant fraction of a wavelength and behaves like a transmission line with GLC performance following the pattern illustrated in Fig. 5-21. At high frequencies, the smooth curves of Fig. 5-20 may be used as an estimate of GLC, but with the realization that maxima and minima exist and that the exact location of each is difficult to predict.

A comparison of Figs. 5-19 and 5-20 shows the advantage of using a balanced signal driver/receiver over an unbalanced system. Parameter lines of 1% and 10% tolerance of unbalance are identified in Fig. 5-20. Comparing Fig. 5-19 and 5-20, it is observed that an overall reduction in GLC is realized by balancing. The reduction is consistent throughout the low-frequency range when both ends of the circuit are grounded ($A/t = $ infinity). However, it is noted that at low frequencies some increase in GLC results by balancing with the circuit floated. For balanced conditions and the circuit floated, the low-frequency GLC performance depends upon the distribution of the impedance unbalance between the two signal leads. The worst-case situation is presented in Fig. 5-20.

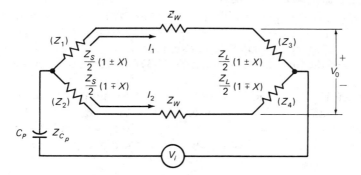

Fig. 5-27. Equivalent circuit to evaluate ground loop coupling at low frequencies for a balanced system with specified percent tolerance of unbalance (X), with circuit floated (ungrounded) at the signal end.

Figure 5-27 illustrates an equivalent balanced circuit floated at the signal end with capacitance C_p introduced as a result of floating. For this circuit, ground-loop coupling at low frequencies can be expressed as follows:

$$\text{GLC} = \frac{V_0}{V_i} = \left[\frac{Z_p}{Z_p + Z_{Cp}}\right]\left[\frac{Z_3}{Z_1 + Z_3} - \frac{Z_4}{Z_2 + Z_4}\right] \tag{5-35}$$

where Z_1, Z_2, Z_3, Z_4, and Z_w assume the same relationships given by equations (5-27) and (5-28). Note that equation (5-35) is equal to equation (5-30) multiplied by the term $[Z_p/(Z_p + Z_{Cp})]$. The impedance Z_p is the equivalent impedance of the two branch circuits in parallel which may vary in range from tens of ohms to a few hundred ohms. The impedance Z_p is given by:

$$Z_p = \frac{(Z_1 + Z_w + Z_3)(Z_2 + Z_w + Z_4)}{Z_1 + Z_w + Z_3 + Z_2 + Z_w + Z_4} = \frac{(Z_1 + Z_3)(Z_2 + Z_4)}{Z_1 + Z_3 + Z_2 + Z_4} \tag{5-36}$$

The approximation in equation (5-36) results from Z_w being much less than other circuit impedances.

Typical values for the capacitance C_p in Fig. 5-27 may range from a few picofarads to a few hundred picofarads. For a ratio $A/t = 100$ cm (default value, corresponding to $C_p = 8.85$ pf), the magnitude of the capacitance impedance Z_{C_p} at a frequency of 1 kHz is approximately 18 megohms, which is considerably greater than Z_p. At a frequency of 1 MHz, $Z_{C_p} = 18$ kilohms, still fairly large compared to Z_p. For large values of Z_{C_p}, equation (5-35) may be approximated by:

$$\text{GLC} = \frac{V_0}{V_i} = \left[\frac{|Z_p|}{|Z_{C_p}|}\right]\left[\frac{Z_3}{Z_1 + Z_3} - \frac{Z_4}{Z_2 + Z_4}\right] \tag{5-37}$$

Equation (5-37) is graphically illustrated in Fig. 5-20 for various values of C_p in terms of the given A/T ratios for the curves provided. The curves are also drawn for values of 1% and 10% unbalance tolerance, X, which determines the values of the impedance in equation (5-37) in accordance with equation (5-27).

Note that Figs. 5-19 and 5-20 apply to the following specific circuit parameters: source and load impedances are matched (i.e., $Z_g = Z_L = 100$ ohms), and cable is an AWG #22 parallel wire pair 10 meters long. Other curves for other parameter values are given in the appendix to this chapter.

Optical Isolators and Fiber Optics.   Figure 5-28 corresponds to Fig. 5-15 and illustrates the installation of an optical isolator (sometimes called an optocoupler or an opto-isolator) at the input of the victim load amplifier. Effectively, the optical isolator introduces a high impedance at low frequencies in the ground-loop path around which the common-mode current (I_{CM}) flows. The compromise of this high impedance results from the input–output capacitance (C_{io}) of the optical isolator, which can range in value from 0.6 pf to 10 pf(1, 8, 12, 15, 17). This capacitance of 3 pf has an impedance of 50 ohms at 1 GHz (1).

To determine the effect of optical isolators on GLC, the charts of Fig. 5-19 and 5-20 can be used. Here, instead of the capacitance-related A/t ratio, the value of C_{io} for the specific optical isolator is substituted. If C_{io} is known (usually specified by the manufacturer), an equivalent A/t ratio can be determined by:

$$\frac{A}{t} = \frac{C_{io}\,(\text{pf})}{8.85 \times 10^{-2}} \quad \text{cm} \tag{5-38}$$

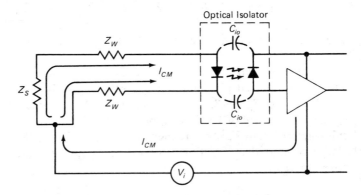

Fig. 5-28. Ground loop coupling reduction using an optional isolator (optocoupler).

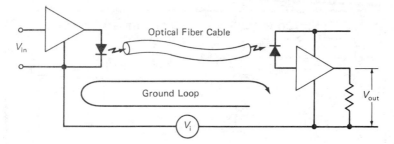

Fig. 5-29. Illustration of fiber optic data link for ground-loop coupling elimination.

The ratio determined by equation (5-38) and the EMI frequency can then be used in Figs. 5-19 and 5-20 to determine GLC (12).

Figure 5-29 illustrates the use of optical fiber cable to essentially introduce a high impedance in a common-mode ground loop as a result of the extremely high resistivity of the optical fiber cable. The use of optical fiber cable and optical couplers for EMI control is covered in detail in reference 1.

Isolation Transformers.   Figure 5-30 illustrates the use of an isolation transformer to reduce GLC at low frequencies. At high frequencies, the transformer performance degrades (i.e., GLC increases), owing to the tranformer primary-to-secondary distributed winding capacitance, C_W.

RF Chokes in Safety Ground Path.   Another EMI control technique for reducing GLC is the use of RF chokes in the box case-to-ground path, as shown in Figs. 5-31a,b (see inset diagram). Here, the objective is to provide a low-impedance path to the 50/60/400 Hz power-mains frequency for shock

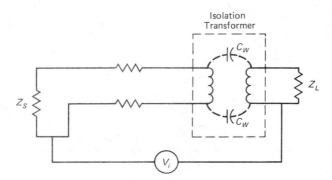

Fig. 5-30. Illustration of ground-loop coupling reduction at low frequencies using an isolation transformer.

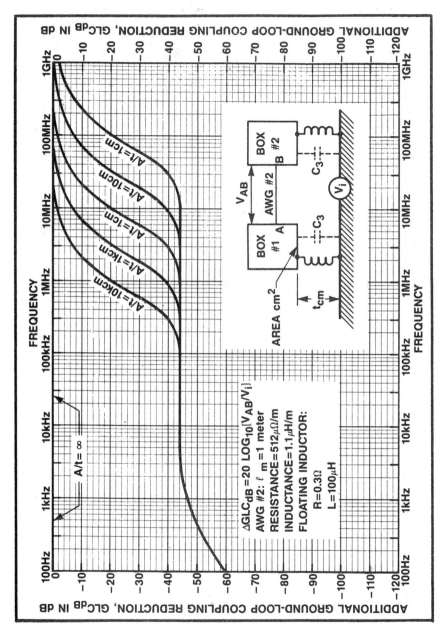

Fig. 5-31a. Improvement in GLC by floating boxes with RF chokes and bonding boxes together with heavy AWG #2 wire (shown for 100 microhenry chokes)

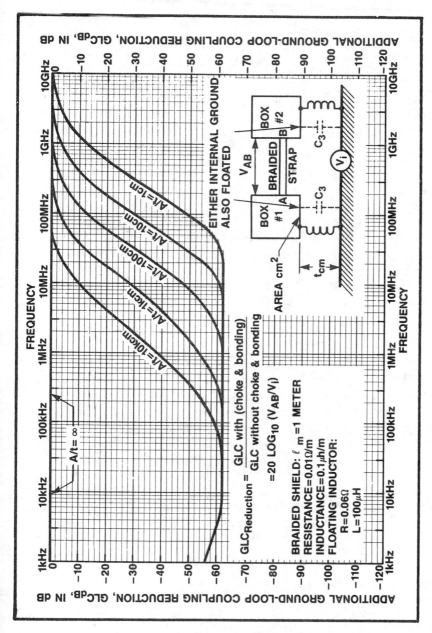

Fig. 5-31b. Additional GLC reduction by floating boxes with RF chokes and bonding them together with large braid.

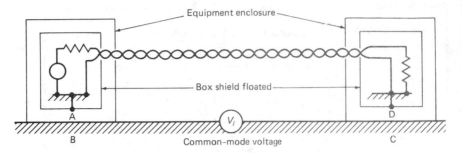

Fig. 5-32. Increasing ground-loop impedance by floating box shield inside equipment enclosure.

hazard control while exhibiting a high impedance around the ground loop at radio frequency. To get the technique to work best, it is also necessary to provide a common-mode choke in the power wires and a low-impedance path between the boxes, such as permitted by a #2 AWG wire or a braided shield grounded to both boxes. This is shown in the insert in Fig. 5-31a. The figure corresponds to a box-to-box cable wire length of 1 meter and must be adjusted for any other length in a given problem (12).

Shield Case Within a Shield.   Another EMI control technique for reducing GLC is to float a box shield inside an equipment enclosure shield, as shown in Fig. 5-32. Here, the signal reference plane can be grounded to the inside shield. The outer shield can also be directly grounded without developing a low-impedance ground loop. The shield-to-shield capacitance A/t ratio now replaces that of a floating signal reference plane, and Figs. 5-19 and 5-20 can once again be used to determine the GLC reduction.

Since the double box shield has the same qualitative effect as floating the signal reference planes, the techniques would not both be used, as they are essentially redundant. Either one would give the improvement shown in Figs. 5-19 and 5-20, but if both were used, only a small (approximately 6 dB) additional improvement would result.

Ferrite Absorbers.   Ferrite beads provide a means of reducing ground loop coupling at high frequencies without introducing significant losses at low frequencies (below 1 MHz) (4, 12, 16). The beads effectively introduce series inductance and effective resistance, providing a high series impedance at high frequency and power absorption of high frequency energy (12).

Ferrite beads may be used to suppress the effects of common-mode (CM) coupling by reducing GLC or to suppress the differential mode (DM). Figure

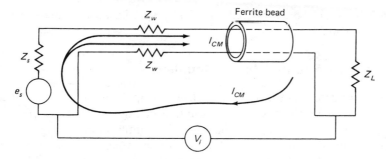

Fig. 5-33. Ferrite bead insertion for suppressing common mode noise, effectively reducing ground-loop coupling.

5-33 illustrates bead insertion for suppressing CM, and Fig. 5-34 for suppressing DM. Note that for CM suppression, both cable conductors must be inserted inside the bead, whereas for DM suppression, only one cable conductor must be inserted (i.e., each wire must have its own bead).

In addition to ferrite beads, ferrite rods and tubing may also be used to reduce GLC. In addition to CM and DM noise suppression, ferrite absorbers are used to damp out parasitic oscillations. These absorbers are also installed in power mains circuits as well as in signal lines (4, 12, 16).

It is important to note that the performance of ferrite absorbers is a function of circuit impedance and frequency (4, 12). For specific applications, it is necessary to determine the proper bead/rod by referring to performance data for specific bead size, material, and so on (16).

Shielded Enclosure. Another approach to reduce the common-mode coupling is to shield the entire area, enclosing all the boxes and interconnecting cables. In essence, this amounts to a shielded-room enclosure. This

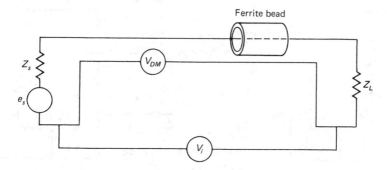

Fig. 5-34. Ferrite bead insertion for suppressing differential mode noise.

approach should be considered as a last resort because of the substantial expense involved.

Radiated Differential-Mode Coupling

This section describes the third of the five principal coupling paths, indicated as field-to-cable differential-mode coupling in Fig. 5-5. This coupling path converts an electric or magnetic field to a differential-mode voltage produced by direct radiation in the circuit loop area of a cable, as shown in Fig. 5-35. This loop area should not be confused with the much larger ground-loop area shown in Fig. 5-15. The differential-mode voltage then appears distributed across the source impedance Z_s, and the input terminals of the victim amplifier or logic circuit (Z_L) to produce the potential EMI threat to the latter. This section presents graphics for both predicting field-to-cable differential-mode coupling (DMC) and EMI control techniques for reducing this coupling.

Differential-Mode Coupling into Balanced Wire Pairs

Equations (5-10) and (5-14) also apply for converting an electric-field strength or a magnetic-flux density into a differential mode voltage. The main difference is that for DM coupling, the electric and magnetic fields couple into the shaded loop area shown in Fig. 5-35, whereas the fields couple into the larger loop area previously shown in Fig. 5-15 for CM coupling.

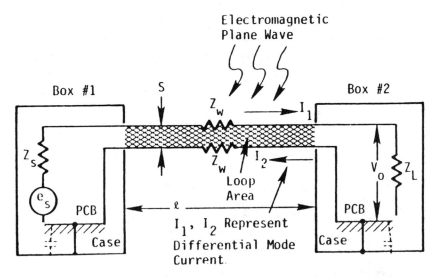

Fig. 5-35. Field-to-cable differential-mode coupling. (*Courtesy of Interference Control Technologies*)

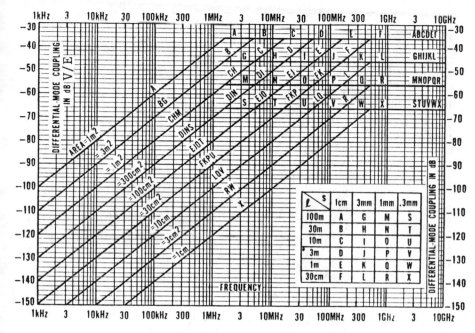

Fig. 5-36. Electric-field differential-mode coupling into box-to-box, cable-loop area.

The differential-mode coupling is plotted in Figs. 5-36 and 5-37 for parallel wire pairs. The parameters in the graphs correspond to the loop dimensions shown in the insert. It is noted that DMC increases with frequency at the rate of 20 dB/decade until the corner frequency condition corresponding to $l = \lambda/2$ is reached. Above this frequency, DMC varies with frequency in accordance with equations (5-10) and (5-14), plotted in Figs. 5-13 and 5-14. Beyond the frequency where the loop length is longer than a half wavelength, the curves in Figs. 5-36 and 5-37 are constructed as illustrated in Fig. 5-16.

To assist the user, the selection of the parameter line is made by choosing the closest l and s loop dimensions in the insert in Figs. 5-36 and 5-37 and reading the corresponding parameter letter. The intersection of the EMI frequency on the X-axis with the applicable parameter line gives the corresponding DMC value on the Y-axis. Note that for determining DMC, the coupling loop height is the separation s between the signal cables.

Differential-Mode Coupling (DMC) in Unbalanced Lines

Unbalanced lines as used herein refer to the coaxial cable family. The physics of DMC for coaxial cables is divided into two separate phenomena: (1) the induced surface current (I_s) on the outer cable conductor, and (2) the voltage induced by I_s, as a result of the coaxial cable transfer impedance (Z_T). The

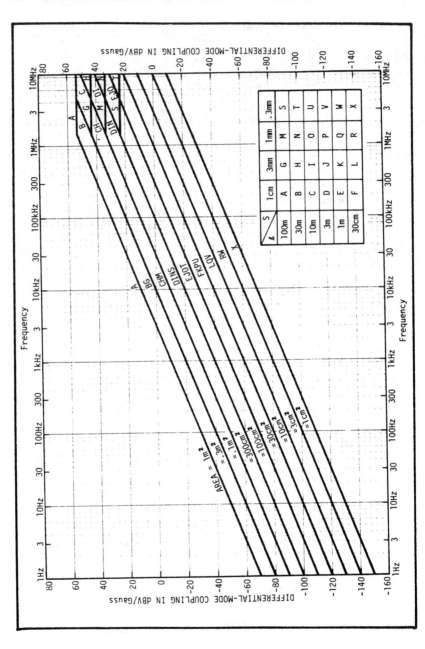

Fig. 5-37. Magnetic-flux density, differential-mode coupling into box-to-box, cable-loop area.

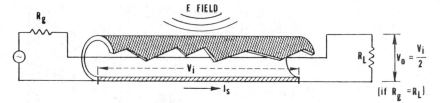

Fig. 5-38. Differential mode coupling into coaxial cables. (*Courtesy of Interference Control Technologies*)

product of $I_s Z_T = V_i$, which is the voltage induced on the inside of the outer coaxial conductor as shown in Fig. 5-38. This voltage V_i is then divided between the generator impedance R_g and the load impedance R_L.

An illustration of coaxial cable DMC follows from Fig. 5-38 (12):

$$\text{DMC} = \frac{V_0}{E} = \frac{kV_i}{E} = k\frac{I_s Z_T}{E} \qquad (5\text{-}39)$$

In terms of dB:

$$\text{DMC}_{\text{dB}} = 20 \log\left[\frac{V_0}{E}\right] = 20 \log\left[\frac{I_s}{E}\right] + 20 \log\left[\frac{V_i}{I_s}\right] + 20 \log\left[k\right]$$

$$(5\text{-}40)$$

where:

I_s/E = ratio of coaxial induced sheath current to ambient electric field
　　　strength (mho-meter)
V_i/I_s = ratio of voltage induced in victim cable per meter of cable length,
　　　to the induced sheath current
$V_i/I_s = Z_T$, transfer impedance of cable (ohm/meter)
　$k = V_0/V_i = R_L/(R_g + R_L)$

From equation (5-40), three quantities are required to determine DMC_{dB} for a coaxial line. The first is a means for determining the induced sheath current, I_s, per unit ambient electric field, I_s/E. This sheath current distribution is a function of the ambient field configuration, frequency, length of cable, cable height above the ground plane, and characteristic impedance Z_0 formed by the cable sheath (or shield, for shielded cables) treated as a single-wire transmission line over a ground plane (6). For a plane-wave illumination, the I_s/E function depends upon the direction of wave propagation relative to cable spatial orientation and the spatial orientation of the elective field vector. For a nonuniform ambient source, the location of the source (i.e., source

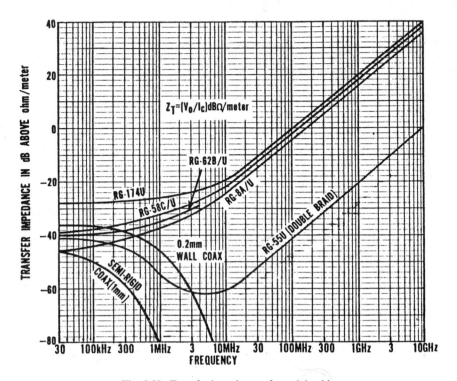

Fig. 5-39. Transfer impedance of coaxial cables.

coordinates) relative to cable coordinates also affects the sheath current distribution, I_s/E (6).

The second quantity in equation (5-40) is the transfer impedance of the coaxial cable, which is a function of the cable type as illustrated in Fig. 5-39 for a variety of specific cables.

The third term in equation (5-40) is the ratio, k, of the differential-mode voltage V_0, at the victim terminals, and the voltage V_i induced in the cable sheath (outer-coaxial conductor). The value of k is determined by the voltage division of V_i between R_g and R_L, where $k = R_L/(R_g + R_L)$ (12).

EMI Control Techniques for Reducing Differential-Mode Coupling

Differential-mode coupling may be reduced by an application of one or more of the following:

- Use twisted wire pairs for balanced parallel lines.
- Use coaxial cable with a lower transfer impedance, Z_T.

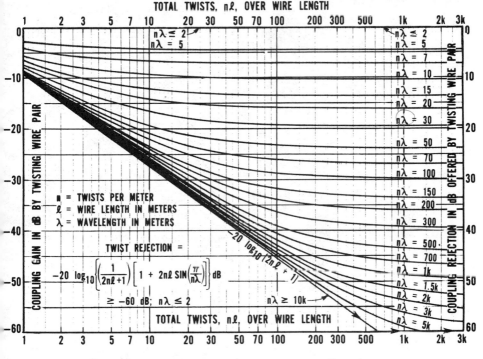

Fig. 5-40. DMC reduction offered by twisting wire pair.

- Add a single braided shield and ground at one end.
- Add a second braided shield and ground at the other end or use triax (for coaxial unbalanced line).
- Add a third braided shield and ground at either end.
- Replace braided shields with solid homogeneous tubular shields or conduit (or use semirigid coax, which has the lowest Z_T in Fig. 5-39).

Twisted Wire Pairs. Figure 5-40 is a plot of the reduction in coupling offered by twisting a wire pair. The X-axis corresponds to the total number of twists over the length of the wire pair, and the Y-axis is the coupling reduction in dB. The parameter line, $n\lambda$, corresponds to the product of the number of twists per meter and the wavelength. Thus, the user selects the applicable parameter line and the total twists on the X-axis. The DMC coupling reduction is then read on the Y-axis at the intersection of the parameter line and the X-axis (12).

Single Braided Shield Over Wire Pair. Figure 5-41 shows the shielding effectiveness of adding a braided shield over a wire pair or harness. It is a plot of the additional DMC reduction in coupling versus the cable length

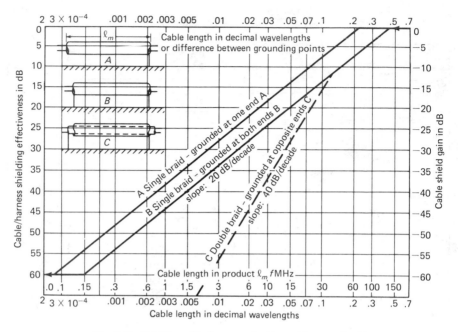

Fig. 5-41. Shielding effectiveness of cable and harness braid.

between ground points in decimal fractions of a wavelength (see upper X-axis). Insert A in the upper left corner is a sketch of the braided shield grounded at either the right or the left end. Curve B corresponds to the situation where the cable is grounded at both ends.

Should the cable be grounded at more than two points, the upper X-axis then corresponds to length between average ground points in decimal wavelengths. The lower X-axis corresponds to the equivalent product of the cable length between grounding points in meters times the frequency in megahertz. It is to be noted that in the upper right corner the added shielding effectiveness of a braid-type cable approaches zero for a significant fraction of a wavelength. Conversely, the figure shows the value truncated at -60 dB to remind the user that the value cannot continue indefinitely because of other coupling paths (12).

Second Braided Shield. Figure 5-41 also shows the effect of adding a second braided shield. While the figure shows the inner braided shield grounded at the left end and the outer braided shield grounded at the right end, this particular configuration is not necessary because they can both be grounded at the same end. The only requirement is that both ends of the same shield not be grounded. For these conditions then, Fig. 5-41 shows that the additional decoupling offered by this double insulated braided shield corresponds to a slope of 40 dB/decade.

Solid Shield or Conduit.   While it is not mentioned in the preceding section, one of the principal reasons for the degradation in performance of braided shielding effectiveness with an increase in frequency shown in Fig. 5-41 is the lack of homogeneity of the braid. Current flowing on the outer surface of the braid also flows on the inner surface of the weave of the individual braid wires. Therefore, there is little net skin effect isolation offered by the shield braid. What isolation does come about results from reflection loss (see Chapter 10). Figure 5-42 shows the reduction in differential mode coupling offered by a solid, homogeneous, copper tubular cable shield which is usually bonded to box cases.

The parameter is the wall thickness, and the Y-axis is the absorption loss in dB offered by the tube. This absorption loss is a function of the wall thickness of the tube as a function of skin depth, δ, given by the expression:

$$\delta = \frac{1}{\sqrt{\pi f \mu \sigma}} \tag{5-41}$$

where

f = noise frequency (Hz)
μ = material permeability (henries/meter)
σ = material conductivity (mho-meter)

The absorption factor is covered in more detail in Chapter 10 under shielding concepts.

Near-Field Cable-to-Cable Coupling

The fourth principal coupling path shown in Fig. 5-5, which is due to the proximity of cables in harnesses, cable trays, raceways, conduits, and so on is often referred to as crosstalk. By capacitive and/or inductive effects, a voltage or current appearing on a source (culprit) cable can result in coupling or the transfer of part of this signal onto an adjacent (victim) cable. The coupling mechanism involves both mutual capacitance and inductance between the culprit and victim wire pairs (12). The undesired emissions can appear as both conducted common and differential mode noise in the victim cable or cable pair. The culprit–victim circuit configuration often includes the presence of a ground plane as part of the return circuit path. By the method of images, this representation can be modified to produce the equivalent coupling between two-wire pairs (12).

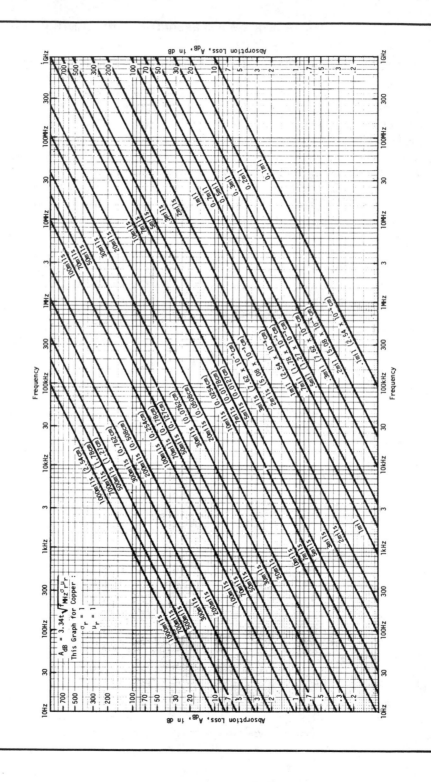

Capacitive and inductive coupling are addressed in Chapter 6. A circuit model (transmission line model) that simultaneously includes both capacitive and inductive effects is also presented in Chapter 6. Each approach has limitations and difficulties. A capacitive-only model or an inductive-only model may give overly simplified results that are not accurate at high frequencies. The transmission line model, however, gives good results over a large frequency band although the equations become rather cumbersome.

After the cable-to-cable coupling is determined, the result is compared to the circuit sensitivity or immunity to determine whether an EMI situation exists.

Power Mains and Power Supply Coupling

This section discusses the fifth coupling path shown in Fig. 5-5, namely, that associated with EMI coupled into and out of power mains and power supplies. It involves conducted emissions on AC or DC unregulated power mains, where the emissions may be harmonics of 50, 60, or 400 Hz power, narrowband signals from intentional emitters, or broadband energy from pulse-type or transient-causing sources. Examples of the latter include differential-mode transients due to other users on the power line or common-mode transients due to radiated energy, such as from a lightning discharge (12).

The coupling of EMI associated with power mains also includes emissions generated by equipment that are coupled into the power mains and potentially can interfere with other equipment on the line. Therefore, this coupling path involves both susceptibility to EMI on the power mains and emissions conducted onto the power mains. (12).

Power Mains Coupling Model.   Figure 5-43 illustrates common-mode (CM) and differential-mode (DM) impedances of both the mains source and the power system, and also illustrates an associated grounding configuration

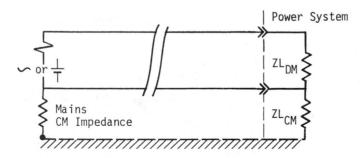

Fig. 5-43. General coupling model. (*Courtesy of Interference Control Technologies*)

(12). Knowledge of actual grounding schemes and associated impedances is important for (12):

- Understanding the predominant coupling paths from the mains to the equipment power system (susceptibility), or from the equipment power system to the mains (emission).
- Calculating CM to DM ratios.
- Selecting adequate filters to suppress emissions and reduce susceptibility.
- Determining system installation overall requirements.

Principal power mains coupling models are depicted in Fig. 5-44, where Z_s represents the mains source impedance, and Z_L the power system's input impedance. The value of the mains impedance is the first consideration regarding the power grounding scheme, source, and load impedance.

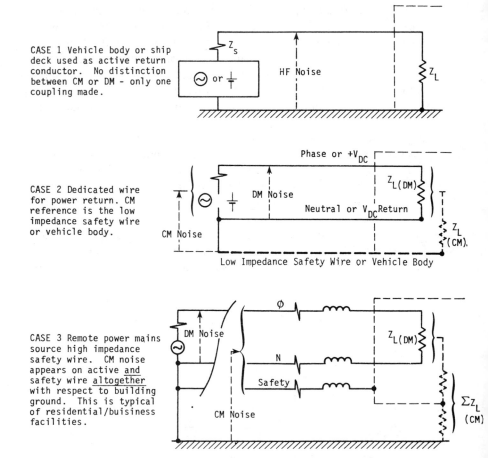

CASE 1 Vehicle body or ship deck used as active return conductor. No distinction between CM or DM - only one coupling made.

CASE 2 Dedicated wire for power return. CM reference is the low impedance safety wire or vehicle body.

CASE 3 Remote power mains source high impedance safety wire. CM noise appears on active and safety wire altogether with respect to building ground. This is typical of residential/buisiness facilities.

Fig. 5-44. Principal power mains coupling models.

From DC to frequencies of a few kilohertz, CM and DM impedances of power mains are roughly equivalent to their DC resistance, which is almost always less than 1 ohm (12). If it is unknown, default values can be used, with the following values being applicable (12):

- Large, high-voltage building service entrance, 3 to 10 milliohms.
- Large, commercial/industrial room panel (100 kVA), 20 milliohms.
- Branch circuit power outlet, 20 milliohms to 1 ohm.

These values represent the total mains impedance including source, plus one phase and return wire at the point of observation.

The values of power mains CM and DM impedances above low frequencies of a few kilohertz depend upon several variables, some of which are (12):

- Frequency.
- Length of mains distribution network.
- Nature and routing of the mains net (wire size, overhead or buried lines, type of conduit, etc.)
- On-line reactive devices (power factor capacitor banks, anti-harmonic chokes, etc.).
- Fluctuating customer load demands, at a given location at a particular time of day.

Figure 5-45 presents general emission and susceptibility models in power mains coupling, and Fig. 5-46 presents a range of aggregated power mains

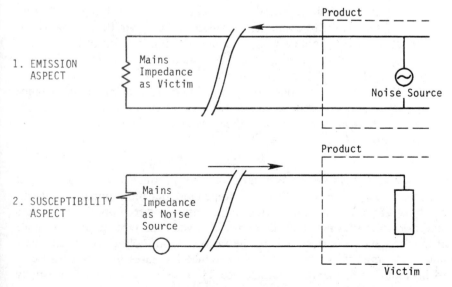

Fig. 5-45. General concept of emission/susceptibility in power coupling.

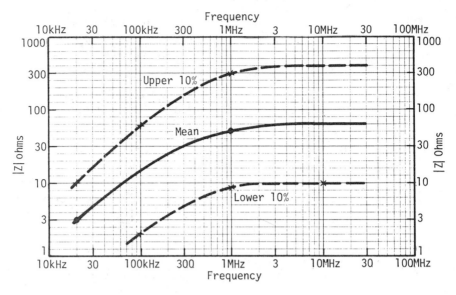

Fig. 5-46. Absolute mains impedances (CM or DM) of aggregate U.S. and European power network (min./max./mean).

impedance values determined for the United States and Europe (12). Referring to Fig. 5-46, the following criteria are recommended (12):

- For emission standards, where the critical aspect is the voltage appearing on AC mains due to equipment emissions, the upper 10% of mains impedance values should be used, so that the highest risk of interference is covered for up to 90% of the representative population of power mains systems.
- For susceptibility considerations, the lower 10% of mains impedance values should be used, to ensure that the worst configuration for susceptibility is covered for 90% of the representative population of product installations (12).

Figure 5-47 illustrates conceptual aspects of product power system input impedance. The value of product input impedance depends upon the front-end circuit of the power system. In the case of input power supply transformers (and assuming that no input filters are involved), the DM input impedance at low frequencies corresponds to the ratio of input voltage over rated current. As frequency increases, DM impedance increases linearly because of wiring and transformer input inductance. The CM input impedance is usually

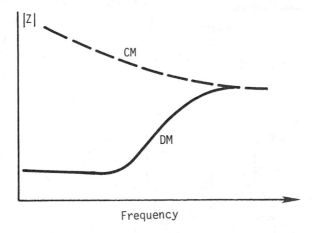

Fig. 5-47. Aspect of power system input impedance.

high at low frequencies, since the power input is floating with respect to the chassis. It decreases with increasing frequency, mainly because of the wiring and PCB-to-chassis and chassis-to-ground capacitance.

To determine CM and DM input impedance accurately, measurements are recommended. If this is impractical, Fig. 5-48 provides default values (12).

Reduction of Power Mains Coupling

Conventional regulators, while effective against short-term and semiperman- ent line voltage fluctuations, are ineffective against short-duration spikes or surges, such as caused by lightning or switching, or high-frequency EMI coupled onto the power mains from broadcast receivers, digital devices, auto- mobile ignitions, and so on. The combination of parasitic coupling paths and slow-acting regulators allows a great deal of unwanted energy to couple into systems from the mains. Thus, a number of options are presented that are useful against this type of coupling.

The reduction of power mains coupling usually is performed with EMI devices especially configured to reduce the coupling from the power mains. Ordinary isolation transformers may be sufficient for reducing common- mode line-conducted interference at 60 Hz, but are totally unsuitable at EMI frequencies of interest. An improvement in isolation transformers is accom- plished with a faraday shield inserted between the primary winding and the secondary winding. Isolation transformers are discussed in Chapter 14, "Power-Line EMC Filters."

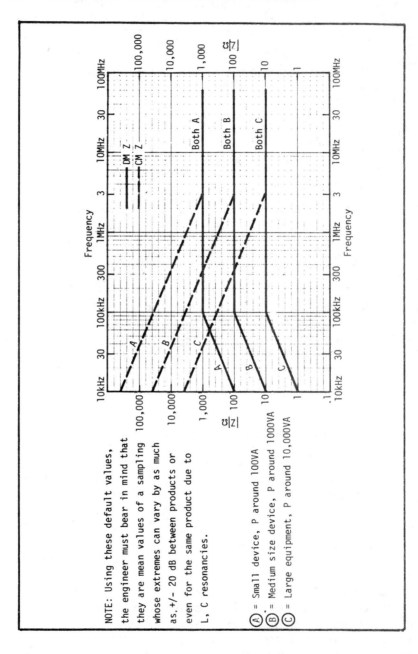

Fig. 5-48. Suggested (default) values of magnitude of Z_{CM} and Z_{DM} for current products.

Other reduction techniques use EMI filters to suppress either common- or differential-mode EMI, or a combination of both. These filters are a network of passive elements arranged in a low-pass configuration. The DM filter is connected from line to line and essentially reflects back to the line-conducted DM noise. The CM filter is connected from both (or all) conductors to the CM ground or reference and shunts the CM noise to ground and away from the equipment input. This topic is also covered in detail in Chapter 14.

Finally, the other types of power mains noise that can affect equipment are due to spikes and surges. The most common types of surge protection elements in use are semiconductor avalanche suppressors, variable resistor elements, and gas discharge tubes. The function of all types are to "clip" fast-rise-time, large-amplitude spikes to a safe level. They are always installed ahead of an EMI filter because the EMI filter is not designed to handle the large amplitudes associated with lightning and switching surges.

The *semiconductor avalanche suppressor* is essentially a large junction diode capable of handling (typically) 1.5 kW of transient power (defined as the peak transient current times the peak transient voltage). The diode acts like a reversed biased zener when conducting and can be obtained with unipolar or bipolar (i.e., like a back-to-back zener) characteristics. It is most often used at levels where the transient energy is relatively low, but where an extremely fast-acting device is needed. It is often inserted "down line" from other transient suppression devices, and most often is used for signal and data line protection.

The variable resistor transient suppressor is often called a *varistor*. It is constructed of a sintered metal-oxide composition (most often a zinc oxide compound). The I-V curve of this type of device resembles that of two paralleled, back-to-back zener diodes. The response time to a transient for these devices is medium, and the energy-handling capabilities are from medium to high for either power or signal-line applications.

The gas discharge tube (simply the gas tube) is a pair of electrodes encased in a gaseous envelope and at low pressure. When the voltage between the two electrodes reaches the breakdown voltage of the gas between the electrodes, the gas undergoes breakdown, and the transient current passes either to ground (CM) or to the other line (DM). The most accurate name for these devices is *surge diverter* because the gas tube diverts the transient energy away from the line rather than absorbing the energy like the semiconductor or varistor transient suppressors. Gas tubes are used both in power line applications (as lightning arresters on phase lines) and in data line applications (often in conjunction with a silicon avalanche suppressor or varistor).

More on transient and surge protection may be found in Chapter 15, "Power Line Isolation Devices."

APPENDIX: GROUND-LOOP COUPLING PREDICTION CURVES

The following figures show additional computed curves of ground-loop coupling or GLC (2). These are presented in addition to Figs. 5-19 and 5-20 which were derived based upon a length of wire of 10 meters. Figure 5-A-1a,b shows GLC for a two-box system with a 1-meter length of wire joining the two boxes. Figure 5-A-2a,b show the·same configuration except that the distance between the boxes has been increased to 100 meters.

These curves, along with Figs. 5-19 and 5-20, give the reader a two-decade span in interconnecting cable length, which is useful in predicting GLC for a variety of configurations.

REFERENCES

1. Georgopoulos, Chris J. *Fiber Optics and Optical Isolators.* Gainesville, VA: Don White Consultants, Inc., 1982.
2. Mardiguian, Michel. Unpublished notes on GLC Measurements conducted at the Don White Consultants, Inc. facilities, 1982.
3. Morse, Philip M., and Fishback, Herman. *Methods of Theoretical Physics,* Parts I and II. New York: McGraw-Hill Book Company, Inc., 1953.
4. Ott, Henry W. *Noise Reduction Techniques in Electronic Systems.* New York: John Wiley & Sons, 1976.
5. Ramo, Simon, and Whinnery, John R. *Fields and Waves in Modern Radio,* Second Edition. New York: John Wiley & Sons, 1953.
6. Smith, A. A., Jr. *Coupling of External Electromagnetic Fields to Transmission Lines.* New York: John Wiley & Sons, 1977.
7. Smythe, W. R. *Static and Dynamic Electricity.* New York: McGraw-Hill Book Company, Inc., 1950.
8. Sorensen, Hans. "*Digital Data Transmission Using Optically Coupled Isolators.*" Hewlett-Packard Application Note 947, June 1982.
9. Stakgold, Ivar. *Boundary Value Problems of Mathematical Physics,* Volume I. Toronto, Ontario, Canada: The Macmillan Company, Collier- Macmillan Canada, Ltd., 1969.
10. Van Bladel, J. *Electromagnetic Fields.* New York: McGraw-Hill Book Company, Inc., 1964.
11. Vance, Edward F. *Coupling to Shielded Cables.* New York: John Wiley & Sons, 1978.
12. White, Donald R. J. *EMI Control Methodology and Procedures,* Third Edition, Second Printing. Gainesville, VA: Don White Consultants, Inc., 1982.
13. White, Donald R. J. *EMI Control in the Design of Printed Circuit Boards and Backplanes.* Gainesville, VA: Don White Consultants, Inc., 1982.
14. White, Donald R. J. *A Handbook on Electromagnetic Shielding Materials and Performance,* First Edition. Germantown, MD: Don White Consultants, Inc., 1975.
15. White, Donald R. J. *EMI Control Methods and Techniques,* A Handbook Series on Electromagnetic Interference and Compatibility, Volume 3. Germantown, MD: Don White Consultants, Inc., 1973.
16. "Bead Balun and Broadband Kit," Engineering Bulletin. Walkill, NY: Fair-Rite Products Corp., March 1981.
17. Optoelectronics Designers Catalog. Hewlett Packard, 1972.

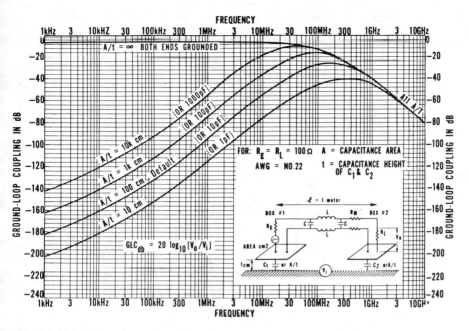

Fig. 5-A-1(a). Design data—ground-loop coupling—unbalanced system. (*Courtesy of Interference Control Technologies*)

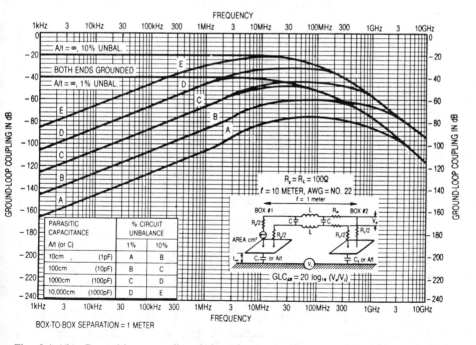

Fig. 5-A-1(b). Ground-loop coupling, balanced system. (*Courtesy of Interference Control Technologies*)

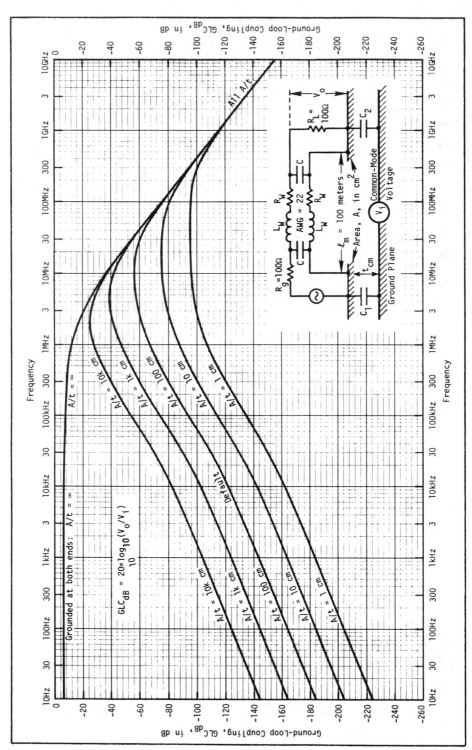

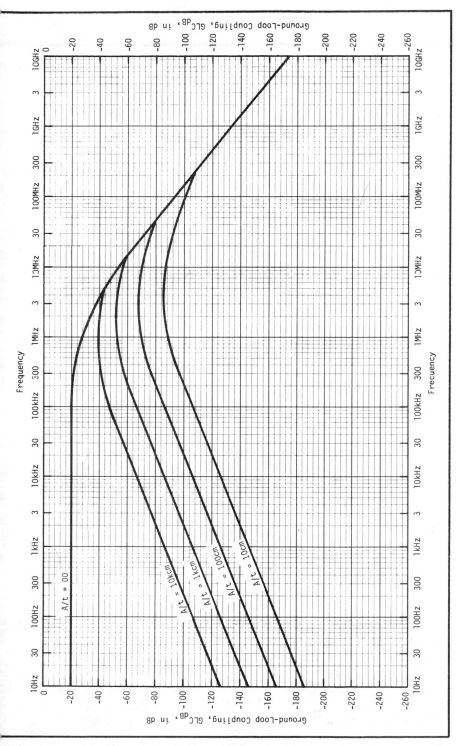

Fig. 5-A-2(b). GLC for balanced circuits having source and load impedance of 100 Ω, Tol = 10% AWG #22 cable length 100 meters.

211

Chapter 6
CABLE WIRING AND HARNESSING

The role of the various types of power and signal cables, as well as the EMI associated with them, continues to be a significant EMC concern. The use of cables permits flexibility in the distribution of electrical power and the transmission of electrical signals required for the operation of various electrical control, communication, and computer systems (1). Since cables are usually routed to accommodate practical operational situations, such as convenient routing paths and equipment location, the EMI environment associated with cables is difficult to quantify and usually varies over a wide range of frequencies and electric and magnetic field amplitudes. Therefore, from both an emission and a susceptibility point of view, cable EMI performance is difficult to predict. In one instance, cables can act as radiating antennas and thereby become a part of EMI radiating sources; or they can act as receiving antennas, thereby becoming a main link in system susceptibility to radiated emissions. Power and signal cables also provide a coupling path for conducted EMI voltages and currents. In addition, cable routing usually finds a variety of cables bundled or harnessed together within a facility, so that two or more cables are in close proximity, causing potential susceptibility to near-field coupling or crosstalk.

WIRING CLASSES AND HARNESSING

Cables may handle 10 kW ($+70$ dBm) of 50/60 or 400 Hz AC power mains supply, or 1 MW ($+90$ dBM) of VHF/UHF peak RF transmitter power. On the low-level side, cables may be used to connect an antenna to a receiver input having a sensitivity level of -120 dBM. This power range thus covers about 200 dB and presents an enormous EMI coupling threat to low-level circuits from high-level emission sources. Therefore, one major aspect of EMI control is to separate wires and cables into similar classes of power-handling and susceptibility levels.

Many specifications attempt to classify types of wiring or cables into four to six groups in order to minimize EMI coupling. However, the typical classification seems to be qualitative and is usually presented in the form of design

**Table 6-1. Air Force wiring classification
(Courtesy of Interference Control Technologies.)**

Class	Identification	Voltage current or power	Frequency
I	DC power circuit	>2 amperes	0
	DC control circuit	<2 amperes	0
II	DC reference circuit	<1 volt or	0
	AF susceptible circuits	<0.2 amperes	
III & IV	AC power circuits	>1 volt or	<400 Hz
	AC reference circuits	>0.2 ampere	<400 Hz
	AF source circuits	>0.2 ampere	<15 kHz
V	RF susceptible circuits		
VI	EMI source circuits	> −45 to −75 dBm	.15 to 5 MHz
		> −75 dBm	5 to 25 MHz
		> −75 to −45 dBm	025 to 1 GHz
		> −45 dBm	>1 GHz
VII	Antenna circuits		

guides. For example, the United States Air Force Systems Command Design Handbook 1-4 suggests the classification listed in Table 6-1 (12).

To determine classification on a more quantitative basis, it may be beneficial to classify wiring and cables into levels of power transmitted, or susceptibility of termination, as applicable. One such classification is achieved by dividing the above 200-dB power-level range into approximately six equal steps of about 30 dB each. Table 6-2 is the result of such a grouping. In Table

**Table 6-2. Wiring Classification by
30-dB power-level groupings.
(Courtesy of Interference Control Technologies.)**

Class	Power range	Identification
A	>40 dBm	High power DC/AC and R-F sources
B	+10 to +40 dBm	Low power DC/AC and R-F sources
C	−20 to +10 dBm	Pulse and digital sources Video output circuits
D	−50 to −20 dBm	Audio and sensor susceptible circuits Video input circuits
E	−80 to −50 dBm	RF and IF input circuits Safety circuits
F	< −80 dBm	Antenna and RF circuits

6-2, the classification has the advantage that:

1. EMI sources and receptors tend to group separately with a power level dichotomy existing at about -20 dBm.
2. Power levels in adjacent wires in a bundle or harness will not likely exceed a 30-dB spread.

The three main concerns to be addressed herein regarding EMI associated with cables are: (1) cables acting as source antennas for radiated emissions, (2) cables acting as susceptibility antennas, and (3) cables in near-field (crosstalk) coupling situations.

CABLES AS RADIATION EMITTERS

Functioning electrical circuits form closed loops of electrical currents. Figure 6-1 conceptually illustrates a current-carrying loop lying in the x-y plane, showing the location and direction of the maximum amplitude of the radiated electric field in the plane of the current loop in the θ-direction. The maximum amplitude of the electric field radiated as shown can be expressed as follows

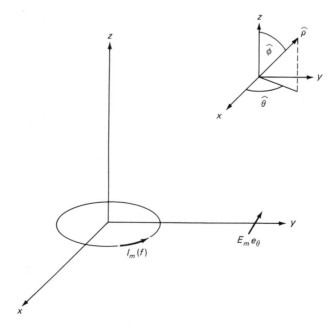

Fig. 6-1. Current carrying loop and associated field.

in vector form (3, 11):

$$E_m = \mathbf{e}_\theta E_m = \mathbf{e}_\theta \left[1.32\sqrt{1 + (\lambda_m/2\pi r_m)^2} \, (f_{\text{MHz}}^2/r_m)ANI_m(f) \right] \qquad (6\text{-}1)$$

where:

λ_m = wavelength in meters = $300/f_{\text{MHz}}$
f_{MHz} = frequency in megahertz
r_m = loop-to-observer distance in meters
A = loop area in cm^2
N = number of loops
$I_m(f)$ = maximum current amplitude in amperes at appropriate frequency
$\mathbf{e}_\theta$ = unit vector in the θ-direction
N = number of coincident conducting loops

The electric field amplitude in equation $(6-1)$ can be written as follows:

$$E_m = K(r_m, f_{\text{MHz}})ANI_m(f) \qquad \text{V/m} \qquad (6\text{-}2)$$

where:

$$K(r_m, f_{\text{MHz}}) = 1.32\sqrt{1 + (\lambda_m/2\pi r_m)^2}\,(f_{\text{MHz}}^2/r_m) \qquad (6\text{-}3)$$

To convert equation (6-2) to dB referenced to 1 microvolt/meter $[\text{dB(uV/m)}]$:

$$\begin{aligned}
E_m\text{dB}(\mu\text{V/m}) &= 20 \log \left[E_m \right] \\
&= 20 \log \left[K(r_m, f_{\text{MHz}}) \right] \\
&\quad + 20 \log \left[A \right] + 20 \log \left[N \right] \\
&\quad + 20 \log \left[I_m(f) \right] \qquad \text{dB}(\mu\text{V/m})
\end{aligned} \qquad (6\text{-}4)$$

For a single current loop ($N = 1$) carrying a current amplitude $I_m(f)$ of 1 ampere, loop area (A) of 1 cm^2, equation (6-4) reduces to E_{m0}, determined as follows:

$$E_{m0}\text{dB}(\mu\text{V/m}) = 20 \log \left[K(r_m, f_{\text{MHz}}) \right] \qquad \text{dB}(\mu\text{V/m}) \qquad (6\text{-}5)$$

Table 6-3 presents values of maximum radiated electric field strength in dB(μV/m) determined from equations (6-3) and (6-5). The loop-to-observer distances selected in Table 6-3 correspond to radiation measurement distances specified in MIL-STD-461, FCC, and VDE regulations (1). If the observation distance r_m is much greater that the loop dimensions, the exact geometric shape of the loop is not a significant factor, and the radiation level is proportional to the loop area.

Table 6-3. Electric and magnetic fields from a straight 1-cm wire and 1-cm sq-loop carrying 1 ampere (for calculating field strengths for CISPR, FCC, and VDE radiated emissions). (Courtesy of Interference Control Technologies.)

FRE-QUENCY	L=1cm; I=1 Ampere E-Field dbμV/m				H-Field dBμA/m				A=1cm sq; I=1 Ampere E-Field dBμV/m				H-Field dBμA/m				FRE-QUENCY
	R=1m	R=3m	R10m	R30m	R=1m	R=3m	R10m	R30m	R=1m	R=3m	R10m	R30m	R=1m	R=3m	R10m	R30m	
10Hz	248	220	188	160	57	38	17	-2	-65	-84	-105	-124	23	-6	-37	-66	10Hz
20Hz	242	213	182	153	57	38	17	-2	-59	-78	-99	-118	23	-6	-37	-66	20Hz
30Hz	239	210	179	150	57	38	17	-2	-55	-75	-95	-115	23	-6	-37	-66	30Hz
50Hz	234	206	174	146	57	38	17	-2	-51	-70	-91	-110	23	-6	-37	-66	50Hz
70Hz	231	203	171	143	57	38	17	-2	-48	-67	-88	-107	23	-6	-37	-66	70Hz
100Hz	228	200	168	140	57	38	17	-2	-45	-64	-85	-104	23	-6	-37	-66	100Hz
200Hz	222	193	162	133	57	38	17	-2	-39	-58	-79	-98	23	-6	-37	-66	200Hz
300Hz	219	190	159	130	57	38	17	-2	-35	-55	-75	-95	23	-6	-37	-66	300Hz
500Hz	214	186	154	126	57	38	17	-2	-31	-50	-71	-90	23	-6	-37	-66	500Hz
700Hz	211	183	151	123	57	38	17	-2	-28	-47	-68	-87	23	-6	-37	-66	700Hz
1kHz	208	180	148	120	57	38	17	-2	-25	-44	-65	-84	23	-6	-37	-66	1kHz
2kHz	202	173	142	113	57	38	17	-2	-19	-38	-59	-78	23	-6	-37	-66	2kHz
3kHz	199	170	139	110	57	38	17	-2	-15	-35	-55	-75	23	-6	-37	-66	3kHz
5kHz	194	166	134	106	57	38	17	-2	-11	-30	-51	-70	23	-6	-37	-66	5kHz
7kHz	191	163	131	103	57	38	17	-2	-8	-27	-48	-67	23	-6	-37	-66	7kHz
10kHz	188	160	128	100	57	38	17	-2	-5	-24	-45	-64	23	-6	-37	-66	10kHz
20kHz	182	153	122	93	57	38	17	-2	1	-18	-39	-58	23	-6	-37	-66	20kHz
30kHz	179	150	119	90	57	38	17	-2	5	-15	-35	-55	23	-6	-37	-66	30kHz
50kHz	174	146	114	86	57	38	17	-2	9	-10	-31	-50	23	-6	-37	-66	50kHz
70kHz	171	143	111	83	57	38	17	-2	12	-7	-28	-47	23	-6	-37	-66	70kHz
100kHz	168	140	108	80	57	38	17	-2	15	-4	-25	-44	23	-6	-37	-66	100kHz
200kHz	162	133	102	74	57	38	17	-2	21	2	-19	-38	23	-6	-37	-66	200kHz
300kHz	159	130	99	70	57	38	17	-2	25	5	-15	-34	23	-6	-37	-65	300kHz
500kHz	154	126	94	66	57	38	17	-2	29	10	-11	-30	23	-6	-37	-65	500kHz
700kHz	151	123	91	63	57	38	17	-1	32	13	-8	-26	23	-6	-37	-65	700kHz

1MHz	148	120	88	61	57	38	17	-1	35	16	-5	-23	23	-6	-37	-64	1MHz
2MHz	142	114	83	58	57	38	18	2	41	22	2	-14	23	-6	-36	-61	2MHz
3MHz	139	110	80	57	57	38	18	5	45	26	6	-8	23	-5	-36	-59	3MHz
5MHz	134	106	77	59	57	38	20	8	49	30	12	0	23	-5	-34	-52	5MHz
7MHz	131	103	76	62	57	39	22	11	52	34	17	6	23	-5	-32	-46	7MHz
10MHz	128	101	75	65	57	39	24	14	55	37	22	12	23	-4	-30	-40	10MHz
20MHz	123	98	81	71	58	42	30	20	62	46	34	24	24	-1	-18	-28	20MHz
30MHz	120	97	84	75	58	45	33	23	66	52	41	31	24	1	-11	-21	30MHz
50MHz	117	99	89	79	60	48	37	28	72	60	49	40	26	8	-2	-12	50MHz
70MHz	116	102	92	82	62	51	40	31	77	66	55	46	28	14	4	-6	70MHz
100MHz	115	105	95	85	64	54	43	34	82	72	61	52	30	20	10	0	100MHz
200MHz	121	111	101	91	70	60	49	40	94	84	73	64	42	32	22	12	200MHz
300MHz	124	115	105	95	73	63	53	43	101	91	80	71	49	39	29	19	300MHz
500MHz	129	119	109	99	77	68	57	48	109	100	89	80	58	48	38	28	500MHz
700MHz	132	122	112	102	80	71	60	51	115	106	95	86	64	54	44	34	700MHz
1GHz	135	125	115	105	83	74	63	54	121	112	101	92	70	60	50	40	1GHz
2GHz	141	131	121	111	89	80	69	60	133	124	113	104	82	72	62	52	2GHz
3GHz	145	135	125	115	93	83	73	63	140	131	120	111	89	79	69	59	3GHz
5GHz	149	139	129	119	97	88	77	68	149	140	129	120	98	88	78	68	5GHz
7GHz	152	142	132	122	100	91	80	71	155	146	135	126	104	94	84	74	7GHz
10GHz	155	145	135	125	103	94	83	74	161	152	141	132	110	100	90	80	10GHz

Notes: Radiation Source-to-Victim Distance Quantized into 4 Distances: 1m, 3m, 10m, 30m. MIL-STD-462 = 1m; FCC = 3m (Class B) & 30m (Class A); VDE = 30m Below 30MHz; Above = 10m For straight wire: E and H = above + 20xlog (LI); L in cm and I in Amperes. For loop wire; E and H = above + 20xlog(AI); A in sq. cm and I in Amperes.

Equation (6-1) forms the basis for the method proposed herein to predict the radiated field levels using the tabulated values in Table 6-3 [or equation (6-5) is used to derive the table quantities] for the reference loop parameters of 1 ampere of current and loop area of 1 cm^2. Once values are extracted from Table 6-3 for specified frequencies and loop-to-observer distances, the additional terms that account for the actual loop area (A), number of loops (N), and current amplitude $I_m(f)$ at the specified frequency (f) must be added to obtain the resultant calculated maximum electric field radiated by N loops in accordance with equation (6-4).

Differential-Mode Radiation

The determination of the current amplitude as a function of frequency, $I_m(f)$, in equations (6-1) through (6-4), usually is not a trivial task. Figure (6-2) illustrates a differential-mode transmission system with a signal source V_s, a source real impedance Z_s, a load real impedance Z_L, and a transmission line (cable, PCB trace, for example) of length L with separation s between conductors and characteristic impedance Z_0.

The loop around which the differential-mode current $I_m(f)$ flows can be classified as either a "small" or a "large" loop, with the reference being the length of the loop, L, compared to the wavelength, λ_m, at a specified frequency. In general, the input impedance Z_{in}, for a transmission line, can be expressed in the familiar ideal transmission line formula, assuming negligible line losses: (14)

$$Z_{in} = Z_0 \frac{Z_L + jZ_0 \tan(\beta L)}{Z_0 + jZ_L \tan(\beta L)} = Z_0 \frac{Z_L + jZ_0 \tan(2\pi L/\lambda_m)}{Z_0 + jZ_L \tan(2\pi L/\lambda_m)} \qquad (6-6)$$

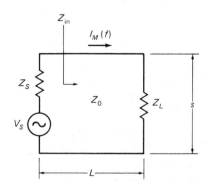

Fig. 6-2. Differential mode transmission system.

where $\beta = 2\pi/\lambda_m$, $j = \sqrt{-1}$. From Fig. 6-2:

$$I_m(f) = \frac{V_{sm}(f)}{Z_s + Z_{in}} \tag{6-6a}$$

For $L \ll \lambda_m (L \leq \lambda_m/10)$, corresponding to low frequencies, equation (6-6) shows that $Z_{in} = Z_L$, and the current $I_m(f) = V_{sm}(f)/(Z_s + Z_L)$, where $V_{sm}(f)$ is the maximum amplitude of the source voltage V_s at frequency f. Also, at low frequencies, the current amplitude is essentially uniformly distributed around the loop, which is characteristic of a "small" loop.

As the frequency of interest increases, the wavelength λ_m decreases, and the circuit length L becomes a significant fraction, or multiple, of λ_m. In this case, the loop becomes a "large" loop (3, 11). Equation (6-6) shows that the input impedance Z_{in} varies with λ_m, and therefore frequency, with corresponding variations in the amplitude of the loop current. As the frequency is varied, the magnitude of Z_{in} varies between maximum and minimum values, with the latter values determined by the relationship between Z_0 and Z_L (5). Also, the current distribution along L is generally in the form of standing waves, with current maxima and minima occurring as a function of position along L. Standing waves are present whenever $Z_L \neq Z_0$. The nonuniform current distribution along L results in directional radiation field patterns, or "lobes," as a function of azimuth angle ϕ around the loop (11). In this instance, the calculation of the radiated field pattern becomes much more difficult.

In general, a transmission system load impedance Z_L will be complex and expressible in the form: $Z_L = R_L + jX_L$. The low-loss, near-ideal transmisson line input impedance in Fig. 6-2 then becomes:

$$Z_{in} = Z_0 \frac{R_L + j[X_L + Z_0 \tan(2\pi L/\lambda_m)]}{Z_0 - X_L \tan(2\pi L/\lambda_m) + jR} \tag{6-7}$$

If Z_L is purely reactive, that is, $R_L = 0$ and $Z_L = jX_L$, then equation (6-7) becomes:

$$Z_{in} = jZ_0 \frac{X_L + Z_0 \tan(2\pi L/\lambda_m)}{Z_0 - X_L \tan(2\pi L/\lambda_m)} \tag{6-8}$$

Equation (6-8) indicates that the input impedance is purely reactive, with poles and zeros occurring at frequencies where:

$$\text{Poles: } Z_0 - X_L \tan(2\pi L/\lambda_m) = 0 \tag{6-9}$$

$$\text{Zeros: } X_L + Z_0 \tan(2\pi L/\lambda_m) = 0 \tag{6-10}$$

The exact (resonant) frequencies at which the conditions of equations (6-9) and (6-10) exist depend upon the relationship between L, λ_m, X_L, and Z_0. This can be critical when determining the loop current amplitude, $I_m(f)$, as an input for determining the radiated field from equation (6-4). If the frequency of interest coincides with a pole indicated (or approximated) by equation (6-9) (antiresonance condition), the impedance Z_{in} is relatively high (theoretically, infinity), the current a minimum (theoretically, zero), and the corresponding radiation low (theoretically, zero) at that frequency. The reverse situation occurs whenever the frequency of interest is close to the conditions of equation (6-10) (series resonance). If the conditions of equations (6-9) and (6-10) are not taken into account, the penalty is likely to be an over-estimation or underestimation of the predicted radiated electric field strength at certain resonant frequencies. When measurements are made to determine compliance with regulatory specifications, the measured radiated field will conform to the actual radiation loop current amplitudes affected by conditions similar to those described by equations (6-9) and (6-10) at resonant frequencies (2).

Digital waveforms are of the pulse and nonsinusoidal periodic types. The former can be represented in the frequency domain by a continuous Fourier spectrum and the latter by a Fourier series corresponding to a line spectrum of harmonic components. If any of these frequencies, determined by Fourier analysis techniques, coincide with the resonant frequencies that result in the conditions described by equations (6-9) and (6-10), the procedure for determining $I_m(f)$ must consider the potential consequences of equations (6-9) and (6-10) (2).

Reduction of Differential-Mode Radiation from Cables

Referring to equation (6-1), the following actions will reduce differential-mode radiation from circuit loops formed by interconnecting cables:

- Keep loop areas (A) small. This usually means keeping cable lengths as short as possible, since the spacing minimum is limited by cable insulation thickness and other construction considerations.
- Operate at the lowest possible frequencies.
- In the time domain, digital waveform rise times should be no shorter than necessary consistent with desired data rates, logic type, and other system requirements.
- Amplitudes of current waveforms should be as low as possible, consistent with logic type and other system design requirements.

In addition to the above considerations, other methods for reducing cable differential-mode radiation include:

- Installing cable shields (1).
- Use of twisted-pair cables, unshielded or shielded (1).
- Use of coaxial and/or triaxial cable.
- Connecting alternate signal returns together (1) (flat ribbon cable).
- Use of fiber optics.

Common-Mode Cable Loop Radiation

Figure 6-3 conceptually illustrates the flow of common-mode (CM) current, I_{CM}, flowing around a loop area of dimensions $A = H \times L$ where H is the average height of a cable above a metal ground plane, metal tabletop, rebar reinforced floor concrete, or possibly the green safety wire; and L is the length of the interconnecting signal cable. As illustrated in Fig. 6-3, the CM radiating loop may be large. The level of the CM radiated field depends upon the loop area $H \times L$ and the amplitude of the current I_{CM} that flows in the ground plane (4, 5, 16).

Reduction of Common-Mode Radiation from Cable Assemblies

The following options are capable of reducing CM radiation (1):

- Reduce the loop area by reducing the cable length (L) and/or the average height above the ground plane (H).
- Reduce the amplitude of the CM current.
- Use a ferrite yoke on the cable close to the connector, where the yoke envelops all cable conductors.
- Use filter-pin connectors to attenuate out-of-band emissions.

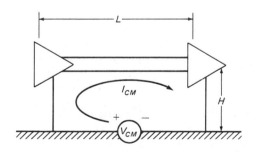

Fig. 6-3. Common mode loop.

CABLE SUSCEPTIBILITY TO RADIATED EMISSIONS

Susceptibility to far-field radiated emissions is covered in Chapter 5 under the discussion of radiation into loops resulting in two coupling paths: common-mode and differential-mode coupling.

Reducing Cable Susceptibility to Radiated Emissions

Reducing Common-Mode (CM) Radiated Susceptibility. Common-mode susceptibility is directly proportional to the loop area $H \times L$ illustrated in Fig. 6-3. To reduce CM susceptibility:

- Reduce the loop area by reducing the cable length (L) and/or the average height above the ground plane (H).
- Reduce ground-loop coupling (see Chapter 5).
- Shield the $H \times L$ loop area (if it is practical to do so).
- Use fiber optics.

Reducing Differential-Mode (DM) Radiated Susceptibility. Differential-mode susceptibility is directly proportional to the loop area $L \times s$ in Fig. 6-2. To reduce DM susceptibility:

- Reduce the loop area, usually by keeping the cable length as short as possible, consistent with other system design considerations.
- Use twisted pair cables, unshielded or shielded (16).
- Install cable shields (16).
- Use EMI-hardened connectors (1).
- Use fiber optics.

CABLE-TO-CABLE COUPLING (CROSSTALK)

This section reviews electromagnetic interference (EMI) situations that result from near-field capacitive and/or inductive coupling (crosstalk) between two or more wires or wire pairs. These phenomena were identified as the fourth coupling path in Fig. 5-5, cable-to-cable coupling. Whenever wires are bundled together in close proximity, such as within a harness, the possibility exists that one of these wires or wire pairs can become a culprit source, and one or more of the others potential victims.

Cable-to-Cable Near-Field Coupling Prediction Model

Models have been developed for the determination of cable-to-cable coupling (crosstalk) resulting from mutual inductance (magnetic field coupling) and mutual capacitance (electric field coupling) between two circuits. A

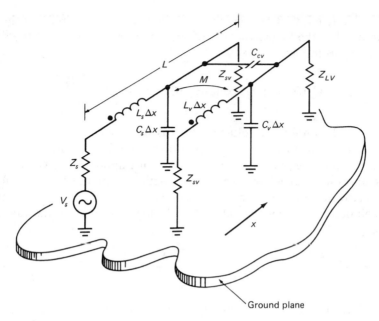

Fig. 6-4. Transmission line model to determine cable-to-cable-coupling.

rigorous analysis for crosstalk prediction was accomplished by C. R. Paul, and the following development presents this analysis and the essential conclusions reached by this effort (6).

Figure 6-4 illustrates an equivalent circuit for a basic three-conductor transmission line, with the ground plane selected as the reference conductor. The transmission-line pair is directed parallel to the x-axis. The total line length is L meters, and the line separation is s meters, but the latter dimension is usually scaled down to millimeters when calculations are made. The line segments are assumed to be perfect conductors (infinite conductivity), and the medium surrounding the conductors is assumed to be lossless and characterized by scalar values for permeability and permittivity. The transmission-line cross section is uniform throughout its length L, and all line voltages are referenced to the ground plane. The left conductor contains the source voltage, V_s, and is considered to be the "culprit" circuit. The culprit circuit also contains a source impedance Z_s, a load impedance Z_{Ls}, a self-inductance per unit length, L_s henries per meter, and a self-capacitance to ground per unit length, C_s farads per meter. The right-hand conductor is the receptor, or victim, circuit, and it contains a source impedance Z_{sv}, a load impedance Z_{Lv}, self-inductance per unit length, L_v henries per meter, and a self-capacitance to ground per unit length, C_v farads per meter. The circuits are inductively and capacitively coupled via the mutual inductance, M, and the mutual capacitance, C_{cv}, per unit length (8).

The objective is to determine the magnitude of the ratio in dB of the voltage induced at the load terminals of the victim circuit, V_{Lv}, and the source voltage, V_s, as follows:

$$\text{Cable-to-cable coupling (CCC)}_{\text{dB}} = 20 \log \left| \frac{V_{Lv}}{V_s} \right| \qquad (6\text{-}11)$$

An exact solution for the voltage ratio has been determined by C. R. Paul by solving the distributed-parameter transmission-line equations for an incremental length of line (7). Equations providing approximate values for the inductances and capacitances per unit length are provided in reference 18 for selected transmission-line cross sections.

A solution for the voltage induced at the load terminals of the victim circuit is (6, 7):

$$V_{Lv} = \frac{j\omega LSV_s}{D} \left[-\frac{MZ_{Lv}}{(Z_{sv} + Z_{Lv})(Z_s + Z_{Ls})} + \frac{C_{cv}Z_{sv}Z_{Lv}Z_{Ls}}{(Z_{sv} + Z_{Lv})(Z_s + Z_{Ls})} \right] \qquad (6\text{-}12a)$$

from which the ratio in equation (6-11) can be determined:

$$\frac{V_{Lv}}{V_s} = \frac{j\omega LS}{D} \left[-\frac{MZ_{Lv}}{(Z_{sv} + Z_{Lv})(Z_s + Z_{Ls})} + \frac{C_{cv}Z_{sv}Z_{Lv}Z_{Ls}}{(Z_{sv} + Z_{Lv})(Z_s + Z_{Ls})} \right] \qquad (6\text{-}12b)$$

where D is the denominator term and is defined by:

$$D = C^2 - S^2\omega^2 t_v t_s \left[1 - k^2 \frac{(1 - a_s a_{Lv})(1 - a_{Ls} a_v)}{(1 + a_v a_{Lv})(1 + a_s a_{Ls})} + j\omega CS[t_v + t_s] \right] \qquad (6\text{-}13a)$$

with:

$C = \cos(\beta L)$

$S = \dfrac{\sin(\beta L)}{L}$

$\omega = $ radian frequency (radians/second)

$\beta = \dfrac{2\pi}{\lambda}$

The coupling coefficient is found from:

$$k = \frac{M}{\sqrt{L_s L_v}} = \frac{C_{cv}}{\sqrt{(C_s + C_{cv})(C_v + C_{cv})}}$$

The time constants of the source and victim circuit are:

$$t_s = \frac{L}{Z_s + Z_{Ls}} L_s + (C_s + C_{cv})(Z_s Z_{Ls}) \qquad (6\text{-}13b)$$

$$t_v = \frac{L}{Z_{sv} + Z_{Lv}} L_v + (C_v + C_{cv})(Z_{sv} Z_{Lv}) \qquad (6\text{-}13c)$$

and the a quantities are found from:

$$a_s = Z_s \sqrt{\frac{(C_s + C_{cv})}{L_s}} = \frac{Z_s}{Z_{0s}}$$

$$a_{Ls} = Z_{Ls} \sqrt{\frac{(C_s + C_{cv})}{L_s}} = \frac{Z_{Ls}}{Z_{0s}}$$

$$a_v = Z_{sv} \sqrt{\frac{(C_v + C_{cv})}{L_v}} = \frac{Z_{sv}}{Z_{0v}}$$

$$a_{Lv} = Z_{Lv} \sqrt{\frac{(C_v + C_{cv})}{L_v}} = \frac{Z_{Lv}}{Z_{0v}}$$

The impedances Z_{0s} and Z_{0v} are the characteristic impedances of the culprit and victim circuits, respectively, each in the presence of (i.e., coupled to) the other circuit. They are determined by:

$$Z_{0s} = \frac{L_s}{(C_s + C_{cv})}$$

$$Z_{0v} = \frac{L_v}{(C_v + C_{cv})}$$

Equations (6-12a) and (6-12b) account for the simultaneous coupling due to mutual inductance and capacitance. These equations can be simplified when the cable coupling is "loose" ($k \ll 1$) and the frequency of interest is such that the length of line is electrically short, say $L < \lambda/10$, and the frequency is "sufficiently low." Under these assumptions, $C = D = S = 1$, and equation (6-12a) can then be simplified to a two-component form (6):

$$V_{Lv} = V_{Lvi} + V_{Lvc} \qquad (6\text{-}14)$$

where V_{Lvi} = voltage induced across the victim load due to inductive coupling, and V_{Lvc} = voltage induced across the victim load due to capacitive coupling.

From low-frequency assumptions and equations (6-12a) and (6-12b), the components of equation (6-14) can be determined, and the inductive and capacitive coupling ratios can be written separately as:

Inductive: $\qquad \dfrac{V_{Lvi}}{V_s} = -\dfrac{Z_{Lv}}{Z_{sv} + Z_{Lv}}(j\omega ML)\dfrac{1}{Z_s + Z_{Ls}}$ $\qquad$ (6-15)

Capacitive: $\qquad \dfrac{V_{Lvc}}{V_s} = \dfrac{Z_{sv}Z_{Lv}}{Z_{sv} + Z_{Lv}}(j\omega C_{cv}L)\dfrac{Z_{Ls}}{Z_s + Z_{Ls}}$ $\qquad$ (6-16)

The expressions in equations (6-14), (6-15), and (6-16) indicate that, *at low frequencies*, the individual components due to inductive and capacitive coupling *may be isolated and superimposed*. Note that in equations (6-15) and (6-16) the self inductances and capacitances of the two circuits do not appear, because of the assumed low frequency where their reactive effects are negligible.

As the frequency of interest increases, the model of equation (6-12) can be adjusted further so that individual inductive and capacitive coupling components of V_{Lv} still can be identified. For this intermediate frequency range, equation (6-12a) can again be expressed in terms of two components, as shown in equation (6-14), with the individual components determined by the following expressions adjusted for a higher frequency range (6):

$$V_{Lvi} = -\frac{Z_{Lv}}{Z_{sv} + Z_{Lv}}(j\omega ML)\frac{V_s}{Z_s + Z_{Ls}}\frac{1}{(1 + j\omega t_s)(1 + j\omega t_v)} \qquad (6\text{-}17)$$

$$V_{Lvc} = \frac{Z_{sv}Z_{Lv}}{Z_{sv} + Z_{Lv}}(j\omega C_{cv}L)\frac{Z_{Ls}V_s}{Z_s + Z_{Ls}}\frac{1}{(1 + j\omega t_s)(1 + j\omega t_v)} \qquad (6\text{-}18)$$

Ratios similar to equations (6-15) and (6-16) can be determined from equations (6-17) and (6-18). For equations (6-17) and (6-18) to apply, the line length must still be electrically short (i.e., $L \ll \lambda$), and the lines weakly coupled (i.e., $k \ll 1$). However, for this intermediate frequency range, the mutual and self inductance and capacitance terms become significant and appear as part of the expressions for t_s and t_v in equations (6-17) and (6-18) [see equations (6-13b) and (6-13c)]. It should also be noted that both self-inductances and self-capacitances appear as part of both the voltage coupled via the mutual inductance (V_{Lvi}) and that due to mutual capacitance (V_{Lvc}).

A superposition technique often employed to calculate cable-to-cable coupling utilizes two separate models: one model includes only the mutual and self-inductances to determine inductive coupling, and another model includes only the mutual and self-capacitances to determine capacitive coupling. The two results are then superimposed to determine overall coupling. This latter superposition approach is shown to produce significant errors at frequencies when $L \geq \lambda/1000$. *An example provided in reference 6 indicates a 49 dB error for $L > \lambda/10$, the superposition technique leading to overly pessimistic coupling.* This superposition technique produces an error of less than or equal to 6 dB for $L < \lambda/1000$ in the same example. This development indicates that this technique is of very limited use (useful only at extremely low frequencies) in determining cable-to-cable coupling. The example in reference 6 also provides experimental data that support equation (6-12) in predicting cable-to-cable coupling.

The dividing line between low frequency, where equations (6-15) and (6-16) apply, and an intermediate frequency range, where equations (6-17) and (6-18) apply, cannot be drawn on the basis of a simple determination because it is a function of the interrelationships of several variables, including circuit geometry, coupling coefficients, and circuit impedances. These interrelationships are contained within the D term of equations (6-12), with D defined in equation (6-13a), and the t and a terms contained within D defined by equations (6-13b) and (6-13c) and subsequent definitions following immediately thereafter. Also, it is noted that D is generally complex, which further complicates calculations. From all indications, except when it can be absolutely determined that only extremely low frequencies are of interest or except for very short runs of cable where the line length L is on the order of a thousandth of a wavelength, the total coupling model specified by equation (6-12) should be used to determine cable-to-cable crosstalk.

Crosstalk Reduction Techniques

A number of measures may be taken to reduce crosstalk between cables. The desired result is to reduce the coupling mechanism, either the mutual inductance, M, or the mutual capacitance, C_{cv}. Since the crosstalk mechanism could also be thought of as a voltage divider occurring via the parasitics between a circuit or circuits, other reduction techniques attempt to reduce coupling by increasing the voltage drop across the source impedances and/or the coupling impedance to effect a decrease in the voltage across the victim load. This discussion will concern the alteration of one or more parameters in the crosstalk equation; so the reader is referred back to equations (6-12b) and (6-13a) and the equation for the coupling coefficient, k.

Some techniques used to reduce one factor in the coupling equation may increase the coupling in another fashion. This discussion will focus on the techniques used to control magnetic- and electric-field coupling that depend on the various parasitic paths and circuit values.

Magnetic-Coupling Reduction Techniques

To reduce the potential for interference that is magnetically coupled to a circuit loop from an adjacent circuit, the following measures can be taken:

1. Reduce the culprit voltage source V_s, or current.
2. Reduce the circuit loop area by reducing either the line length, L, or the wire height, h (separation), or both. Usually, L cannot be reduced significantly, since it is presumed that as short a cable run as possible was used in the first place. Figure 6-5 illustrates the evolution of reducing circuit loop area.

    The size of a loop gives a relative indication of how much interference can couple into it. The large loops tend to have a high self-inductance L_s and L_v, and low wire to ground capacitance C_s and C_v, which tends to increase coupling. The "worst case" (maximum loop area) is illustrated in Fig. 6-5a. By placing the insulated circuit wire directly over its ground plane, as shown in Fig. 6-5b, the loop area is significantly reduced due to a much smaller h. A far better practice is to achieve Fig. 6-5b by using a dedicated ground return, as shown in Fig. 6-5c, to avoid common-mode impedance coupling problems. A best practice is to twist the dedicated ground return with its outgoing wire.(9) The twist tends to make local environmental EMI contributions cancel because the induced voltage in each incremental twist area is approximately equal and opposite to that of its neighbor.

3. Separate (isolate) victim circuits and potential sources as much as possible. This reduces both C_{cv} and M.
4. Operate at lower frequencies, if possible, to increase the reactance of C_{cv} and M.
5. Design digital waveforms, or select logic, for longest possible practical rise-times and fall-times. Short rise-times lead to a broad EMI frequency spectrum.
6. Use magnetically shielded cables in which the relative permeability of the shield $\mu_r \gg 1$ to reduce the mutual inductance M by shunting the flux away from the victim wire.
7. Operate sensitive circuits using differential amplifiers so that common-mode, induced currents in both leads will tend to cancel.
8. Twist each signal wire together with its return to reduce DM radiation and pickup.

Some additional remarks regarding the above EMI-control techniques for reducing magnetic coupling in wires and cables are in order. When a dedicated return wire is used (best practice), as shown in Fig. 6-5 (c and d), the

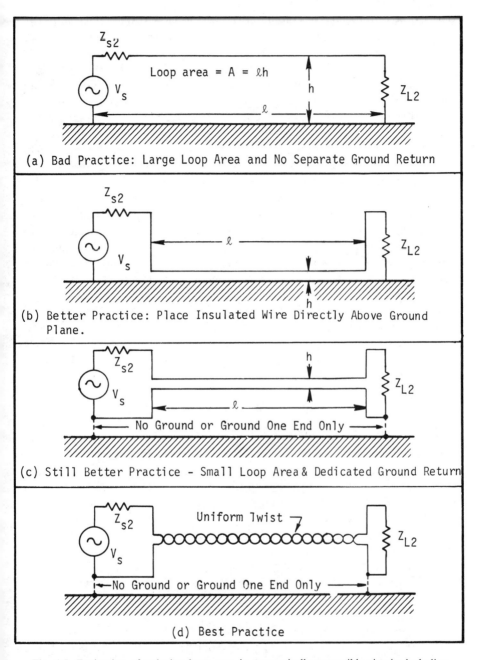

Fig. 6-5. Evaluation of reducing loop area in magnetically-susceptible circuits including circuit grounding practice.

topic of how to ground the circuit surfaces. This was discussed at length in Chapter 5. For the present discussion, the first question raised is whether circuit grounding is needed at all. If grounding is required, then it should be done at one point only, as shown in Fig. 6-5 (c and d). Should both ends of the cable or both sending and receiving circuits be grounded, then an alternative return path for the signal current has been introduced. Now, current return will go through both the return twist and the ground plane. Since the latter offers a much lower impedance path than the twisted wire, nearly all of the current will follow this route, and the return twisted wire is effectively out of the circuit. Thus, the effect is the equivalent circuit of Fig. 6-5a with its large circuit loop area.

In Chapter 5, sections on radiated common-mode and ground-loop coupling and radiated differential-mode coupling discuss control of the ground-loop area, use of ferrites to reduce common-mode current, and the techniques of floating (ungrounding) circuits, boxes, and cabinets. Other techniques to reduce ground-loop coupling include the use of RF chokes, optical isolators, isolation transformers, and balanced differential drivers and receivers. Many of these techniques are also useful in reducing cable-to-cable crosstalk.

Chapter 5 also covers control of the cable wire loop area for parallel-wire lines and the transfer impedance for coaxial lines; twisting parallel wires; and use of single, double, and triple braided shields, as well as solid tubular shields and conduits. Similarly, many of these techniques are useful in reducing cable-to-cable crosstalk.

Cable Shielding of Magnetic Fields

The list of magnetic coupling reduction techniques (above) indicates that further magnetic-field decoupling can be achieved by using a magnetic shield over the wire. Typical tin-plated, copper-braided shields, used for electric-field decoupling purposes (shielding) at higher frequencies, will have little to no effect on magnetic coupling, since the shield has a relative permeability of $\mu_r = 1$, which is the same as that of air. However, where magnetic shielding is desired, as in either secure systems or susceptible equipments operated in the presence of strong magnetic fields, a ferrous-based cable shield may be used. The culprit magnetic field will be reduced at the equivalent victim loop area because of the low reluctance path offered by the shield, as shown in Fig. 6-6.

The degree of magnetic shielding, SE_{dB}, offered by a permeable cylindrical sheath placed over either a wire pair of a harness at low frequencies is (5):

$$SE = 20 \log \left(1 + \frac{\mu_r t}{2R} \right) \text{dB}, \qquad (6\text{-}19)$$

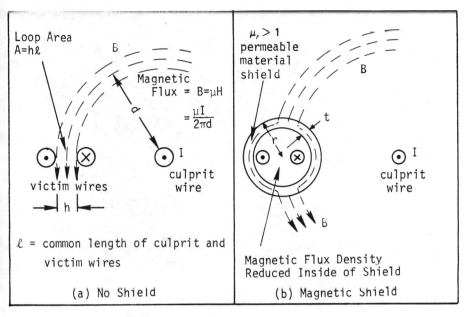

Fig. 6-6. Reducing magnetic flux loop area by using permeable magnetic shield. (*Courtesy of Interference Control Technologies*)

where:

$$\mu_r = \text{relative permeability of the magnetic sheath}$$
$$t = \text{thickness of the cylindrical sheath}$$
$$R = \text{outer radius of the sheath}$$

Equation (6-19) applies for low frequencies (say, $f < 10$ kHz). Further, the equation only applies for transverse fields, such as that shown in Fig. 6-6b, and not for longitudinal fields.

Equation (6-19) indicates that the shielding effectiveness is reduced for thin-wall cylindrical sheaths. Since magnetic-field shielding is due to the mechanism of absorption, proper absorption of the magnetic field depends on the number of *skin depths* to which the thickness of the shield corresponds. A skin depth is that distance inside a material to which a field decays to $1/e$ of its value at the surface (see Chapter 10). For example, if a $1/16 =$ inch-wall iron pipe ($\mu_r \simeq 1000$) having a 1-inch radius is used around a large harness in a fixed installation, the magnetic shielding effectiveness is only about 30 dB. If a 2-mil (0.002″) iron tape had been wrapped around a 1-inch-radius harness with a 50% overlap per pitch, the shielding effectiveness would only be about 3 dB. Thus, materials of higher permeabilities and/or greater thicknesses are necessary for good shielding performance.

Fig. 6-7. Double-knitted wire mesh tape shield (*Courtesy of Chomerics*)

A number of tapes made of highly permeable material are commercially available that can be used for magnetic-field shielding of a wire harness. Figure 6-7 shows one such tape available in 1-inch widths. Based on a tin-coated, copper-clad steel, double-wire mesh with about 90% air space, the equivalent relative permeability of the tape is about 100 per 0.015-inch of tape thickness. Table 6-4 shows the low-frequency shielding effectiveness of tapes for harnesses of different radii when they are wrapped as a bandage, advancing one-half layer per turn.

Table 6-4 also shows the typical shielding effectiveness expected from other highly permeable, solid-foil tapes when wrapped with a 50% overlap per turn. The permeability is a function of field strength, degree of annealing, and other parameters. Netic tape offers less shielding than Co-Netic, but because of saturation in high fields, it is used above 5 gauss, and Co-Netic is used below 5 gauss.(17) The corrugated flexible tube shown in the table is not a tape but is shown because of interest in its high shielding properties. Figure 6-8 shows a typical corrugated flexible harness shield used in EMP environments.

Table 6-4. Low-frequency magnetic shielding effectiveness of tapes. (Courtesy of Interference Control Technologies.)

MAGNETIC TAPE	RADIUS OF WIRE HARNESS				
	r = 2.5mm	r = 6.4mm	r = 12.7mm	r = 25.4mm	r = 50.8mm
TECKNIT MAGNETIC TAPE	30 dB	21 dB	16 dB	10 dB	4 dB
NETIC, 0.1mm TAPE	32 dB	24 dB	18 dB	12 dB	6 dB
CO-NETIC, 0.1mm TAPE	45 dB	37 dB	31 dB	25 dB	19 dB
CO-NETIC, 0.25mm TAPE	52 dB	44 dB	38 dB	32 dB	26 dB
HYPERNOM, 0.1mm TAPE	46 dB	38 dB	32 dB	26 dB	20 dB
CORRUGATED HYPERNOM 0.2mm	NA	49 dB	43 dB	37 dB	31 dB

Another form of cable wire harness shield is zipper tubing. When fabricated from tin-coated, copper-clad steel, its magnetic shielding properties are similar to the shielding effectiveness listed in Table 6-4 for Tecknit. Zipper tubing distinguishes itself by permitting a rapid installation and removal, and is useful when checking out R & D systems, test arrangements, and quick-fixes in the field. It is otherwise relatively heavy, bulky, and expensive.

Capacitive Coupling Reduction Techniques

Parameters in the crosstalk development that contribute to capacitive coupling are the mutual coupling capacitance, C_{cv}, and the wire to ground capacitances. These are controllable by varying wire spacing and length. Other

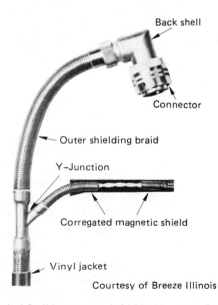

Back shell

Connector

Outer shielding braid

Y–Junction

Corregated magnetic shield

Vinyl jacket

Courtesy of Breeze Illinois

Fig. 6-8. Typical flexible corrugated shields. (*Courtesy of Breeze-Illinois*)

control techniques follow the same rationale developed for the inductive coupling case. The following coupling reduction guidelines are based upon reducing one or more of these parameters. Suggestions for reducing electric field coupling include:

1. Reduce the source voltage, V_s, or the frequency in the culprit circuit.
2. Increase the wire to ground capacitances, C_s and C_v, by running the wires close to a ground plane. This is necessary in many instances to achieve good impedance matching.
3. Separate the circuits as much as possible to reduce the mutual coupling capacitance.
4. Reduce the coupling capacitance by using a shorter length of wire. The capacitance will also be decreased if the wires are not parallel to each other for long runs of wire. The minimum capacitance results when the wires cross over at a 90° angle.
5. Reduce the capacitance by shielding—a Faraday shield can be constructed around either or both of the wires. One popular method of running interconnect lines between PCBs and backplanes utilizes miniature coaxial lines that to some degree, perform as shielded cables.
6. Decrease the circuit impedance in the victim circuit. A decrease in the victim circuit impedance will increase the voltage across the coupling capacitance, leading to a decreased voltage across the victim load impedance. However, this action will increase magnetic field coupling.
7. Crosstalk in digital circuits often is a result of improper impedance matching that causes reflections and ringing on the line; so investigate careful matching of circuit impedances.
8. Use fiber optics to eliminate the problem of capacitive and inductive crosstalk.

Shielded Wires

Figure 6-9 shows (a) the victim and culprit wires end-on. Wire #2 of the culprit circuit now has a metal shield around it with the shield bonded to the ground. The mutual capacitance, C'_{cv}, now exists between wire #1 and the shield of wire #2 (of the victim circuit). A larger capacitance, C_{c2}, exists between wire #2 and its shield. The equivalent circuit is shown in part (b) of the figure. C'_{cv}, C_{c2}, and the shield-to-ground bond impedance, Z_b, essentially form a high-pass T-filter resulting in reduced coupling between wire #1 and wire #2. The low inductance and resistance of this bond are necessary to maintain effective high-frequency isolation. The ratio of $Z_b/X_{C'}$ where X_c is the reactance of the capacitance between the wires results in the voltage-divider-action improvement offered by the Faraday shield.

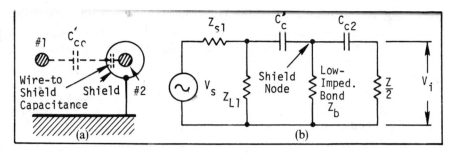

Fig. 6-9. Equivalent circuit of wire shield to reduce electric-field coupling. (*Courtesy of Interference Control Technologies*)

Figure 6-10 shows (a) both wires encircled by a shield and (b) the resultant equivalent circuit. An increase in isolation between the two wires is effected by an increase in high-pass filtering that is highly dependent on the shield-to-ground bond impedances, Z_b. The voltage-dividing action takes place in two stages and yields an overall decoupling improvement of about $Z_b^2/X_{C1}X_{C''}$ or approximately twice the dB of a single-wire shield (Fig. 6-9).

Other Shielded Wire Configurations

Besides the alternatives discussed for magnetically shielding wires, there are other ways to shield closely spaced wires, such as various ribbon cable configurations utilizing ground planes that are built into the ribbon and effectively provide shielding for the signal conductors.

Some shielding benefit can be derived using regular ribbon cable if each signal wire is run alongside a ground or guard wire. The electric flux lines from each signal wire go to ground, rather than from signal line to signal line, providing isolation from crosstalk. This arrangement also lowers the

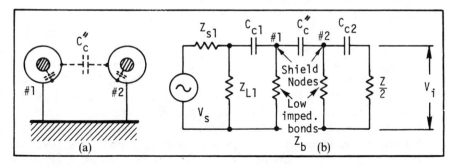

Fig. 6-10. Equivalent circuit with both conductors shielded. (*Courtesy of Interference Control Technologies*)

impedance of the signal lines, which is important in high-speed digital applications (Chapter 17). Less isolation is provided by placing a ground wire for every two signal lines, even less for every three signal lines, and so on.

Multiconductor ribbon cable can also be shielded by running a flexible ground plane under the ribbon cable. Again, the electric flux lines tend to go to ground, increasing the capacitance from the signal lines to ground, lowering the impedance, and resulting in a decrease in crosstalk. Ribbon cable is available that has such a shield built into the cable configuration. Other shield configurations and discussions are found in Chapter 5.

Shield Grounding

The topic of shield grounding is quite complex. For the electric and far-field case, the frequencies of EMI are generally higher than the magnetic field case, and problems result when the length of the shield becomes a significant fraction of a wavelength. Thus, the prediction of the response of a circuit with a shield grounded in a certain manner is very difficult. (13, 18) This difficulty arises because of the variation in circuit impedances, parasitics, and geometry that can occur even in two seemingly identical circuits.

Because of this difficulty in circuit parameter prediction, the following general guidelines will be presented with the warning that, as with many EMI problems, other configurations may perform better, and that the best configuration may be the result of trial and error.

A suitable shield ground configuration may be determined by examining the ground-loop coupling response as presented in Chapter 5. Under certain conditions, the shield ground will influence the GLC response. Under these conditions, it is recommended that both the GLC and the coupling due to the shield connection (called shield coupling) be calculated, and the more severe coupling path be addressed first (16).

The problem associated with shield coupling (SC) is illustrated in Fig. 6-11. The number of possible ground connections is shown, but what is of interest is the two possible shield ground connections, S_1 and S_2. When the shield is grounded at only one end, the CM current, I_s, flows through the shield. Since the common-mode source impedance, Z_s, is much less than that of the shield, Z_{sh}, the magnitude of the common-mode current is not affected by the presence of the shield. The shield impedance constitutes only a drain current path for the common-mode current, and is not in the path of the generator impedance, Z_g, or the load impedance, Z_L. Therefore, the common-mode current has no effect on the undesired load voltage, V_0.

This situation changes when the shield is grounded at both ends.(10) Since the shield has a finite (i.e., non-zero) transfer impedance, defined as the voltage induced on the inside of the shield due to current flowing on the outside of the shield, an EMI voltage develops that can couple onto the signal line.

This voltage utimately affects the victim input. This transfer impedance is defined as:

$$Z_t = \frac{V_l}{I_s} \tag{6-20}$$

where V_l is the coupled voltage drop down the wire line per 1-meter length of wire. This voltage approaches the voltage induced on the inside of the shield above a few kHz due to the shield-to-wire mutual inductance. Also, Z_t is normalized to a 1-meter length of cable, unbalanced circuits, and $Z_g = Z_L$.

The differential EMI voltage V_0 is then calculated from:

$$V_0 = \frac{lV_l Z_L}{(Z_g + Z_L)} C_b \tag{6-21}$$

where l is cable length in meters and C_b is decoupling due to the unbalance of the circuit, in percent. If this parameter is unknown, the following can be taken as default values:

- Ordinary, low-cost shielded twisted pair: 10%
- Good-quality twinax: 3%

If both boxes are grounded, then the full common-mode voltage develops across the shield, and $I_{sh} \simeq V_i/Z_{sh}$. The shield coupling is found as follows:

$$SC = \frac{V_0}{V_i} = \frac{lZ_t Z_L C_b}{lZ_{sh}(Z_g + Z_L)} = \frac{Z_L C_b}{Z_{sh}} \frac{Z_L}{(Z_g + Z_L)} \tag{6-22}$$

In order to determine whether SC is significant, both GLC and SC must be calculated and either the larger of the two is selected, or they are added together coherently (16).

For balanced, shielded transmission lines, the following approximations can be used (based on RG-22/U with 3% unbalance):

$$Z_t = (3 + j2f_{MHz}) \text{ millimho/m} \tag{6-23}$$

$$Z_{sh} = (3 + j3000f_{MHz}) \text{ millimho/m} \tag{6-24}$$

Figure 6-12 shows SC in dB as a function of frequency for three different qualities of cable shield and two different cases of load and generator impedance. For comparison, GLC is also shown for a 10-meter wire length and unbalanced and 1% balanced circuits.

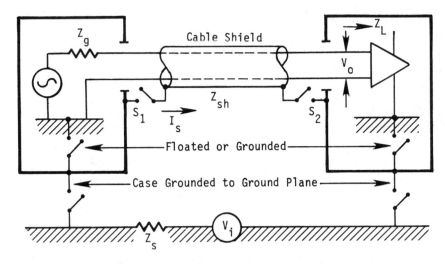

Fig. 6-11. Cable shield interaction and grounding conditions. (*Courtesy of Interference Control Technologies*)

It is evident that SC is significant at lower frequencies, and the grounded shield degrades GLC. Thus, for $l \ll \lambda$, the shield should generally be floated. The values of SC apply only for shielded cables with quality comparable to those of Fig. 6-12. Shields with lower transfer impedance will, of course, yield less shield coupling than shown in Fig. 6-12.

REFERENCES

1. Don White Consultants Engineering Staff. "The Role of Cables and Connectors in the Control of EMI." *EMC Technology*, Volume 1, Number 3, July 1982.
2. Keenan, R. K. *Digital Design for Interference Specification*. Vienna, VA: TKC Publications, 1983.
3. Kraus, J. D. *Electromagnetics*. New York: McGraw-Hill Book Co., 1953.
4. Ott, Henry W. *Noise Reduction Techniques in Electronic Systems*. New York: John Wiley & Sons, 1976.
5. Ott, Henry J. "Ground-A Path for Current to Flow," *EMC Technology*, Volume 2, Number 1, January–March, 1983.
6. Paul, C. R. "On the Superposition of Inductive and Capacitive Coupling in Crosstalk Prediction Models." *IEEE Transactions on Electromagnetic Compatibility*, Volume EMC-24, Number 3, August 1982.
7. Paul, C. R. "Solutions of the Transmission Line Equations for Three Cable Conductor Lines in Homogeneous Media." *IEEE Transactions on Electromagnetic Compatibility*, Volume EMC-20, Number 1, February 1978.
8. Paul, C. R., and Feather A. E. "Computation of the Transmission Line Inductance and Capacitance Matrix from the Generalized Capacitance Matrix." *IEEE Transactions on Electromagnetic Compatibility*, Volume EMC-18, Number 4, November 1976.

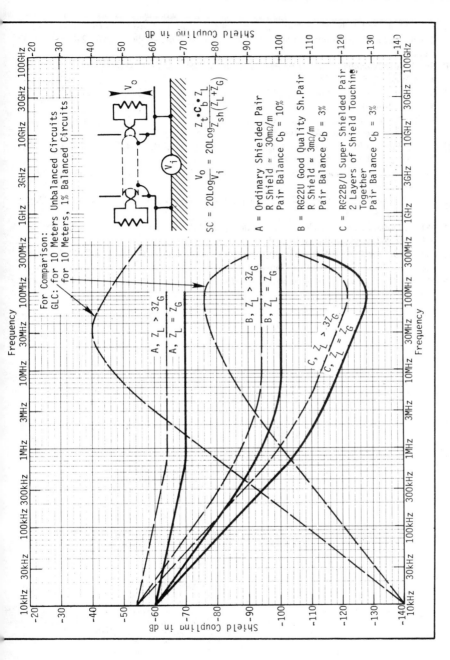

Fig. 6-12. Shield Coupling (SC) vs. ground loop coupling (GLC). (*Courtesy of Interference Control Technologies*).

9. Paul, C. R., and Jolly, M. B. "Susceptibility of Crosstalk in Twisted-Pair Circuits to Line Twist." *IEEE Transactions on Electromagnetic Compatibility*, Volume EMC-24, Number 3, August 1982.

10. Smith, A. A. *Coupling of External Electromagnetic Fields to Transmission Lines*. New York: John Wiley & Sons, 1977.

11. Stutzman, Warren L., and Thiele, Gary A. *Antenna Theory and Design*. New York: John Wiley & Sons, 1981.

12. United States Air Force Systems Command, Design Handbook 1-4, *Electromagnetic Compatibility*, Third Edition. Andrews Air Force Base, MD, 1975.

13. Vance, Edward F. "Cable Grounding For the Control of EMI." *EMC Technology*, Volume 2, Number 1, January–March, 1983.

14. Weeks, Walter L. *Transmission and Distribution of Electrical Energy*, New York: Harper and Row, 1981.

15. Weeks, Walter L. *Electromagnetic Theory for Engineering Applications*. New York: John Wiley & Sons, 1964.

16. White, Donald R. J. *EMI Control Methodology and Procedures*, Third Edition, Gainesville, VA: Don White Consultants, Inc., 1982.

17. White, Donald R. J. *A Handbook on Electromagnetic Shielding Materials and Performance*. Germantown, MD: Don White Consultants, 1975.

18. White, Donald R. J. *Electromagnetic Compatibility Handbook*, Volume 3. Gainesville, VA: Don White Consultants, 1973.

Chapter 7
CONNECTORS

A companion to either a wire cable or a harness, a connector is an assembly of mating contacts that links a cable to or separates it from another cable or other equipment. There may be anywhere from one to several hundred individual wire-pin and/or coaxial sheaths making simultaneous contact via a connector. Individual pin contacts are embedded in insulating material to isolate them from one another within a connector and to prevent their coming in contact with bare hands. While in a linked or engaged position, the connector should provide a low-impedance conducting path for all internal wires and a low-impedance bond whenever an outer shell is used, such as for providing electrical continuity for a cable harness shield (1). Connector design and assembly also determine if, and how, a cable shield is to be grounded.

CONNECTOR CONTACT PROBLEMS

EMI problems associated with connectors are similar to those manifested by mechanical switches. One principal problem is poor contact which may result directly in arcing, or in overheating that eventually may lead to arcing. Poor connector contact also invites driven-circuit voltage variations due to contact impedance modulation of the driving-current source. Common-impedance coupling from outside sources can exist in connector grounding paths. Improperly shielded connectors or poor cable–connector–equipment enclosure contact can invite radiated emission penetration or leakage through resulting apertures. (3)

Contact Impedance

The geometry of the connector may cause EMI because of discontinuity in its structure. For example, shell discontinuities can create unshielded loops and unwanted localized lumped impedances, possibly at sensitive locations in circuits. All connectors usually become limiting factors in circuits operating at high frequencies, especially when homogeneous (solid-wall tubing) cable shields or coaxial cables are used. At RF, manufacturers rate connectors in terms of the resulting voltage standing wave ratio (VSWR) versus frequency, which is a measure of the impedance match between the connector and the cable to which it is attached. While a well-designed connector will not cause interference per se, design deficiencies may result in considerable EMI to other circuits due to either radiation coupling or conductor path variations. The best design combines low spring pressure at the point of conductor contact and good peripheral (360°) electrical contact between the connector backshell and cable shield(s).

In the link position, contacts should maintain a low-impedance bond. Therefore, testing of connectors to detect and isolate high-impedance contacts requires:

- That the applied test voltage be about one-tenth the normal working voltage, so that it does not break down the normal film and tarnish accumulation.
- Use of the highest test frequency to which the driven loads may be susceptible. It is sufficient to conduct DC resistance measurements when the loads are audio and low-frequency.
- That the connector be submitted to low-level, low-impedance test, and then to an environmental test of mechanical forces, wear, and corrosion.

A common cause of faulty connector contacts is damage during mating. Adequate contact floating should be ensured to permit insertion without binding or scoring and to prevent wedging by correct pin layout. Properly placed guide pins will reduce bending, gouging, and abrasion due to misalignment. With properly designed mechanical guides, pin contact alignment occurs without trial-and-error scraping of pins across the female contact to find the alignment position. Protective coverings should extend over the male pins to reduce pin damage. Pin overdesign for an extra low length-to-diameter ratio provides mechanical ruggedness. Inexpensive protective plastic caps should be provided for use during handling and storage to prevent pin damage. Potting the back ends of connectors decreases entry of moisture, fumes, contaminants, and foreign objects. Clamps prevent wires from being pulled and twisted from their contacts. Good contact pressure should be ensured over a long period of time by the use of low-fatigue, high-resilience spring materials, such as beryllium–copper alloy.

Contact Finishes

Low-impedance contact can occur in either of two ways: (1) simple contact under pressure, which breaks or wipes away low or nonconducting film or tarnish when pins are pressed together, or (2) breakdown contact, in which the film on the pins is not ruptured by connector mating. In the latter situation, minute arcing forms channels of molten metal resulting in a low-resistance contact. A capacitive effect can exist if the voltage level between the pins is not sufficient to cause the dielectric oxide film on the contact surfaces to break down. High-resistance contact can be produced by the formation of some highly resistive metal oxides on the surface of the connectors after the contact bond is formed. This can occur whenever the contact pressure is inadequate, resulting in a loose bond that can then be penetrated by moisture or corrosive vapors with the formation of oxides.

Contact plating is used to mitigate the above problems. Plating increases tarnish and corrosion resistance. Gold is ideal for this purpose. Certain hard gold alloy platings are preferred because of their electrical conductivity, corrosion resistance, and wearability. As an underplate for the gold alloy, 0.1 mil of ductile nickel (elongation of not less than 5%) provides the best overall combination from a performance standpoint for a sustained period of time.

Non-hermetically sealed connectors contain a copper-based alloy, while hermetically sealed connectors usually consist of an iron–nickel alloy. The gold underplating for hermetic seals should be copper over the iron–nickel substrate, followed by a plating of nickel. Gold should not be plated directly to the copper plating because it will diffuse into the copper, possibly causing seal leakage and corrosion (1).

The finer the microfinish of the contact mating surfaces, the better the corrosion resistance characteristics and the less the insertion and withdrawal forces. Microfinish thicknesses of 10 micro-inches (0.254 micron) are achievable (1).

Cable Shield/Connector Interface Leakage

Radiation leakage can occur either at the cable shield–connector–equipment enclosure interface due to an improper shield–connector or connector–enclosure bond, or through the connector itself if it is improperly designed. The cable shield and connector should be viewed as a cable system assembly composed of electrically and mechanically compatible elements, rather than as separate entities. Also, the cable system is part of a closed electromagnetic barrier, as illustrated in Fig. 7-1 (2). For effective EMI control, it can be concluded that shielded cable systems should be closed, whether or not they are grounded (2).

A completely closed barrier consists of a closed surface that prevents both radiated and conducted interference from passing through. As illustrated in Fig. 7-1a, a barrier usually consists of a metal (or metallized) equipment enclosure or case; filters on signal and power conductors; wire mesh screens or conductive coatings over apertures in the enclosure for viewing windows for meters, cathode ray tube screens, cooling vents, and so on; and cable shields and connector assemblies (2). Figure 7-1b illustrates a cable shield and connector assembly as an element of a closed barrier against radiated fields from sources external to the barrier, demonstrating the necessity for

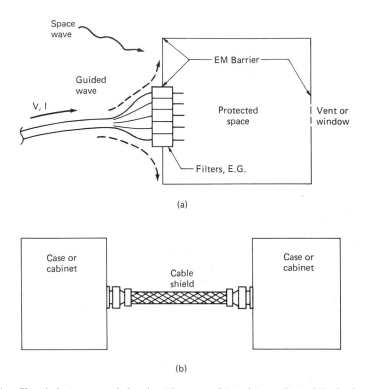

(a)

(b)

Fig. 7-1a. Closed electromagnetic barrier. (*Courtesy of Interference Control Technologies*)
Fig. 7-1b. Cable shield and connector assembly as elements of a closed electromagnetic barrier.

making both the cable shield and the equipment enclosure electrically continuous around both ends of the connector perimeter to provide barrier complete shielding integrity. The cable-shield-to-connector shielding integrity can be obtained by using a connector with a metallic backshell that makes continuous contact around the entire 360° backshell-shield perimeter. The connector-to-enclosure shielding integrity is usually obtained via the two metallic connector sections, one of which is attached to the metallic (or metallized) enclosure and the other to the cable-shield assembly, with the two sections joined together via a male–female thread assembly or an appropriate connector clamp.

CONNECTOR BACKSHELLS

Connector types may be divided into three classes: (1) low-frequency single and twin-conductor connectors, (2) low-frequency multi-pin connectors, and (3) high-frequency unbalanced-line (coaxial, triaxial, and quadrax cable) connectors. There may be a division between low and high frequency anywhere from 100 kHz to 10 MHz. This section discusses multi-pin connectors and the shielding of their outer shells to reduce radiation leakage and penetration. Coaxial cable connectors are discussed in a later section.

Multi-pin connectors generally have an external shield that slips over the harness at the connector and is secured to the conductor termination or mating shell. The backshell, as it is called, serves as a form of cable or harness mechanical retainer and strain distributor and provides a 360° peripheral shielded configuration around the cable or harness assembly at the wire–connector interface. The backshell also serves to terminate (i.e., ground) the cable or harness shield to a compatible connector housing or another mating connector shell assembly. Thus, a good multi-pin connector is one in which the shielding effectiveness of the mated connector equals or exceeds that of an equal length of the interconnecting cable shield.

In an otherwise adequate shield, induced RF currents that are conducted along cable shields may be coupled to the system wiring at the point of improper cable termination, that is, at the cable–connector interface. When a shield is properly terminated, the entire periphery is grounded to a low-impedance reference that minimizes RF potentials at the termination. This method results in a bond superior to that obtainable by the use of epoxy or other synthetic-conducting material.

Figure 7-2 illustrates a permanent termination of the cable shield to a connector. Here, the outer shield is made continuous with the connector backshell by a soldered or metal-formed bond. Spring fingers are used to carry the shell continuity to the mating connector. The illustration also shows the through-path for unshielded individual connectors. When more than one shielded inner conductor must be routed through a single cable and connector, the technique suggested in Fig. 7-2b is employed to preserve individual internal-wiring shielding. The internal coaxial shields should never be pulled back, twisted, and then bonded to the outer connector sheath; that is, no portion of the coaxial shield should be broken before it is bonded to the connector shell. Individual shields for connections that are routed through multi-pin coaxial connectors should be terminated individually in the manner illustrated in Fig. 7-2b.

Figure 7-2 indicates that the cable shield is permanently secured to the connector shell. While offering the best bond, this practice is not particularly cost-effective in manufacturing time. Methods of quick mechanical compression bonding of the cable braid to the shell have been developed by the EMC connector manufacturers. Many such connector varieties permit rapid assembly, require no special tools, are reparable in the field, and permit environmental sealing. They are available in both permeable and non-permeable-base materials to shield against both H and E-fields or E-fields only.

Figure 7-3 illustrates typical adaptors and backshells used for overall cable shield termination, and Figure 7-4 shows adaptors for individual termination of shielded wires using the connector shell as a ground point.

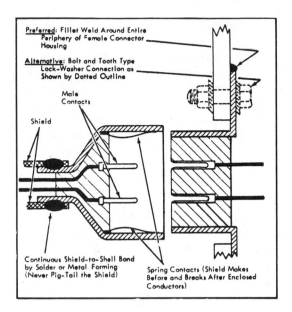

(a) Individual Conductors Are Unshielded

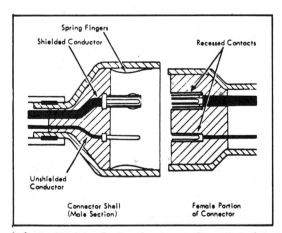

(b) Individual Conductors Are Shielded

Fig. 7-2. Shield termination for electrical connectors. (*Courtesy AFSC Design Handbook, DH* 1-4 *EMC*)

Fig. 7-3. Adaptors and backshells for overall cable shield termination. (*Courtesy of Glenair, Inc.*)

Fig. 7-4. Adaptors for individual termination of shielded wires using connector shell as ground. (*Courtesy of Glenair, Inc.*)

TERMINATION OF INDIVIDUAL WIRE SHIELDS

When a cable harness contains many individual shielded wires, in which each shield acts as a Faraday cage, continuity through the mating connector interface is obtained via an individual pin for each shield. This suggests that for noncoaxial pin connectors twice as many pin-receptacle contacts are necessary for individual shielded-wire cables. To cut down on the number of extra pin contacts necessary for shield continuity, a technique of daisy-chaining is sometimes employed where the harness contains many wires carrying signals from DC up to about 1 MHz. In the daisy-chain technique, a single dedicated pin is not used for the continuity of each individual wire shield in the assembly, but, rather, one pin may carry up to five individual wire shield connections.

The daisy chain practice is to peel back the outer braid of each shielded wire and connect these braids in groups of five by an insulated wire looping from outer shield to outer shield in a daisy-like manner. The final wire from the shield group goes to a separate dedicated feed-thru pin. While this practice compromises Faraday shielding between short lengths of resulting unshielded wires at high frequencies, reliance is made upon the outer cable connector backshell for overall shielding at the cable–connector interface. Cables carrying signals above 1 MHz should not use this practice; in fact, daisy-chaining is considered to be obsolete because multi-coaxial pin connectors are now available to give better performance.

One alternative to the daisy-chain technique is the halo-ring technique, in which individual shielded wires in a harness must have a common shield

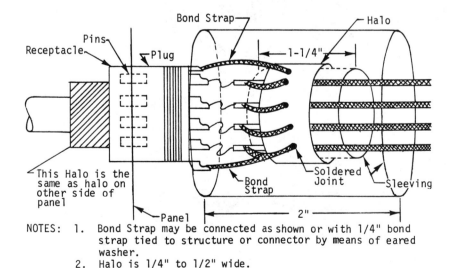

NOTES: 1. Bond Strap may be connected as shown or with 1/4" bond strap tied to structure or connector by means of eared washer.
2. Halo is 1/4" to 1/2" wide.

Fig. 7-5. Bonding ring or halo as connector for terminating shields in a harness.

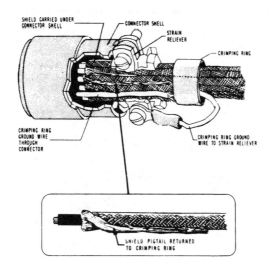

Fig. 7-6. Crimping-ring technique for terminating shields.

ground at the connector. Here, a cylindrical conductor (the halo) is used as shown in Fig. 7-5 to ground all applicable shields through one or more connector pins. Where final termination is to exist at an equipment housing, shield halos should be bonded to the ground plane by 3.8 cm or less of a 6.4-mm- to 12.7-mm-wide, tin-plated, copper strap.

The halo technique is acceptable only when relatively few shielded wires are involved. A preferred method where cost implications become important is to use a collectively crimped peripheral ring, as illustrated in Fig. 7-6, for all wire shields exclusive of those intentionally operated as either individual coaxial cables or low-level audio shielded leads. The collective crimping ring uses two ground wires. Connect one wire from the ring to the connector shell where design permits. The other wire is carried through the connector. Figure 7-8 shows what the resulting outer shield grounding configuration would look like.

The best-performing method to use, but a relatively expensive one to manufacture, is the interlacing-strap method, shown in Fig. 7-7. It is used for a common shield ground in multishielded wires in harnesses that have a large number of individual internal shields. The interlacing strap should be at least 6.4 mm wide by .25 mm thick, and bonded securely to the connector as shown in the figure. The strap follows the wiring system ground shown in Fig. 7-8.

Where multishielded wires are to protect audio-susceptible circuits, they should be grounded at one end only, as shown in Fig. 7-9. Individual twisted-wire-pair shields should each be insulated from other pairs to prevent undesired grounding, and the shield should never be used as a signal return.

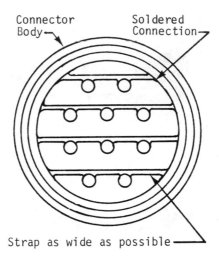

Fig. 7-7. Interlacing technique for terminating shields.

IRIS Concept Adapters

A development in cable shield terminations is the "IRIS Concept." IRIS-shaped devices permit uniform pressure to be concentrically applied to either resilient or rigid surfaces. In addition, the IRIS-shaped devices can be constructed from metals, for RF shielding, bonding, or grounding; and/or from various elastomers for environmental or pressure sealing requirements.

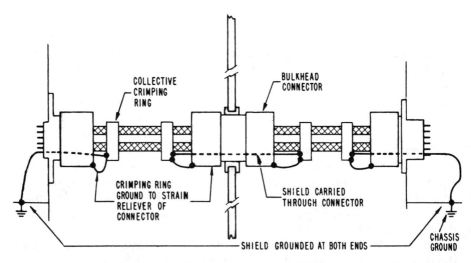

Fig. 7-8. Wiring system termination of shielded wires. (See Fig. 7-6 for details of collective crimping.) (*Courtesy of Interference Control Technologies*)

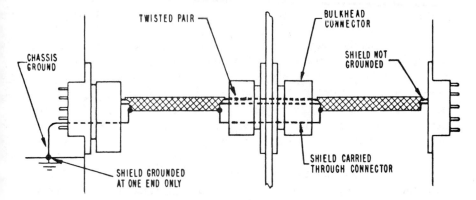

Fig. 7-9. Termination of shielded audio susceptible wiring. (*Courtesy of Interference Control Technologies*)

When pressure is applied to the IRIS, the aperture becomes concentrically smaller. The concentrically applied pressure is transferred by multiple point contacts to any solid, flexible, or resilient cable, tube, or conduit. Surface distortion is minimal, and the pressure can be maintained at a fixed value or varied over a wide range. Figure 7-10 is a sectional view of an IRIS Concept adapter.

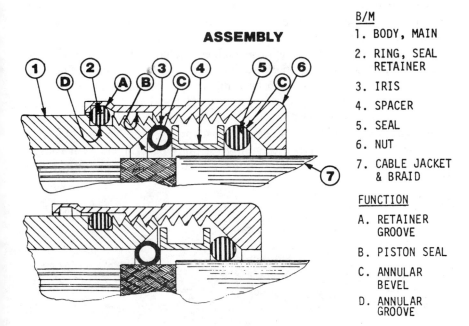

ASSEMBLY

B/M
1. BODY, MAIN
2. RING, SEAL RETAINER
3. IRIS
4. SPACER
5. SEAL
6. NUT
7. CABLE JACKET & BRAID

FUNCTION
A. RETAINER GROOVE
B. PISTON SEAL
C. ANNULAR BEVEL
D. ANNULAR GROOVE

Fig. 7-10. Sectional view of an IRIS concept adapter offering circumferential connection to cable shield. (*Courtesy of Interference Control Technologies*)

FILTER-PIN CONNECTORS

Filters offer significant possibilities for controlling conducted interference. Generally, EMI filters are employed as lumped elements in various portions of circuits and input–output wiring of equipments. In recent years, however, filters have been miniaturized to such small sizes that some can now be built into the cable-pin assembly. Figure 7-11 illustrates one type of miniature multi-pin connector employing π-type filters in each pin. Because of the limitation of the obtainable shunt capacitance and series inductance that can be constructed in the pin, filters of this small type, typically about 3.2 mm × 9.6 mm in size, exhibit little or no attenuation below 1 MHz. Typical attenuation offered by these filter pins in a 50-ohm system is about 20 dB at 10 MHz and up to 80 dB at 100 MHz.

Another filter-pin connector of a somewhat larger body dimension is designed to carry 5 amperes. Thus, for low, DC working voltages, capacitances up to about 1 nf are achievable in the larger pins. Figure 7-12 shows some typical dimensional data of these connectors and associated insertion loss versus frequency. Many of these filters exhibit cutoff frequencies of the order of 100 kHz when measured in a 50-ohm system per MIL-STD-220A.

Filter pins are available for use in a variety of connector shapes including the miniature and subminiature rectangular connector. Three types of contacts are available for each contact position: filter contacts, power contacts, and grounded contacts. The contacts can be intermixed in any arrangement for maximum circuit flexibility. Not only do filter connectors have the same layout pattern and contact spacing as their equivalent nonfiltered connectors, but they are also intermatable and intermountable with them. The basic difference between a filtered and an unfiltered connector is that the filtered connector is approximately 6 mm longer or deeper than its unfiltered counterpart. The filter pin connectors have a large capacitance to the filter shell or ground, and that can be a serious disadvantage or even preclude their use. The capacitance varies, depending on the type of filter, from roughly 500 pf to as much as 150 nf. The designer of circuits requiring critical wave

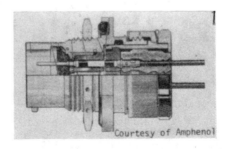

Fig. 7-11. Typical miniature multi-pin filter connector. (*Courtesy of Amphenol*)

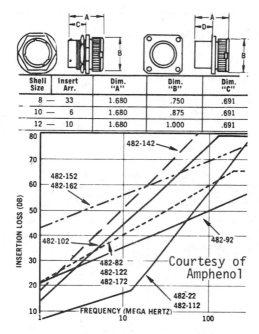

Fig. 7-12. EMI filtering connectors and insertion loss.

shapes should be well aware of the magnitude of the capacitance to ground that these filter pins present to the circuit.

Recent developments in filter pin connectors have pushed the whole frequency attenuation down considerably while still maintaining relatively small sizes. It is now possible to obtain a 20 dB attenuation at 1 MHz and 60 dB attenuation at 300 MHz in subminiature rectangular connectors. The cost for this low-frequency filtering, however, is substantially higher than with the normal or regular filter pins. Figure 7-13 illustrates typical attenuation

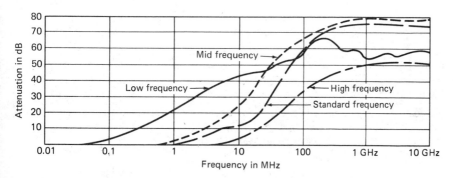

Fig. 7-13. Typical performance of filter pins. (*Courtesy of Interference Control Technologies*)

Table 7-1. Coaxial connector selection chart.*

Characteristics	SMA	SSMA	SMB	SMC	TNC	BNC	SNC	SCC	UHF	HN	LT
COUPLING Threaded	X	X		X	X		X		X	X	X
Bayonet						X		X			
Quick Disconnect			X								
CABLE ATTACHMENT Crimp	X	X	X	X	X	X	X	X	X	X	
Clamp	X		X	X	X	X	X	X	X	X	X
Solder	X	X	X	X	X				X		
IMPEDANCE 50 Ohm	X	X	X	X	X	X	X	X		X	X
70 Ohm			X	X	X	X	X				
FINISH (BODY)											
Passivated Stainless Steel	X	X	X	X	X		X				
Gold Plated	X	X	X	X							
Silver Plated				X	X	X	X	X	X	X	X
Nickel Plated				X	X	X	X	X	X	X	X
SIZE 0.320" Maximum		X	X	X							
0.400" Maximum	X										
0.625" Maximum					X	X					
0.827" Maximum							X	X	X	X	
1.500" Maximum											X
VOLTAGE RATING 250 VRMS		X	X	X							
500 VRMS	X				X	X	X				
1500 VRMS								X	X	X	
5000 VRMS											X
SEALING Moisture	X	X		X	X	X	X	X		X	X
MAXIMUM OPERATING FREQUENCY											
0.3 GHz									X		
2.5 GHz										X	
4.0 GHz			X			X					X
10.0 GHz				X							
11.0 GHz					X		X	X			
12.4 GHz	X	X									
15.0 GHz						X					
18.0 GHz	X	X									
VSWR AT MAXIMUM OPERATING FREQUENCY											
1.25:1		X					X	X			
1.30:1		X			X	X	X	X			
1.35:1								X	X	X	X
1.40:1	X										
1.50:1			X								
1.60:1				X							
1.70:1				X							
QPL MIL-C-39012	X		X	X	X	X	X	X	X		
BODY MATERIAL											
Beryllium copper	X	X									
Stainless Steel	X	X			X		X				
Brass			X	X	X	X	X	X	X	X	X
Aluminum							X			X	

*Courtesy of Electronic Products Magazine, May 15, 1972; pp. 148.

254

of filter pin connectors as a function of frequency for each of the four types of filter pins.

COAXIAL CONNECTORS

For applications above 10 kHz, and more typically above 10 MHz, connectors are of the unbalanced-line, coaxial type in order to mate with coaxial cables. Crosstalk between wires at high frequencies is due to electric-field coupling. To provide both a return-wire path and a shield at and above HF, coaxial lines and connectors are used. Some balanced, parallel lines are used at HF/VHF. Connectors of this type maintain a 360° low-impedance integrity of the outer shield through the connector interface to the mating connector assembly. A low-impedance shield is extremely important because this impedance exists in the return-wire path of the associated coaxial cable. Thus, the outer cable shield impedance of the termination is of paramount consideration in the performance of coaxial connectors.

One significant EMI problem of coaxial connectors is the impedance mismatch in a 50-ohm, 72-ohm, or other cable characteristic impedance. Impedance mismatch is rated in terms of maximum voltage standing wave ratio (VSWR) versus frequency. The maximum signal amplitude variation as a function of VSWR is the amplitude of the VSWR per se. For example, depending upon the length of cable, a connector rated with a VSWR of 2:1 at a particular frequency could exhibit at 6 dB peak-to-peak variation in signal or EMI amplitude. Thus, connector VSWR becomes very important, especially at frequencies for which an associated cable length approaches or exceeds 1/8 wavelength.

Specifying the correct coaxial connector for a specific requirement will simultaneously accomplish most EMI considerations. To do this, Table 7-1 may be used. To use the table, work down the characteristics in the left-hand column, noting the connector series that have the desired characteristics and progressively eliminating those series that do not. When the checklist is completed, the selection will have been narrowed down to types that will fulfill the requirements (1).

REFERENCES

1. Don White Engineering Staff. "The Role of Cables and Connectors in the Control of EMI." *EMC Technology*, Volume 1, Number 1, July 1982.
2. Vance, E. F. "Cable Grounding for the Control of EMI." *EMC Technology*, Volume 2, Number 1, January 1983.
3. White, Donald R.J. *Electromagnetic Compatibility Handbook*, Volume 3. Gainesville, VA: Don White Consultants, 1973.

Chapter 8
GROUNDING AND BONDING
FOR SAFETY CONTROL

The subject of grounding has appeared several times in the preceding chapters. For example, in Chapter 5, on intrasystem EMI prediction and analysis, grounding had a major role relative to both common-mode coupling and ground-loop coupling. Chapter 6 covered the topics of cables and harnesses and the part played by the shield, including its grounding. Chapter 7 involved connectors, and showed that the shielding effectiveness of the connector was often predicted on the impedance of its bond (ground) to the frame. There are individual grounding requirements for buildings, shielded enclosures, equipments, and filters. Grounding appears deceptively straightforward, but it is one of the least understood EMC subjects. Improper grounding is a major contributor to intrasystem EMI problems (1, 3, 5).

There are two reasons for grounding devices, cables, equipments, racks, and systems. The first reason is both to prevent a shock hazard, in the event that wiring or component insulation within an equipment frame or housing develops accidental breakdown, and to protect against lightning damage. The second reason for grounding is to reduce EMI due to either electric field or magnetic flux coupling, as well as common impedance coupling. The EMI reasons for grounding are covered in earlier chapters and in Chapter 9, on bonding and corrosion control. This chapter deals with shock and lightning hazards.

SHOCK HAZARD CONTROL

AC power distribution in private homes, buildings, hospitals, and industrial sites in the United States is governed by local and national codes. The National Fire Protection Association (NFPA) publishes and issues the National Electrical Code, which provides standards on wiring and other electrical devices (4). One requirement is shown in Fig. 8-1, which illustrates control of the shock hazard to personnel and equipment, as well as the fire threat, should a pole transformer develop a fault.

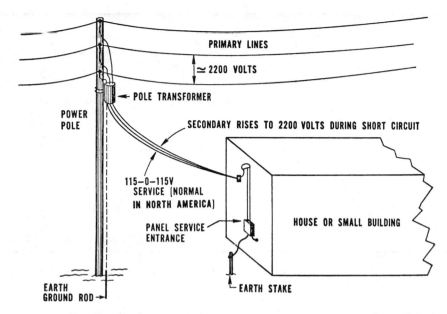

Fig. 8-1. Shock hazard to personnel and insulation fire threat when transformer shorts. (*Courtesy of Interference Control Technologies*)

In Fig. 8-1, a 2200-volt distribution line is shown. To provide service to the small building or house, the power pole transformer reduces the primary voltage to the 115/230-volt service. The standard practice is to ground the neutral at the secondary of the transformer by a conductor running directly down the power pole to the earth connection, as shown. It is also observed that the three-wire service comes down from the masthead to the panel service entrance. There, the neutral is grounded in accordance with the National Electrical Code requirements. This, then, provides redundant grounding of the neutral (viz., back at the transformer pole and at the service entrance panel). The reason for the redundancy is to give protection against the eventuality that either of the earths becomes accidently removed. In essence, then, if the primary 2200 volts were to be accidently shorted over to the secondary neutral, this practice would prevent the neutral from rising to approximately 2200 volts. The higher voltage, of course, could burn out insulation, create fire hazards, and develop a far more serious shock hazard. Thus, earthing is required of the neutral and panel service entrance to reduce these hazards.

Macroshock Hazard and Control

Figure 8-2 illustrates a typical 115/230-volt service at the point of panel service entrance. A safety wire, sometimes referred to as the green wire, is also established at the point of neutral earthing. Thus, this system requires a

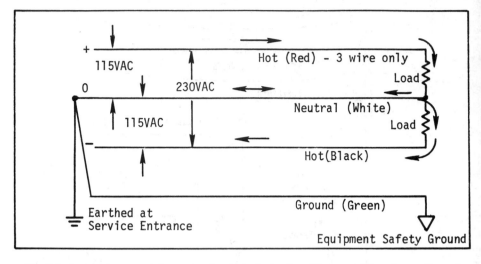

Fig. 8-2. Standard two- and three-wire electrical wire loading. (*Courtesy of Interference Control Technologies*)

three-wire distribution for each circuit. One wire is made up of the 115-volt hot line, the second wire is the white neutral, and the third is the green safety wire. Also, for 230-volt AC services, the two high sides, which are 180° out of phase, are distributed together with the safety wire. Theoretically, no return circuit current should pass through the safety wire. If the high side were accidently short-circuited to the metallic equipment frame, the frame could rise 115 volts above ground reference. If someone were touching that frame, as shown in Fig. 8-3, an alternate path of current would flow through the individual's hand. It would continue through the body and out of either (1) the other hand, if it were touching some other reference such as ground, or (2) the soles of shoes touching a grounded floor and thence to the building ground. For this condition, if the 50/60 Hz current is approximately 2 mA, it represents the threshold of perception in most individuals. Between 10 and 20 mA, a condition of "lock-on" or "no-let-go" could exist, in which the person could be gradually electrocuted. If the current reached values of the order of 75 mA, flowing through the body including the chest cavity, it could prove fatal by developing heart fibrillation. In essence, this condition is a heart attack in which the synchronism of the heart pump and valves is out of phase and loads down the heart, causing it to stop. Resuscitation must be initiated immediately (2).

The foregoing situation, as shown in Fig. 8-3a, is referred to as macroshock hazard. This can be avoided by grounding with a third wire (viz., the green wire to the equipment frame) through a modern AC power cord as shown in Fig. 8-3b Here, if a short circuit were to develop, the green wire would carry the fault current, and the circuit breaker would be tripped to protect

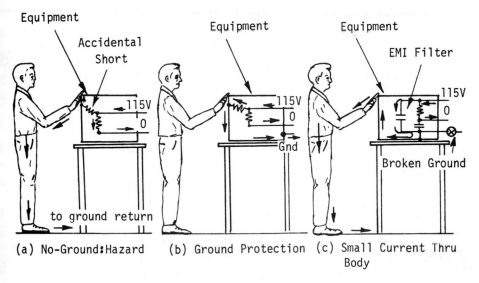

Equipment

Accidental
Short

115V

0

to ground return

(a) No-Ground:Hazard

Equipment

115V

0

Gnd

(b) Ground Protection

Equipment

EMI Filter

115V

0

Broken Ground

(c) Small Current Thru
Body

Fig. 8-3. Safety and shock hazards with/without equipment grounds and EMI filters. (*Courtesy of Interference Control Technologies*)

the user of the equipment. If the fault current were insufficient to trip the breaker, the frame would no longer be at 115 volts because the fault current would be returned through the green wire. Therefore, if a user were to touch the equipment frame as shown in the illustration, there would be no shock hazard as long as the integrity of the safety wire return was preserved.

To help reduce either development of or susceptibility to EMI in equipment of all types, it is common practice today to use EMI filters in the power line at the point of entry to the equipment, as illustrated in Fig. 8-3c. Here, either capacitors or multi-element filters are placed from both hot and neutral lines to the green safety wire to bypass common-mode EMI. The National Electrical Code previously limited the use of such capacitors to 0.1 µf at 60 Hz, corresponding to a leakage reactance current of 5 mA. If an individual touched an equipment frame and the frame were not grounded, the maximum current through the body would be limited to 5 mA. Increasingly, many such filters are being used in different equipment. If a number of devices with filters, each having the potential of putting 5 mA of reactive current back on the line, were all connected to a single (safety wire) line, a shock hazard could develop due to current buildup. Consequently, in more recent years, the electrical codes have limited the maximum capacitance to 0.01 µf so that the leakage reactive current would not exceed 0.5 mA per equipment. It should be noted that the green safety wire carries all of the conducted EMI that is returned to the power line by the equipment, thereby generating essentially common-mode inteference.

Microshock Hazard and Control

Another kind of shock hazard exists when several devices with EMI filters are operated from the same circuit, or a single device develops a short to the equipment frame. Both situations involve potential microshock hazard to patients in hospitals, clinics, and medical centers when catheters via elec-

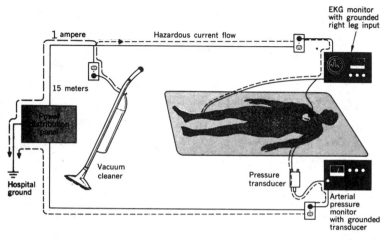

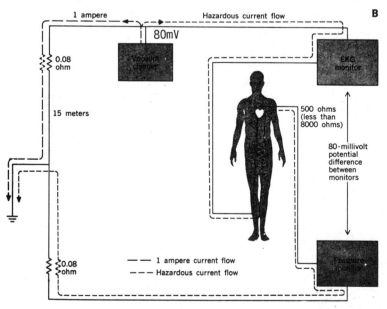

A—Case in whi a vacuum cleaner is plugg into a wall outlet on sa circuit as the EKG monit The cleaner has a thr wire power cord, with third wire grounding outer case. But the windir of the motor are expose dust (often damp), wh provides a good path for eventual "winding-to-out case" short circuit. Becau this kind of short circu makes the case rise to line voltage, the case grounded to protect operator. In this examp the vacuum cleaner has completely failed, but I developed a fault suffici to permit 1 ampere to f down the ground wire a back to the power distri tion panel. B—In this ana sis of the incident, if power distribution pane 15 meters distant and power wiring is 12 gauge, 15 meters of ground w have 0.08 ohm of resistan The hazard here is that faulty appliance caused ference in ground poten between two devices a allows a possibly lethal c rent to flow through patient.

Fig. 8-4. Microshock hazard in hospitals. (*Courtesy of IEEE Spectrum*)

trodes from EKG, arterial pressure monitors, and similar biophysical instruments are in direct contact with the heart. In such a situation, the level of hazardous current can be of the order of a few microamperes.

When a high impedance leakage or direct short develops in a piece of equipment, such as a vacuum cleaner sharing a common green safety ground with the medical instruments, a substantial current may flow in the ground wire. As Fig. 8-4 illustrates, this fault current will divide, with most of the current returning directly to the power distribution panel via the safety ground return. However, some current follows another path directly through the patient, as shown. Should this current reach values of the order of 50 μA, it could be fatal to the patient. Even if the level of current were not fatal, it could disrupt the indications of the sensitive measuring equipment.

A number of precautions can be taken to prevent the kind of problem illustrated. A primary measure involves monitoring the fault current in the safety ground. The objective is to ensure that a high rather than a low impedance exists. If the level of current should increase above some predetermined value, say 1 mA, then an alarm or monitor system could be designed and installed to indicate the presence of a shock hazard whenever this value was exceeded.

Another measure to prevent microshock hazard is to prohibit the use of common-mode filters between either the high side or neutral and the safety line. Differential-mode filters can be installed between the high and low sides of the power line with no direct connection to the green safety wire. Other protective measures include the use of fiber optics and optical isolators to provide a very high impedance path in all critical monitoring leads, including catheter lines. Power, where necessary, is developed at high frequencies by switching regulators to permit the use of isolation transformers for developing high impedance paths. Also, if it is at all possible, a dedicated branch circuit should be specified to provide power to sensitive equipment, with no other devices connected to this circuit.

Ground fault interrupters (GFIs) are also required by the National Electrical Code to protect circuits in certain shock high-risk areas, such as circuits with outside outlets at ground level. A GFI is designed to open the branch circuit if the current difference between the supply line and return path exceeds 5 ± 1 mA (4).

LIGHTNING HAZARD CONTROL

Another aspect of grounding involves protection against lightning hazards to buildings, their contents, and personnel. In one sense, lightning may be regarded as a greater threat than other types of electrical shock. This is so primarily because of the energy levels discharged during a stroke. A typical lightning stroke may have a path length of about 3 km and carry about

Table 8-1. Lightning parameters. (Courtesy of Interference Control Technologies.)

Parameter	90%	50%	10%	Maximum observed	Number of observations
Crest current	2 to 8 kA	10 to 25 kA	40 to 60 kA	230 kA	4,150
Current rate of rise per pulse (10 to 90% crest value)	2 kA/µs	8 kA/µs	25 kA/µs	50 kA/µs	40
Total stroke duration	0.01 to 0.1 s	0.1 to 0.3 s	0.5 to 0.7 s	1.5 s	100
Duration of a single pulse in a stroke	0.1 to 0.6 ms	0.5 to 3.0 ms	20 to 100 ms	400 ms	150
Time interval between end of one pulse and start of next pulse	5 to 10 ms	30 to 40 ms	80 to 130 ms	500 ms	525
Time between start of pulse and half crest value on decay side	10 to 25 µs	28 to 42 µs	52 to 100 µs	120 µs plus	425
Time to crest for a single pulse	0.3 to 2 µs	1 to 4 µs	5 to 7 µs	10 µs	45
Number of pulses in an individual stroke	1 to 2	2 to 4	5 to 11	34	500
Time for atmosphere to recharge after a stroke so that another stroke will be produced	—	20 s	—	—	—

30,000 amperes. The stroke duration corresponding to the 50% height is about 30 microseconds. Thus the power discharged during the stroke can achieve values of approximately 10^{13} watts or an enery of about 300 megajoules. Table 8-1 provides a statistical distribution of lightning parameters (2).

Lightning energy corresponds to several orders of magnitude above typical burnout levels of most electrical and electronic circuitry, with solid-state circuitry burnout levels being as low as 1 microjoule. It is seen, then, that great efforts are necessary to protect electrical and electronic equipment from catastrophic burnout. To a lesser extent, measures must be taken to protect buildings against damage. Consequently, one of the main precautions involves actions to properly ground structures, that is, to divert as much of the lightning stroke energy as possible to the earth's energy sink via proper grounding techniques.

Earth Potential Gradients

Figure 8-5 illustrates one possible situation, involving lightning burnout of electronic equipment in a shack located near the foot of a tower carrying microwave antennas. Here, a lightning stroke of 20,000 amperes is assumed to strike the top of the tower. This current goes down the tower structure,

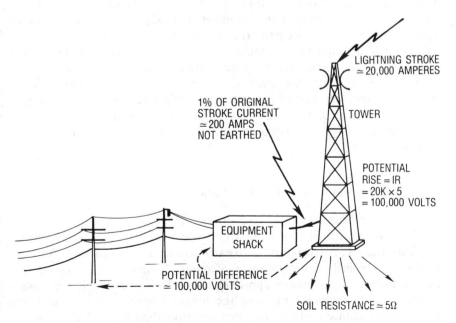

Fig. 8-5. Lightning burnout to communications equipment. (*Courtesy of Interference Control Technologies*)

which is well sinked (grounded) to earth. In the example shown, it is presumed that the impedance from the base of the tower to a local reference point in the earth achieves a value of 5 ohms. Under this condition, then, during the lightning stroke, 20,000 amperes would flow throughout this 5 ohms to create a potential difference of 100,000 volts between the base of the tower and the local earth near the shack.

This potential difference suggests the possibility that the tower potential referenced to a neutral earthed at the left end of the shack would be 100,000 volts relative to cables coming in from the tower at the right end of the shack. This high voltage could, if not properly controlled, burn out much of the electrical and electronic equipment located inside the shack. While it will be shown later that lightning arresters and surge supressors are heavily used to limit this voltage rise, grounding still is one of the ultimate means of dissipating the lightning-stroke energy. The arresters and suppressors provide low impedance paths for the lightning-generated currents to be diverted to ground.

Umbrella Effect

Major measures available for the control of lightning and protection of structures located near a strike are: (1) to design a system that will direct the stroke (with a high degree of probability) to an object that is designed to absorb the energy or divert it to ground and is unlikely to be damaged in the process; (2) to provide a low impedance path from the termination of the lightning stroke (i.e., point of attachment) to ground; and (3) to assure that the impedance of the conductor termination to ground will be as low as possible.

Figure 8-6 illustrates the areas of influence resulting from the height of the highest structure relative either to other structures or to nearby terrain. In the illustrations shown, statistically 99% of the lightning strokes in an area equal to $9\pi h^2$ should strike the highest structure of height h above the nearby terrain (2).

Low-Impedance Down-Conductors

The next consideration, then, is to provide a low impedance path (down-conductor) from the highest point where lightning is likely to strike down to the location where connections to earth will be made. Because a lightning stroke has a typical pulse duration (τ) of about 30 microseconds, and because the rise-time (τ_r) of the pulse is approximately 0.5 microsecond, it is necessary that the down-conductor carrying the lightning stroke current offer low values of impedance up to a frequency corresponding to a value of $1/\pi\tau_r$ or about 637 kHz. To offer such a low impedance path, it is necessary that the conductors either have a very large cross section or provide a large peripheral

STROKES TO A STRUCTURE ON LEVEL TERRAIN — AREA OF INFLUENCE OF MAST

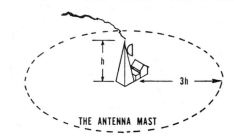

STROKES TO A STRUCTURE ON A PROMINENT
ELEVATION — EFFECTIVE AREA OF INFLUENCE OF MAST

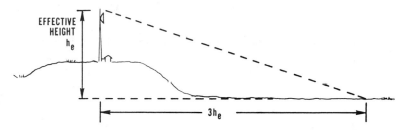

Fig. 8-6. Areas of influence of structures on level terrain and on prominent elevations. (*Courtesy of Interference Control Technologies*)

dimension relative to the cross-sectional area. This means, for example, that heavy round cables are less desirable than flat pieces of sheet metal or straps because of the greater inductance per unit length. This topic is covered in greater detail in Chapter 9.

Even though the preceding discussion makes it apparent that the reactance—not the resistance—of the grounding conductors may be the limiting factor in obtaining a low-impedance path to ground, nearly all the classical literature deals with grounding-grid and grounding-rod resistances. This topic is covered in detail later.

Conductor–Earth Interface Requirements

The resistivity of earth varies with many soil types, temperature, moisture content, and other factors. Perhaps an average soil might be defined as one having a resistivity on the order of 50 ohm-meters. The resistivity of a copper conductor, on the other hand, is 1.7×10^{-8} ohm-meter. Thus, average soil is about 3×10^9 poorer in conductivity than copper. While the earth, taken as a large mass, is a good sink to dissipate the energy of the lightning stroke, nearly all the significant impedance in the path of a lightning stroke is that due to the conductor–earth interface impedance.

The problem under consideration is illustrated in Fig. 8-7. Figure 8-7a shows a cylindrical conductor of radius a and height h buried in the earth. A current of I amperes flows into the top of the cylindrical conductor and then is dissipated into the earth, flowing radially outward through the cylinder surface and into the surrounding earth. The radial current flow pattern is illustrated in the top view of the conductor shown in Fig. 8-7b. If the current flowing through the bottom of the cylinder is negligible ($h \gg a$), then

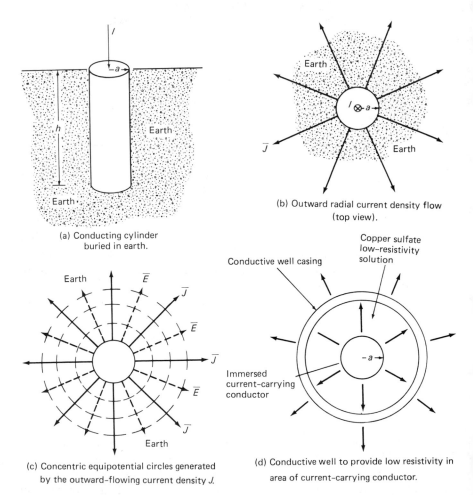

(a) Conducting cylinder buried in earth.

(b) Outward radial current density flow (top view).

(c) Concentric equipotential circles generated by the outward-flowing current density J.

(d) Conductive well to provide low resistivity in area of current-carrying conductor.

Fig. 8-7. Potential gradient developed in homogeneous earth surrounding a buried cylindrical current-carrying conductor.

the current density at the cylinder surface can be approximated by:

$$J = \frac{I}{2\pi ah} \quad \text{amperes/m}^2 \tag{8-1}$$

where $2\pi ah$ is the cylindrical surface area. Assuming that the earth in the vicinity of the conductor is homogeneous and of resistivity ρ, then the radially directed electric field produced in the earth at a radial distance r ($r \geq a$) from the cylinder axis is approximated by the point relationship between the current density J and electric field E in a resistive medium:

$$E = J\rho = \frac{I\rho}{2\pi rh} \quad \text{volts/m} \tag{8-2}$$

Equation (8-2) is illustrated in Fig. 8-7c. If the current I is the result of a "constant-current" lightning discharge, then equation (8-2) is an expression for the potential gradient produced in the earth, and it is a maximum at $r = a$, where the current density is a maximum:

$$E_{\text{max}} = \frac{I\rho}{2\pi ah} \quad \text{volts/m} \tag{8-3}$$

The objective is to minimize the potential gradient, or electric field E_{max}, given by equation (8-3). This can be accomplished by making the conductor radius a and height h as large as practically possible, which essentially increases the conductor–earth contact surface area; or by reducing the resistivity ρ of the earth.

Figure 8-7d provides an illustration of a ground well designed to effectively increase the conductor–earth contact surface area or decrease the earth resistivity in the conductor vicinity. The conductor in the ground well is immersed in a watertight conductive casing containing a copper sulfate water solution. The water solution resistivity as much lower than that of the surrounding earth, thus lowering the overall maximum electric field or potential gradient in the region of maximum current density at the conductor surface. Typically, the ratio of the well casing diameter to that of the ground rod is about 10:1.

Another approach to increasing the effective conductor–earth interface area is to immerse the conductors in concrete that is in contact with the earth. Although concrete is a poor conductor relative to copper, it is a good conductor relative to some soil types. Figure 8-8 shows one scheme for attaining this in which the ratio of the areas of the concrete footing per unit length to that of the conductor may now be on the order of 100 to 1. This results in a substantial reduction of the conductor–earth interface impedance over typical well casing situations discussed above.

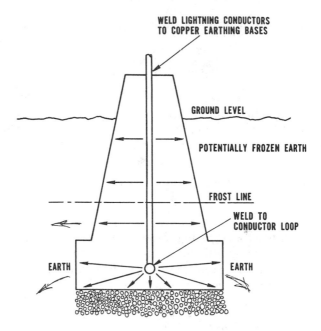

Fig. 8-8. Conductor immersed in concrete footing. (*Courtesy of Interference Control Technologies*)

The relative effectiveness of the scheme shown in Fig. 8-8 will depend upon the depth of the frozen earth. The resistivity of frozen earth is considerably higher than that of earth at a temperature above the freezing level. It is thus desirable to assure that there is an appropriate conductor–earth contact area below the freezing level. Other factors affecting the scheme shown in Fig. 8-8 include the aggregate content of the concrete (stone is a poor conductor relative to concrete) and compactness of the backfilled soil around the concrete (2).

Another scheme for reducing the earthing resistance of lightning down-conductors is to weld the down-conductors to the reinforcing bars (rebars) of the concrete foundation. The rebars are welded to each other after they have been put in place. This is called a concrete-encased rebar grounding system or Ufer ground, and it has been found to perform much better than ordinary ground-rod and ground-ring systems in areas of poor soil conductivity.

STRUCTURAL GROUNDING

The preceding section discussed two non-EMC reasons for grounding: (1) to prevent a shock hazard to personnel should an electrical equipment frame or housing develop a dangerous voltage to frame due to accidental breakdown (short-circuit) of wiring or components; and (2) to protect a building and its contents from lightning-stroke damage by providing a very low impedance path from the point of lightning contact to earth. Chapter 5 discussed an EMC reason for grounding, namely, to provide a reference voltage for all electrical and electronic systems and equipment in order to avoid a shift in operating voltage levels and to prevent circulating ground-current loops resulting in common-mode impedance coupling.

A building ground of the neutral and safety (green) wire at the service entrance may exhibit an impedance to earth as low as 0.5 ohm. During an electrical storm, a typical lightning stroke from a charged-cloud source, having a potential to ground of about 100 megavolts, may result in a current of about 30,000 amperes. Assuming a negligible impedance for the lightning arrester and building-girder structure, the neutral ground point then is suddenly elevated to 15,000 volts (30,000 amps × 0.5 ohm earth-ground impedance). Were it not for the control of this ground impedance and the line-to-neutral surge protectors, many electrical and electronic equipments connected to the phase-(black)-line-to-neutral would be burned out whenever struck by lightning.

The next section reviews the topic of structural grounding with regard to earth grounds and soil impedance, ground rods and grids, structure frame, power grounds, safety and instrument grounds, and corrosion control in grounding and bonding.

Earth Ground and Soil Impedance

The installation and grounding of power mains servicing a facility in the United States are governed by local and national codes. The National Fire Protection Association (NFPA) National Electrical Code (4) provides standards for wiring and other electrical devices.

Another aspect of the safety problem involves providing a low-impedance path to earth in order to protect personnel and equipment during lightning strokes in an electrical storm. When an ionized column of lightning strikes a building, it will most likely seek out higher-elevation, sharp surfaces, and the current will follow low-impedance paths to earth. This means that a building's steel structure and/or internal wiring become natural paths for high lightning current to follow. If the steel structure were not well earthed, the majority of the stroke current could follow the building wire path. In this case, the wiring could burn out, as could connected equipment, input filters, transformers, motors, and so on. Therefore, the object here is to earth the building structure better than the AC power ground, and to connect the structural girders to lightning air terminals (lightning rods) on top of the building so that lightning surges are diverted away from the building wiring and equipment.

Another aspect of the earth-ground situation is to recognize that no local earth region has zero impedance between any two points. Hence, circulating ground currents produce potential drops between these points, which can result in common-mode impedance coupling (see Chapter 5). Apart from the earth impedance problem discussed in a later section, a non-zero potential earth ground, which should be avoided as a reference for EMI purposes, results from earth pollution due to both AC and DC ground currents. Two examples are (2):

1. Gas distribution lines that are kept at -0.8 volt with respect to a reference electrode buried in earth. While modern gas lines are insulated in earth and at the meter services, occasionally current leakage into earth occurs, due to leaks in the insulation called "holidays."
2. AC power distribution systems, where a significant portion of the 60 Hz unbalanced current is returned through earth. Ground currents on the order of 1000 amperes have been measured at substation locations. Two ground rods driven into the earth 50 cm apart can produce a voltage sufficient to provide power for 100-watt bulbs connected to the ground rods. While a typical building ground is not located at a substation, considerably smaller but significant unbalanced currents can flow through the local earth to cross-modulate 60 Hz and associated RF contamination upon the EMI grounding complex.

The above serves to illustrate that lightning and power-safety earth grounds can conflict with EMI control reference earth grounds unless a special effort is made to assure that no significant common-impedance, earth-ground coupling path exists between the two ground systems; that is, that a true single-point ground exists.

Ground Rods and Grids for Power and Safety Grounds

This section discusses techniques to establish earth grounds and potential reference plane systems that are compatible with requirements of various structure types and use. Sample problems are presented to illustrate the use of grounding criteria. Complementary techniques, necessary for the effective implementation of ground planes, are described in detail. The following terms are used in subsequent discussions (5):

- *Ground rods*: rods constructed of highly conductive metals that are driven into the earth and bonded to metallic or structural masses above ground to reduce the development of potentials that may prove hazardous to personnel and equipment.

- *Earth-ground grid meshes*: meshes constructed of highly conductive materials that are bonded together at all junctions, installed at or below the earth's surface, and bonded to metallic or structural masses above the earth to complement ground rods.
- *Reference-plane, ground-grid mesh*: highly conductive mesh construction above earth grade level that serves the primary purpose of providing a low-impedance, zero-potential, signal reference plane for grounding electromagnetic shields and electronic equipment.

Earth-grounding systems and reference-plane systems serve two distinct functions and therefore are presented separately. Earth-grounding systems are not ordinarily sufficient voltage reference planes because of the relatively high earth resistance obtainable between ground rods and between two points within earth-grid meshes. Good results are obtained from reference planes by connecting such planes at a single point to earth ground. Thus, undesirable earth-loop currents are isolated from the reference plane by this single-point ground practice.

The underground water table and buried pipelines have historically proved to be excellent media for connecting structural steel to earth ground because they provide a large surface area exposed to the earth. It hass been standard practice for many years to bond above-ground metallic structures to water and gas pipes by means of copper bonds or copper ground rods. However, copper, in its various forms, creates an undesirable coupling of dissimilar metals. When in contact with iron or steel, copper acts as a cathode to accelerate corrosion of the less noble metal (see Chapter 9). Corrosion between a copper bond and the less noble metal can increase the bond impedance to such an extent that it may be essentially ineffective.

The corrosion of underground and above-ground piping systems results in costly maintainance; so water utility and other companies are resorting to coated pipes or nonconductive pipes and couplings, which will eliminate this traditional, widely used method of grounding. This presents a significant problem in establishing effective, economical grounds, as it is becoming necessary to install dedicated grounding systems for nearly all new buildings.

Ground Rods and Earth-Ground Grid Meshes. Because of the increasing need for dedicated grounding systems, the use of ground rods and meshes has become necessary. The following discussions apply where water and gas pipes are either inaccessible or prohibited for use as grounds; that is, where ground rods and meshes are to be used in lieu of utility pipes to satisfy National Electrical Code requirements. [Refer to the current edition of the National Electrical Code (4), section 250-81.]

National Electrical Code requirements are generally adhered to for industrial and commercial usage. However, various private concerns and branches of the Armed Services prescribe techniques to be used to provide earth grounding of structures and equipments by using ground rods and meshes. Copper or copper-clad aluminum ground rods are commonly used because copper has a long life when buried in the earth and has the best conductivity of commercially available metals.

Design of Grounding System Utilizing Ground Rods. Ground resistances are computed either by application of the concepts of field theory or by the average-potential method. Although the latter method is not exact from the standpoint of applied physics, it furnishes fairly accurate results and is readily adaptable to most practical problems. In recent years, it has been accepted as the only practical means of solving problems of a complex nature.

The grounding resistance of many closely spaced parallel ground rods is expressed as (2):

$$R = \frac{\rho}{2\pi nL}\left[\ln\left(\frac{4L}{b}\right) - 1 + \frac{2kL}{\sqrt{A}}(\sqrt{n} - 1)^2\right] \qquad \text{ohms} \qquad (8\text{-}4)$$

where:

R = grounding resistance in ohms
ρ = soil resistivity in ohm-centimeters
L = length of each rod in centimeters
b = radius of rods in centimeters
n = number of equally spaced rods within area A
A = area of rod coverage in square centimeters
k = coefficient explained below

Coefficient k in equation (8-4) is obtained from the expression $(\rho/\pi) \times (k/\sqrt{A})$ for the resistance of a horizontal thin plate. When L approaches infinity, equation (8-4) approaches this value. For square and rectangular plates at different depths, coefficient k is plotted as in Fig. 8-9. In most practical cases, grids or rod beds are buried to depths much less than $\sqrt{A}$ so that the coefficient k for the surface level usually holds with sufficient accuracy.

The determination of soil resistivity as a function of locality where measured, and depth of ground-rod penetration, cannot be practically realized with a high degree of accuracy. The effect of various terrain considerations upon soil resistivity and the accuracy of various methods of measuring soil resistivity are discussed in the next section.

Rearranging equation (8-4) into units more frequently used yields (2):

$$R = \frac{0.52\rho}{nL}\left[\ln\left(\frac{2L}{3b}\right) - 1 + \frac{2kL}{\sqrt{A}}(\sqrt{n} - 1)^2\right] \qquad \text{ohms} \qquad (8\text{-}5)$$

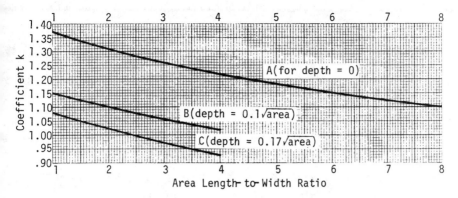

Fig. 8-9. Coefficient k (see text). (*Courtesy of Interference Control Technologies*)

where:

ρ = soil resistivity in ohm-meters
L = rod or pipe length in feet
b = rod or pipe diameter in inches
n = number of parallel rods
A = area in square feet between rods at farthest outside position

Equation (8-5) is used to calculate ground resistance as a function of number of evenly spaced rods, area of coverage, and depth of earth penetration. Resultant data are shown in Fig. 8-10. Calculations have been based upon a value of soil resistivity equal to 50 ohm-meters. Earth resistance resulting from soil resistivities different from this value are:

$$R_1 = R \frac{\rho^1}{\rho} \qquad (8\text{-}6)$$

where:

R = grounding resistance obtained from Fig. 8-10
ρ = 50 ohm-meters (reference)
ρ^1 = measured value of soil resistivity

Whenever grounding resistance calculations are made, sufficient tolerances should be added to allow for: (1) increase in grounding resistance due to age and corrosion, (2) resistance of rods and tie cables, (3) bonding resistance resulting from structure to rod and/or tie cables, and (4) variations in soil resistivity. To obtain the decreased resistance, change: (1) depth of penetration, (2) number of rods, and/or (3) area of coverage. A 50% tolerance is often used.

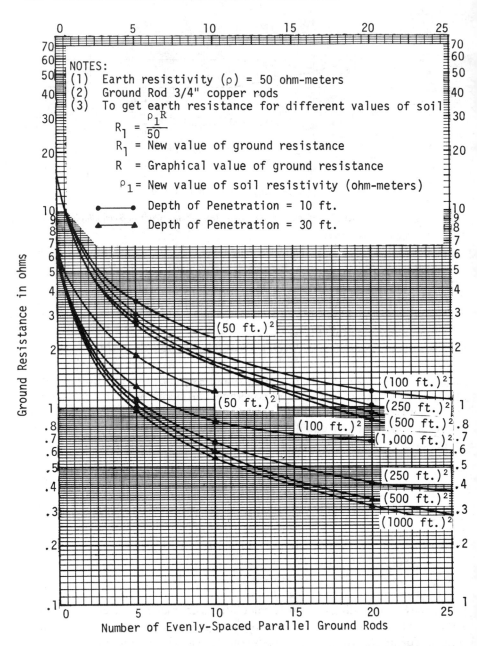

NOTES:
(1) Earth resistivity (ρ) = 50 ohm-meters
(2) Ground Rod 3/4" copper rods
(3) To get earth resistance for different values of soil

$$R_1 = \frac{\rho_1 R}{50}$$

R_1 = New value of ground resistance

R = Graphical value of ground resistance

ρ_1 = New value of soil resistivity (ohm-meters)

●———● Depth of Penetration = 10 ft.

▲———▲ Depth of Penetration = 30 ft.

Ground Resistance in ohms

Number of Evenly-Spaced Parallel Ground Rods

Fig. 8-10. Ground resistance vs. number of evenly spaced parallel ground rods for various areas of coverage and depths of earth penetration.

Materials to be Used in Ground Rods and Coatings. Copper ground rods are commonly used for grounding purposes because of their high conductivity and high corrosion-resistance properties. Figure 8-11 illustrates the physical makeup of existing ground rods that have proved to be both effective and capable of being driven to substantial depths. Such rods come in a variety of sizes, but the 3/4-inch (1.9-cm)-diameter rod appears to be the most compatible with electrical and mechanical requirements.

The factors that contribute to corrosion must be considered in the selection of compatible ground rods: (1) in most cases, both water and oxygen are necessary for corrosion; (2) the initial rate of corrosion is comparatively rapid, slowing as protective films form; (3) surface films are important in controlling the rate and distribution of corrosion; (4) an increased rate of motion increases corrosion in water; and (5) dissimilar metals in contact in the presence of an electrolyte accelerate corrosion of the metal higher in the electrochemical series.

While copper is reasonably corrosion-resistant, it is anodic (higher in the electrochemical series) when compared to the position of some metals used underground for piping and construction purposes. To reduce copper-rod corrosion, it is thus desirable to coat copper-grounding rods with a material that: (1) is less anodic than copper to metallic objects located in the same underground vicinity; (2) is of high electrical conductivity; (3) posseses high resistance to corrosion; and (4) is galvanically compatible with the base metal.

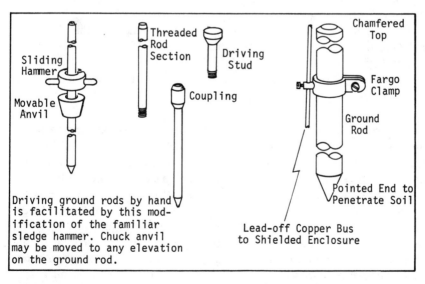

Fig. 8-11. Physical characteristics of typical grounding rods. (*Courtesy of Interference Control Technologies*)

Table 8-2. Metal corrosivity as a function of soil resistivity. (Courtesy of Interference Control Technologies.)

Resistance in ohms/cm^3	Severity of galvanic effects
Less than 400	Extremely severe
400–900	Very severe
900–1,500	Severe
1,500–3,500	Moderate
3,500–8,000	Mild
8,000–20,000	Slight

Tin coatings on copper and copper alloys are normally anodic to the base metal. The tin–copper alloy layer, formed in coating by hot dipping, is cathodic to tin and may be slightly cathodic to copper, and, in general, the corrosion of tinned copper is essentially corrosion of the tin. The function of tin coatings is usually to provide, between copper and the material in question, a layer that, if corroded at all, will yield as innocuous a corrosion product as possible.

The advantage obtained from deep-driven ground rods is realized because of decreased soil resistivity and increased volume of earth associated with deep rods. However, as indicated in Table 8-2, metal corrosiveness increases as soil resistivity decreases, thus imposing stringent requirements on the corrosion-resistance properties of ground rods.

Two promising groups of alloys that may be valuable in the prevention of galvanic corrosion, are the austenitic irons (SDTMA-439, Type D2) and austenitic stainless steel of the 18% chromium and 8% nickel variety. However, additional research is required to evaluate the effectiveness of such materials for use in conjuction with grounding.

Method of Connecting Ground Rod to Structure and Grid Mesh. Figure 8-12 illustrates preferred techniques for connecting ground rods to structures and grid meshes. The portion of the ground wire making contact between the Joslyn washer and the base shoe should be tinned to reduce the effects of galvanic corrosion. All bolts and nuts should be securely tightened to maintain contact pressure and prevent bond impedance deterioration with age and wear. Bond contact areas should be coated with a moisture-proof coating after installation is complete, and the coating should be capable of maintaining its physical properties over an extended period of time. The cover that is placed over a portion of the ground rod extending above the surface of the earth may be of a nonconducting medium as long as it remains waterproof over time. Such covers should be removable or provide

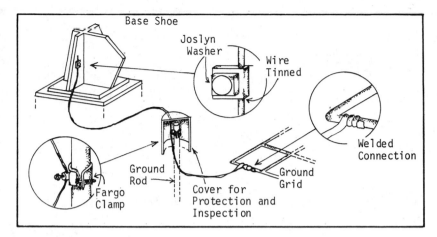

Fig. 8-12. Method for connecting ground rods to structure and grid mesh. (*Courtesy of Interference Control Technologies*)

access to facilitate periodic inspection of the ground-rod connection and measurement of earth resistance. The bonding cable can be a #4/0 or solid 0.63-cm copper wire, or of a larger size. Techniques displayed in Fig. 8-12 can be applied to all types of building structures with steel frames and base shoes.

Design of Grounding System Utilizing Ground-Grid Mesh.

Ground-grid meshes are often required to complement rod beds or to be used separately when deep-driven rods are impractical because of soil and terrain considerations. The following formula is used to calculate the ground resistance of an earth ground-grid mesh (5):

$$R = \frac{\rho}{\pi L}\left[\ln\left[\frac{2L}{a^1}\right] + k_1\left(\frac{L}{\sqrt{A}}\right) - k_2 \right] \qquad \text{ohms} \qquad (8\text{-}7)$$

where:

ρ = soil resistivity in ohm-centimeters
L = total length of all connected conductors in centimeters
$a^1 = \sqrt{(a \times 2z)}$ for conductors buried at a depth of z cm
$a^1 = a$ for conductors on earth surface
a = conductor radius in centimeters
A = area covered by conductors, in square centimeters
k_1, k_2 = coefficients explained below.

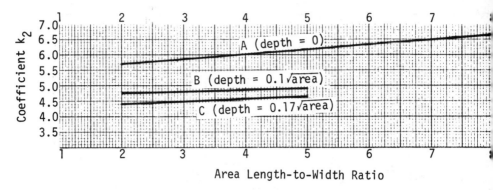

Fig. 8-13. Coefficient k vs. area shape and depth. (*Courtesy of Interference Control Technologies*)

The coefficient k_1 is the same as k used in equations (8-4) and (8-5). Coefficient k_2 has been calculated for loops encircling areas of the same shape and depth as used for calculating k_1 (see Fig. 8-9). Calculated results for k_2 are shown in Fig. 8-13.

Equation (8-7) is rearranged into units more frequently used:

$$R = \frac{1.045\rho}{L}\left[\ln\left(\frac{2L}{a}\right) + k_1 \frac{L}{\sqrt{A}} - k_2\right] \quad \text{ohms} \quad (8\text{-}8)$$

where:

a = conductor diameter in inches × depth in feet
L = total length of all connected conductors in feet
ρ = soil resistivity in ohm-meters

Resistance has been calculated as a function of foundation or area of grid coverage and number of grids per side using #4/0 copper cable. The resistivity of the earth was assumed to be 50 ohm-meters for calculation purposes. Coefficients k_1 and k_2 vary only slightly with area, and the error introduced by using calculated data for both square and rectangular grids is negligible. Calculated data are presented graphically in Fig. 8-14.

Grounding resistance afforded by buried grid meshes can be reduced significantly by increasing both the number of grids and the area of grid coverage. Data have been extrapolated from Fig. 8-14 and replotted in Fig. 8-15 to show ground resistance as a function of area of grid mesh coverage, for a single loop and for 30 grids per side.

Use of such graphical presentations in developing optimum design criteria for earth grid meshes has shown that: (1) an increasing reduction in ground

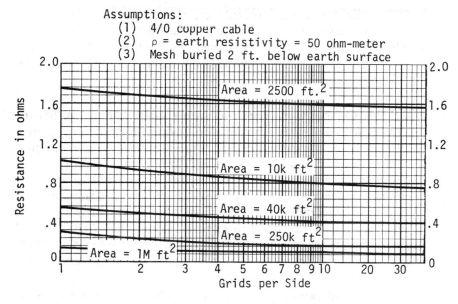

Assumptions:
(1) 4/0 copper cable
(2) ρ = earth resistivity = 50 ohm-meter
(3) Mesh buried 2 ft. below earth surface

Fig. 8-14. Ground resistance vs. number of grids in mesh as a function of total mesh area. (*Courtesy of Interference Control Technologies*)

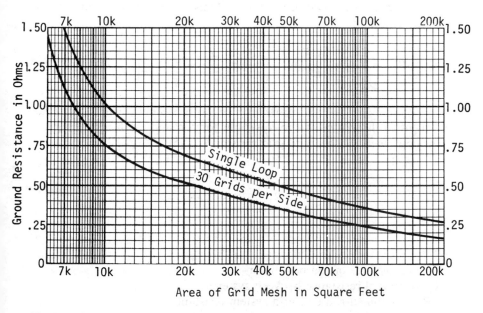

Fig. 8-15. Ground resistance vs. grid mesh. (*Courtesy of Interference Control Technologies*)

resistance is realized by using increased areas of grid coverage up to approximately 2286 m² (90,000 ft²); (2) beyond 2286 m², maximum ground resistance reduction is realized by use of a number of grids; and (3) the average reduction of ground resistance resulting from additional grids is approximately 0.2 ohm.

Method for Connecting Earth Ground Grid Mesh to Structure. Figure 8-16 illustrates a preferred technique for connecting an earth ground-grid mesh to a structure. The end of the bond cable should be tinned where connection is made with the structure's base shoe in order to reduce effects of galvanic corrosion. A double connection is made between the bond cable and the grid mesh to reduce the possibility of bond deterioration with age and wear. All connections should be wrapped, welded, and covered with a protective coating, as indicated. All structure base shoes should be connected in a similar manner to the grid mesh. Techniques illustrated in Fig. 8-12 should be followed when grid meshes are used in conjunction with ground rods.

Method for Approximating Combined Ground Resistance of Mesh and Ground Rods. In many cases it may be necessary to use a combination of ground rods and a grid mesh below ground to obtain a sufficiently low ground resistance. Figure 8-17 illustrates how a combination of a grid mesh and ground rods might be physically implemented. The mutual resistance between the two grounding systems can be approximated by the following equation:

$$R_{12} = R_{21} = \frac{\rho}{\pi L} \left[\ln \left(\frac{2L}{L_1} \right) + k_1 \frac{L}{\sqrt{A}} - k_2 + 1 \right] \qquad \text{ohms} \qquad (8\text{-}9)$$

where $R_{12} = R_{21}$ = mutual resistance of both systems, with parameters equivalent to those used in the equations for rod and mesh resistance.

The combined rod bed and grid resistance is (5):

$$R = \frac{R_{11}R_{12} - R_{12}{}^2}{R_{11} + R_{22} - 2R_{12}} \qquad (8\text{-}10)$$

where:

R_{11} = resistance of grid alone
R_{22} = resistance of rod bed alone
R_{12} = mutual resistance between systems
R = combined rod and grid system resistance

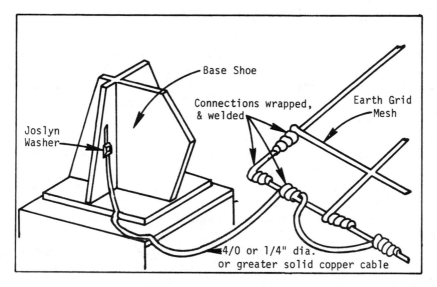

Fig. 8-16. Method for connecting earth ground-grid mesh to structure. (*Courtesy of Interference Control Technologies*)

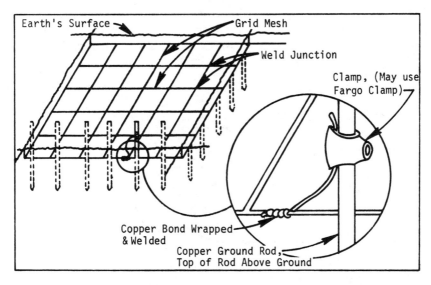

Fig. 8-17. Typical combination of ground rods and grid mesh. (*Courtesy of Interference Control Technologies*)

The reduction in ground resistance achieved by adding rods to a grid will hardly warrant the extra cost. Yet, there are points in favor of such an arrangement: (1) it ensures practically constant ground resistance near the earth's surface where soil resistivity may fluctuate due to extreme climatic conditions, and (2) rods used to provide a reliable ground source and grid serve as a safety measure to equalize fault potentials over the earth's surface.

Soil Impedance. Ground rod and grid mesh criteria were developed in the preceding paragraphs on the assumption that a sufficiently low earth resistivity could be realized for effective implementation. In extremely rocky or frozen soil, deep penetration of ground rods is impractical. In such cases, ground grids might be used, but in regions subjected to extreme climatic variations, earth resistivity will vary considerably, causing resistance changes in shallow buried grid meshes. In various localities such as dry sandy soils, earth resistivity may be extremely high regardless of the depth of ground rod penetration. In such situations, other techniques may be utilized to obtain the necessary low-ground resistance: (1) impregnation of soil with a salt solution, (2) immersion of grid or plate in nearby water sources and connection of such grounding media to structures to be grounded, and (3) utilization of underground piping systems.

Resistivity Variations as a Function of Soil Type. The ground resistance of any type of electrode that may be used is directly proportional to ρ, the resistivity of the soil. Consider a metallic hemisphere of radius a, which is buried flush with the surface of the earth, as shown in Fig. 8-18. If the

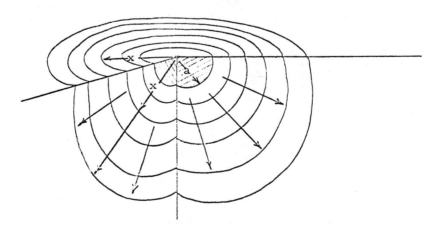

Fig. 8-18. Distribution of current about a metallic hemispherical ground electrode. (*Courtesy of Interference Control Technologies*)

Table 8.3 Earth Resistivity
of Different Soils

R & ρ Soil Fills	Resistance* Ω 5/8" x 5' rods			Resistivity in Ω/cm³		
	Avg.	Min.	Max.	Avg.	Min.	Max.
Ashes, Cinders, Brine Waste	14	3.5	41	2,370	500	7,000
Clay, Shale, Gumbo, Loam	24	2	98	4,060	340	16,300
Same, with varying proportions of sand & gravel	93	6	800	15.8k	1,020	135k
Gravel, sand, stones with little clay or loam	554	35	2,700	94k	59k	458k

*Bureau of Standards Technical Report No. 108

resistance of the electrode itself is neglected, the resistance of the earth connection, R, will be that offered to the current flow through the soil volume immediately surrounding the electrode (5):

$$R = \rho \int_a^\infty \frac{dx}{2\pi x^2} = \frac{\rho}{2\pi a} \qquad (8\text{-}11)$$

Earth resistivity varies as a function of soil type, temperature, and moisture content. Table 8-3 shows data related to soil type. From this table, it is noted that a grounding system that is entirely adequate in clay soil will be almost worthless in sandy soil.

Resistivity Variations as a Function of Moisture Content. Soils that are relatively good conductors under normal moisture content become good insulators when such content is low. Figure 8-19 and Table 8-4 show the variation of soil resistivity with moisture content for various soil types. For most soil types, a moisture content of 30% will result in a sufficiently low resistivity.

Resistivity Variations as a Function of Temperature. Soils that have sufficient moisture to be good conductors at normal temperatures will become ineffective below freezing due to increased resistivity. This is illustrated in Fig. 8-20 and Table 8-5.

Resistivity Variation Due to Salt Content. Soil resistivity varies as a function of the salt content of the soil. Figure 8-21 and Table 8-6 show the effects of salt content upon reduction of the resistivity of various types of soil.

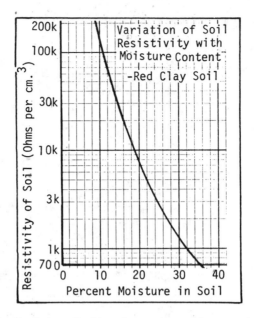

Fig. 8-19. Resistivity of red clay soil with moisture content. (*Courtesy of Interference Control Technologies*)

Table 8-4. Effect of moisture content on the resistivity of soil.* (Courtesy of Interference Control Technologies.)

Moisture content by wt.	Resistivity in Ω/cm^3	
Soil type	Top soil	Sandy loam
0%	$< 1,000 \times 10^6$	$< 1,000 \times 10^6$
2.5%	250,000	150,000
5%	165,000	43,000
10%	53,000	18,500
15%	19,000	10,500
20%	12,000	6,300
30%	6,400	4,200

* "An Investigation of Earthing Resistances," by P. J. Higgs, *IEE Journal*, Vol. 68, p. 736, February, 1930.

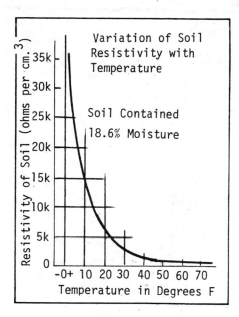

Fig. 8-20. Resistivity of red clay soil with temperature. (*Courtesy of Interference Control Technologies*)

Table 8-5. Effect of temperature on the resistivity of soil. (sandy loam, 15.2% moisture)* (Courtesy of Interference Control Technologies)

Temperature		Resistivity in Ω/cm^3
°C	°F	
20	68	7,200
10	50	9,900
0 (water)	32	13,800
0 (ice)		30,000
−5	23	79,000
−15	14	330,000

* "Lightning Arrester Grounds, Parts I, II, and III," by H. M. Towne, *General Electric Review*, Vol. 35, pp. 173, 215, and 280, March, April and May, 1932.

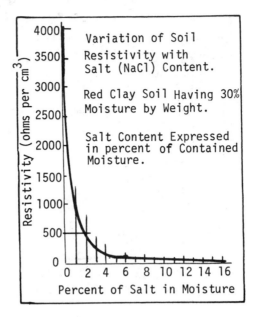

Fig. 8-21. Effect of addition of salt upon resistivity of red clay. (*Courtesy of Interference Control Technologies*)

Table 8-6. The effect of salt content on the resistivity of soil. (Sandy loam, moisture content 15% by weight, temperature 17°C).* (Courtesy of Interference Control Technologies)

Added salt weight	Resistivity in Ω/cm^3
0%	10,700
0.1%	1,800
1.0%	460
5%	190
10%	130
20%	100

* "An Investigation of Earthing Resistances," by P. J. Higgs, *IEE Journal*, Vol. 68, p. 736, Feb. 1930.

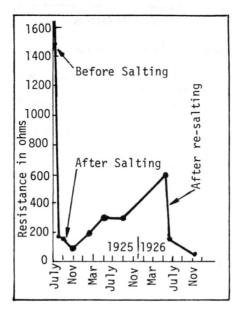

Fig. 8-22. Changes in resistivity of a ground connection. (*Courtesy of Interference Control Technologies*)

The soil can be artificially treated with salt to increase conductivity. In Fig. 8-22, test data show resistance variations as a function of time and salting for a specific ground connection.

Methods for Measuring Soil Resistivity. It is necessary to know, at least approximately, the value of soil resistivity existing in an area where a structure requiring an earth ground is to be built. Previous discussions and resultant criteria have been based upon the assumption that an approximate value of soil resistivity will be available for design purposes.

After grounding systems have been implemented, techniques must be available for ensuring that resultant systems produce the specified grounding resistance. Ground resistance can be expected to fluctuate as a result of climatic changes, age, and wear, and must be periodically checked to ensure compliance with stipulated specifications.

Although ground resistance test requirements vary from situation to situation, various techniques are available that are compatible with the requirements of all situations. These techniques are identified as: (1) the fall-of-potential method, (2) the four-point-array method, and (3) the radioactivewave-propagation method.

REFERENCES

1. Denny, Hugh W. *Grounding for the Control of EMI.* Gainesville, VA: Don White Consultants, Inc., 1983.
2. Hart, William C., and Malone, Edgar W. *Lightning and Lightning Protection.* Gainesville, VA: Don White Consustants, Inc., 1979.
3. Ott, Henry W. *Noise Reduction Techniques in Electronic Systems.* New York: John Wiley & Sons, 1976.
4. Schram, Peter J., Ed., *The National Electrical Code Handbook*, Quincy, MA: National Fire Protection Association. 1983.
5. White, Donald R. J. *An Electromagnetic Compatibility Handbook*, Volume 3. Gainesville, VA: Don White Consultants, Inc., 1973.

Chapter 9
GROUNDING AND BONDING
FOR EMI CONTROL

Chapter 5 presents a number of considerations regarding grounding in discussing the broader subject of intrasystem EMI prediction and control. The main consideration is to avoid the creation of low-impedance ground loops associated with resulting common-mode EMI. For the control of EMI, wiring, circuits, printed circuit boards, equipment racks, and so on, in some instances should not be grounded i.e., should not be connected to a designated system potential reference or "ground plane"; or (usually) they are grounded at a single point. Circuit designs and layouts should consider maintaining high ground-loop impedances to minimize circulating common-mode currents that may flow into circuits containing sensitive victim loads. Floating is often especially effective at low frequencies. The proper grounding of faraday shields, which are installed to protect wiring (cable) harnesses, reduce isolation transformer interwinding capacitance, and protect filters, is necessary to permit a voltage-dividing filtering action to reduce the voltage induced by an ambient electric field or electromagnetic plane wave. Also, the susceptibility of equipment to electrostatic discharge requires proper grounding techniques to assure that the discharge energy will not find its way into sensitive circuits and components.

This chapter focuses on grounding for the reduction of EMI. In the present complex and sophisticated aerospace and microprocessor era, it is necessary that such a mundane-sounding subject as grounding be thoroughly understood because the success or failure of modern systems depends upon proper grounding practice based upon a systems approach.

It is the opinion of the authors that the topic of grounding is not better understood for at least the following two reasons:

- Shock and safety control requirements existed long before the electronics and high frequency era, and traditional grounding techniques were developed to satisfy those requirements.
- Often, a conflict exists between the requirements for safety grounds, including specifications in electrical codes, and EMI control.

A systems approach to signal and safety grounding is required to assure that all variables are accounted for, and that safety and EMI control requirements are both satisfied.

EQUIPOTENTIAL GROUND PLANE

In its purest (ideal) sense, an equipotential ground plane consists of a plane region wherein all points within the plane are at the same potential. In the practical world, this cannot be realized, since no material exists that has an infinite conductivity, or zero resistivity, with which to construct the plane. Under most practical circumstances, a ground "plane," ground "bus," chassis ground, or the like consists of an identifiable metallic structure within a circuit or equipment configuration that is used as a potential reference, or "zero" potential level. Under practically all circumstances, electrical conduction currents will be induced to flow within the metallic structure, referred to as the potential reference. Since the metal used to construct the reference structure (e.g., copper, aluminum, steel, etc.), has a finite conductivity (or non-zero resistivity), the currents flowing through this resistivity result in the generation of potential differences between points within the ground plane (or along the length of a ground bus).

Figure 9-1 illustrates potential contours produced by currents flowing within a ground plane or metallic surface structure (1). Currents $i_1, i_2, i_3, \ldots i_n$, flowing within the ground plane structure produce potential (voltage) contours $V_1, V_2, V_3, \ldots V_n$. As illustrated in Figure 9-1, a voltage exists between points A and B equal in magnitude to $(V_1 - V_4) = I_G Z_{AB}$, with I_G an equivalent ground current, and Z_{AB} an equivalent impedance, between the points A and B.

One of the consequences of finite ground plane impedance is the probability of generating common-impedance, common-mode coupling, whenever a multipoint grounding scheme is used. This corresponds to the first of the five principal coupling paths described in Chapter 5.

Common-Impedance Coupling

Figure 9-2 illustrates two circuits, with circuit element subscripts 1 and 2, respectively, that share the same ground plane as a return path. The ground plane impedance is indicated by Z_G. In this instance, the two circuits can mutually interfere with each other as follows: If Kirchhoff's voltage law is applied around the loop formed by circuit #1, including the ground plane return path, the following equation is obtained:

$$V_{s1} = I_1(Z_{s1} + Z_{L1} + Z_G) + I_2 Z_G \qquad (9\text{-}1)$$

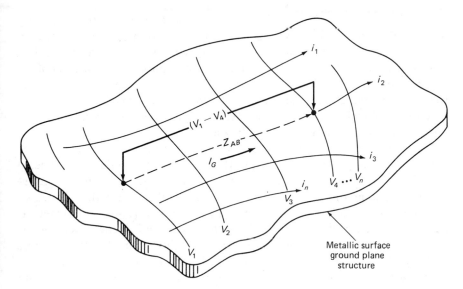

Fig. 9-1. Potential contours produced by currents flowing within a ground plane or metallic surface structure.

where the term $I_2 Z_G = V_{N1}$ represents a noise voltage in circuit #1 induced by the current I_2 flowing through the ground plane impedance Z_G. Similarly, around circuit #2, the following equation is obtained:

$$V_{s2} = I_2(Z_{s2} + Z_{L2} + Z_G) + I_1 Z_G \qquad (9\text{-}2)$$

where the term $I_1 Z_G = V_{N2}$ represents a noise voltage in circuit #2 induced by the current I_1 flowing through the ground plane impedance Z_G.

The situation illustrated in Fig. 9-2 and described immediately above is often encountered in practice. The critical question concerns the level of the induced noise due to the common-impedance coupling in comparison to the

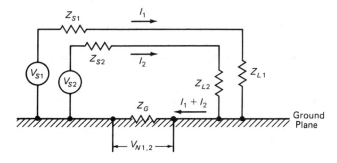

Fig. 9-2. Common impedance coupling between circuits.

victim circuit susceptibility. If the induced noise V_N is below the sensitivity of an analog victim or below the noise immunity of a digital circuit, then presumably no problem exists. However, in a complex circuit configuration, ground plane currents can be generated by many sources, and the level of the induced noise may be difficult to predict. Sharing a common return path between two circuits should only be done when it is determined that an acceptably low probability of EMI exists. An obviously desirable condition is to assure that the impedance Z_G is as low as possible.

The foregoing problem, illustrated by Fig. 9-2, results because both circuits are "multipoint" grounded; that is, both ends of both circuits are connected at different points within the ground plane. Figure 9-3 illustrates common-impedance, common-mode coupling when a signal circuit with a dedicated return path is grounded to a ground plane at two points (1, 4). In this instance, a ground-plane current I_G flowing through impedance Z_G generates a voltage $V_i = I_G Z_G$, which appears simultaneously as a common-mode voltage across both signal circuit lines at points B and C. This results in the flow of common-mode currents I_1 and I_2, as shown. Since I_1 and I_2 flow through circuit branches of unequal impedance, a differential-mode noise voltage V_0 is developed across the load Z_L (2, 3). An EMI situation exists if V_0 exceeds circuit susceptibility (analog sensitivity or digital noise margin or immunity). In Figure 9-3, the path around $ABCD$ is referred to as the "ground loop" around which common-mode currents can circulate. An EMC objective is to assure that the impedance around this loop is as high as possible to minimize the amplitude of circulating common-mode currents. In Chapter 5, this is referred to as reducing ground-loop coupling (GLC) (4).

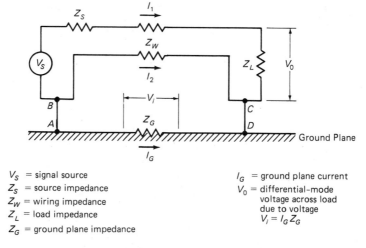

V_S = signal source
Z_S = source impedance
Z_W = wiring impedance
Z_L = load impedance
Z_G = ground plane impedance

I_G = ground plane current
V_0 = differential-mode
   voltage across load
   due to voltage
   $V_i = I_G Z_G$

Fig. 9-3. Common impedance coupling for a circuit grounded to a ground plane at two points.

SINGLE POINT AND MULTIPOINT GROUNDING

The rationale for single point versus multipoint grounding has been widely developed, with different possible configurations available for both schemes. For a complex system, the final grounding arrangement usually ends up a hybrid system, composed of subsets of both single and multipoint grounds. The demarcation for determining whether to use single or multipoint grounding ultimately depends upon the EMI and/or signal frequency (1-4).

Single Point Grounding

Figure 9-4a illustrates three interconnected equipment boxes of a system with a multipoint grounding configuration to a ground plane or ground bus. A current I_G induced by a variety of sources flows within the ground plane, inducing voltages V_{G1} and V_{G2} between points A and B and B and C as shown. As a consequence, points A, B, and C are not at the same potential, and the conditions illustrated by Fig. 9-2 and 9-3 could be created, resulting in an EMI situation. Figure 9-4b illustrates the same equipment as in Fig. 9-4a with the three equipment boxes all grounded at the single point B, thus preventing common impedance coupling due to the current I_G. It is assumed that points A', B', and C' are essentially at the same potential as point B. This is essentially true at low frequencies. However, as the frequency of the EMI or signal increases, Fig. 9-4b electrically behaves more like Fig. 9-4c, with the parasitic distributed inductances, capacitances, and frequency-dependent lead resistances (due to skin-effect) introducing resonant and anti-resonant effects. In Fig. 9-4c, points A', B', and C' are not at the same potential as point B at high frequencies, but each point is generally at some indeterminate potential with respect to B. Thus single point grounding appears desirable and acceptable at low frequency as long as the effects of parasitic element impedances are negligible.

Multipoint Grounding

Whenever the signal or EMI frequency is relatively high, or fast logic is used, a multipoint grounding scheme must be used, at least in part of the circuit (3). In a multipoint grounding system, connections to ground are made via the shortest route from the circuit node that is to be grounded to the reference ground, with the latter usually being the chassis, dedicated ground bus, or printed circuit board (PCB) ground plane (3, 4). The connecting wires between each circuit and the ground plane should be as short as possible to minimize their series impedance.

The facts are that a single point grounding scheme operates best at low frequencies, and a multipoint ground often behaves best at high frequencies. If the overall system, for example, is a network of audio equipment, with

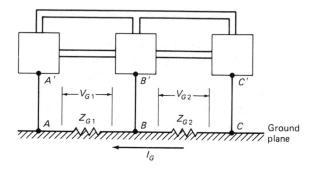

(a) Multipoint grounding—common-mode voltages V_{G1} and V_{G2} induced.

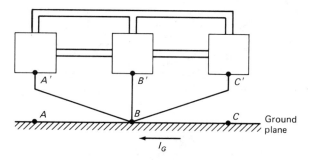

(b) Single point grounding.

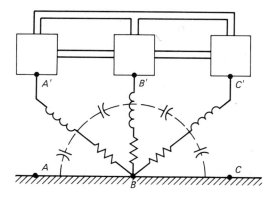

(c) Single point ground configuration at high frequencies.

Fig. 9-4. Multipoint and single point grounding.

many low-level sensors and control circuits behaving as broadband transient noise sources, then the high-frequency performance is irrelevant (assuming that no high frequency spurious or parasitic responses are present, such as caused by audio rectification), since no receptor responds to frequencies above audio-frequency. Conversely, if the overall system is a receiver complex consisting of high-frequency (VHF–UHF range) tuners, amplifiers, and displays, then low-level, low frequency performance is irrelevant. Here, multipoint grounding applies, and interconnecting, unbalanced coaxial lines are used (3).

The above dichotomy of audio versus VHF/UHF systems makes clear the selection of the correct single point or multipoint grounding approach. The problem then narrows down to one of defining where low and high frequency crossover exists for any given subsystem or equipment. In part, the answer involves the relationship between the shortest wavelength (highest significant operating frequency) of the low-level (most sensitive) circuits relative to the longest length of wire required to complete a single point grounding system. If, at the highest frequency of interest, that is, at the shortest wavelength determined by $\lambda_m = 300/f_{MHz}$ meters, the longest wire connection required is $L_m > \lambda_m/20$, then a multipoint ground scheme should be used (2, 3). Another rule of thumb is suggested by Ott (3): at frequencies below 1 MHz, a single point ground is preferred; and at frequencies above 10 MHz, use a multipoint ground. For frequencies between 1 MHz and 10 MHz, a single point grounding scheme may be used to avoid common impedance coupling, provided that the length of the longest ground conductor is less than 1/20th of a wavelength.

Hybrid Grounding

The matter of single point versus multipoint grounding discussed in the previous sections is based upon the criterion of the length of the longest conductor compared to 1/20th wavelength. For low-frequency operation and small circuit dimensions compared to a wavelength, use single point grounding. For high frequency and large circuit dimensions compared to a wavelength, use multipoint grounding. For transitional situations, one or the other may perform better. Hybrid grounds perform best in circuits where portions operating at low frequency use single point grounding while high-frequency portions use multipoint grounding, all connected in a "ground-tree" fashion (3).

In general, regarding mathematical modeling and prediction of EMI as a function of grounding, it is extremely difficult to accomplish this for other than (1) relatively simple system configurations and (2) single point grounding. The number of undeterminable significant parasitic reactances in complex systems, or at high frequencies for all systems, renders modeling of a grounding situation difficult.

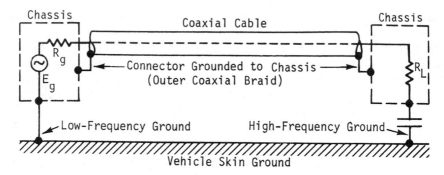

Fig. 9-5. Low-frequency ground current loop avoidance with high-frequency ground. (*Courtesy of Interference Control Technologies*)

The term "hybrid ground" is sometimes used in two somewhat different senses: (1) when a grounding scheme either appears as a single point ground at low frequencies and multipoint at high frequencies; (2) when a system grounding configuration simultaneously employs both single point and multipoint grounds. Each of these is illustrated below.

Figure 9-5 shows a low-level video circuit in which both the sensor and driven circuit chassis must be grounded to the skin of a vehicle (not necessarily by choice), and the coaxial cable shield is grounded to the chassis at both ends through its mating connectors. A low-frequency ground current loop would be generated, were it not for the high impedance of the capacitor. At high frequencies, the low impedance of the capacitor assures that the cable shield is grounded at the load end of the circuit to protect the Faraday-shield effect. Thus, this circuit simultaneously behaves as a single point ground at low frequencies and a multipoint ground at high frequencies.

A different kind of example is shown in Fig. 9-6, in which all the computer and peripheral frames must be grounded to the power system green

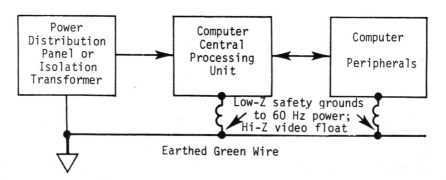

Fig. 9-6. Safety ground with high-frequency isolation. (*Courtesy of Interference Control Technologies*)

wire for safety purposes (shock-hazard protection) in accordance with the National Electrical Code. Since it is recognized that the green wire generally contains significant electrical noise trash, this code conflicts with the desire to float the computer system ground from the noisy green wire ground. Thus, one or more isolation coils (RF chokes) of about 1 mH value are used to provide a low-impedance (less than 0.4 ohm) safety ground at AC 50/60 Hz power line frequencies, and RF isolation (of the order of 1000 ohms) in the 50 kHz to 1 MHz spectrum containing significant energy generated by computer pulses. This inductor helps keep induced transient and EMI noise in the green wire from entering the computer supply voltage logic busses (1). Essentially, this approach may be viewed as the "dual" of the approach illustrated in Fig. 9-5.

ELECTRICAL BONDING

Electrical bonding refers to the process in which components or modules of an assembly, equipment, or subsystems are electrically connected by means of a low-impedance conductor. The purpose is to make the structure electrically homogenous with respect to the flow of RF currents. This mitigates EMI-producing electrical potential differences among metallic parts.

An example of the importance of bonding to reduce EMI is shown in Fig. 9-7, which illustrates how the effectiveness of a filter can be nullified by improper bonding. In that example, the contact resistance of a poor bond does

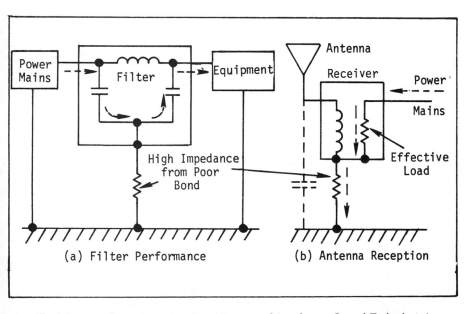

Fig. 9-7. Two effects of poor bonding. (*Courtesy of Interference Control Technologies*)

not provide the low-impedance path necessary for shunting to ground the interference currents coming from the power mains. These currents now flow through the filter capacitors and onto other components that were to have been protected. In Fig. 9-7b, the receiver is not well bonded to a common ground plane reference for both the antenna and the power mains return. Thus, RF currents appearing on the power mains share a common impedance path at the bond with RF signals picked up by the antenna.

Equivalent Circuits of Bonds

A low-impedance path is possible only when the separation of the bonded members is small compared to the wavelength corresponding to the highest frequency of the EMI being considered, and the bond strap or wire material is a good conductor. This was discussed previously. At high frequencies, structural members may behave as transmission lines whose impedance can be inductive or capacitive and of varying magnitudes, depending upon geometrical shape and frequency.

Figure 9-8 is the equivalent electrical circuit of a bond strap assembly. The circuit contains the series combination of the strap inductance and its two-part resistance: the resistance due to the finite conductance of the strap metal material, and the contact resistance at both ends of the strap. The

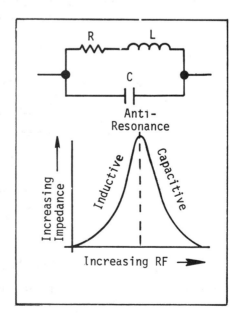

Fig. 9-8. Equivalent circuit of bond strap and its impedance. (*Courtesy of Interference Control Technologies*)

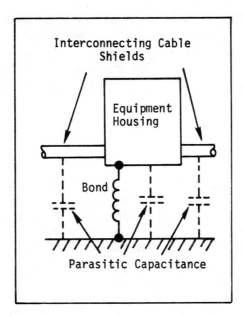

Fig. 9-9. Bond impedance and installation parasitics. (*Courtesy of Interference Control Technologies*)

shunt capacitance is comprised of the parasitic capacitance of the strap and the capacitance between the equipment boxes, conduits, and so on, and the ground reference. This capacitance and bond strap self-inductance form a parallel anti-resonant circuit, resulting in the adverse impedance response shown in Fig. 9-8. At frequency f_0, the circuit is parallel resonant, with the effective impedance across the strap assembly reaching some maximum value.

There is little correlation between the DC resistance of a bond and its RF impedance. The measured RF impedance of artificial bonds, per se, such as jumpers, straps, rivets, and so on, is not a reliable indication of bonding effectiveness as a function of frequency in an actual installation. Here, the artificial bond is in parallel with the various members to be bonded, and the total impedance includes various parallel paths over which RF conductive or displacement currents may flow. Thus, a bond strap of low inductance combines with the capacitance of the installation, as shown in Figs. 9-8 and 9-9, to form a high-impedance, anti-resonant circuit at some frequency.

Types of Bonds

The best-performing electrical bond consists of a permanent, directly connected metal-to-metal contact area, such as provided by welding, brazing, sweating, or swagging. Though adequate for most purposes, the best soldered joints have appreciable contact resistance and cannot be depended upon for the most satisfactory type of bonding. Semipermanent joints, such as provided by bolts or rivets, can provide effective bonding. However, the eventual loosening of the joints and possible relative motion of the joined members will likely reduce the bonding effectiveness by introducing a varying joint contact impedance.

Bonding Hardware

Star or lock washers should be used with bolt or lock-thread bonding nuts to ensure the continuing tightness of a semipermanent bonded joint or members. Figure 9-10 shows one recommended arrangement. Star washers are especially effective in cutting through protective or other insulating coatings on metal such as anodized aluminum or unintentional oxides or grease films developed during periods between maintenance.

Joints that are press-fitted or joined by screws of the self-tapping or sheet-metal type cannot be relied upon to provide low-impedance RF paths. Among other considerations, these screws are made on a screw machine in which a jet of coolant oil is used, and the threads thus may contain some residual oil in spite of a degreasing bath. Often, there is a need for relative motion between members that should be bonded, as in the case of shock mounts. A flexible metal strap can be used as a bonding agent, as shown in Fig. 9-11. This strap must satisfy both the mechanical, shock-mounting properties and the electrical requirements of the bond.

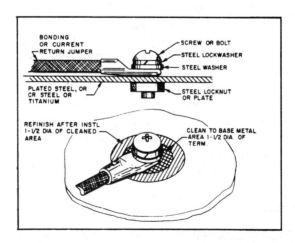

Fig. 9-10. Bonding connections. (*Courtesy of Interference Control Technologies*)

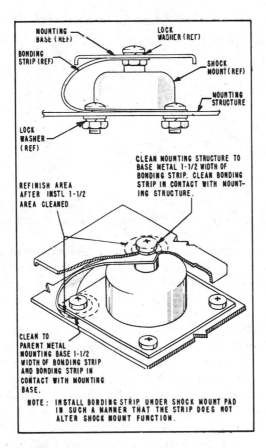

Fig. 9-11. Bonding shock mounts. (*Courtesy of Interference Control Technologies*)

Bonding Jumpers and Bond Straps

Bonding jumpers are short, rectangular or round, either braided or stranded conductors for application where interference frequencies to be grounded are below about 10 MHz. They are frequently used in low-frequency devices where the development of static charges must be prevented. They are also used to provide good electrical continuity across tubing members and associated clamps, such as shown in Fig. 9-12. In their application, the clamps should not be relied upon for continuity because of tubing surface finishes, grease films, and/or metal oxides.

To provide a low-impedance path at RF, it is necessary to minimize both the series self-inductance and residual capacitance of a bond assembly, in order to maximize the parasitic resonant frequency. Since it is difficult to

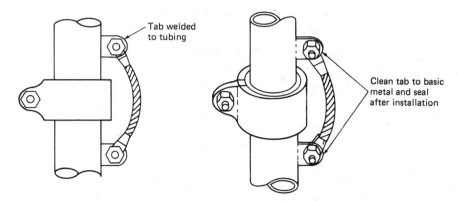

Fig. 9-12. Bonding connections. (*Courtesy of Interference Control Technologies*)

change the residual capacitance of the strap and mounting, self-inductance becomes the main controllable variable. Thus, rectangular cross-section bond straps are preferable to round wires of equivalent cross-sectional areas.

Bond straps are either solid, flat, metallic conductors or a woven braid configuration where many conductors are effectively in parallel. Solid metal straps are generally preferred for the majority of applications. Braided or stranded bond straps are not generally recommended because of several undesirable characteristics. Oxides may form on each strand of nonprotected wire and cause corrosion. Because such corrosion is not uniform, the cross-sectional area of each strand or wire will vary throughout its length. The nonuniform cross-sectional areas (and possible broken strands of wire) may lead to generation of EMI within the cable or strap. Broken strands may act as efficient antennas at high frequencies, and interference may be generated by intermittent contact between strands.

Solid bond straps are also perferable over stranded types because of lower series self-inductance. The direct influence of bond-strap construction on RF impedance is shown in Fig. 9-13, where the impedances of two bonding straps and on Number 12 wire are compared as a function of frequency. The relatively high impedance at high frequencies illustrates that there is no adequate substitute for direct metal-to-metal contact. A rule of thumb for achieving minimum bond strap inductance is that the length-to-width ratio of the strap should be a low value, such as 5:1 or less. This ratio determines the inductance, the major controllable factor in the high-frequency impedance of the strap.

Rack Bonding

Equipment racks (48 cm or other sizes) provide a convenient means of maintaining electrical continuity between such items as rack-mounted chassis,

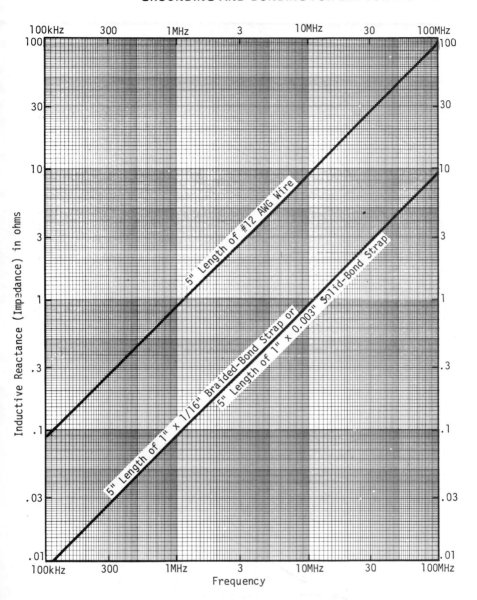

Fig. 9-13. Impedances of wire braided and solid-bond straps. (*Courtesy of Interference Control Technologies*)

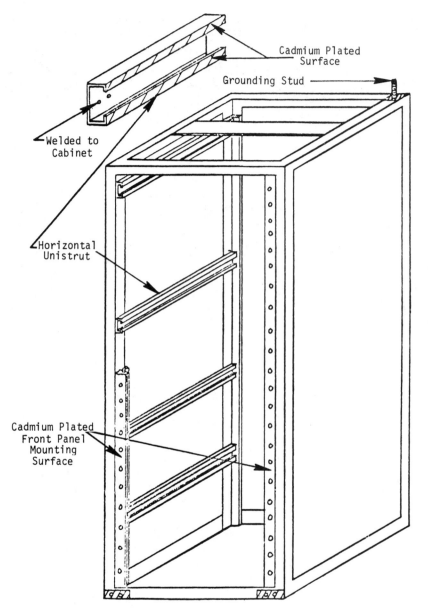

Fig. 9-14. Cabinet bonding modifications. (*Courtesy of Interference Control Technologies*)

panels, and ground planes. They also serve as an electrical inter-tie for cable trays. A typical equipment cabinet, with the necessary modifications to provide such bonding, is shown in Fig. 9-14. Bonding between the equipment chassis and rack is achieved through the equipment front panel and rack right-angle brackets. These brackets are grounded to the unistrut horizontal slide that is welded to the rack frame. The lower surfaces of the rack are treated with a conductive protective finish to facilitate bonding to a ground-plane mat. The ground stud at the top of the rack is used to bond a cable tray, if used, to the rack structure, which is of welded construction.

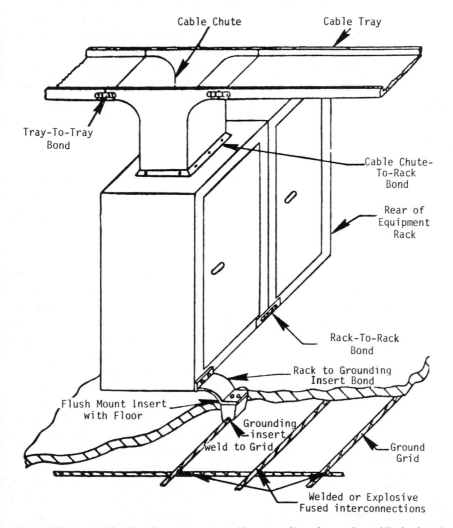

Fig. 9-15. Typical cabinet bonding arrangements. (*Courtesy of Interference Control Technologies*)

Figure 9-15 illustrates a typical bonded installation. Cable trays are bonded together, and the cable tray is bonded to the cable chute. The cable chute is bonded to the top of the cabinet; the cabinet is bonded to the flush-mounted grounding inset (which is welded to the ground grid); and the front panel of the equipment is bonded to the rack or cabinet front-panel mounting surface. Nonconductive finishes are removed from the equipment front panel before bonding. The joint between equipment and cabinet may serve a dual purpose: that of achieving a bond, and that of preventing interference leakage from the cabinet if the joint is designed to provide shielding. If such shielding is a requirement, conductive gaskets should be used around the joint to ensure that the required metal-to-metal contact is obtained. If equipment is located in a shock-mounted tray, the tray should be bonded across its shock mounts to the rack structure. Connector mounting plates should use conductive gasketing to improve chassis bonding. If chassis removal from the rack structure is required, a one-inch-wide braid with a vinyl sleeving should be used to bond the back of the chassis to the rack. The braid should be long enough to permit withdrawal of the chassis from the rack.

CORROSION AND CONTROL

This section discusses the effects of corrosion on bond contact impedance and methods of corrosion control.

Corrosion

When two metals are in contact (bonded) in the presence of moisture, corrosion may take place through either or both of two chemical processes. The first process is termed galvanic corrosion, and develops from the formation of a voltaic cell between two dissimilar metals in contact with the moisture present to form an electrolyte. The degree of resultant corrosion depends on the relative positions of the metals in the electrochemical (sometimes called electromotive) series. Part of this series is shown in Table 9-1, with the metals listed at the top of the table corroding more rapidly than those at the bottom. If the potential between two metals differs appreciably in this series, such as between aluminum and copper (2.00 volts difference), the resulting electromotive force will cause a continuous ion stream with a significant accompanying decomposition of the more active (higher in the electrochemical series or less noble) metal as it gradually goes into solution.

The second chemical corrosion process is termed electrolytic corrosion. While this process also requires two metals in contact through an electrolyte, the metals need not have different electrochemical activity; that is, the contact junction can be between bodies of the same material. In this case, decomposition is attributed to the presence of local electrical currents which, for

Table 9-1. Electrochemical series. (Courtesy of Interference Control Technologies.)

Metal	EMF (volts)	Metal	EMF (volts)
Magnesium	+2.37	Lead	+0.13
Magnesium alloys		Brass	0.0
Beryllium	+1.85	Copper	−0.34
Aluminum	+1.66	Bronze	
Zinc	+0.76	Copper–nickel alloys	
Chromium	+0.74	Monel	
Iron or steel	+0.44	Stainless steel	
Cast iron		Silver solder	
Cadmium	+0.40	Silver	−0.80
Nickel	+0.25	Graphite	
Tin	+0.14	Platinum	−1.20
Lead-tin solders		Gold	−1.50

example, may be flowing across a metallic junction as a result of using a structure as a power-system ground return.

Since mating bare metal to bare metal is essential for a satisfactory bond, a frequent conflict arises between bonding and finishing specifications. For EMI control, it is preferable to remove the finish where compromising bonding effectiveness would otherwise occur. Certain conductive coatings, such as alodine, iridite, and Dow #1, or protective-metal platings such as cadmium, tin, or silver, need not generally be removed. Most other coatings, however, are nonconductive and destroy the concept that a bond must offer a low-impedance RF path. For example, anodized aluminum appears to the eye to be a good conductive surface for bonding, but, in reality, it is covered with an insulated coating. Figure 9-16 shows the effect of various protective coatings on the electric-field shielding effectiveness of aluminum and magnesium, illustrating degradation in shielding effectiveness due to the application of various finishes. The shielding effectiveness is referenced, in dB, to the shielding effectiveness of the untreated metal. Thus, the shielding effectiveness at 5 MHz for a metal treated with DOW #7, for instance, is 25 dB worse than the shielding effectiveness of the untreated metal. The shielding effectiveness is a function of the coating conductivity (see Chapter 10).

Corrosion Protection

The most effective way to avoid the adverse effects of corrosion is to use metals low in the electrochemical activity table, such as tin, lead, or copper. This is not generally practical in the design of many structures (e.g., aircraft) because of weight and cost considerations. Consequently, the more active,

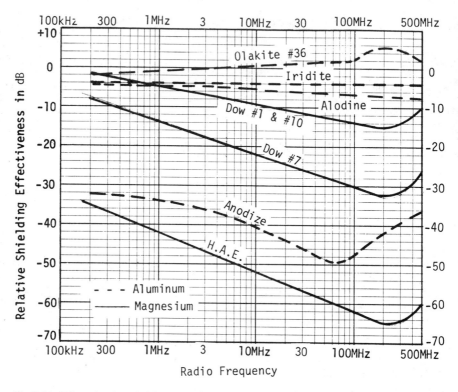

Fig. 9-16. Effect of various finishes on shield effectiveness of aluminum and magnesium. (*Courtesy of Interference Control Technologies*)

lighter metals, such as magnesium and aluminum, are employed, although stainless steel has been used in many missile programs.

Joined metals should be close together in the activity series if excessive corrosion is to be avoided. Magnesium and stainless steel form a galvanic couple of high potential (about 3 volts) that tends toward a rapid corrosion of the magnesium. Where dissimilar metals must be used, select replaceable components such as grounding jumpers, washers, bolts, or clamps, rather than structural members as the object likely to corrode. Thus, the object with the smaller mass should be of the higher potential in the electrochemical series (e.g., steel washers could be used with brass structures).

When members of the electrolytic couple are widely separated in the activity table, it is often practical to use a plating, such as cadmium or tin, to help reduce the dissimilarity. Sometimes, it is possible to electrically insulate metals with organic and electrolytic finishes and seal the joint against moisture to avoid corrosion. One solution to electrolytic corrosion is to avoid the use of the structure or housing for a power-ground return.

Considering the broad aspects of corrosion, both galvanic and electrolytic activity, the problem can be eliminated either by removing the metal-to-metal interface or by applying an effective seal to prevent moisture penetration. For example, some ships and aircraft are constructed of fiberglass skins. Alternatively, if the metal-to-metal interface is covered with a moisture-proof coating, the electrolyte will be prevented from forming. During maintenance, where the metallic parts must be temporarily separated, the moisture seal is usually broken. During reassembly, the old seal material should be removed and a new coating applied.

REFERENCES

1. Denny, Hugh W. *Grounding for the Control of EMI*. Gainesville, Virginia: Don White Consultants, Inc. 1983.
2. White, Donald R. J. *EMI Control Methods and Techniques*. A Handbook on Electromagnetic Interference and Compatibility, Volume 3. Gainesville, VA: Don White Consultants, Inc. 1977.
3. Ott, Henry, W. *Noise Reduction Techniques in Electronic Systems*. New York: John Wiley & Sons. 1976.
4. White, Donald, R. J. *EMI Control Methodology and Procedures*, Third Edition, Second Printing. Gainesville, Virginia: Don White Consultants, Inc. 1982.

Chapter 10
SHIELDING THEORY AND MATERIALS

Shielding basics and information on materials are presented in this chapter. These topics include field theory and near- and far-field definitions that lead to concepts such as wave and metal barrier impedances, absorption losses, reflection loss, and overall shielding effectiveness. As proper shielding design is necessary in many instances to achieve EMC objectives, (3) this chapter is of paramount importance to this *Handbook*.

FIELD THEORY

This section presents some relations of magnetic, electric, and electromagnetic fields as pertinent background for understanding and applying field theory to shielding concepts.

The electric (E_θ and E_r) and magnetic (H_ϕ) field components in spherical coordinates are found by solving Maxwell's equation about a small dipole (doublet) that is electrically excited in an oscillatory manner, and whose length D is considerably less than a wavelength ($D \ll \lambda$). The solutions for the E and H fields are (7, 9):

$$E_\theta = \frac{Z_0 I D \pi \sin \theta}{\lambda^2} \left[-\left(\frac{\lambda}{2\pi r}\right)^3 \cos \phi - \left(\frac{\lambda}{2\pi r}\right)^2 \sin \psi + \left(\frac{\lambda}{2\pi r}\right) \cos \psi \right] \quad (10\text{-}1)$$

$$E_r = \frac{2 Z_0 I D \pi \cos \theta}{\lambda^2} \left[\left(\frac{\lambda}{2\pi r}\right)^3 \cos \psi + \left(\frac{\lambda}{2\pi r}\right)^2 \sin \psi \right] \quad (10\text{-}2)$$

$$H_\phi = \frac{I D \pi \sin \theta}{\lambda^2} \left[\left(\frac{\lambda}{2\pi r}\right)^2 \sin \psi + \left(\frac{\lambda}{2\pi r}\right) \cos \psi \right] \quad (10\text{-}3)$$

where:

$Z_0 =$ free space impedance (for $r \gg \lambda/2\pi = \sqrt{\mu/\varepsilon} = 120\pi = 377$ ohms)
$I =$ current in a short wire (dipole)
$D =$ length of short wire

θ = zenith angle to radial distance r
λ = wavelength corresponding to frequency $f = (c/\lambda)$
r = distance from short wire dipole to measuring or observation point
$\psi = (2\pi f/\lambda) - \omega t$
ω = radian frequency = $2\pi f$
t = time = $1/f$
$c = 1/\sqrt{\mu\varepsilon} = 3 \times 10^8$ m/sec

It is assumed at this juncture that the wave propagating from the oscillatory dipole is normally incident upon a barrier surface; that is, the barrier surface is normal to the radial spherical coordinate, r, as illustrated in Fig. 10-1.

From a shielding point of view, the component of the Poynting vector of interest is that which accounts for the propagating wave energy density normal to the barrier surface. This normal component, given by the vector cross product $\bar{E}_\theta \times \bar{H}_\phi$, accounts for the energy density of the wave component that penetrates the barrier surface. The component $\bar{E}_r \times \bar{H}_\phi$ represents the wave energy that propagates in a direction parallel to the barrier surface without penetrating this surface. Therefore, the field vectors of interest are $\bar{E}_\theta$ and $\bar{H}_\phi$, and impedance relationships between these components are developed.

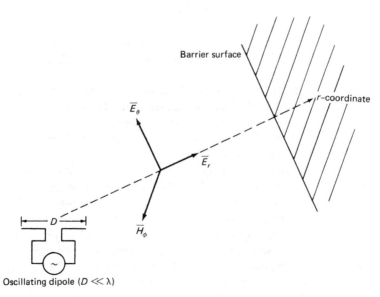

Fig. 10-1. Electromagnetic wave propagating outward from an oscillating dipole and impinging upon a barrier surface.

Several observations can be made about the near and far fields from equations (10-1) and (10-3):

1. When the coefficient multiplier is unity, $(\lambda/2\pi r) = 1$, in the electric field and the magnetic field terms, all coefficients of either the sine or cosine terms within the brackets are unity and equal. Thus, when $r = \lambda/2\pi$ (about one-sixth wavelength), this distance corresponds to the transition field condition or boundary between the near field and far field for electromagnetic compatibility considerations (7, 8).

2. When $r \gg (\lambda/2\pi)$ (far field conditions), only the last term of equations (10-1) and (10-3) is significant. For this condition, the wave impedance $Z_0 = E_\theta/H_\phi = 377$ ohms. This is called the radiation field (plane waves), and both E_θ and H_ϕ are in time phase although they are in directional quadrature.

3. When $r \ll (\lambda/2\pi)$ (near field conditions), only the first term of each equation is significant. For this condition, the wave impedance $E_\theta/H_\phi = Z_0(\lambda/2\pi r)$. Note that the wave impedance is now $\gg Z_0$, the free space impedance, since $(\lambda/2\pi r) \gg 1$. This is sometimes called an electric field or a high impedance field; that is, the impedance is high relative to plane wave impedance. It is also the case where the induction field and E_θ and H_ϕ are both temporally and spatially in quadrature.

4. Had the oscillating source not been a small straight wire or a dipole exhibiting high circuit impedance, the first term appearing in equations (10-1) and (10-2) would vanish, and a similar first term would appear in equation (10-3). For this condition, the wave impedance in the near field $E_\theta/H_\phi = Z_0(\lambda/2\pi r)$. Note that the wave impedance is now $\ll Z_0$, for $(\lambda/2\pi r) \gg 1$. This is sometimes called a magnetic field or a low impedance field, that is, low impedance relative to Z_0, the plane wave (radiation) impedance.

Figure 10-2 illustrates observations (1) through (3) above for the amplitude of each of the electric field terms in equation (10-1). Note that the quasi-stationary field is the largest term in the near field, and the induction term is the next largest, whereas the radiation term is the largest in the far field. For $r = (\lambda/2\pi)$, the coefficients of all three terms are equal.

Figure 10-3 illustrates preceding observations (3) and (4) in the near or induction field (5). Situation (a) in Fig. 10-3 is a short monopole or straight wire in which the circuit current is low (the conduction current in the wire is the same as the displacement current flowing in the capacitance from wire to ground or, in the case of a wire pair, from one end of the wire to the lower return side). Consequently, the source or circuit impedance $= V/I$ is a high impedance. As previously shown, the wave impedance in the near field is also high relative to 377 ohms. The associated electric field attenuates more rapidly $(1/r^3)$ with an increase in distance than the magnetic field $(1/r^2)$ in the induction region (cf. equation 10-1 and 10-3). Thus, the wave impedance

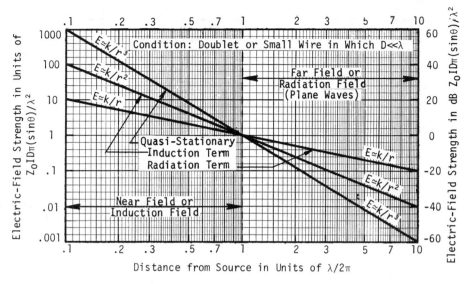

Fig. 10-2. Electric-field strength vs. source distance. (*Courtesy of Interference Control Technologies*)

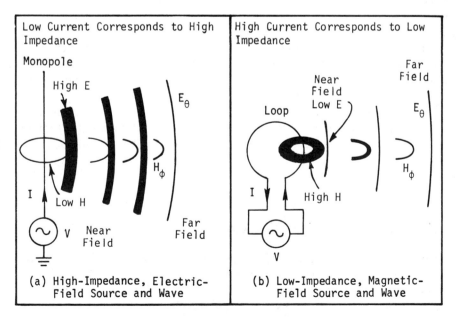

Fig. 10-3. Conceptual illustration of field strength vs. source type and distance. (*Courtesy of Interference Control Technologies*)

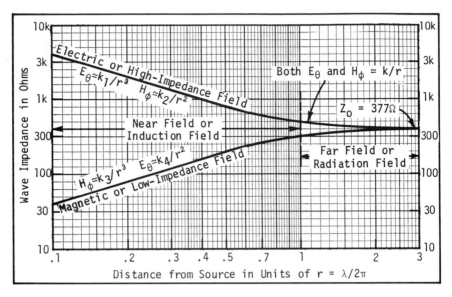

Fig. 10-4. Wave impedance as a function of source distance. (*Courtesy of Interference Control Technologies*)

decreases linearly (20 dB per decade) with an increase in distance, asymptotically approaching $Z_0 = 377$ ohms in the far or radiation field (for $r \gg \lambda/2\pi$). This relation is depicted in Fig. 10-4.

The converse is true in Fig. 10-3, wherein a low impedance source or circuit creates a low impedance wave (i.e., a magnetic field). This wave impedance increases linearly (20 dB per decade) with distance and asymptotically approaches 377 ohms in the far field.

When the dimension D of an emission source (e.g., a wire or an antenna) becomes an appreciable fraction of a wavelength, the expressions presented in equations (10-1) to (10-3) no longer apply. Different relations then must be used, since the curvature of the arriving wave front no longer permits the incremental elements of an antenna or radiating source to be in phase. This situation is described in Chapter 2 of Reference 7.

WAVE AND METAL IMPEDANCE

The previous section showed that the electromagnetic wave impedance is a function of the impedance of the source circuit generating the wave, the distance, r, from the source-to-metal barrier measuring point, and the frequency. Summing up, the wave impedance, Z_w, is:

$$Z_w = E/H = kZ_0 = k\sqrt{\mu/\varepsilon} = 120\pi k = 377k \text{ ohms} \qquad (10\text{-}4)$$

where:

$k \simeq \lambda/2\pi r$ for constant current source (very high impedance) and
$r \leq \lambda/2\pi$ (10-5)

$k \simeq 2\pi r/\lambda$ for constant voltage source (very low impedance) and
$r \leq \lambda/2\pi$ (10-6)

$k \simeq 1$ for any network impedance and $r \geq \lambda/2\pi$ (10-7)

$\mu = \mu_0\mu_r$ = permeability of medium

μ_0 = absolute permeability of free space (air) = $4\pi \times 10^{-7}$ henries/meter

μ_r = permeability of medium relative to free space (air)

$\varepsilon = \varepsilon_0\varepsilon_r$ = permittivity of medium

ε_0 = absolute permittivity of free space (air) $\simeq 8.84 \times 10^{-12}$ farads/meter

ε_r = permittivity of medium relative to free space (air)

provided that the length, l, of the emission generating element of the source is $l \ll \lambda$. Thus, in the far field, the plane wave impedance, 377 ohms, is independent of the circuit source impedance, but in the near field, the wave impedance can never exceed $377(\lambda/2\pi r)$ nor can it be less than $377(2\pi r/\lambda)$.

The above development raises the question about just how the source generating circuit and wave impedances are related in the near field. This relation is important because the shielding effectiveness in the near field is presented in terms of a high impedance source (electric fields) and low impedance source (magnetic fields). Use of electric field sources may give rise to optimistic results (relatively high shielding effectiveness), and use of magnetic sources gives rise to pessimistic results (relatively low shielding effectiveness). The former could result in an underdesign condition, and the latter could result in overdesign of any shielded circuit, box, equipment, cabinet, and so forth (7).

The development of a discrete relation between circuit, Z_c, and wave, Z_w, impedance in the near field is beyond the scope of this *Handbook*. However, the following mathematical relations *are* suggested for all conditions in which the circuit dimensions $l \ll \lambda$:

$$Z_w = \frac{Z_0\lambda}{2\pi r}, \text{ for } Z_c > \frac{Z_0\lambda}{2\pi r} \geq Z_0 \qquad (10\text{-}8)$$

$$Z_w \simeq Z_c, \text{ for } \frac{Z_0\lambda}{2\pi r} \geq Z_c \geq Z_0 \qquad (10\text{-}9)$$

$$Z_w \simeq Z_0, \text{ for } Z_c = Z_0 \text{ or } \frac{\lambda}{2\pi r} \geq 1 \qquad (10\text{-}10)$$

$$Z_w \simeq Z_c, \text{ for } Z_0 > Z_c \geq \frac{Z_0 2\pi r}{\lambda} \qquad (10\text{-}11)$$

$$Z_w \simeq \frac{Z_0 2\pi r}{\lambda}, \text{ for } Z_0 > \frac{Z_0 2\pi r}{\lambda} > Z_c \qquad (10\text{-}12)$$

Equations (10-8) through (10-12) are plotted in Fig. 10-5 for several values of Z_c, including common transmission-line impedances of 50, 100, 300, and 600 ohms. For these common transmission line impedances and the above conditions, neither a very high nor very low impedance conditions exists when $r \ll \lambda/2\pi$.

As discussed later, sometimes it is more useful to present equations (10-4) through (10-7) in terms of frequency, f, rather than wavelength. Since:

$$f\lambda_m = c = (1/\sqrt{\mu_0 \varepsilon_0}) = 3 \times 10^8 \text{ m/sec} \qquad (10\text{-}13)$$

or:

$$f_{\text{MHz}} = 300/\lambda_m = \text{frequency in MHz} \qquad (10\text{-}14)$$

Equations (10-5) through (10-7) then become:

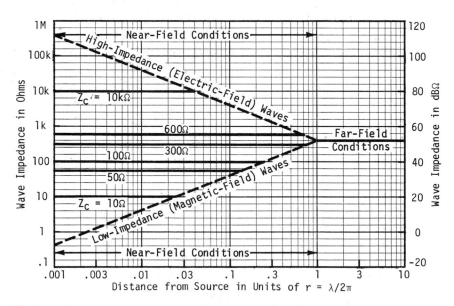

Fig. 10-5. Wave impedance for several circuit impedances. (*Courtesy of Interference Control Technologies*)

For a high impedance circuit in the near field:

$$Z_w = \frac{Z_0 \lambda}{2\pi r} = \frac{18,000}{r_m f_{MHz}} \quad \text{ohms} \tag{10-15}$$

For a low impedance circuit in the near field:

$$Z_w = \frac{Z_0 2\pi r}{\lambda} = 7.9 r_m f_{MHz} \text{ ohms} \tag{10-16}$$

For free-space far-field conditions:

$$Z_w = 120\pi = 377 \text{ ohms} \tag{10-17}$$

In terms of the more general condition of any circuit impedance as presented in equations (10-8) through (10-12), the generated wave impedance is:

$$Z_w \simeq 18,000/r_m f_{MHz}, \text{ for } Z_c \geq 18,000/r_m f_{MHz} \tag{10-18}$$

$$Z_w \simeq Z_c, \text{ for } 18,000/r_m f_{MHz} \geq Z_c \geq 7.9 r_m f_{MHz} \tag{10-19}$$

$$Z_w \simeq 7.9 r_m f_{MHz}, \text{ for } 7.9 r_m f_{MHz} \geq Z_c \tag{10-20}$$

Metal Impedance

All homogeneous materials are characterized by a quantity known as the intrinsic impedance of the material, which is defined by (7):

$$Z_i = \sqrt{\frac{j\omega\mu}{\sigma + j\omega\varepsilon}} \tag{10-21}$$

where:

$\omega = 2\pi f$ radians/sec

f = frequency in Hz

μ = permeability of the material = $\mu_0 \mu_r$

μ_0 = absolute permeability of free space (air) = $4\pi \times 10^{-7}$ henries/meter

μ_r = permeability of material relative to free space

σ = conductivity in mhos/meter

ε = permittivity of material = $\varepsilon_0 \varepsilon_r$

ε_0 = absolute permittivity of free space = $1/(36\pi \times 10^9) \simeq 8.84 \times 10^{-12}$ farads/meter

ε_r = permittivity of material relative to air

As an electromagnetic wave propagates through the material, the impedance of the wave, $Z_w = E/H$, approaches the value Z_i [see equation (10-21)].

For air, conductivity is extremely small; that is $\sigma \ll \omega\varepsilon$. Thus, the intrinsic impedance of equation (10-21) for air becomes:

$$Z_a = \sqrt{(\mu/\varepsilon)} = \sqrt{(\mu_0/\varepsilon_0)}$$
$$= \sqrt{(4\pi \times 36\pi \times 10^2)} = 120\pi = 377 \text{ ohms} \qquad (10\text{-}22)$$

In constrast to air and low-loss dielectrics, metals are defined as materials for which the conductivity is extremely high relative to air, such that $\sigma \gg \omega\varepsilon$. Thus, the intrinsic impedance of equation (10-21) for a metal becomes (7):

$$Z_m = \sqrt{\frac{j\omega\mu}{\sigma}} = \sqrt{\frac{j2\pi f\mu}{\sigma}}$$

$$= (1 + i)\sqrt{\frac{\pi f\mu}{\sigma}}, \text{ for } \sigma \gg \omega\varepsilon \text{ and } t \gg \delta \qquad (10\text{-}23)$$

$$|Z_m| = \sqrt{(2 \times 10^3)}\sqrt{(\pi\mu f_{MHz}/\sigma)} \text{ ohms/sq., for } f \text{ in MHz} \qquad (10\text{-}24)$$

where:

    t = thickness of the metal in meters
    δ = skin depth of metal

The intrinsic impedance of air (equation 10-22) is a purely resistive constant, whereas the intrinsic impedance of a metal contains both a resistive and an inductive component. Consequently, Z_m depends upon both the permeability and the conductivity of the metal.

Equation (10-24) may be expressed in terms relative to copper:

$$|Z_m| = 369\sqrt{\mu_r f_{MHz}/\sigma_r} \text{ } \mu\text{ohms/sq.} \qquad (10\text{-}25)$$

where:

    $\sigma = \sigma_c\sigma_r$
    σ_c = conductivity of copper = 5.8×10^7 mhos/meter
    σ_r = conductivity of metal relative to copper

Equation (10-25) is plotted in Fig. 10-6 for various metals (7).

The barrier impedance of metals is sometimes expressed in terms of skin depth:

$$|Z_m| = \left|\frac{(1 + j)}{\sigma\delta}\right| = \frac{\sqrt{2}}{\sigma\delta} \quad \text{ohms/sq.} \qquad (10\text{-}26)$$

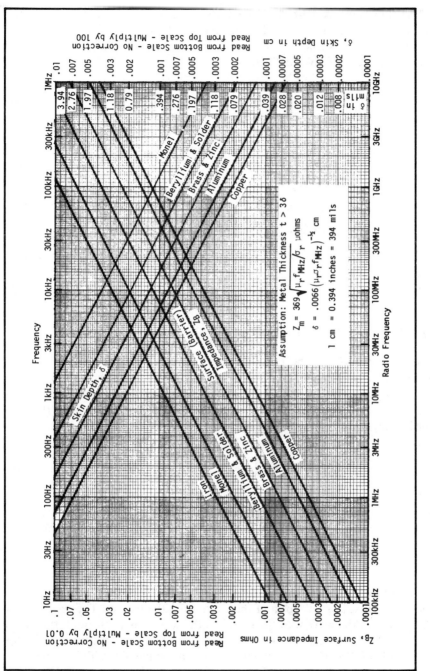

Fig. 10-6. Surface impedance and skin depth of various metals vs. frequency. (*Courtesy of Interference Control Technologies*)

The skin depth is defined as the surface thickness of a metal, at any frequency, at which the current has decayed to $1/e$ (37%) of the amplitude of the current at the surface of the metal. By combining equations (10-23) and (10-26), the skin depth is defined:

$$\delta = \frac{\sqrt{2}}{\sigma Z_m} = \frac{1}{\sigma}\sqrt{\frac{\sigma}{\pi f \mu}} = \frac{1}{\sqrt{\pi f \mu \sigma}} \gg t \qquad (10\text{-}27)$$

$$\delta = \sqrt{\frac{10^{-3}}{\pi f \mu \sigma}} \text{ for frequency in MHz} \qquad (10\text{-}28)$$

$$\delta = 0.0066/\sqrt{f_{\text{MHz}}} \text{ cm for copper } (\mu_r = \sigma_r = 1) \qquad (10\text{-}29)$$

$$\delta = 2.6/\sqrt{f_{\text{MHz}}} \text{ mils for copper } (\mu_r = \sigma_r = 1) \qquad (10\text{-}30)$$

$$\delta = 0.0066/\sqrt{\mu_r \sigma_r f_{\text{MHz}}} \text{ cm for any metal} \qquad (10\text{-}31)$$

$$\delta = 2.6/\sqrt{\mu_r \sigma_r f_{\text{MHz}}} \text{ mils for any metal} \qquad (10\text{-}32)$$

Equations (10-31) and (10-32) are plotted in Fig. 10-6 for various metals.

SHIELDING EFFECTIVENESS

This section develops the relations for a figure of merit that describes the performance of a shield in reducing the electromagnetic energy either impinging upon a potentially susceptible victim or exiting from an interfering source. This figure of merit is called *shielding effectiveness (SE)*. It is defined as (7):

$$SE_{\text{dB}} = 10\log_{10}\left[\frac{\text{incident power density}}{\text{transmitted power density}}\right] \qquad (10\text{-}33)$$

where incident power density = power density at a measuring point before a shield is in place, and transmitted power density = power density at the same measuring point after a shield is in place.

Equation (10-33) is defined as a loss so that the shielding effectiveness is always positive.

The amount of attenuation offered by a shield depends upon three mechanisms. The first is a reflection of the wave from the barrier. The second is an absorption of the wave into the metal as the wave passes through the metal. The third is a re-reflection that takes place as the wave encounters the opposite side of the shield after passing through the metal (7). This happens an infinite number of times as the wave bounces about inside the metal, but only the first re-reflection component will be discussed here.

In terms of field strength, equation (10-33) may also be defined as long as the two fields are measured in the same medium having the same wave impedance:

$$SE_{dB} = 20 \log_{10} \left(\frac{E_b}{E_a} \right) \quad \text{for electric fields} \qquad (10\text{-}34)$$

$$SE_{dB} = 20 \log_{10} \left(\frac{H_b}{H_a} \right) \quad \text{for magnetic fields} \qquad (10\text{-}35)$$

where:

E_b = electric field strength before shield is installed
E_a = electric field strength after shield is installed
H_b = magnetic field strength before shield is installed
H_a = magnetic field strength after shield is installed

Figure 10-7 shows the conceptual mechanism for determining shielding effectiveness. It is detailed in Fig. 10-8 (7) and may be explained as follows: The incident field strength (an electric field is illustrated) is considered to be normalized. The first mechanism of shielding effectiveness is due to reflection that occurs because of the impedance mismatch at the air–metal interface.

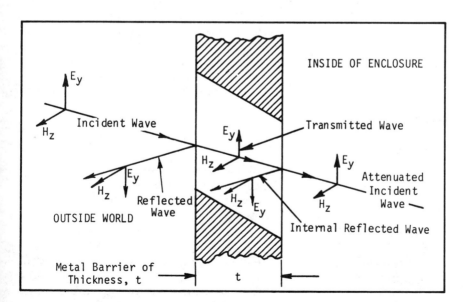

Fig. 10-7. Representation of shielding phenomena for plane waves. (*Courtesy of Interference Control Technologies*)

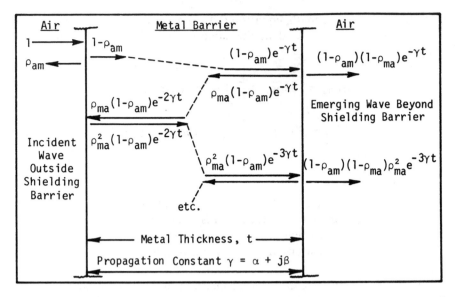

Fig. 10-8. Geometry of metal basrier used in explaining shielding effectiveness (see text). (*Courtesy of Interference Control Technologies*)

The normalized reflected field is:

$$p = \frac{1-K}{1+K} = -1 \text{ for } K \gg 1 \tag{10-36}$$

$$= 0 \text{ for } K = 1 \tag{10-37}$$

$$= +1 \text{ for } 0 \geq K \ll 1 \tag{10-38}$$

where:

$$K = Z_w/Z_m \tag{10-39}$$

Z_w = wave impedance as defined in equation (10-4)
Z_m = metal impedance as defined in equation (10-23)

and:

$$p_{am} = \frac{1-K_{am}}{1+K_{am}} \tag{10-40}$$

= reflection coefficient at the air-to-metal interface (i.e., wave propagation from air medium into metal)

$$p_{ma} = \frac{K_{ma} - 1}{K_{ma} + 1} \tag{10-41}$$

= reflection coefficient at the metal to air interface (i.e., wave propagation from metal medium into air), in which case the roles of Z_w and Z_m are interchanged, whereby K_{ma} becomes $1/K_{am}$ in equation (10-40) and equation (10-41).

The K term may be thought of as the voltage standing wave ratio (VSWR) for $Z_w/Z_m \geq 1$ in equation (10-39). When $Z_w/Z_m \leq 1$, the VSWR concept still applies if equation (10-40) is defined as equation (10-41), in which case $Z_m/Z_w \geq 1$.

The relative transmitted field, Γ_{am}, just inside the left edge of the metal-to-air barrier material is:

$$\Gamma_{am} = 1 - p_{am} \tag{10-42}$$

This field undergoes an attenuation in traversing the thickness of the metal barrier, the second mechanism of shielding effectiveness, that is essentially due to ohmic losses in the metal and turns the associated losses into exothermic heat. The arriving field results in a lower field strength emerging at the inside of barrier, and is described in equation (10-43):

$$\Gamma_{AR} = \Gamma_{am}e^{-\gamma t} = e^{-(\alpha + j\beta)t}\Gamma_{am} = e^{-(\alpha + j\beta)t}(1 - p_{am}) \tag{10-43}$$

where:

γ = propagation constant = $\alpha + j\beta$
α = attenuation constant
β = phase constant = $2\pi/\lambda$
t = metal thickness

The third mechanism of shielding effectiveness is due to re-reflection. The re-reflected relative field strength, $\Gamma_{RR'}$ at the inside right edge of the metal-to-air barrier of Fig. (10-18) is:

$$\Gamma_{RR} = p_{ma}\Gamma_{AR} = e^{-\gamma t}p_{ma}(1 - p_{ma}) \tag{10-44}$$

where p_{ma} = metal-to-air reflection coefficient (equation 10-41).

The relative transmitted field, $\Gamma_{RT'}$ to the right, just outside the metal barrier is:

$$\Gamma_{RT} = 1 - p_{ma}\Gamma_{AR} = e^{-\gamma t}(1 - p_{am})(1 - p_{ma}) \tag{10-45}$$

Equation (10-45) is the shielding effectiveness expressed as a gain (a negative number) when $\gamma \gg 1$.

When all fields are combined, the relations may be expressed in terms of the impedance ratio of the air–metal and metal–air interfaces:

$$\Gamma_T = e^{-\alpha t}\left(\frac{2K}{1+K}\right)\left(\frac{2}{1+K}\right)\left[1 - \left(\frac{K-1}{K+1}\right)^2 e^{-2\gamma t}\right]^{-1} \tag{10-46}$$

$$\Gamma_T = e^{-\alpha t}\frac{4K}{(1+K)^2}\left[1 - \left(\frac{K-1}{K+1}\right)^2 e^{-2\gamma t}\right]^{-1} \tag{10-47}$$

where $e^{-\alpha t}$ = absorption component (A), $4K/(1+K)^2$ = reflection term (R), and the term in the brackets is the re-reflection term (B).

Expressing equation (10-47) as a loss (i.e., shielding effectiveness) rather than a gain, and converting it to decibels, there results the following expression:

$$SE_{dB} = 20 \log_{10}(1/\Gamma_T)$$

$$= 20 \log_{10}\left[e^{\alpha t}\left[\frac{(1+K)^2}{4K}\left(1 - \left(\frac{K-1}{K+1}\right)^2 e^{-2\gamma t}\right)\right]\right]$$

$$= A_{dB} + R_{dB} + B_{dB} \tag{10-48}$$

where:

$$\text{Absorption loss, } A_{dB} = 8.686\alpha t \tag{10-49}$$

$$\text{Reflection loss, } R_{dB} = 20 \log_{10}(1+K)^2/4K \tag{10-50}$$

The re-reflection correction is expressed as:

$$B_{dB} = 20 \log_{10}(1 - [(K-1)^2/(K+1)^2]e^{-2\gamma t}) \tag{10-51}$$

The topic of shielding effectiveness will now be examined in further detail. The three loss components to be examined are: (1) the absorption loss, (2) the reflection loss, and (3) the re-reflection loss correction (7).

Absorption Loss

Equation (10-49) may be expanded:

$$A_{dB} = 8.686\alpha t = 8.686t\sqrt{\pi f \mu \sigma}$$

where:

$$\gamma = \alpha + j\beta = \sqrt{j\omega\mu(\sigma + j\omega\varepsilon)} \qquad (10\text{-}52)$$

$$\gamma = \sqrt{j\omega\mu\sigma} \text{ since } \sigma \gg \omega\varepsilon \text{ for metals} \qquad (10\text{-}53)$$

$$\gamma = (1 + j)\sqrt{\pi f\mu\sigma} \qquad (10\text{-}54)$$

$$\alpha = \beta = \sqrt{\pi f\mu\sigma} \text{ for metals} \qquad (10\text{-}55)$$

If equation (10-52) is defined in terms of t in mils (thousandths of an inch) for English units and in terms of centimeters for the metric system, and if f is in megahertz, equation (10-52) becomes:

$$A_{dB} = 3.338 t_{mils}\sqrt{f_{MHz}\mu_r\sigma_r} \text{ dB, English units} \qquad (10\text{-}56)$$

$$A_{dB} = 1314.3 t_{cm}\sqrt{f_{MHz}\mu_r\sigma_r} \text{ dB, metric units} \qquad (10\text{-}57)$$

where μ_r and σ_r are the permeability and conductivity relative to copper.

Figure 10-9 shows the absorption loss as a function of frequency for a number of commonly used shielding materials. Note that the absorption loss is independent of wave impedance.

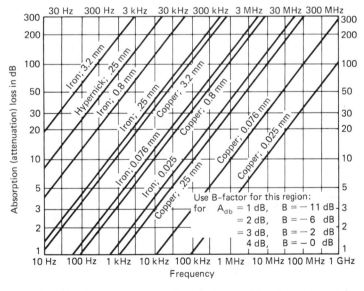

Fig. 10-9. Shielding absorption (A_{dB}) loss vs. frequency, material and thickness (due to attenuation by penetration: independent of wave impedance). (*Courtesy of Interference Control Technologies*)

Reflection Loss

The contribution of reflection loss to the overall shielding effectiveness is due to an impedance mismatch at the metal–barrier interfaces. Thus, it is useful to substitute the impedances of equation (10-39) by their equivalents from equation (10-4) for Z_w and equation (10-23) for Z_m:

$$K = \frac{Z_w}{Z_m} = \frac{k\sqrt{\mu_0/\varepsilon_0}}{(1+j)\sqrt{\pi f \mu/\sigma}} \qquad (10\text{-}58)$$

where:

$$k = (\lambda/2\pi r) = 1/2\pi r f \sqrt{\mu_0 \varepsilon_0} \text{ for high impedance, } E \text{ fields} \qquad (10\text{-}59)$$

$$k = 2\pi r/\lambda = 2\pi r f \sqrt{\mu_0 \varepsilon_0} \text{ for low impedance, } H \text{ fields} \qquad (10\text{-}60)$$

$$k = 1 \text{ for far fields, } r \geq (\lambda/2\pi) \qquad (10\text{-}61)$$

Combining equation (10-58) through (10-61) yields:

$$k = 1/2\pi r f \varepsilon_0 \sqrt{2\pi f \mu/\sigma} \text{ for high impedance fields} \qquad (10\text{-}62)$$

$$k = r\sqrt{2\pi f \sigma \mu_0/\mu_r} \text{ for low impedance fields} \qquad (10\text{-}63)$$

$$k = 1/\sqrt{2\pi f \mu_r \varepsilon_0/\sigma} \text{ for plane waves (far field)} \qquad (10\text{-}64)$$

The reflection loss term is equation (10-50) may be expanded in terms of the wave and metal barrier impedances:

$$R_{dB} = 20 \log_{10} (1 + K)^2/4K \qquad (10\text{-}65)$$

$$R_{dB} \simeq 20 \log_{10} (K/4) = 20 \log_{10} (Z_w/4Z_m) \text{ for } K \gg 1 \qquad (10\text{-}66)$$

Equations (10-62) through (10-64) may be substituted into equation (10-66) to yield:

Electric (high impedance) fields:

$$R_{dB} \simeq 152.0 - 10 \log_{10} (\mu_r f_{MHz}{}^3 r_{ft}{}^2/\sigma_r) \text{ dB, English units} \qquad (10\text{-}67)$$

$$R_{dB} \simeq 141.7 - 10 \log_{10} (\mu_r f_{MHz}{}^3 r_m{}^2/\sigma_r) \text{ dB, metric units}$$

where;

r_{ft} = source to barrier distance in feet
r_m = source to barrier distance in meters

Magnetic (low impedance) fields:

$$R_{dB} \simeq 64.3 - 10 \log_{10} (\mu_r/f_{MHz}\sigma_r r_{ft}^2) \text{ dB, English units} \quad (10\text{-}69)$$

$$R_{dB} \simeq 74.6 - 10 \log_{10} (\mu_r/f_{MHz}\sigma_r r_{m}^2) \text{ dB, metric units} \quad (10\text{-}70)$$

Plane waves (far field conditions):

$$R_{dB} = 108.1 - 10 \log_{10} (\mu_r f_{MHz}/\sigma_r) \text{ dB, both units} \quad (10\text{-}71)$$

(independent of r as long as $r \gg (\lambda/2\pi) \simeq 47.8/f_{MHz}$).

Figure 10-10 (a, b, and c) shows calculation of the reflection loss versus frequency for the three types of waves: plane, high impedance, and low impedance.

When the concept of circuit impedance is introduced rather than simply high or low impedance waves (cf. equations 10-8 through 10-20), the shielding effectiveness in the near field becomes:

$$R_{dB} \simeq 20 \log_{10} \left[\frac{Z_c}{4 \times 369 \times 10^{-6}\sqrt{\mu_r f_{MHz}/\sigma_r}} \right] \quad (10\text{-}72)$$

$$R_{dB} = 56.6 - 10 \log_{10} (\mu_r f_{MHz}/\sigma_r Z_c^2) \quad (10\text{-}73)$$

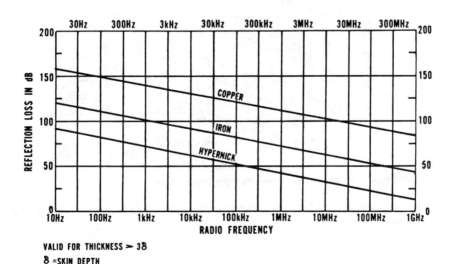

VALID FOR THICKNESS > 3δ
δ =SKIN DEPTH

Fig. 10-10a. Reflection loss R_{dB} of plane waves vs. radio frequency. (*Courtesy of Interference Control Technologies*)

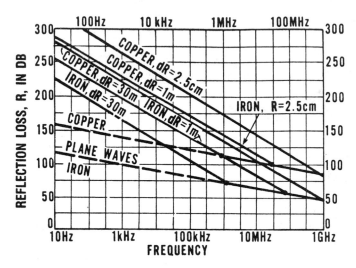

Fig. 10-10b. Reflection loss $*(R_{dB})$ of electric fields vs. frequency. (*Courtesy of Interference Control Technologies*)

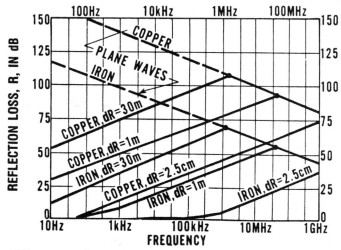

*Permeability assumed constant with frequency

Fig. 10-10c. Reflection loss (R_{dB}) of magnetic fields vs. frequency. (*Courtesy of Interference Control Technologies*)

or:

$$R_{dB} = \text{Eqs. (10-67) and (10-68), whichever is less for } E \text{ fields.}$$

$$R_{dB} = 56.6 - 10 \log_{10} (\mu_r f_{MHz}/\sigma_r Z_c^2) \tag{10-74}$$

or:

$$R_{dB} = \text{Eqs. (10-69) and (10-70), whichever is less for } H \text{ fields.}$$

$$R_{dB} = \text{Eq. (10-71) for plane waves.}$$

Re-reflection Correction

Equation (10-51) may be expanded in terms of the wave and metal-barrier impedance (see equation 10-58):

$$B_{dB} = 20 \log_{10} \left[1 - (K - 1)^2/(K + 1)^2 e^{-2\gamma t} \right] \tag{10-75}$$

$$B_{dB} = 20 \log_{10} \left[1 - \left(\frac{K - 1}{K + 1} \right)^2 10^{-.1 A_{dB}}(\cos 0.23 A_{dB} - j \sin 0.23 A_{dB}) \right] \tag{10-76}$$

$$B_{dB} \simeq 20 \log_{10} \left[1 - e^{2(\alpha + j\beta)t} \right], K \gg 1 \tag{10-77}$$

$$B_{dB} \simeq 20 \log_{10} \left[1 - e^{2t\sqrt{\pi f \mu \sigma}} e^{2jt\sqrt{\pi f \mu \sigma}} \right] \tag{10-78}$$

where A_{dB} is given in Equations (10-56) and (10-57).

Shield Performance Graphs

Shielding effectiveness may be plotted as a function of frequency in which the parameters are: source to shield distance; field type (E field, H field, or or plane waves); shield metal and its properties; and metal thickness. Two such graphs are shown in Figs. 10-11 and 10-12 for a source to-metal distance of 1 meter (7).

Figure 10-11, developed for copper, may be used for any of the good conductor family (e.g., aluminum, brass, gold) with only a few decibels difference in performance. Note that the performance of the shield to E fields is everywhere equal to or better than plane waves, and the performance of the shield to plane waves is everywhere equal to or better than for H fields. The E and H field performance curves approach the plane wave performance curves at the corner frequency of 48 MHz, corresponding to $\lambda/2\pi$ for a distance of 1 meter. The region in which the E fields and plane waves correspond to an increase in shielding effectiveness with an increase in frequency is the beginning of the absorption loss region. This region occurs where the metal thickness is on the order of one skin depth ($t \simeq \delta$).

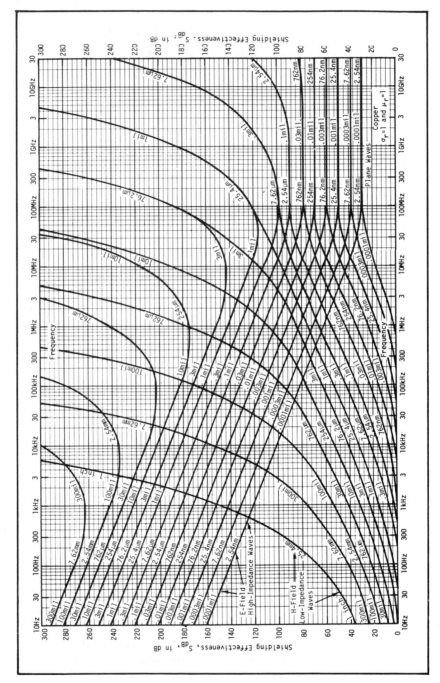

Fig. 10-11. Shielding effectiveness of copper vs. frequency for source-to-metal distance of 1 m. (*Courtesy of Interference Control Technologies*)

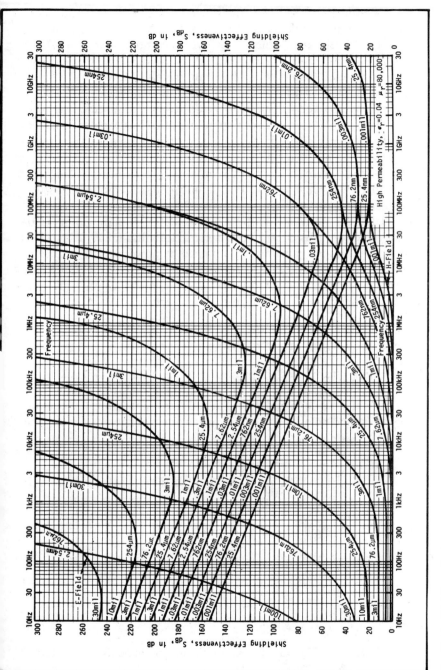

Fig. 10-12. Shielding effectiveness of high permeability vs. frequency for source-to-metal distance of 1 m. (*Courtesy of Interference Control Technologies*)

It is also seen in Fig. 10-11 that the shielding effectiveness of household aluminum foil, having a thickness of about 40 µm, corresponds to at least 130 dB to both E-field and plane waves over the entire spectrum. On the other hand, this same foil is transparent to H fields ($SE = 0$ dB) at power mains frequencies of 50 and 60 Hz. Thus, thin metals of low relative permeability do not perform as a magnetic shield at low frequencies.

In order to obtain a good shield to H fields at low frequencies, it is necessary to choose a metal thickness corresponding to at least one or more skin depths, as shown in Fig. 10-11. For example, to obtain about 40 dB of shielding effectiveness for a 1-meter distance at power mains frequencies, copper or aluminum would have to be about 1 cm thick! To avoid such thick and heavy construction, the high-permeability family of metals should be used for shielding against magnetic fields.

Figure 10-12 shows the shielding effectiveness of high-permeability materials for a source-to-metal distance of 1 meter. Note that for H fields, 40 dB of shielding effectiveness can now be obtained at power mains frequencies with a metal thickness of about 0.5 mm. Further, the H-field shielding effectiveness flattens out at very low frequencies and becomes constant right down to DC. This is the region discussed in the next section.

Low Frequency Magnetic Shielding Effectiveness

Overall shielding effectiveness is presented in equation (10-48) and is applicable for the far-field region and for and E and H fields in the near-field region for any thickness of metal (7). A substantial simplification of equation (10-48) results at and below very low frequency for magnetic materials when t/δ, K, and $2\delta t \ll 1$:

$$SE_{dB} = 20 \log_{10}\left[\frac{1}{4K}[1 - (1 - 4K)(1 - 2\gamma t)]\right] \qquad (10\text{-}79)$$

$$SE_{dB} = 20 \log_{10}\left[\frac{1}{4K}[1 - (1 - 4K - 2\gamma t)]\right] \quad \text{for } 8K\gamma t \ll 4K \text{ and } 2\gamma t$$

$$\qquad (10\text{-}80)$$

$$SE_{dB} \simeq 20 \log_{10}(1 + \gamma t/2K) \qquad (10\text{-}81)$$

where:

$$\gamma = (1 + j)\alpha = \sqrt{2\pi f \mu \sigma} \qquad (10\text{-}52)$$

$$K = Z_w/Z_m = r\sqrt{2\pi f \sigma \mu_0/\mu_r} \text{, for magnetic fields} \qquad (10\text{-}63)$$

Substituting equations (10-52) and (10-63) into equation (10-81) yields:

$$SE_{dB} \simeq 20 \log_{10}\left(1 + \frac{t\sqrt{2\pi f\mu_0\mu_r\sigma}}{2r\sqrt{2\pi\sigma\mu_0/\mu_r}}\right) \qquad (10\text{-}82)$$

$$SE_{dB} \simeq 10 \log_{10}\left[1 + \left(\frac{\mu_r t}{2r}\right)^2\right], \quad \text{for } t/\delta, K, \text{ and } 2\gamma t \ll 1 \qquad (10\text{-}83)$$

Equation 10-83 is a useful and simple expression for determining the shielding effectiveness for very low frequency and DC magnetic fields. It is equally useful for determining the shielding of a wire pair, coaxial line, or harness having a magnetic shield of thickness t and radius r. If a wire or harness shield is used, it is necessary to be certain that the effective relative permeability, μ_r, is used for braided shields, since much of the shield may be air or some other material.

Example

Determine the low frequency magnetic shielding effectiveness of a tin-plated, copper-clad steel braid in which tin = 3%, copper = 37%, and steel = 60%. The braid is 80% material and 20% air by volume and has an average thickness of $t = 0.020$ inch. The braided shield, when slipped over a harness, has an average radius of 0.50 inch:

$$\mu_r = (M_{de}MM_{de}u_{MM} + NMM_{de} \times 1) + A_{de} \times 1 \qquad (10\text{-}84)$$

$$\mu_r \simeq M_{de}MM_{de}u_{MM}, \text{ since } NMM_{de} \ll MM_{de}u_{MM} \qquad (10\text{-}85)$$

where:

M_{de} = decimal equivalent of total material by volume

MM_{de} = decimal equivalent of magnetic material by volume, referenced to total material

u_{MM} = permeability of magnetic material

NMM_{de} = decimal equivalent of nonmagnetic material by volume, referenced to total material

A_{de} = decimal equivalent of air by volume = $1 - M_{de}$

Equation (10-85) for the above example becomes:

$$\mu_r = 0.80 \times 0.60 \times 1000 = 480$$

Thus, the low frequency magnetic shielding effectiveness is:

$$SE_{dB} = 10 \log_{10}\left[1 + \left(\frac{480 \times 0.02}{2 \times 0.50}\right)^2\right] = 19.7 \text{ dB}$$

Performance Degradation

All preceding discussions assume that the shielding material is both homogeneous and large in planar dimensions such that neither leakages nor edge effects take place. Leakage through slots, apertures for switches and I/O cables, resonance cavity effects, and the effects of boundary discontinuities on the field distribution will result in a compromised shield performance. The shielding effectiveness expressed by equation (10-48) may then be conceptually redefined to account for nonideal shield performance:

$$SE'_{dB} = A_{dB} + R_{dB} + B_{dB} - \text{Leakage effects} - \text{Standing wave effects}$$

$$(10\text{-}86)$$

where:

A_{dB} = absorption loss, Eq. (10-49) or (10-56)
R_{dB} = reflection loss, Eq. (10-50) or (10-66)
B_{dB} = re-reflection correction, Eq. (10-51) or (10-75)

Leakage effects in equation (10-86) may be identified as due to one or more of the following situations, which exist in any practical, real-life shield. The leakage effects are due to the following types of leakage paths:

- Seams:
  Mating members
  Screws
  Crimps
  Welds
  Brazes
  Silver solder
  Soft solder
- Doors
- Cover plates
  Access panels
  Drawers
- Vents:
  Ventilation
  Air conditioning
  Heating

- Holes and apertures for:
  Connectors
  Fuses
  Power lines
  Signal and control cables
  Fiber optic lines
  Potentiometers
  Viewing windows
  Status indicators
  Push buttons
  On/off switches
  Control switches
  Convection cooling
- Nonhomogeneous areas:
  Screens
  Shield braids
  Meshes
  Thin areas

Degradation due to standing waves in equation (10-86) involves resonance effects at higher frequencies where enclosures act as microwave cavities. This results in areas or regions within a shielded enclosure that exhibit poorer performance (i.e., lower shielding effectiveness).

Another type of degradation occurs due to field reinforcement at corners and sharp edges in the enclosure that results in high field levels in the vicinity of abrupt metal transitions. Reinforcement is due to the "piling up" of the waves at these discontinuities. Some of this effect can be alleviated by making welds and corners as smooth and as round as possible.

Equation (10-86) is expressed in conceptual form rather than in explicit form. The method of combining effects due to shield leakages (L_{dB}) and the base metal shielding effectiveness (SE_{dB}) (without these effects) is (7):

$$SE'_{dB} = -20 \log_{10}\left[\log_{10}^{-1}(-SE_{dB}/20) + \sum_{i=1}^{n} \log_{10}^{-1}(-L_{idB}/20) \right]$$

(10-87)

Note that equation (10-87) coherently adds all leakage effects together and coherently adds these results with SE_{dB}. This is a worst case, since the phases may range from nearly all in phase (coherent at low frequencies) to approaching random phase (incoherent at high frequencies). If coherent addition did not take place (i.e., effects combined randomly or incoherently), the three "20s" in equation (10-87) would be replaced by three "10s." However, usually one or a few leakage paths predominate so that coherent worst-case combination is justified.

Example

Suppose that the shielding material, per se, of an enclosure results in a shielding effectiveness of 110 dB at some frequency, but the leakages separately result in shielding effectiveness at each leakage point of: (1) filter/connector panel = 101 dB, (2) vent leakage = 92 dB, (3) door leakage = 88 dB, and (4) shield panel seam leakage = 83 dB. Calculate the total leakage shielding and the combined shielding effectiveness of the enclosure.

Application of the second half of equation (10-87) yields a total leakage shielding effectiveness of 76.8 dB. Applying all of equation (10-87) gives $SE_{dB} = 76.6$ dB. Thus, in this example, it does not particularly matter what the basic enclosure material performance is, as long as it is about 10 dB better than the lowest or least leakage point. This may be illustrated by applying equation (10-87) for several different values of SE_{dB} in the above example having a total leakage shielding effectiveness of 76.8 dB.

Basic enclosure material SE_{dB}	Shielding for total leakages L_{dB}	Combined shielding SE'_{dB}
60	76.8	58.8
70	76.8	66.7
80	76.8	72.2
90	76.8	75.1
100	76.8	76.2
110	76.8	76.6
120	76.8	76.7
130	76.8	76.8
≥ 150	76.8	76.8

The tabulation serves to demonstrate the importance of controlling leakages in shielded enclosures. Usually, but not always, a leakage results in a poorer performance than that of the base metal (one exception is for low frequency magnetic fields where the phase shift through the leakage can add out of phase with the penetration of the base shield metal, resulting in a reduced field amplitude inside the enclosure). Leakages and the techniques for controlling them in shielded housing designs are discussed in Chapter 11.

In summary, the shielding effectiveness of a six-sided homogeneous box may be regarded as the upper limit that approaches the analytical relations and design graphs presented in this chapter. Where overall shielding effectiveness is relatively low, say less than about 40 dB, there usually exists a good correlation between theory and practice. When shielding effectiveness is in the range of 40 to 100 dB, the correlation between theory and practice may range from good to poor depending upon (1) the degree of departure from a homogeneous enclosure, (2) the frequency, and (3) the degree to which the leakage points have been shored up. Where correlation is not good, the measured values are almost always below theoretical values. For theoretical SE_{dB} values in excess of about 100 dB, correlation between theory and practice is generally poor because there are so many leakage points in practical enclosures.

For magnetic materials ($\mu_r > 1$), the relations in this chapter apply for the stipulated permeability, μ_r, condition only. The permeability varies with both magnetic field strength (or flux density) and frequency, especially above several kilohertz. Above about 1 MHz, the relative permeability of most magnetic materials has already become or approaches unity. Thus, the applicability of magnetic materials should be thoroughly investigated, and these materials should be used unless for shielding against low frequency magnetic fields.

SHIELDING MATERIALS

The previous section contains the development of shielding effectiveness relations for homogeneous materials with shielding characteristics that must be independent of frequency and field strength or other saturating effects (i.e., they are linear). These equations apply, with restrictions, to pseudo-homogeneous metals such as conductive paints, coatings, depositions, flame-spray processes, and the like. Except at lower frequencies, they do not apply for small-aperture materials such as screens, meshes, and metalized textiles, all of which are described by different expressions.

This section summarizes the properties of homogeneous metals, shields with apertures, and pseudo-homogeneous metals.

Homogeneous Metals

Many of the metals that the designer might expect to encounter are listed, with their associated conductivity and permeability, in Table 10-1 (7). The fourth column, entitled "Product $\sqrt{\sigma_r \mu_r}$," is a ranking of the latent absorption loss of metals relative to copper. For nonmagnetic materials except silver, all $\sqrt{\sigma_r \mu_r}$ values relative to copper are less than 1.0. Thus, nonmagnetic materials provide relatively poor absorption loss. All magnetic materials, on the other hand, have relative absorption loss values exceeding 2 and are relatively good absorbers of energy at low frequencies, compared to nonmagnetic materials. On the other hand, since the relative permeability degrades with increasing frequency, magnetic materials offer absorption losses less than those of most nonmagnetic materials at higher frequencies (above approximately 100 kHz).

For thin metal thickness (thicknesses much less than a skin depth) the absorption loss is negligible (for the shielding effectiveness is based on absorption alone), and the choice of metal, whether magnetic or nonmagnetic, is unimportant. However, when the metal is not thin compared with skin depth, the absorption loss becomes significant. Apart from cost and other considerations, Cols. 2 and 3 are useful in choosing a metal for shielding or simply in extracting applicable σ_r and μ_r values for use elsewhere in this *Handbook*.

The fifth column of Table 10-1, entitled "Quotient $\sqrt{\sigma_r / \mu_r}$," is a ranking of the latent reflection loss of metals relative to copper. Expressed in dB form in Col. 6, the nonmagnetic materials outperform nearly all magnetic materials for reflection loss performance. For thin metal thickness, the reflection loss is the only important term for overall shielding effectiveness. For this condition, then, the designer would select a nonmagnetic material of high conductivity.

For RF shielding of windows and viewing apertures, such as glass or plastic substrates, t is measured in nanometers (nm) and is a tiny fraction of skin depth. Here the more stable (lower in the electrochemical series) and more conductive metals would be selected, such as gold.

In dealing with magnetic shielding materials, the question arises of the numerical value of relative permeability to be employed in the calculation of shielding effectiveness, as discussed in the previous section. The values of relative permeability, μ_r, given in the literature are often applied as if they

Table 10-1. Relative conductivity and permeability of Metals. (Courtesy of Interference Control Technologies.)

(1) Metal	(2) Relative conduct. σ_r	(3) Relative permbly. μ_r @ ≤ 10 kHz	(4) Product $\sqrt{\sigma_r \mu_r}$ $A = k_1\sqrt{\sigma_r \mu_r}$	(5) Quotient $\sqrt{\sigma_r/\mu_r}$ $R = k\sqrt{\sigma_r/\mu_r}$	(6) Relative reflection R_{dB}
1. Silver	1.064	1	1.03 dB	1.3	+0.3 dB
2. Copper (Solid)	1.00	1	1	1	0
3. Copper (Flame Spray)	0.10	1	0.32	0.32	−10.0
4. Gold	0.70	1	0.88	0.88	−1.1
5. Chromium	0.664	1	0.81	0.81	−1.8
6. Aluminum (Soft)	0.63	1	0.78	0.78	−2.1
7. Aluminum (Tempered)	0.40	1	0.63	0.63	−4.0
8. Aluminum (Household Foil, 0.6 mil)	0.53	1	0.73	0.73	−2.8
9. Aluminum (Household Foil, 0.6 mil)	0.61	1	0.78	0.78	−2.1
10. Aluminum (Flame Spray)	0.036	1	0.19	0.19	−14.4
11. Brass (91% Cu, 9% Zn)	0.47	1	0.69	0.69	−3.3
12. Brass (66% Cu, 34% Zn)	0.35	1	0.52	0.52	−5.7
13. Magnesium	0.38	1	0.61	0.61	−4.3
14. Zinc	0.305	1	0.57	0.57	−4.9
15. Tungsten	0.314	1	0.56	0.56	−5.0
16. Beryllium	0.33	1	0.53	0.53	−5.5
17. Cadmium	0.232	1	0.48	0.48	−6.3
18. Platinum	0.17	1	0.42	0.42	−7.6
19. Tin	0.151	1	0.39	0.39	−8.2
20. Tantalum	0.12	1	0.33	0.33	−9.6
21. Lead	0.079	1	0.28	0.28	−11.0
22. Monel (67 Ni, 30 Cu, 2 Fe, 1 Mn)	0.041	1	0.20	0.20	−13.9

23. Manganese	0.039	1	0.20	0.20	−14.1
24. Titanium	.036	1	0.19	0.19	−14.4
25. Mercury (Liquid)	0.018	1	0.13	0.134	−17.4
26. Nichrome (65 Ni, 12 Cr, 23 Fe)	0.0012	1	0.035	0.035	−29.2
27. Supermalloy	0.023	100,000	53.7	0.0005	−65.4
28. 78 Permalloy	0.108	8,000	29.4	0.0037	−48.7
29. Purified Iron	0.17	5,000	29.2	0.0058	−44.7
30. Conetic AA	0.031	20,000	28.7	0.0011	−58.8
31. 4-79 Permalloy	0.0314	20,000	25.1	0.0013	−58.0
32. Mumetal	0.0289	20,000	24.0	0.0012	−58.4
33. Permendur (50 Cu, 1-2V, & Fe)	0.247	800	14.1	0.018	−35.1
34. Hypernick	0.0345	4,500	12.5	0.0028	−51.1
35. 45 Permalloy (1200 anneal)	0.0384	4,000	12.4	0.0031	−50.2
36. 45 Permalloy (1050 Anneal)	0.0384	2,500	9.80	0.0039	−48.1
37. Hot-Rolled-Silicon Steel	0.0384	1,500	7.59	0.0051	−45.9
38. Sinimax	0.0192	3,000	7.59	0.0025	−51.9
39. 4% Silicon Iron (Grain Oriented)	0.037	1,500	7.45	0.0050	−46.1
40. 4% Silicon Iron	0.029	500	3.81	0.0076	−42.4
41. 16 Alfenol	0.01113	4,500	7.13	0.0016	−56.0
42. Hiperco	0.069	650	6.70	0.010	−39.7
43. Monimax	0.0216	2,000	6.57	0.0033	−49.7
44. 50% Nickel Iron	0.0384	1,000	6.20	0.0062	−44.2
45. 45-25 Perminvar	0.091	400	6.03	0.015	−36.5
46. Commercial Iron (0.2% Impure)	0.17	200	5.83	0.29	−30.7
47. Cold-Rolled Steel	0.17	180	5.53	0.031	−30.2
48. Nickel	0.23	100	4.70	0.047	−26.6
49. Stainless Steel (1 Cu, 18 Cr, 8 Ni, & Fe)	0.02	200	2.00	0.010	−40.0
50. Rhometal (36 Ni)	.019	1,000	4.36	0.0044	−47.2
51. Netic 53-6	.172	300	7.18	0.024	−32.4

were a fixed material parameter, nonvarying under different electromagnetic ambient conditions. Unfortunately, the value of relative permeability for materials classified as magnetic materials ($\mu_r > 1$) invariably changes with the magnetic ambient and frequency. A misapplication of the shielding effectiveness relations, such considering μ_r to be a fixed quantity, can lead to an overly optimistic prediction of shielding effectiveness. Thus, a realistic evaluation of SE_{dB} depends upon utilization of the correct value of μ_r.

An accurate assessment of the correct value of μ_r to be used in calculating shielding depends on several factors. Figure 10-13 shows the magnetization curve and hysteresis curve of a typical material (7). In this figure various characteristics of the material are defined. H_c is the coercive force or the value of applied reverse magnetic field strength required to drive the magnetic induction (flux density) to zero. B_r is the residual magnetism or the value of magnetic induction retained by the material after a saturation value of induction, B_m, has been achieved. B_m is produced in the material from the application of a magnetic field strength, H_m. After the magnetic field is reduced to zero, the value of flux intensity, B_r, remains in the material.

Relative permeability, μ_r, is defined as the ratio of magnetic induction, B, to applied magnetic field strength, H. Examination of the figure shows the definition of the three different relative permeabilities, μ_r. These are: μ_0, the initial permeability; μ_{max}, the maximum permeability; and μ_Δ, the incremental

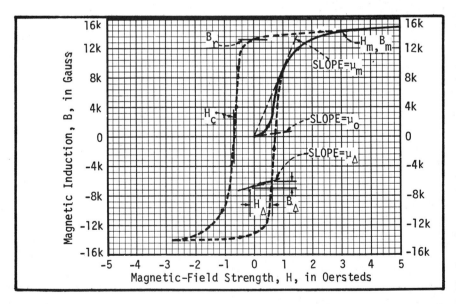

Fig. 10-13. Magnetization curve (solid) and hysteresis loop (dotted). Some important magnetic quantities are illustrated. (*Courtesy of Interference Control Technologies*)

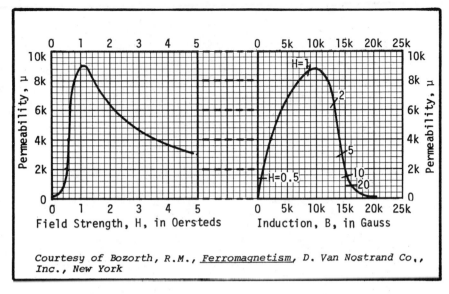

Courtesy of Bozorth, R.M., *Ferromagnetism*, D. Van Nostrand Co., Inc., New York

Fig. 10-14. Permeability curves of iron, with μ plotted against H and B, I and B-H are also used as abscissae (*Courtesy of Interference Control Technologies*)

permeability. These three values of permeability can vary by more than an order of magnitude (or more than 10 dB of SE_H for reflection loss and by any amount for absorption loss). Since magnetic SE_{dB} values are usually low at low frequencies, this can result in a significant error.

The incremental permeability, μ_Δ, is of particular interest in this discussion. Appearing in the figure as a minor loop on the hysteresis curve, it shows the effect of a low-level signal applied in addition to the hysteresis loop generation signal. At the single point shown, the effect of the incremental permeability is to produce a flux density in opposition to that of the hysteresis loop generation signal. This represents the situation when a magnetic field strikes a permeable shield.

Figure 10-14 shows the relative permeability, μ_r, as a function of applied magnetic field strength and induction for iron (7). This clearly portrays the dependence of permeability upon the magnetic environment, a factor that must be considered in evaluating μ for the computation of SE_{dB}.

Removing the hysteresis curve while retaining the presence of the externally applied magnetic signal results in the situation shown in Fig. 10-15, where the small loops generated by applied fields are shown on the magnetization curve. Note that μ_Δ will vary with applied field, in that the *B-H* slope varies, which defines the incremental permeability, μ_Δ.

The situation experienced by a shield material in real life is, of course, even more complex. Shields are only used in the presence of alternating fields.

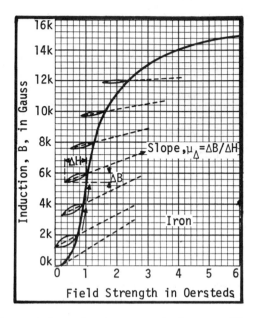

Fig. 10-15. Minor hysteresis loops shown on magnetization curve. (*Courtesy of Interference Control Technologies*)

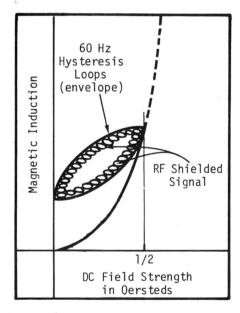

Fig. 10-16. Portrayal of real world situation. (*Courtesy of Interference Control Technologies*)

Thus, one typical situation would be a shield existing in the earth's magnetic field ($\simeq 0.5$ Oersted, where 1 Oersted $= 10^3/4\pi \simeq 79.58$ ampere-turns/meter) which constitutes a DC bias, with typically a 60 Hz H-field superimposed, causing hysteresis effect and possibly necessitating the shielding of vulnerable signal wires. As shown in Fig. 10-16, this situation allows the estimation of an effective permeability, μ_e, to be used for computation for SE_{dB}:

$$\mu_e \simeq \mu_\Delta \simeq 2\mu_0 = 0.1\mu_m$$

In the more general situation, several effects take place, including eddy current losses which are a function of (1) magnetic field strength, (2) frequency, (3) distances to the metal barrier, (4) metal thickness, and (5) metal permeability and resistivity. Thus, it is concluded that unless all conditions are known, it is impossible to define an equivalent μ_r and to directly use the μ_r values in Table 10-1 and the equations for magnetic materials.

The relative magnetic perrmeability approaches unity above a few hundred kilohertz. In addition, Fig. 10-17 shows the behavior of μ_r versus magnetic flux density at 60 Hz. About all one might conclude from this is that below saturation and at very low frequencies, the permeability can increase up to about an order of magnitude above its rated low-level values listed in Table 10-1.

At and above medium frequencies (300 kHz to 3 MHz), $\mu_r \rightarrow 1$ for all conditions. Thus the shielding effectiveness curves presented in the appendix to the chapter apply only for the stated values of μ_r used.

When the maximum magnetic flux density may be exceeded, a double shield containing either a nonmagnetic metal or a higher-saturation magnetic metal should face the more hostile magnetic field source. For hostile emissions coming from outside a box or cable shield, this means that the first layer of protective metal should face the outside, whereas if the emissions originate within, the protective metal should face the inside.

Availability and Applications

Several of the metals in Table 10-1 are available off the shelf in sheet stock form from thicknesses of about 1/64th inch (0.4 mm) or less to about 1/8 inch (3.2 mm) or more. Metals having thicknesses less than 1/64th inch are sometimes regarded as foils. Many of the high-permeability metals come in foil thicknesses ranging from about 1 mil (25.4 μm) to 10 mils (254 μm). They are usually available in both sheet and tape form. The foil stock is also available in the form of adhesive-backed foil and in rolls.

The thinner nonmagnetic foils, whose thickness is of the order of a few mils, are widely used for RF shielding. While there are many stories spread about how aluminum foil saved the day, (One mil of household foil can produce 80 dB of shielding effectiveness to plane waves and electric fields in excess of

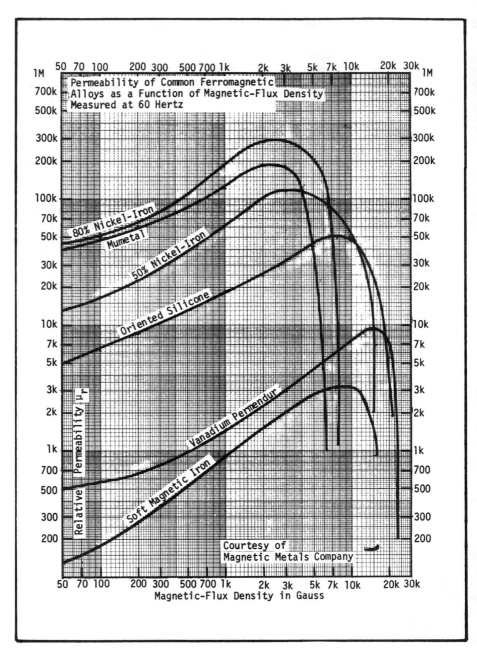

Fig. 10-17. Relative permeability vs. magnetic-flux density. (*Courtesy of Interference Control Technologies*)

1 GHz) its performance against low frequency magnetic fields is very poor (i.e., it is nearly transparent).

Metal foils are used in a number of ways. One rather interesting application involves metal-foil wallpaper for converting an entire room into a limited shielded enclosure. It exhibits considerably less shielding effectiveness than a commercial shielded enclosures, but it does have the advantage of lower cost. These materials have been used on occasion to construct shielded chambers from thousands of square feet in surface area down to equipment-box size. Metal-foil wallpaper must be used in conjunction with other materials such as pressure-sensitive compounds. They also require taking a number of measures to reinstate lost shielding integrity around doors, windows, AC power mains penetrations, heating/air conditioning ductwork, and the like.

Metal-foil wallpaper usually comes in thickness of 2 to 3 mils (51 to 76 μm), and is made of either aluminum or copper foil or special stainless steel foil of relatively high conductivity and high permeability. Representative shielding effectiveness is 25 to 40 dB for magnetic fields at 200 kHz, 80 to 100 dB for electric fields from 200 kHz to 10 MHz, and 60 to 80 dB for plane waves above 400 MHz when measured in accordance with MIL-STD-285.

Sometimes, thin foils with thicknesses on the order of 1 mil are bonded to (metallized on) a plastic base such as 5 to 10-mil Mylar. They can be used as an air-inflatable structure that performs as a shielded enclosure. Such applications are discussed in Section 1.1.3 of Volume 2 of reference 7.

Shielding Effectiveness of Apertures

A practical equipment box shield is violated by various apertures. The necessity of having holes in the shield for power cords, push buttons, control shafts, cooling, and so on, dictates that the amount of shielding compromise by a hole should be considered. Specific solutions to these problems are discussed in Chapter 11. This section presents some analytical relations that will allow the shielding degradation due to single and multiple apertures to be calculated (1). The shielding effectiveness of a shield with apertures is calculated as follows:

$$SE_{dB} = A + R + B + K_1 + K_2 + K_3 \qquad (10\text{-}88)$$

where:

SE = shielding effectiveness
A = aperture absorption loss
R = aperture reflection loss
B = correction term for aperture reflections
K_1 = correction term for a number of openings
K_2 = correction term for field penetration at low frequencies
K_3 = correction term for coupling between closely spaced holes

Absorption Loss, A. The basic absorption term is derived from waveguide theory, at frequencies below the cutoff frequency of the waveguide, the field suffers attenuation as it passes through the waveguide/aperture. The absorption offered by a waveguide to frequencies below cutoff is:

$$A_c = 32 \, L/D, \text{ for circular apertures} \qquad (10\text{-}89)$$

$$A_r = 27.3 \, L/W, \text{ for rectangular apertures} \qquad (10\text{-}90)$$

where:

L = length of the hole
D = diameter of the circular hole
W = greatest width of the rectangular hole

Reflection Loss, R. The reflection loss is arrived at in a manner similar to that for the reflection loss of a solid shield. That is, the reflection loss depends on the impedance mismatch at a boundary. Thus, the impedance mismatch is derived from the characteristic impedance of the incident wave and the characteristic impedance of the aperture. The reflection loss is found from:

$$R = 20 \log_{10} \left[J/4 + 1/2 + 1/4J \right] \qquad (10\text{-}91)$$

where:

$J = Z_a/Z_w$, the ratio of the characteristic impedance of the aperture to the characteristic impedance of the incident wave

$Z_{ar} = j\omega\mu_0 W/\pi$, the characteristic impedance a rectangular aperture

$Z_{ac} = j2\pi f\mu_0 D/3.682$, the characteristic impedance of a circular aperture

$Z_{wh} = j\omega\mu_0 r$, the impedance of the conductor between apertures for magnetic fields

$Z_{we} = j/\omega\varepsilon_0$, the characteristic impedance of the conductor between apertures for electric fields

ω = frequency in radians/sec
f = frequency in Hz
r = interference source to shield distance in millimeters
μ = magnetic permeability of shield material
ε = dielectric constant

Aperture Reflection Loss Correction, B. When the absorption component of the shielding effectiveness is less than 10 dB, then the correction term applies:

$$B = 20 \log_{10} \left[1 - \frac{(J-1)^2}{(J+1)^2} 10^{-(A/10)} \right] \text{dB} \qquad (10\text{-}92)$$

Multiple Aperture Correction, K_1. The shielding effectiveness will be degraded by a number of apertures in the shield. The formula for the multiple aperture correction term is expressed as a loss in the total shielding effectiveness relation:

$$K_1 = -10 \log_{10} an \qquad (10\text{-}93)$$

where:

a = area of all apertures
n = number of holes per square area

The product ($a \times n$) gives the total number of holes number of holes, and thus the SE degradation due to the total number of holes is accounted for.

Correction Factor for LF Field Penetration, K_2. At low frequencies, the skin depth of the metal is large, and thus the fields penetrate the conductor and emerge on the opposite side of the shield. An empirical relation was derived to calculate the field penetration as follows:

$$K_2 = -20 \log_{10} \left[1 + \frac{35}{1.15(\pi d^2 f \sigma \mu)} \right] \qquad (10\text{-}94)$$

where:

σ = conductivity of material
c_w = conductor width between holes
d = wire diameter

Note: When calculating the shielding effectiveness of screening, substitute c_w for d^2.

Coupling Between Holes, K_3. The mutual coupling between holes tends to increase the impedance of the entire aperture configuration, leading to an increase in the total shielding effectiveness. This coupling is:

$$K_3 = 20 \log_{10} \left[1/\tanh (A/8.686) \right] \qquad (10\text{-}95)$$

where A is the absorption term calculated above.

Aperture Design. A number of techniques are available that can be used to increase the shielding effectiveness of apertures in a box shield. These include using honeycomb for cooling; using a wire screen over cooling apertures; designing switch and control shaft openings as waveguides beyond cutoff; and using separate box shields to enclose switches. (See Chapter 11.)

Fig. 10-18. Surface resistance of copper vs. volume resistivity or various metal thickness. (*Courtesy of Interference Control Technologies*)

Pseudo-homogeneous Metals

Metals that lack homogeneity but do not have holes, slits, or other apertures, whether small or large, are called pseudo-homogeneous metals (PHM) in this *Handbook*. Examples of PHM include conductive paints and coatings and products of the flame spray process of metallizing an insulator. Another technique used to give electromagnetic shielding properties to plastics consists of adding metal or conductive particles to the plastics before molding; this also yields pseudo-homogeneous material. PHM may have thin areas because of the problem of adequate quality control inherent in a process (i.e., the process does not lend itself to homogeneity). Consequently, PHM may have in a theoretical shielding effectiveness that is good to poor with respect to measured results.

The electrical properties of pseudo-homogeneous conductive coatings are usually measured either in units of surface resistance per unit thickness or in units of volume resistivity, absolute conductivity, or occasionally conductivity relative to copper. The surface resistance, R_{sde}, is measured in ohms/square and is related to volume resistivity, ρ:

$$R_{sde} = 100\rho_{\text{ohm-m}}/t_{cm} \text{ ohms, metric system} \tag{10-96}$$

$$= 39,370\rho_{\text{ohm-m}}/t_{mils} \text{ ohms, English system} \tag{10-97}$$

where t = thickness of the surface coating.

The volume resistivity is related to conductivity, σ, and surface resistance in the following manner at DC:

$$\sigma_{\text{mhos/m}} = 1/\rho_{\text{ohm-m}} = 100/R_{sde}t_{dm} \tag{10-98}$$

$$= 39,370/R_{sde}t_{mils} \tag{10-99}$$

Finally, the conductivity relative to copper (σ_r for copper = 1; $\sigma_{cu} = 5.8 \times 10^7$ mhos/m) is:

$$\sigma_r = \sigma_{\text{metal}}/\sigma_{cu} = 1.72 \times 10^{-8}/\rho_{\text{ohm-m}} \tag{10-100}$$

$$= 1.72 \times 10^{-6}/R_{sde}t_{cm}, \text{ metric} \tag{10-101}$$

$$= 6.79 \times 10^{-4}/R_{sde}t_{mils}, \text{ English} \tag{10-102}$$

Equations (10-98) and (10-99) are plotted in Fig. 10-18 to yield surface resistance versus volume resistivity with metallized coating thickness as a parameter. Equations (10-101) and (10-102) are plotted in Fig. 10-19 to present coating conductivity relative to copper versus surface resistance with thickness as a parameter. This figure is especially useful because it allows

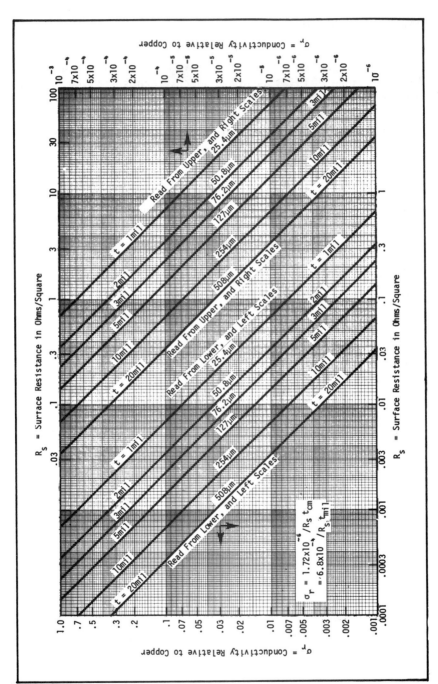

Fig. 10-19. Conductivity relative to copper vs. surface resistances. *(Courtesy of Interference Control Technologies)*

the development of an equivalent relative conductivity, σ_r, that can be used in the many expression of shielding effectiveness throughout this *Handbook*.

CONDUCTIVE COATINGS

Flame Spray Process

Flame spraying of metals is a process by which a surface is metallized by using a special metallizing gun that vaporizes a metal onto the surface. This process includes thermo spraying and plasma spraying as well as flame spraying.

In flame spraying, a wire rod of the copper, aluminum, or other metal is fed into a special metallizing gun. This gun also accepts a gas such as oxy-acetylene or oxypropane and compressed air. The wire rod is drawn through the gun and nozzle and melted in the oxygen/fuel gas flame and atomized by a blast of compressed air that carries the metal particles to the surface to be metallized. This is illustrated in Fig. 10-20.

Any metal (with the exception of tungsten) obtainable in a uniform wire or rod size can be sprayed using special guns. Flame spraying speeds depend on a number of factors. Using 15 psi (5 kg/cm^2) of acetylene pressure, for example, 12 pounds (26 kg) of aluminum, 29 pounds (64 kg) of copper, 18 pounds (40 kg) of monel, or up to 16 pounds (35 kg) of steel can be deposited per hour. Higher spraying speeds can be achieved using propane gas. Typically, metal wire and rod diameters from 18 B&S gauge (0.91 mm) to 3/16 inch (4.76 mm) diameter can be accommodated in the special spray guns.

Since the flame (about 6000°F = 3000°C), which is located in the nozzle, does not contact the surface, materials such as plastics and fiberglass can

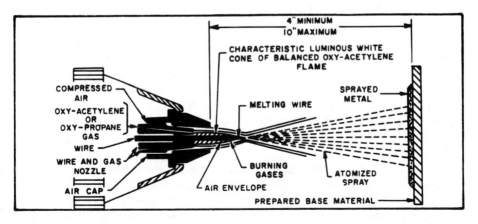

Fig. 10-20. Flame spray wire metalizing gun. (*Courtesy of Interference Control Technologies*)

be sprayed without damage. However, air pockets in the deposition can develop, and, depending upon the process control, the metallized surface may range from a sandpaper-type finish to a relatively smooth one. Chemical analysis indicates that oxides of the base metal deposited on the surface are low.

The flame-sprayed deposits of copper and aluminum listed in Table 10-1 exhibit conductivities relative to their own base metals of about 10% for copper and 6% for aluminum. This is so because most of the deposit is air pockets or tiny voids in the deposited surface; that is, the basic metal density is low. Consequently, it takes about 10 times as much for copper and 17 times for aluminum of the base-metal thickness to yield the same conductivity as the base metal. For thin deposits there is also danger of the formation of many small-area voids during the flame spray process.

A variation of the flame spray process, called Electrospray by Wall Comonoy Corp., uses an electric arc gun. Used for higher-volume metallizing operations, it works faster than the conventional method and costs somewhat less to operate. The 7000°F (3900°C) temperature of the arc melts a wire rod faster and deposits particles having higher heat and greater fluidity. There is some danger, however, that plastic substrates can be deformed if the flame spray dwells too long on a surface.

The electric arc spray rate is claimed to be three to five times faster than that of oxyacetylene equipment, and bond strength is greater, as welded adhesion is added to the usual mechanical bond. This latter process, of course, cannot be used with plastics because a minimum speed across the surface must be maintained, and the resultant heat would be enough to melt the plastic. Good bonds can be achieved with less than ideal surface preparation. Since hotter particles mean denser coatings, coating oxide content is reduced compared to conventional methods.

A different type of flame spray process, called ThermoSpray by Metco, Inc., accepts the metal to be sprayed in powder form. The powdered aluminum, copper, or other metal is held in a hopper atop the gun and is gravity-fed into the gun where it is picked up by oxyacetylene or a hydrogen gas mixture and carried to the gun nozzle. Here, it is melted almost instantly and carried to the surface being sprayed by means of a siphon-jet arrangement at the gun nozzle. Because the burning gases extend well beyond the nozzle, use of the ThermoSpray process necessitates that a required minimum distance to plastic base materials be observed very carefully and the speed of the spray gun across the surface of the plastic be high to prevent deforming of the plastic.

Plasma spraying utilizes a technique involving the application of high-frequency electrical energy to ionize an inert gas. This creates a higher energy level that generates temperatures as high as 30,000°F (16,650°C). The primary benefit of such high temperatures is that the melting point of the material being sprayed is reached quickly. Thus, the metal need not remain in the hot zone as long as for other processes, such as those using lower-temperature oxyacetylene flames. A shorter heating time means the metal can be propelled through the plasma arc faster and reaches the substrate with less heat loss and with greater impact. These two advantages give plasma coatings a greater density and higher bond strengths than other processes yield.

The plasma gun generates its high temperatures without combustion, so even the most reactive materials can be melted with little or no oxidation and without altering the original chemical composition to any significant degree. Also, the spraying can be done within a protective-atmosphere chamber, to further protect the sprayed material. The elimination of oxides produces a more cohesive coating that will finish to a better surface and protect itself better in service than other coatings.

Most metals can be sprayed onto almost any material. The substrate need not be heated over 300°F (most materials are unaffected using 400°F as a maximum), so heat distortion problems are eliminated.

Conductive Paints

Conductive paints (also called conductive coatings) are made by Chomerics, Emerson and Cuming (E & C), Technical Wire Products (Tecknit), Acheson Colloids, and others. They are usually lacquer, elastomeric, silicone resin, vinyl, acrylic, or latex base, and require careful surface preparation. Some conductive paints require an overcoat for protection. To assure good electrical and mechanical reliability, surface preparation of plastics, wood, ceramics, and other base materials requires the removal of all greases, waxes, oils, dirt, mold, and foreign matter until the surface is water-break free. Most conductive paints may be applied by one or more of the following processes: dipping, spraying, silk screening, roll coating, or brushing. Several conductive coating are also available in aerosol spray cans.

Silver lacquer paint (e.g., E & C Eccocoat CC-2) is a highly conductive silver particulate and organic resin formulation. In most cases, a single spray coat on a reasonably nonporous surface using air drying is adequate to produce a surface resistivity of about 0.1 ohm/sq. for a 1 mil (25.4 µm) coating thickness. This corresponds to a volume resistivity of 0.04 ohms-m ($\sigma =$ 25 mhos/m, or $\sigma_r = 0.0042$; see Figs. 10-18 and 10-19). Oven curing, if possible, gives improved conductivity (7).

A silver-filled elastomeric coating (e.g., E & C Eccocoat CC-4) is also highly conductive, and a film will stretch over 100%. Thus, it is particularly useful where the substrate to which it is applied is subject to flexing or stretching. Air drying typically results in a surface resistance of about 50 milliohms/sq. for a 1 mil coating thickness. This corresponds to a $\sigma_r = 0.014$. Oven curing results in a surface resistance of about 1 milliohm/sq. or $\sigma = 0.68$. It provides a resistance to salt spray and has an operating temperature range of $-65°F$ $(-54°C)$ to $250°F$ $(121°C)$. Performance under extended use at temperatures of $250°F$ is excellent.

The silver base silicone surface coating (e.g., E & C Eccocoat CC-10) exhibits both high conductivity and high operational temperature properties. The surface resistance is 40 milliohms/sq. for a 1 mil coating thickness ($\sigma_r = 0.39$). It can be used continuously at a temperature of $600°F$ ($315°C$). When applied to a substrate that will withstand even higher temperatures, the surface coating can be fused to the substrate. Moisture and chemical resistance is good.

Conductive paints, consisting of a pure silver filler in a vinyl copolymer (e.g., Tecknit 72-00026), result in a conformal coating having surface resistances of about 10 milliohms/sq. for a 1 mil coating. It will stretch, compress, and deflect with its substrate while maintaining its basic electrical properties. The operating range is from $-65°F$ ($-54°C$) to $250°F$ ($121°C$).

The conductive acrylic coatings developed for the plastic industry (e.g., Tecknit 72-00025) exhibit a surface resistance of about 50 milliohms/sq. for a 1 mil coating. The variation depends upon the percent solids used, the base material, the deposition depth, and so on. In preparation for spraying, the surface is typically cleaned with either alcohol or naphtha. Should environmental, decorative, and/or dielectric insulation protection be required on conductive acrylic paint, the overcoat must be a nonsolvent paint. Typical overcoats exhibit dielectric strengths of 300 V/mil (12 kV/mm) and have temperature ranges that match the conductive coating: $-65°F$ ($-54°C$) to $200°F$ ($93°C$) (7).

A number of different conductive surface coatings, called Electrodag, are made by Acheson Colloids. They range in surface resistance from 15 milliohms/sq./mil for more expensive, highly reflective applications, to less expensive, 200 ohms/sq./mil for low shielding effectiveness use. A particular coating is selected based on the abrasion resistance; surface hardness; adhesion to plastics, glass, or ceramics; chemical and solvent resistance; and service temperature. While conventional paint spraying is the usual application, brushing or dipping techniques may also be used. Figure 10-21 illustrates a typical application technique.

Shielding Effectiveness of Conductive Coatings

A common-wall attenuation test setup is used to measure the shielding effectiveness of conductive panels to ascertain relative shielding effectiveness performance (2). The test setup consists of two shielded enclosures sharing a common wall in which is placed the flat rectangular panel constructed of the material to be tested, whether it be a conductive coating on plastic or a metallized plastic panel. This test setup is shown in Figure 10-22.

Fig. 10-21. Application of conductive coating to interior of plastic equipment enclosures. (*Courtesy of Chomerics*)

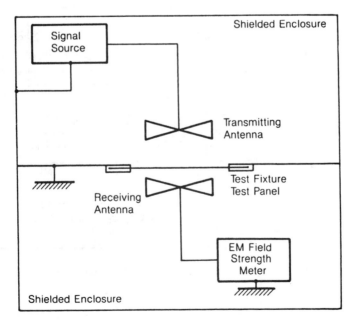

Fig. 10-22. Test set up using dual chamber test method. (*Courtesy of Interference Control Technologies*)

Table 10-2. Relative shielding effectiveness of different conductive coatings.

Material	Thickness mils	Sheet resistance ohms/se. @ 1 mil	Attenuation dB
Sheet polycarbonate	1/8 inch	—	0
Aluminum Sheet	1/8 inch	0	64–80
Silver paint	1.5 mil	0.01	54–70
Silver graphite	0.2/1.0 mil	0.01/100	54–77
Copper	2.0 mil	8.0	20–54
Copper/graphite	2.0/2.0 mil	8.0/100	27–62
Graphite	1.0 mil	100	11–60

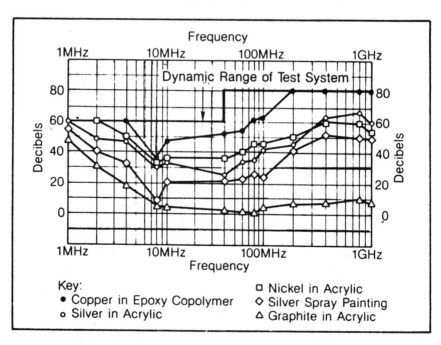

Fig. 10-23. Shielding effectiveness (selected materials). (*Courtesy of Interference Control Technologies*)

The best agreement between theory and measurement occurs for a large-size test sample in place in the common wall. In one of the rooms a variable-frequency RF transmitter is connected to a transmitting antenna to produce a radiated EM field. The other room contains a receiving antenna connected to an EMI receiver or spectrum analyzer. As the frequency of the transmitter is varied, the amount of attenuation through the test sample is measured from the readings taken from the receiver.

In one such test, different conductive coatings were sprayed onto plastic panels and tested (2). For reference, a plastic panel was installed and tested for which the shielding effectiveness was 0 dB. It must be noted that the measured attenuation of the various coatings is *relative*, and the performance of any shield depends on the presence and treatment of apertures, the quality of the application of the conductive coating, the frequency of the incoming field, and the specific type of field (high or low impedance) to be shielded for.

The results of the test on the conductive coatings are shown in Table 10-2 (2), and show the average attenuation in dB over the test frequency range of 1 to 10 GHz. Note the role that conductivity plays in the shielding effectiveness performance.

The performance of other metallized plastics is shown in Fig. 10-23 using the same test procedure as before, but with a number of conductive-particle-filled plastics being tested (5).

Conductive Coatings and Electrostatic Discharge

Conductive coatings, applied over plastic cabinetry, can be an effective solution to the problem of electrostatic discharge (2). The purpose of the coating is to divert ESD currents away from the inside of the cabinet which contains sensitive circuitry. However, the use of a very highly conductive coating can tend to be counterproductive because of the secondary EMI generated by the ESD as it flows through the cabinet. The very fast rise-time of the ESD may be preserved if it flows through a conductive cabinet, resulting in radiated EMI. If the conductive coating has some resistance, the ESD will be somewhat damped and the rise-time slowed. Thus, the choice of conductive coating is a tradeoff between required shielding effectiveness and ESD suppression.

APPENDIX: TESTING AND MIL-STD-285

MIL-STD-285, dated 25 June, 1956, is entitled "Military Standard Attenuation Measurements for Enclosures, Electromagnetic Shielding for Electronic Purposes, Method of" (4, 6). There are several points mentioned below regarding MIL-STD-285 and its use:

- E-field or high impedance waves, where $r < (\lambda/2\pi)$ and $Z_w \gg 377$ ohms, are measured at a distance of only 12 inches (30.5 cm) from the shielding barrier interface. This results in much higher shielding effectiveness (optimistic values) than would exist at greater distances. Measurements are made at least at 200 kHz, 1 MHz, and 18 MHz and generally at other frequencies below 30 MHz.
- H-field or low impedance waves, where $r < (\lambda/2\pi)$ and $Z_w \ll 377$ ohms, are measured at a distance of only 12 inches (30.5 cm) from the shielding barrier interface. This results in much lower shielding effectiveness (pessimistic values) than would exist at greater distances. Measurements are made at least at 200 kHz and generally at other frequencies below 1 MHz.
- Plane waves or far-field measurements, where $r > (\lambda/2\pi)$ and $Z_w = 377$ ohms, are made at a distance of 6 feet (1.8 meters) or more (where $r \geq 2\lambda$) from the shielding barrier interface. Here, distance is unimportant as long as $r > (\lambda/2\pi)$, with proper attention paid to the measurement range available. Measurements are made at least at 400 MHz and generally at several other frequencies above 100 MHz.
- Initially, measurements usually are made out in the open with no barrier between transmitting and receiving antennas; then, they are repeated with the same orientation and separation of antennas, but with the shielding barrier located in between the transmitting and receiving antennas. This give erroneous readings for E-fields because the antennas are detuned due to capacitive loading. Antenna probes, whose dimensions are small compared to 12 inches (30.5 cm), will not have this problem. H-field antennas are farady-shielded.
- The measurement range suggested in MIL-STD-285 can be substantially increased by using power amplifiers or power oscillators on the transmitting side and narrowband tuned amplifiers and/or correlators with fiber optic reference paths on the receiving side.
- For gaskets, wire screens, woven mesh, and other metal thicknesses that are very small compared to the shielded enclosure metal thickness, an 18-inch (45.7-cm) or 24-inch (61-cm) square aperture is usually cut in the enclosure wall of an internally partitioned shielded enclosure. Shielding effectiveness (insertion loss) measurements are made by taking the difference between field strength measurements when no test sample is used and those obtained when the test sample is in place.

Discussions regarding test methods and procedures are beyond the scope of this chaprter. However, it is informative to amplify the above comments regarding measurement errors. For example, suppose there is a known source of interference in a real problem at a distance of 10 meters against which one desires to be shielded. How much shielding effectiveness is required when manufacturers' data sheets are reported in the language of MIL-STD-285 at a distance of 12 inches (30.5 cm)?

The answer to the above question is found by developing correction data based on extrapolating the results of the MIL-STD-285 data in order to fit the user's distance. Such extrapolation data are a function of many variables and would require many graphs. However, the problem may be delimited by certain assumptions so that quantitative correction data can be presented here for user application.

For conditions under which the reflection loss is significant (i.e., $K \geq 10$), equation (10-48) becomes:

$$SE_{dB} \simeq 20 \log_{10} (0.707Kt/\delta) \text{ for } t/\delta \ll 1$$

$$SE_{dB} \simeq 20 \log_{10} (e^{(t/8)}K/4) \text{ for } t/\delta \gg 1$$

where:

$$K = Z_w/Z_m$$
$$Z_w = kZ_0 = k120\pi$$

and:

$$k \simeq (\lambda/2\pi r) \text{ for } E \text{ fields}$$
$$k \simeq (2\pi r/\lambda) \text{ for } H\text{-fields}$$
$$k \simeq 1 \text{ for plane waves}$$

When the above equations are applied for MIL-STD-285 for any two distances, an error is developed that is a function of k alone. Thus, for any measurement distance, r_m, and any user applied distance, r_u, the correction in shielding efficiency, ΔSE_{dB}, becomes:

$$\Delta SE_{dB} = 20 \log_{10} (r_m/r_u) \text{ for } E\text{-fields} \qquad (10\text{-}103)$$

$$\Delta SE_{dB} = 20 \log_{10} (r_u/r_m) \text{ for } H\text{-fields} \qquad (10\text{-}104)$$

$$\Delta SE_{dB} = 0 \text{ for plane waves} \qquad (10\text{-}105)$$

in which it is understood that both r_m and r_u are in the near field.

When one of the distances is in the near field and the other in the far field, the equations for the correction in shielding effectiveness become:

$$\Delta SE_{dB} = 20 \log_{10} (2\pi r_m/\lambda) \text{ for } E\text{-fields,}$$
$$r_m \text{ in near field and } r_u \text{ in far field} \qquad (10\text{-}106)$$

$$\Delta SE_{dB} = 20 \log_{10} (\lambda/2\pi r_u) \text{ for } E\text{-fields,}$$
$$r_u \text{ in near field and } r_m \text{ in far field} \qquad (10\text{-}107)$$

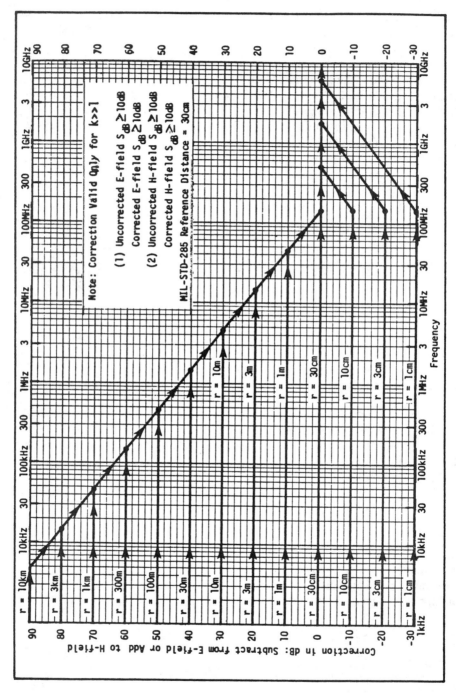

Fig. 10A-1. Correction in shielding effectiveness to convert MIL-STD-285 results to another distance. (*Courtesy of Interference Control Technologies*)

$$\Delta SE_{dB} = 20 \log_{10} (2\pi r_u/\lambda) \text{ tor } H\text{-fields},$$
$$r_u \text{ in near field and } r_m \text{ in far field} \qquad (10\text{-}108)$$

$$\Delta SE_{dB} = 20 \log_{10} (\lambda/2\pi r_m) \text{ for } H\text{-fields},$$
$$r_m \text{ in near field and } r_u \text{ far field} \qquad (10\text{-}109)$$

Equations (10-106) through (10-109) are plotted in Fig. 10-A-1 for MIL-STD-285 at a distance of $r_m = 0.305$ meter (12 inches) and any user distance, r_u, as shown.

Example

A manufacturer's literature states that as per MIL-STD-285 a metallized silicone elastomer offers at least 90 dB of shielding effectiveness to E-fields at 10 MHz, 30 dB of shielding effectiveness to H-fields at 10 MHz, and 70 dB to plane wave above VHF ($f > 300$ MHz). Determine the likely shielding effectiveness at a distance of 10 meters.

From Fig. 10-A-1, the correction to E-fields at $r_u = 10$ meters is -23 dB and to H-fields is $+23$ dB. Thus, for a 10-meter distance SE_{dB} (E-field) = 67 dB and $SE_{dB}(H$-field) = 53 dB. Note that the 10-meter distance is in the far field ($r_u > \lambda/2\pi$) at 10 MHz, since the $r_u = 10$ meters line in Fig. 10-A-1 ends at 4.8 MHz ($r_u = \lambda/2\pi$). Thus, there is no correction for SE_{dB}(plane waves) = 70 dB.

REFERENCES

1. Air Force Systems Command, Design Handbook for Electromagnetic Compatibility, 1-4, Series 1-0, March 1984.
2. Coniglio, James J. "Basics for Conductive Coating Users," Interference Technology Engineers Master. Plymouth Meeting, PA: R & B Enterprises, 1980.
3. Kendall, C. "Boundary Conditions: Valid Factors in Shielding Analysis." Record of the 1976 IEEE International Symposium on Electromagnetic Compatibility, Volume 76, 76-CH-1104-9 EMC, July 13–15, 1976, Washington, DC.
4. Lee, J. D. "MIL-STD-1377 Vs. MIL-STD-285 Microwave Shielding Effectiveness Measurements," Record of the 1975 IEEE Electromagnetic Compatibility Symposium, Volume 75, 75-CH1002-5, October 7–9, 1975, San Antonio, TX.
5. Mauriello, A. J. "Selection and Evaluation of Conductive Plastics." EMC Technology, Volume 3, Number 4. Gainesville, VA: Don White Consultants, Inc., 1984.
6. MIL-STD 285, "Method of Attenuation Measurements for Enclosures, Electromagnetic Shielding for Electronic Test Purposes."
7. White, Donald R. J. A Handbook on Electromagnetic Compatibility, Volume 2, Gainesville, VA: Don White Consultants, 1973.
8. White, Donald R. J. EMI Control Methodology and Procedures, Third Edition. Gainesville, VA: Don White Consultants, Inc., 1982.
9. White Donald R. J. Electromagnetic Shielding, Materials and Performance, Gainesville, VA: Don White Consultants, Inc 1972.

Chapter 11
SHIELDING INTEGRITY PROTECTION

The previous chapter discussed the subjects of shielding theory and materials. With the exception of low-frequency magnetic field shielding, it was shown that it is quite simple to obtain more than 100 dB of shielding effectiveness across the entire spectrum from DC to daylight for electric and electromagnetic waves. However, since any practical enclosure has apertures, the theoretical shielding is never obtained, because of the resultant loss of integrity. This chapter discusses the loss of shielding integrity, how the integrity can be reclaimed, and practical applications to shielded boxes, chassis and equipments, cabinets, rooms, and vehicles.

INTEGRITY OF SHIELDING CONFIGURATIONS

The attenuation offered by materials to electric, magnetic, and electromagnetic waves described in the previous chapter is achieved theoretically. In practice, however, this attenuation is not often achieved because a shielded enclosure or housing is not completely sealed. In other words, nearly any practical application of shielding has necessary penetrations and apertures of one kind or another. Some examples of such penetrations and apertures include:

- Cover plates and access cover members
- Meter windows
- Windows for viewing digital or other displays
- Potentiometer shafts
- Cooling apertures
- Power-line and signal-lead connectors
- Indicator lamps
- Push buttons
- On–off switches
- Fuses

Thus, it is not uncommon to find the plane-wave attenuation of a basic shield material to be 120 dB, for example, while the actual enclosure will exhibit 50 dB in the VHF/UHF portion of the spectrum. Here, leakage through one or more of the above aperture types compromises the integrity of the basic shielding material.

From a mathematical modeling point of view, either of two approaches may be used: (1) compile a data bank on the shielding effectiveness (attenuation vs frequency) of equipment materials and configurations and choose that closest to the specimen under examination, or (2) compute the shielding effectiveness based on an inventory of data listed in the above 10 items and use worst-case coupling. Since no significant shielding data bank on equipment has ever been accumulated and reported on, as suggested in the first approach, the second approach is generally used.

Bonding Seams and Joints

Loss of RF shielding integrity across the interface of clean mating material members is a main reason why shielding effectiveness is compromised. Here, the resistivity of the interface and/or the permeability may be much lower, than the theoretical, depending on the type of interface bond used. Thus, resulting material interfaces may be classified into two types: physically inhomogeneous and physically homogeneous.

A physically inhomogeneous interface bond results when sheilding members are directly connected by screws, rivets, spot welds, and the like. The interface connection is not continuous, and a bowing or waviness effect results between connected members. This in turn develops slits or gaps that lead to radiation or penetration at frequencies where the gap or slit dimension approaches 0.01λ. The attenuation A in dB (A_{dB}) at such a gap is assumed to follow the waveguide-beyond-cutoff criteria:

$$A_{dB} = 0.0046 L f_{MHz} \sqrt{(f_c/f_{MHz})^2 - 1} \text{ dB} \qquad (11\text{-}1)$$

where:

L = gap depth in inches for overlapping members or the thickness of the material for butting members

f_{MHz} = operating frequency in MHz

f_c = cutoff frequency of gap in MHz

f_c = $5900/g$ for rectangular gap $\qquad (11\text{-}2)$

f_c = $6920/g$ for circular gap $\qquad (11\text{-}3)$

g = largest gap transverse dimension in inches

When $f_c \gg f_{MHz}$, equation (11-1) becomes:

$$A_{dB} = 0.0046 L f_c = 27L/g \text{ dB for rectangular gap} \qquad (11\text{-}4)$$

$$A_{dB} = 32L/g \text{ dB for circular gap} \qquad (11\text{-}5)$$

There are a number of techniques available for reducing electromagnetic emission leakage or receptor penetration of a shielded specimen. If members are joined by screws or rivets, equation (11-4) shows that A_{dB} may be significantly increased by using more screws or rivets per linear dimension of the interface, owing to the reduction in the potential gap, g, between the screws and rivets. Figure 11-1 shows a joint shielding effectiveness as a function of screw spacing for the indicated parameters (1). Also note the improvement due to the application of a typical EMI mesh gasket (see Chapter 12).

Other techniques available for reducing the leakage in a physically inhomogeneous mating member bond involve attempting to eliminate or reduce the inhomogeneity. Figure 11-2 illustrates some of these approaches. Where members do not have to be disengaged or separated, a continuous seam weld around the periphery of the mating surfaces is preferred. This type of weld is not critical, provided it is continuous and has no weld pin holes. One

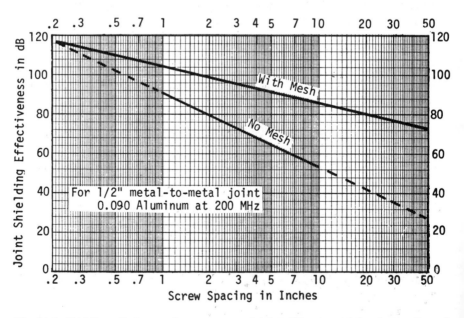

Fig. 11-1. Shielding effectiveness for screw-secured joints. (*Courtesy of Interference Control Technologies*)

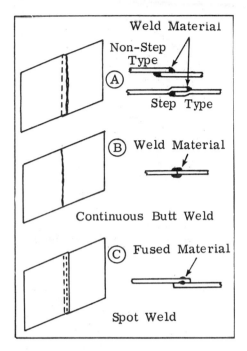

Fig. 11-2. Illustration of permanent welded shield seam configurations. (*Courtesy of Interference Control Technologies*)

exception involves the departure of the weld filler material characteristics, such as the conductivity and permeability, from the characteristics of the basic shield member material. Hence, the conductivity and/or permeability of the weld filler material may be significantly different from that of the base material, resulting in the degradation of shielding effectiveness. The seam weld technique is of questionable value when used with the more exotic magnetic materials, which must be annealed before assembly. Here, welding will destroy the specific properties that the annealing produced, and they are often restored by reannealing.

An alternative technique, shown in Fig. 11-3, is the overlap seam. All nonconductive material (e.g., paint, rust, coating, etc.) must be removed from the mating surfaces before they are crimped. Crimping must be performed under sufficient pressure to ensure positive contact between all mating surfaces.

Shield members, such as cover and access plates for equipment alignment or maintenance, may have to be separated from time to time; so none of the above techniques is acceptable. Whenever a temporary but good bond is required, this is the role of RF gasketing material such as finger stock or resilient mesh. The topic of gaskets is covered later in this chapter.

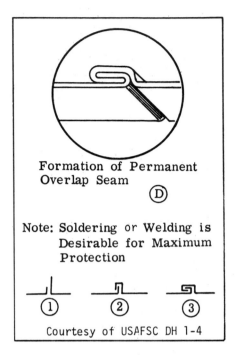

Fig. 11-3. Illustration overlap seam shield construction. (*Courtesy of Interference Control Technologies*)

Ventilation Openings

Most shielded housings or enclosures require either convection or forced-air cooling. Since associated openings will compromise the integrity of the basic shield material, a suitable electromagnetic mask must be installed that will provide substantial attenuation at RF while not significantly impeding the mechanical flow of air. Two approaches are possible: screened covers and honeycomb aperture covers. As explained in the next section, screens are inexpensive approaches to this problem but are limited in shielding effectiveness and tend to disrupt the smooth flow of air, thereby lowering the efficiency of the cooling system. Thus, a honeycomb material is generally used because it provides higher shielding effectiveness and maintains a streamlined, and therefore more efficient, flow of air. This can impact the power requirements of the cooling system.

In typical honeycomb construction, illustrated in Fig. 11-4, the hexagonal elements ("hexcells") use the waveguide-beyond-cutoff characteristic to accomplish the desired shielding effectiveness. Representative honeycomb panels are shown in Fig. 11-5. Equation (11-1) indicated the expected attenuation. However, for honeycomb the shielding effectiveness at frequencies

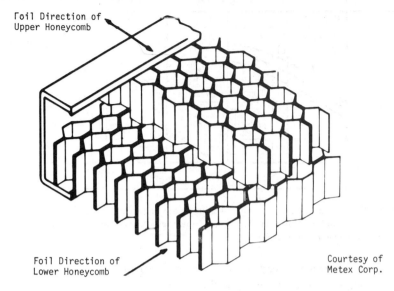

Foil Direction of Upper Honeycomb

Foil Direction of Lower Honeycomb

Courtesy of Metex Corp.

Fig. 11-4. Typical honeycomb construction. (*Courtesy of Metex Corp.*)

Fig. 11-5. Representative honeycomb configuration. (*Courtesy of Chomerics*)

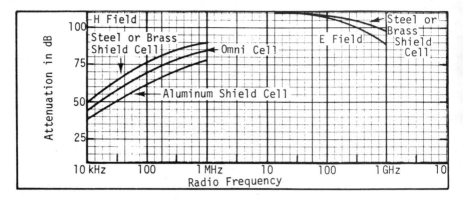

Fig. 11-6. Typical shielding effectiveness of honeycomb vent covers. Both the honeycomb and mesh covers are mounted over the ventilation opening with gasketing material. (*Courtesy of Interference Control Technologies*)

well below cutoff is reduced by the total number of waveguide elements, N, in the panel, since the emerging field from each hexcell coherently combines with its neighbor. Therefore, for the honeycomb ventilation covers, the shielding effectiveness can be approximated by the expression:

$$A_{dB} = 27L/g - 20 \log_{10}[N] \qquad (11-6)$$

Figure 11-6 illustrates typical attenuation (shielding effectiveness) performance of different honeycomb configurations. However, it must be noted that the performance of honeycomb does not follow equation (11-6) for low frequency magnetic fields. For low frequency magnetic fields, other ventilation opening shielding techniques must be employed, such as a "swiss cheese" steel panel with ventilation holes.

Sometimes, it is necessary to provide reduction or removal of air-intake dust in the ventilation process, which honeycomb construction will not do. Thus, an additional woven-wire mesh shield screen must be fabricated and installed across the honeycomb opening. The shielding mesh medium can be either dry or wetted with an oil coating for more effective dust removal. (See Fig. 11-7.)

When ventilation cover panels are used for convection cooling, it is often common practice to employ a number of perforations in the panel rather than to use honeycomb or wire screen. Holes are punched into the panel with a die that also cuts the cover panel. For this situation, the shielding effectiveness, A_{dB}, is:

$$A_{dB} = \frac{kL}{g} + 20 \log_{10}\left(\frac{C}{D}\right)^2 \qquad (11-7)$$

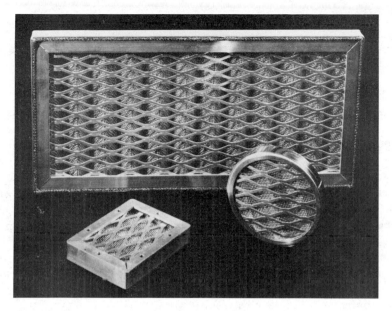

Fig. 11-7. Representative screen mesh ventilation for air filtering and EMI shielding. (*Courtesy of Chomerics*)

where:

$k = 27$ for square perforations

$k = 32$ for circular perforations

L = thickness of cover panel in inches (or cm)

g = width of square perforations or diameter of circular perforations in inches (or cm)

C = center-to-center spacing of perforations in inches (or cm)

D = length of aperture side for squares or diameter for circular apertures in inches (or cm)

If the cover plate perforations are not equally spaced, then C^2 in equation (11-7) may be replaced by $C^2 = A/N$, where A = area of aperture = D^2, and N is the number of perforations or holes. For this situation, equation (11-7) becomes:

$$A_{\text{dB}} = \frac{kL}{g} - 20 \log_{10} \left[\frac{D^2 N}{A} \right] \qquad (11\text{-}8)$$

$$A_{\text{dB}} = \frac{kL}{g} - 20 \log_{10} \left[N \right] \qquad (11\text{-}9)$$

Viewing Apertures

Another requirement that compromises the integrity of the basic shield material is the need for viewing panel meters, digital displays, scopes, and other types of status monitors or read-out presentations contained inside the shielded housing or enclosure. This is accomplished by either a laminated-screen window or a conductive optical substrate.

Screen Windows.   A shielded window must be used to prevent RF penetrations. The screen configuration may be one of two types: (1) similar to a screen room or ordinary window screen in which the individual screen elements or wire strands are welded together at the crossover junctions, or (2) made of fine knitted wire laminated between two layers of acrylic or glass. Screens are characterized by metal-thickness-to-air-gap-area ratios on the order of 0.05 (may range from 0.01 for chicken wire to more than 0.1 for some screens and approximately 0.5 for some metallized textiles). The air-gap spacings, g, may vary from 0.51 mm to about 5.1 cm. The shielding effectiveness model that approximately describes the performance of the screen family in the far field (plane waves) is:

$$SE_{dB} = 20 \log_{10} \left[\frac{\lambda/2}{g} \right] dB, \text{ for } g < \lambda/2 \qquad (11\text{-}10)$$

$$SE_{dB} = 0 \text{ for } g \geq \lambda/2 \qquad (11\text{-}11)$$

where:

λ = wavelength
g = air gap dimension in same units as λ

The screen shielding effectiveness is essentially due to reflection loss with no appreciable loss due to absorption.

Equation (11-10) may be expressed in terms of frequency instead of wavelength, using the following relationships for the English and metric systems of units:

$$SE_{dB} = 20 \log_{10} \left[5906/g_{in} f_{MHz} \right] dB, \text{ for } g_{in} < 5906/f_{MHz} \qquad (11\text{-}12)$$

$$SE_{dB} = 0 \text{ for } g_{in} \geq 5906/f_{MHz} \qquad (11\text{-}13)$$

$$SE_{dB} = 20 \log_{10} \left[15{,}000/g_{cm} f_{MHz} \right] dB, \text{ for } g_{cm} < 15{,}000/f_{MHz} \quad (11\text{-}14)$$

$$SE_{dB} = 0 \text{ for } g_{cm} \geq 15{,}000/f_{MHz} \qquad (11\text{-}15)$$

where g_{in} and g_{cm} are the screen gap spacing in inches and centimeters, respectively.

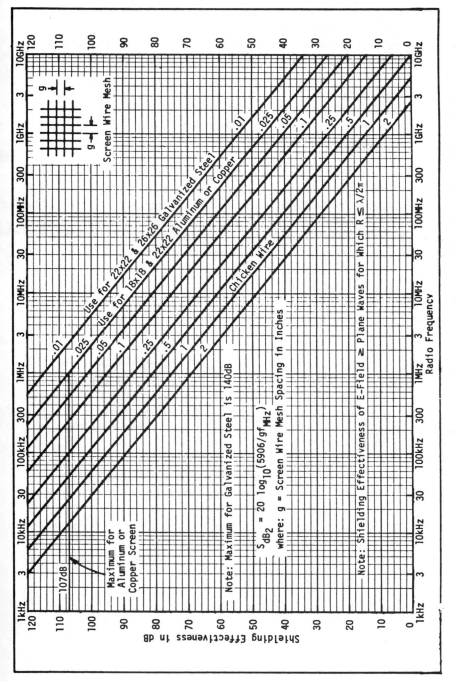

Fig. 11-8. Shielding effectiveness of screen wire to plane waves. (*Courtesy of Interference Control Technologies*)

As the EMI frequency decreases to low values, SE_{dB} does not continue to rise at 20 dB/decade indefinitely for far-field conditions. Measurements indicate that the attenuation or shielding effectiveness for copper and aluminum screen reaches a maximum value of about 110 dB at frequencies of $5\frac{1}{2}$ decades below cutoff frequency, with cutoff frequency corresponding to $g = \lambda/2$. For galvanized steel, the maximum approaches 140 dB at these same low frequencies. To support this theory, it appears that when the screen mesh gap, g, becomes a very tiny fraction of a wavelength (i.e., 10^{-5} to 10^{-7}), the material approaches a homogeneous medium from a macroscopic point of view.

Equations (11-12) through (11-15) are plotted in Fig. 11-8. The figure shows that for plane waves the shielding effectiveness of screen mesh becomes relatively small above a few GHz, whereas it is very significant below 1 MHz. The shielding effectiveness to near-field ($R \ll \lambda/2\pi$) electric fields would be greater than the values given in Fig. 11-8, whereas to near-field magnetic fields it would be less. It is assumed here that the screen mesh wires are making good electrical contact with each other at the crossover junctions by welds or other comparable bonding processes.

The second type of screen window is made of fine knitted wire mesh laminated between two layers of acrylic or glass as illustrated in Figs. 11-9 and 11-10. The wire may be monel with typical sizes of 0.05 mm diameter (one opening per mm) or 0.11 mm diameter (one opening per 2 mm). This corresponds to a low-shadow area (15–20% blockage giving good visibility). Typical shielding effectiveness is shown in Fig. 11-11. This approach is becoming less popular than that of the conductive optical substrate described below because of the less aesthetic aspects of the former. Furthermore, under some conditions, screen window exhibits undesired diffraction-grating optical viewing problems.

Conductive Optical Substrate Windows. Another approach is available for providing shielding across apertures through which either optical viewing or other light transmission is necessary. This approach involves the use of a conductive window, a technique in which a thin film of metal is vacuum-deposited on an optical substrate. These conductive window designs, such as shown in Fig. 11-12, are evolved by establishing some or all six basic design parameters, as applicable:

- Window material
- Conductive coating
- Optical coating and finishes
- Reticle requirements
- EMI gasketing
- Framing and mounting

Fig. 11-9. Shielding screen window for installation over face of cathode ray tube. (*Courtesy of Chomerics*)

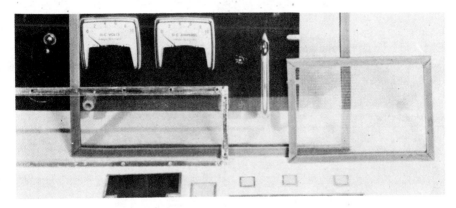

Fig. 11-10. Representiative shield screen windows for viewing.

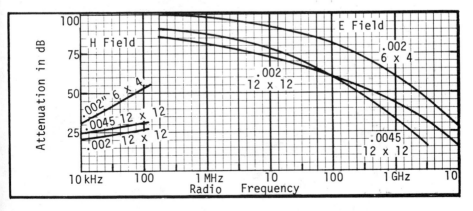

Fig. 11-11. Shielding effectiveness of shield screen windows.

Fig. 11-12. Typical conductive optical viewing panels.

Most plastic and glass optical panel materials are suitable as substrates for the application of a conductive metal coating. A conductive coating can be applied to almost any solid substrate, making it conducive for use as an EMI shield, switch element, filter, or other active low current carrying device. Acceptable substrates are those that will not outgas in a high vacuum. A quick test may be made by checking the substrate for odor. If there is none, it is not likely to outgas. The commonly accepted, more standard materials

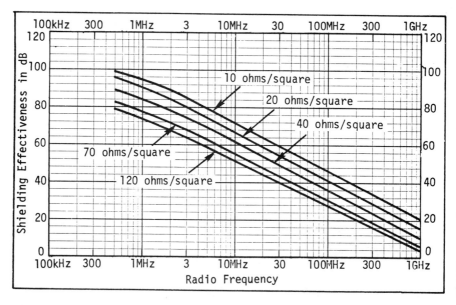

Fig. 11-13. Shielding effectiveness of conductive glass. (*Courtesy of Interference Control Technologies*)

are: glass, acrylic, polycarbonate, and fluorocarbon plastics. The substrates may be clear or colored as required by the application. There are no restrictions on substrate thickness. Curved or three-dimensional parts generally can be coated.

Most thermosetting and thermoplastic substrates have minute surface scratches produced in their normal manufacture. The application of the coating will inherently make these scratches more apparent, although actual user experience indicates that no functional problem will arise. The following list illustrates a sample of the large selection of common substrate materials suitable for conductive coating:

- Glass, plate
- Glass, single strength
- Glass, float
- Glass, tempered

- Glass, laminated, PVB film, safety
- Glass, quartz
- Crystals, ruby
- Crystals, quartz
- Vycor[1]
- Pyrex[1]
- Lexan[2]

- Plexiglass, thermoplastic acrylic[3]
- Plexiglass, transparent, colorless
- Plexiglass, frosted, colorless
- Plexiglass, colored: yellow, amber, gray, bronze, green, red, blue
- Homalite, thermosetting plastic[4]
- Kapton[5]
- Mylar[5]
- Abcite, coated acrylic[5]
- Polycarbonate
- Self-extinguishing plexiglass
- Fluorocarbons

Trademarks of: (1) Corning, (2) General Electric, (3) Rohm and Hass, (4) Homalite, and (5) DuPont. In the plastic substrate group, the most scratch-resistant materials are Abcite followed by Homalite.

Polarized filter laminate finishes are available for contrast improvement. Coatings are unaffected by application of laminated circular polarizers. Translucent or frosted finishes, with a rough surface, are available. They are best employed on the side opposite the conductive face. They can only be used for display or rear projections or where the object is extremely close to the window surface. Anti-reflective, vacuum-deposited coatings may be applied to windows before coating.

Figure 11-13 illustrates typical shielding effectiveness vs. frequency (2) for different film coating thicknesses on glass measured in surface resistance units of ohms/square. Since the film thickness is deposited in microns, little contribution to attenuation comes from absorption loss. Accordingly, reflection loss is the principal medium of attenuation. Above about 1 MHz, the loss decreases with an increase in frequency at a rate of approximately 20 dB per decade and becomes negligible above about 1 GHz.

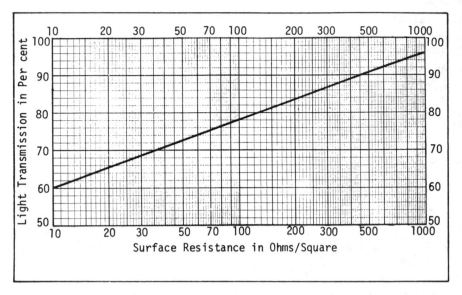

Fig. 11-14. Light transmission of conductive glass. (*Courtesy of Interference Control Technologies*)

Light transmission vs. surface resistance for the above conductive glass is shown in Fig. 11-14. Transmission values of 60 to 80% correspond to resistances of about 10 to 100 ohms/sq. Thus, the values shown in Fig. 11-13 may be compared with the attenuation data of the shield screen depicted in Fig. 11-11 to determine relative performance for specimens of comparable size (area). The shield screen is seen to be everywhere superior to the conductive glass in shielding effectiveness, as shown in Table 11-1, in which the difference becomes greater with increasing frequency. In reviewing the commercial product literature, it is noted that the performance of conductive optical substrates for surface resistivities on the order of 10 ohms/sq. most closely approximates that of the screen. Measurements were made by different

Table 11-1. Comparison of shielding effectiveness of screen and conductive glass windows. (Courtesy of Interference Control Technologies.)

Frequency	Shield screen	Conductive glass	Superiority of shield screen
1 MHz	98 dB	74–95 dB	3–24 dB
10 MHz	93 dB	52–72 dB	21–41 dB
100 MHz	82 dB	28–46 dB	36–54 dB
1 GHz	60 dB	4–21 dB	39–56 dB

observers, using different test setups on different specimens, with expected variations observed.

Thus, it is concluded that if significant VHF and UHF attenuation is required for viewing apertures, shield-screen windows should be used. If the aesthetics or other considerations do not permit that, conductive glass cannot be relied upon to provide significant RF attenuation to E-fields much above 30 MHz.

Control-Shaft Apertures

Another aperture class that compromises the shielding integrity of an equipment housing or instrument panel is that resulting from shafts of potentiometers, tuning dials, and control devices. Generally, an external metallic front panel or housing is either drilled or punched with sufficient clearing tolerance for the control shaft to result in a leaky aperture. The inside wall of the panel hole forms an outer conductor to a coaxially situated internal control shaft (i.e., the inner conductor). In other words, potential EMI can enter or exit through this effective short-length coaxial line, and the extended shaft beyond the panel acts as either as a pickup or a radiating antenna.

To preserve the shielding integrity of otherwise leaky control-shaft situations, one method of minimizing the degradation of shielding effectiveness is to design a supporting bushing extender to act as a circular waveguide-beyond-cutoff attenuator (cf. equations 11-1 and 11-5). For 100 dB attenuation in a circular waveguide, the length (L) of the waveguide must be somewhat more than three times its diameter ($L/g > 3$ in equation 11-5). Figure 11-15 shows an acceptable use of a metal tube bonded to the wall containing the clearance aperture for control shafts.

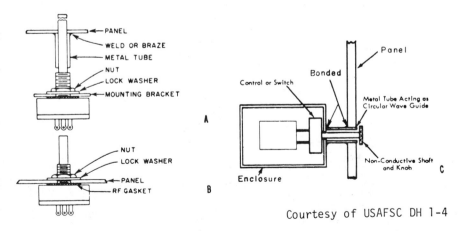

Courtesy of USAFSC DH 1-4

Fig. 11-15. Use of circular waveguide in a permanent aperture for control-shaft EMI leakage control. (*Courtesy of Interference Control Technologies*)

If the preceding situation were implemented without regard to the control-shaft properties and relations to the added metal tube, little improvement might result for typical metal shafts. This situation corresponds to a low-impedance coaxial line in which an intervening dielectric may result from contaminants such as oil films or oxides. To preclude this, one of two techniques is followed: (1) replace the metallic control shaft with a nonconductive shaft as shown in Fig. 11-15, or (2) use a cylindrical-shim EMI gasket between the shaft and tube. The latter method does not require modification of existing control shafts.

Indicator Buttons and Lamps

Some instruments or equipments require the use of push buttons, status indicator buttons, and/or indicator lamps. These devices also provide another compromise of shielding integrity by virtue of the required apertures in a front panel or housing. Two techniques are available to mitigate the EMI leakage through such devices:

1. Encase them in a shielded compartment behind the front panel when they are mounted as shown in Fig. 11-16. Feed-through capacitors or filter-pin conductors are used for hard wiring from outside the compartment to the buttons or indicator lamps, since conducted EMI could exist on either side of the barrier.

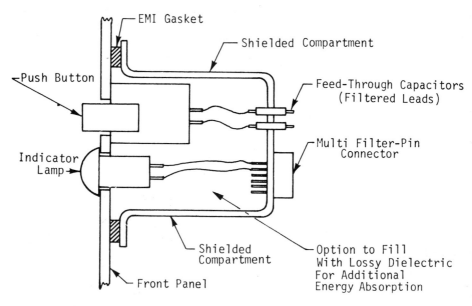

Fig. 11-16. Shielded and filtered compartment technique to restore shielding integrity of button and lamp apertures. (*Courtesy of Interference Control Technologies*)

2. Use special EMC designed hardware where such devices are mounted directly to a front panel. Examples of this include wire-mesh indicator lamps (which look like miniature photo-flash bulbs). This mesh serves to reflect and absorb entering or exiting EMI energy.

EMC GASKETS

This section discuss a very important class of techniques used to restore shielding integrity when loss is due to seams and joints where other than permanent fastening methods are permitted.

Gasketing Theory

Gaskets are employed for either temporary or semipermanent sealing applications between joints or structures; for example:

Temporary RF sealing applications:
- Securing access doors to enclosures, cabinets, or equipments.
- Mounting cover plates or removal panels for equipment maintenance, alignment, or other purposes.

Semipermanent RF sealing applications:
- Mounting either screen or conducted glass windows to housings containing electrical or electronic test equipments.
- Mounting honeycomb and other ventilation covers to enclosures, cabinets, or equipment.
- Securing parallel members of an equipment housing to a frame structure using machine screws.

All gaskets of the non-spring-finger stock type, whether they seal EMI or higher-pressure fluid, make a container dunk-proof, or simply keep forced ventilating air from escaping at a door-to-cabinet joint, conform to the unavoidable irregularities of the mating surfaces of a joint. Some examples are:

- The joint between a garden hose and water faucet.
- Housing for an emergency radio or beacon to be dropped into the sea.
- The joint between the cover and enclosure of a radar pulse modulator.

In each example the joint has two relatively rigid mating surfaces, and neither surface is perfectly flat or conforms to the same geometric contour as the other. When the surfaces are mated without a gasket, even high closing forces will not cause the two surfaces to mutually seal. Resultant gaps will allow leaks to exist. A gasket resilient enough to conform to the contour of both surfaces under reasonable force, however, will eliminate these leaks. Thus, in most cases, the least expensive way to obtain a tight, continuous, metal-surface-to-metal-surface joint (watertight, oil-tight or EMI-tight) is to make the mating surfaces conform to the same geometric contour within reasonable tolerances, and then to add a gasket to compensate for the resulting gaps between the two surfaces.

Joint Unevenness. The degree of misalignment or misfit of the mating surfaces is commonly called joint unevenness, and is designated ΔH in Fig. 11-17a. It is the maximum separation between the two surfaces when they are just touching, or making contact, at some points of surface conformity. If the surfaces are not rigid, then the joint unevenness also includes any additional separation between the two surfaces due to joint distortion when pressure is applied.

Figure 11-17b shows the same joint with a gasket installed. The dashed lines indicate the gasket height, H_g, before compression. The compressed minimum gasket height, H_{min}, occurs at a point where the surfaces would touch without a gasket. Compressed maximum gasket height, H_{max}, is at the point of maximum joint separation. Thus, joint unevenness of the mating surface is:

$$\text{Joint unevenness} = \Delta H = H_{max} - H_{min} \qquad (11\text{-}16)$$

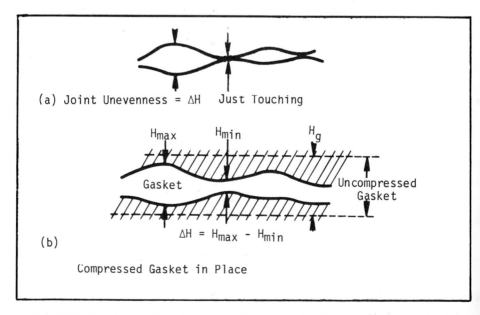

Fig. 11-17. Description of joint unevenness. (*Courtesy of Interference Control Technologies*)

Required Compression Pressure. Three factors determine the required compression pressure on a gasket: its resiliency, the minimum pressure required for a seal, and the total joint unevenness.

Resiliency. Resiliency is the amount by which a gasket compresses per unit of applied compression pressure. Resiliency is usually expressed in percent of original (uncompressed) gasket height divided by pressure in pounds per square inch (psi). A soft gasket would compress more than a hard gasket with the same applied pressure. Stated another way, a soft gasket requires less pressure than a hard gasket to compress at the same percentage of gasket height. For example, a sponge neoprene gasket might compress 10% under an applied compression pressure of, say, 6 psi, but a solid neoprene gasket would require 40 psi for the same 10% deflection, as shown in Fig. 11-18.

Minimum Pressure for Seal. A gasket must at least make contact at the point of maximum separation between mating surfaces; that is, $H_{max} < H_g$ in Fig. 11-17. Actually, the pressure at this point must be the stated minimum amount required for the particular gasket type to assure an EMI seal. Thus, the pressure at the H_{max} point must be high enough to prevent leakage. For EMI gaskets, this minimum pressure, P_{min}, is determined by the pressure required to break through corrosion films and to make a suitable low-resistance contact. P_{min} is typically about 20 psi, but can be as small as 5 psi, depending upon gasket resiliency.

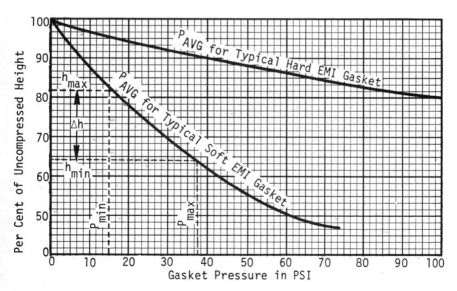

Fig. 11-18. Typical hard and soft EMI gasket height vs. pressure relations. (*Courtesy of Interference Control Technologies*)

Average Pressure. The average pressure applied to the gasket must also be large enough to compress the overall gasket so that the difference between the minimum height and the maximum gasket height (determined by P_{min} from the previous paragraph) is equal to the joint unevenness, that is, $\Delta H = H_{max} - H_{min}$, as previously presented in equation (11-16). In general, the average pressure should equal or exceed that corresponding to the average compressed gasket height, H_{avg}:

$$H_{avg} = (H_{max} + H_{min})/2 \qquad (11\text{-}17)$$

and:

$$P_{avg} = (P_{min} + P_{max})/2 \qquad (11\text{-}18)$$

The required compression force, F, in units of pounds, may be calculated from P_{avg} by determining the surface area of the gasket to be sandwiched between the mating members:

$$F = P_{avg} \times A \text{ pounds} \qquad (11\text{-}19)$$

where A = gasket area in square inches or other units.

Required Gasket Height. To obtain the required EMI seal from a gasketed joint, the gasket height must meet these criteria:

- The pressure at the point of maximum joint separation (H_{max}) must correspond to the minimum pressure needed to obtain the required EMI seal.
- The difference between maximum and minimum compressed heights of the gasket must equal the joint unevenness of the mating surfaces.

If the average pressure available to compress the gasket is P_{avg}, the maximum pressure, P_{max}, is obtained from equation (11-18):

$$P_{max} = 2P_{avg} - P_{min} \qquad (11\text{-}20)$$

where P_{min} is a selected value of minimum pressure required to obtain a satisfactory gasket seal.

The percentages of uncompressed height corresponding to P_{min} and P_{max} in Fig. 11-18 are H_{max} and H_{min}, respectively. To calculate the required uncompressed gasket height, H_g, as a dimension:

$$H_g = \frac{\Delta H_{inches}}{\Delta H_{decimal}} \qquad (11\text{-}21)$$

Thus, the required height is the actual joint unevenness in inches divided by the joint unevenness expressed in decimal equivalent of percent gasket compression (see Fig. 11-18).

Compression Set. Some gaskets do not return to their original uncompressed height after release of compression. This behavior is called compression set. It may be visualized by assuming that the lower curve shown in Fig. 11-18 applies for a particular soft gasket. When compression pressure is removed, the gasket returns to a lesser height whose properties might look somewhat like the upper curve in Fig. 11-18 (this is exaggerated for illustrative purposes).

The importance of compression set depends upon how the gasket is to be used; three classes of use are now defined:

- *Class A, permanently closed:* Compression set is unimportant because the gasketed component will in all probability neve be removed.
- *Class B, repeated identical open-close cycles* (c.g., hinged door, or symmetrical covers): Here, compression-set problems are marginal and must be determined in accordance with specific applications.
- *Class C, completely interchangeable* (complete freedom to reposition gasket on repeat cycles; e.g., round gasket in waveguide): Since the compression-set height at a point of maximum compression may end up being less than the minimum compressed height, no contact at all would result between gasket and mating surfaces at this point. For Class C uses, do not reuse gaskets with compression set limits; instead, use a new gasket.

Gasket Types and Materials

There exists a plethora of EMI gasket types, shapes, binders, and materials. In fact, the profusion of gaskets is so great that it is likely to be confusing to all but those who specify or use them with some degree of regularity. Suppliers recognize this, so they produce creditable application notes and design and order guides.

For convenience of this discussion and comparison, EMI gaskets are divided into the following types: (1) knitted-wire mesh, (2) oriented immersed wires, (3) conductive plastics and elastomers, (4) spring finger stock, and (5) pressure-sensitive, foam-backed foil. The last two types are different from the first three and operate on significantly different principles. A brief summary of each is presented below, followed by a comparison of all the types.

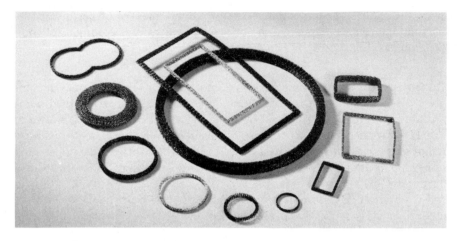

Fig. 11-19. Typical knitted wire mesh gaskets. (*Courtesy of Chomerics*)

Knitted-Wire Mesh Gaskets.

Figure 11-19 shows some examples of knitted-wire mesh gaskets. They are made from resilient, conductive, knitted-wire and somewhat resemble the outer jacket of a coaxial cable. Nearly any metal that can be produced in a fine-wire form can be fabricated into these EMI gaskets. Typical materials used are: monel; aluminum; silver-plated

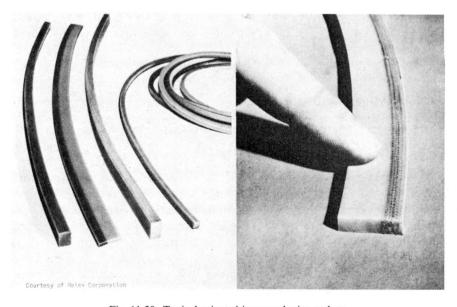

Courtesy of Metex Corporation

Fig. 11-20. Typical oriented immersed-wire gaskets.

brass; and tin-plated, copper-clad steel. These gaskets may employ an air core, or, for maximum resiliency, they may use a spongy neoprene or silicone core. Cross sections may be round, rectangular, or round with fins for mounting. They are generally applied to shielding joints having a periphery greater than 4 inches and cross sections between 0.063 inch and 0.75 inch.

Oriented-Immersed-Wire Gaskets. Figure 11-20 shows some examples of oriented immersed-wire gaskets. They are made with a myriad of fine parallel, transverse-conductive wires whose parallel impedance across the gasket interface is very low. Each convoluted wire is insulated from its neighbor, at a density of about 100 wires per square inch. Typical materials used are monel or aluminum embedded in either a solid silicone (hard gasket) or a sponge silicone (soft gasket) elastomer. This gasket provides a simultaneous EMI and pressure seal. The embedded wires protrude a few mils on either side to assist in percing any residual grease/oil film and oxide on the surface of the mating members, a characteristic that is especially good where aging and subsequent maintenance may result in a panel member that is no longer clean and degreased. Cross sections available range from $0.125'' \times 0.125''$ to $0.625'' \times 0.5000''$, and come in any length.

Conductive Plastics and Elastomer Gaskets. Figure 11-21 shows some examples of conductive plastic and elastomer gaskets. They are often made with a myriad of tiny silver balls or silvered brass balls immersed in a

Fig. 11-21a. Typical conductive elastomer gaskets. (*Courtesy of Chomerics*)

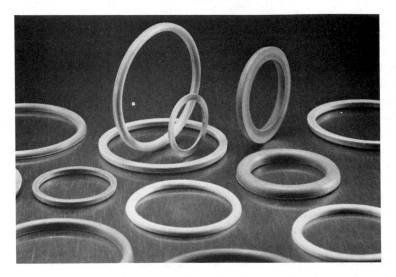

Fig. 11-21b. Typical conductive elastomer gaskets. (*Courtesy of Chomerics*)

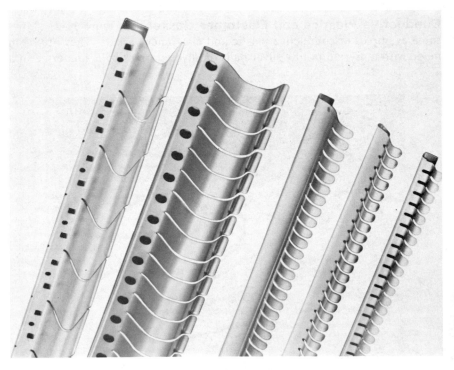

Fig. 11-22. Typical spring finger strip gaskets. (*Courtesy of Instrument Specialties Co., Inc.*)

silicone rubber or vinyl elastomer binder and carrier, the gasket providing a simultaneous EMI and hermetic seal. Offering volume resistivities from 0.001 to 0.01 ohm-m and useful over a wide range of temperatures, these gaskets are available in sheets, die cuts, molded parts, and extruded shapes. Some versions are operable down to cryogenic temperatures. They offer low closing pressures, low compression set and maintenance, and long life.

In recent years it was discovered that most of these gaskets do not give reliable performance when subjected to vibrations. Because the joint unevenness varies over the phase of the vibration cycle, the pressure on the gasket will vary. When the compression pressure goes below the minimum amount to effect an electrical seal, the joint will become leaky. Therefore, although these gaskets provide adequate shielding under mechanically static conditions, some degrade in performance under vibration.

Spring-Finger Stock Gaskets.   Figure 11-22 shows some examples of beryllium–copper, spring-finger gaskets stamped into different configurations. Basically, gaskets similar to these were introduced over 30 years ago, and were the first type of EMI gasket to appear on the market. Since limited elastomer technology existed in the 1940s, joint unevenness was accounted for by a gasket configuration consisting of a line of individual fingers, each capable of flexing a different amount. For shielded enclosures, cover plates, and other heavy-duty applications where frequent access was required, this type of gasket was, and still is, widely used. Recent design changes, shown in Fig. 11-22, make this type of gasket quite flexible and competitive with other types. New spring-finger contact strips offer self-adhesive backing to eliminate older mechanical fastening methods. They are available in a wide variety of sizes and shapes. Principal disadvantages are related to the tendency of the fingers to oxidize and to break off.

Pressure-Sensitive, Foam-Backed Foil Gaskets.   Another type of gasket differing from the above is a beryllium–copper foil backed by a highly compressible neoprene foam. The foam side, containing a synthetic rubber pressure-sensitive adhesive, is applied to cover plates. When placed over an electronics package containing shielded compartments, the foam-backed foil assumes the irregularities of the compartment heights, including outside plates, to result in a continuous EMI seal. This $\frac{1}{16}$-inch (1.59-mm) gasket is available in sheet widths to 6 inches (15 cm), and may be die-cut. EMI shielding effectiveness of 90 dB to electric fields is claimed over the 1 kHz to 10 GHz frequency spectrum.

Comparison of Gasket Types and Materials.   The profusion of different gasket types and materials (well over 1000 variations) is confusing to the design or specification engineer who has the responsibility of selecting

Table 11-2. Comparison of gasket types & materials.
(Courtesy of Interference Control Technologies.)

Comparison Factors / Gasket Types			Knitted Wire Mesh	Oriented Immersed Wires	Conductive Plastics and Elastomers	Spring Finger Stock
Available Forms			Strips, Jointless Rings	Strips & Sheets. Jointless Rings Die-cut Shapes	Strips & Sheets. Die-Cut, Molded, Extruded Shapes	Strips
Size	Periphery		>4"			Any
	Cross Section	Min	0.063"			
		Max	0.750"			
Type of Seal	EMI only		Good to Excellent	Good	Also seals Hermetically	Good to Excellent
	EMI plus Hermatic		NA	Fair to Excellent	Good to Excellent	NA
Conductive Material			Silver Plate, Monel, Aluminum, Steel SN/CU/FE	Monel, Aluminum	Many Tiny Silver Balls	Beryllium-Copper
Binder or Core Material			Rubber, Air Core, Neoprene, Silicone Sponge	Solid & Sponge Silicone	Silicone or Plastic	NA
Temperature Range			Limited to Core	-70oF to 500oF	-100oF to +400oF	-65°F to 100°F
Available Gasket Heights			.062" to 0.500	.062" to 1.000"	0.020" to 0.160"	.062" to 0.400
Joint Unevenness Accommodations			.020" to 0.160"	.010" to 0.100"	0.003" to 0.030"	.035" to 0.250"
Compression Height Range						7:1
Compression Pressure			5 psi to 100 psi	20 to 100 psi	20 to 100 psi	
EMI Shielding Performance	10 kHz (H)		25-30dB	>45dB	>35dB	>10dB
	10 MHz		>100dB	>100dB	>100dB	>120dB
	1 GHz		>90dB	>90dB	>95dB	>100dB
	10 GHz				>70dB	>100dB

one or more best candidates for a particular application. Accordingly, Table 11-2 presents a comparison of some of the best principal characteristics of EMI gaskets. No one type is the best for all applications. For example, those gaskets that are relatively low in cost tend to have a relatively higher volume resistivity, resulting in a less impressive shielding effectiveness. Some gaskets are designed to operate down to cryogenic temperatures and some up to 500°F, but none can be singularly applied throughout this range. Since there are several different methods of mounting, gaskets are available in sheets and strips, die cuts, molded shapes, and extruded forms. At the risk of gener-alizing, conductive plastics and elastomers seem to offer the widest range of applications and price.

Gasket Selection. EMI gasket selection involves making suitable matches and tradeoffs between (1) available EMI gasket materials and their characteristics (see Table 11-2), and (2) performance requirements of equip-ment and design constraints of mating surfaces. Gasket mounting (hence selection) involves a number of alternatives. A suggested guide-line is: *When-ever the need for shielding arises, determine the most cost-effective way to obtain an electrically continuous, high-conductivity surface or path across the seam, aperture, etc., to be sealed.*

In selecting one or more suitable EMI gaskets for sealing mating surfaces, gasket characteristics, application requirements and constraints, and price are the major considerations. Application requirements are usually stated in the form of equipment performance specifications. They include the amount of required shielding, pressure sealing, and environmental considerations (e.g., temperature, salt spray, ambient pressure, and corrosive material). Appli-cation constraints are usually imposed by equipment housing design. They include space available, compression force, joint unevenness, contact surface characteristics, and attachment possibilities (geometrical configuration).

The important *matches and tradeoffs* between application requirements and constraints on the one hand and gasket characteristics and price on the other are:

- Gasket height and compressibility must be large enough to compensate for joint unevenness under the available force.
- The gasket must be capable of providing the required EMI sealing and hermetic sealing (when applicable) when compressed by the available force.
- There must be sufficient space for the gasket within the design limitations of the application.
- The gasket must be attached or positioned by means that are compatible with the joint design.

- The metal portion of the EMI gasket must be sufficiently corrosion-resistant and electrochemically compatible with the mating surfaces.
- The EMI gasket must meet the temperature and other environmental needs of the equipment specifications.

Gasket manufacturers and suppliers provide design guide tables to help the user select the gasket that most nearly meets the application requirements and constraints.

The price range for EMI gaskets is wide. Cost depends upon type, size, quantity and performance. A number of suppliers will provide samples of their products at no charge. Others will supply a development kit of different sizes and shapes for a nominal cost.

Gasket Mounting. A number of methods are used to position the gasket to a metal mating surface: (1) hold-in slot, (2) pressure-sensitive adhesive, (3) bonding non-EMI mounting portion of gasket, (4) conductive adhesive, (5) bolt-through bolt holes, and (6) special attachment methods. Each method is summarized below.

Hold-in Slot. This method is recommended if the slot can be provided at relatively low cost, such as in a die casting. All solid elastomer materials, which embody the gasket material, are essentially incompressible. These products appear to compress because the material flows while it maintains a constant volume. Therefore, when these products are used in a slot, extra cross-sectional area must be allowed for the material to flow axially. At least a 10% extra volume, more if possible, is recommended, as shown in Fig. 11-23a.

Pressure-Sensitive Adhesive. This method of mounting is often the least expensive for attaching EMI gasket materials. Installation costs are substantially reduced with only a slight increase in gasket cost over a material without adhesive backing. Most sponge-elastomer materials are used for applications that do not require any hermetic sealing. The adhesive-backed rubber portion of this material serves only as an inexpensive attachment for the EMI portion.

Bonding Non-EMI Portion. Many good nonconductive adhesives are now available to bond an EMI gasket in position by applying the adhesive to the non-EMI portion of the gasket. The gasket can be insulated from the mating surfaces by a nonconductive material, and it is often a good way of mounting EMI gaskets. This method is shown in Fig. 11-23b.

The designer specifying nonconductive adhesive attachment must include adequate warnings in applicable drawings and standard procedures for production personnel. These cautions state that adhesive is to be applied only to the portion of the gasket material that is not involved with the EMI gasketing function. Experience indicates that installation workers, either

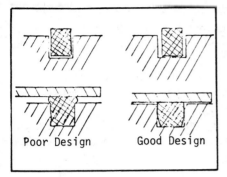

(a) Making Allowance for Solid Elastomer Gasket Flow

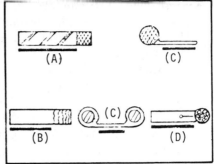

(b) Areas where non-conductive or dry-back adhesive can be used

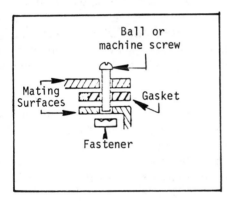

(c) Bolt Through Bolt Holes

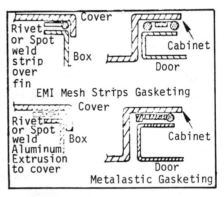

(d) Special Mounting Methods

Fig. 11-23. Different methods of mounting gaskets. (*Courtesy of Interference Control Technologies*)

through carelessness or a misguided desire to do a better job, will apply the nonconductive adhesive to the entire gasket including the EMI gasket portion. It is not uncommon to hear: "This gasket would hold better if I glued all of it rather than half of it." This action can completely degrade the EMI performance.

Conductive Adhesive. Conductive adhesive are available to mount EMI gaskets. However, the following cautions should be observed:

- Most conductive adhesives are hard and incompressible. Thus, if too much adhesive is applied, and it is allowed to soak too far into the EMI gasket material, the compressibility will be destroyed. Irregularly applied adhesive also has the effect of increasing joint unevenness.
- The volume resistivity of the adhesive should be .01 ohm-cm or less, preferably .001 ohm-cm.

- Most conductive adhesives do not bond well to either neoprene or silicone. This is why all products that have conductive paths in an elastomer are rated poor for conductive adhesive bonding by the manufacturers.
- Applying a $\frac{1}{8}$-inch (≈ 3 mm) to $\frac{1}{4}$-inch (≈ 6 mm) diameter spot of conductive adhesive every 1 to 2 inches (2.5 to 5.0 cm) is preferred over a continuous bead.
- Conductive epoxies will attach the gasket permanently. Thus, the removal of an EMI gasket without destroying it is almost impossible if conductive epoxy is used.

Bolt-through Bolt Holes. This is a very common and inexpensive way to hold gaskets in position, as shown in Fig. 11-23c. For most products, providing bolt holes involves only a small initial tooling charge. There is generally no extra cost for bolt holes in the piece price of the gasket. Bolt holes can be provided in the fin portion of EMI strips or in rectangular-cross-section EMI strips if they are sufficiently wide, such as over 3/8 inch (≈ 10 mm).

Special Attachment Means Provided. The knitted mesh fins provided on some versions of EMI strips and the aluminum extrusions in aluminum gasketing were designed to attach these products, as illustrated in Fig. 11-23d. The mesh fins can be clamped under a strip of metal held down by riveting or spot welding, or the mesh fins can be bonded with an adhesive or epoxy. The aluminum extrusions of aluminum gasketing can also be held in position by riveting or bolting.

EMI gaskets should be positioned so that they receive little or no pressure due to sliding motion when being compressed. This is illustrated in Fig. 11-24. The EMI gasket shown in Fig. 11-24a is subject to sliding motion when the door is closed. This may cause it to tear loose or to wear out

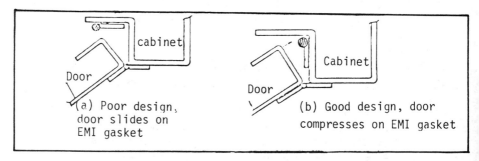

Fig. 11-24. Proper method of mounting gasket in cabinet door well. (*Courtesy of Interference Control Technologies*)

quickly. In Fig. 11-24b, the gasket is subject to almost pure compression forces. The latter is the preferred installation.

EMC SEALANTS

This section discusses another form of EMC shield integrity protection in the form of conductive epoxies and caulking.

Conductive Epoxies

Conductive epoxies are used to join, bond, and seal two or more metallic mating surfaces. The silver-epoxy resins replace soldering and other bonding techniques and cure at room temperatures. The conductive epoxy adhesive and solder family are used in the following applications:

- Electrical connections to:
  - Heat-sensitive components
  - Ferrites
  - Capacitor slugs
  - Integrated circuits
- Connecting electroluminescent panels
- Forming bus bars or strips on conductive glass
- Bonding flanges to waveguides
- Bonding waveguide sections
- Bolting holes and fasteners on electronic enclosures
- Joining dissimilar metals
- Sealing integrated circuit (IC) packages against moisture and EMI
- Repair of printed circuits
- Interconnecting conductive-metal gaskets
- Field repairs to circuits
- Permanent seam shielding
- Sealing EMI shields

Conductive Epoxy Preparation and Curing.   The conductive epoxies are easily mixed on a volumetric basis, eliminating much time and equipment that would otherwise be necessary for weighing. Most expoxies can be prepared with either equal volumes or equal weights of the components. They are formulated with mixed viscosities that produce a light, creamy paste to make application with standard dispensing equipment reasonably easy and fool-proof. Typical cure times are specified by the manufacturer and can vary from one day at room temperature to 30 minutes at 200°F.

Typical Properties of Conductive Epoxies. Depending upon the type of silver-epoxy resin used, typical volume resistivity will range from 0.001 to 0.02 ohm-cm. Operating temperature range is about $-80°F$ to $+250°F$. Shear strength is about 1200 psi, and tensile strength varies with type, but averages about 2500 psi. It exhibits excellent moisture resistance. The cured specific gravity is about 2, suggesting its relatively light weight for many payload weight-limited applications.

Conductive Caulking

Conductive caulking is used to shield and seal two or more metallic mating members principally held together by other mechanical means. Silver particles are suspended in resin to provide conductive sealing. Conductive caulking is used in the following applications:

- Caulking EMI-shielded shelter panels
- Caulking EMI-tight cabinets and enclosures
- Improving joint and seam integrity of electronic enclosures
- Protecting mating members of shielded conduits
- EMI-sealing and grounding bulkhead panel fittings
- Moisture-sealing of mating members
- Adhering metal-foil tape to shielded room joints
- Repairing damaged conductive gaskets
- Sealing ends on wire-mesh gaskets
- Caulking fasteners, panels, and handles

Conductive Caulking Preparation and Use. The conductive caulking compounds, as with any EMI sealant and bond, require that surfaces be thoroughly degreased and cleaned of oxide coating. The caulking may be applied with conventional caulking guns and dispensing equipment, such as small bead-orifice syringes. Hand application with a spatula or putty knife may be used. The caulking is free of any corrosive binders. It is used at room temperatures, and most caulking will not cure (is permanently nonsetting). This feature permits easy disassembly and cleaning of caulked parts for movement or maintenance.

Typical Properties of Conductive Caulking. Depending upon the type of silver resin used, typical volume resistivity will range from 0.005 to 0.02 ohm-cm. Operating temperature ranges from $-80°F$ to $+400°F$. Moisture resistance is excellent. The final specific gravity is about 1.8, suggesting its relatively light weight for many payload weight-limited applications.

CONDUCTIVE GREASE

Conductive grease is not a member of the EMI gasket and sealant family discussed in this chapter. However, it is related to the extent that one of its functions is to provide a low-resistivity contact path between mating members. Here, relative linear or rotary motion may be required between mating members, or they may engage and disengage more often than in most EMI gasketing applications, except for finger stock used in shielded enclosures.

Conductive Grease Preparation and Use

Conductive grease is provided commercially in squeezable tubes in a variety of other containers. It is used on power substation switches and in suspension insulators to reduce EMI noise. It also reduces make-break arcing and pitting of the sliding metal contact surfaces of switches, and fills in pitted areas with silve-silicone. In addition, normally closed switches are prevented from sticking due to corrosion or icing. The grease is designed to maintain a continuous electrical path between contact surfaces which must be free to move. These include ball-and-socket connections of power insulators, which, if allowed to arc, can generate EMI. Conductive grease is designed to maintain low-resistance electrical contact and thereby maintain equipment operating over extended environmental conditions, helping to deliver continuous electrical service.

Conductive grease is used on the contacting surfaces of circuit breakers and knife blade switches. It reduces localized overheating or hot spots, in turn maintaining the mechanical spring properties of blades and the current rating of the switch or breaker at the original-equipment specified level. Lubricating with conductive grease prevents freeze-up in operating equipment and permits restoration of marginal or discarded breakers to rated capacity.

Typical Properties of Conductive Grease

Conductive grease is a low-resistivity, silver-silicone grease that contains no carbon or graphite fillers. The material will maintain its electrical and lubricating properties over a broad environmental range. It is stable at high and low temperatures, resistant to mosture and humidity, inert to many chemicals, and unaffected by ozone and radiation. A conductive grease is usually a viscous paste that can be applied at elevated operating temperatures to vertical or overhead surfaces without dripping or running.

Typical volume resistivity of conductive grease is about 0.02 ohm-cm. The typical operating temperature range is from $-65°F$ to $+450°F$. Conductive grease provides excellent moisture resistance and does not corrode metals. Its pot life is unlimited, and any that is unused may be returned to the container for future applications.

REFERENCES

1. Lassiter, Homer A. "Low Frequency Shielding Effectiveness of Conducted Glass." *IEEE Transactions on Electromagnetic Compatibility*, Volume EMC-6, Number 2, July 1964.
2. White, Donald R. J. *Electromagnetic Compatibility Handbook*. Volume 3. Gainesville, VA: Don White Consultants, Inc., 1973.

Chapter 12
EMI-SHIELDED HOUSINGS

The preceding two chapters covered the subject of shielding. Chapter 10 discussed shielding theory and materials. It was shown that for other than low-frequency magnetic fields, it is relatively easy to obtain more than 100 dB shielding effectiveness across the spectrum for nearly any metallic solid surface. The shielding problem then develops from the fact that practical enclosures have apertures and penetrations that compromise the effectiveness of the basic shield material. Thus, shielding effectiveness of a housing could be reduced to 60 dB or less because of loss of enclosure integrity.

Chapter 11 covers methods for restoring shielding effectiveness of an enclosure with apertures by design and/or retrofit. The discussion covers such topics as how to compensate for the degradation of shielding effectiveness using techniques applicable to seams and joints, ventilation covers, access plates, gaskets, and so on, as well as conductive epoxies and caulking.

This chapter brings the foregoing material together in the form of practical applications to shielded enclosures. It presents shielding techniques applicable to shielded compartments for chassis and equipments, cabinets, shielded rooms and enclosures, huts, and vehicles. Configurations discussed range from enclosures of relatively small physical size (e.g., shielded compartments) to large-size complexes (shielded rooms and buildings).

SHIELDED COMPARTMENTS

In terms of physical size, some of the smallest shielding configurations are shielded compartments for amplifiers, filters, and the like, where either stage-to-stage or input–output crosstalk is to be kept below a specified amount. Input–output crosstalk often affects the gain of an overall amplifier or the out-of-band attenuation of a filter. Individual component shields are discussed in Chapter 16.

To illustrate the above, one example of compartment shielding problems involves capacitive coupling from stage to stage in which the role of the

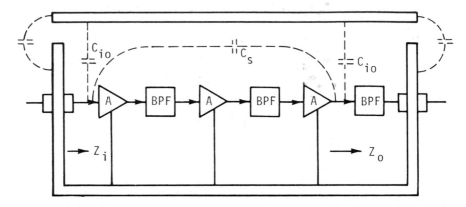

Fig. 12-1. Box shield for high-gain amplifier showing two paths or output-input feedback. (*Courtesy of Interference Control Technologies*)

shield is to provide a faraday cage to eliminate crosstalk. Figure 12-1 shows an I-F amplifier and interstage band-pass filter. The need to shield the amplifier from the outside world was recognized by the designer; hence the outer shield. The need to internally isolate input and output was also recognized; hence the compartments or stage septums. The figure shows a floating, nongrounded, metallic cover plate. Thus, a portion of the high-level output voltage, e_o, is capacitively coupled back to the amplifier's input voltage, e_i. The amplifier is unstable and may oscillate when:

$$e_o\left(\frac{Z_i}{2/\omega C_{io}}\right) = .5e_o Z_i \omega C_{io} > e_i \tag{12-1}$$

or:

$$\frac{e_o}{e_i} = A > \frac{2}{Z_i \omega C_{io}} \tag{12-2}$$

where:

$.5C_{io}$ = input–output capacitance
$\omega = 2\pi f$, where f = operating frequency
Z_i = input impedance
A = overall amplifier gain

For the situation shown in Fig. 12-1, the ungrounded cover plate serves to intensify the intra-amplifier EMI coupling problem, rather than help.

When the cover plate shown in Fig. 12-1 is grounded to the outer housing, the feedback capacitance is transferred to ground where it can do no harm,

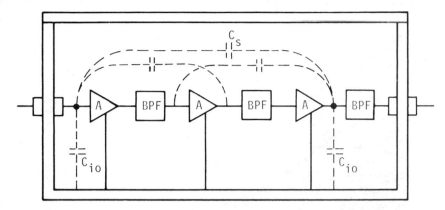

Fig. 12-2. Box shield cover plate in position to eliminate larger feedback path of Fig. 12-1. (*Courtesy of Interference Control Technologies*)

as shown in Fig. 12-2. However, a small residual capacitance, C_s, between input and output still remains. If equation (12-2) is satisfied, the amplifier may still oscillate. Thus, compartment shields are added, as shown in Fig. 12-3, to further decouple stages by the introduction of additional faraday cages.

To assure the full effectiveness of Fig. 12-3, it is necessary to bond the top of the compartment septums to the cover plate, and this is accomplished by gasketing. Figure 12-4 shows one method used to achieve this. Here, a conductive epoxy adhesive is applied to the cover plate, over which is placed the EMI mesh gasket (see Chapter 11). With the gasket held in place, the septum height variations indicated in Fig. 12-3 are accomodated.

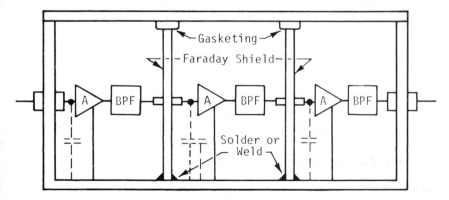

Fig. 12-3. Box-shield with compartment septum shields to ground only interstage feedback. (*Courtesy of Interference Control Technologies*)

Fig. 12-4. EMI gasketed cover plate with nonconductive outer gasket for environmental protection. (*Courtesy of Chomerics*)

Figure 12-5 shows another shielded compartment used in the construction of power-line filters. The purpose of both the shielded box and the compartments is similar to that discussed in previous paragraphs, namely, to preclude feedback from input to output and between stages by capacitive coupling. However, in this case the metallic configuration is constructed of galvanized steel or other permeable material in order to: (1) prevent magnetic coupling from interstage inductors by providing a low-reluctance path for residual magnetic fields from the toroids, and (2) mitigate magnetic coupling (penetration) through the box housing to the outside world. Note the use of separate shields at both ends of the filter to further prevent direct input–output crosstalk and radiation coupling to feed and load wires.

SHIELDED EQUIPMENT AND CABINETS

In terms of physical size, the next level up from the shielded compartment includes shielded equipment and cabinets or racks. This level may be regarded as an ensemble of many individual electrical and electronic building blocks to result in a single housing.

Shielded Boxes and Equipment

Typical examples of chassis, or equipment-level, shielded housing include electronic test instruments, biomedical equipment, mobile transceivers, hi-fi amplifiers, and minicomputers.

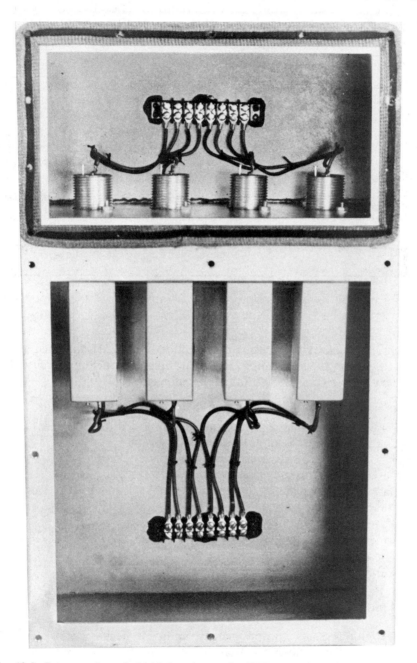

Fig. 12-5. Cutaway view of shielded enclosure for EMI power line filter. (*Courtesy of Lectromagnetics*)

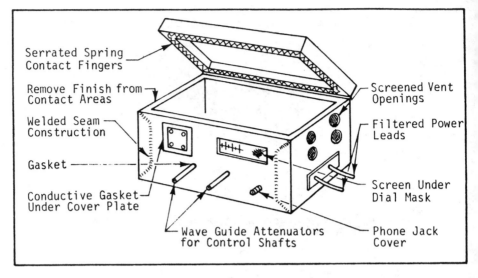

Fig. 12-6. An EMI hardened shielded equipment case. (*Courtesy of Interference Control Technologies*)

Figure 12-6 shows the housing construction of an EMI-hardened equipment case. Note the application of several topics on shielding discussed in the previous two chapters and other control techniques:

- Base shielding material:
  - Permeable or nonpermeable depending upon EMI environment
  - Continuous seam-weld construction
- Shielded integrity protection:
  - Gasketed hinged top
  - Gasketed cover plates
  - Viewing window screen or conductive glass
  - Waveguide attenuators for control shafts
  - Screened vent openings
  - Unused connector caps
- Power-line filters

Shielded Cabinets

This section represents a further increase in the physical size of shields from the chassis or box-size equipment of the previous section to the console size discussed here. Shielded cabinets, such as a 48-cm rack, single or double-bay console, or boxes occupying a volume on the order of 1 m³ are used as examples of this size enclosure.

Fig. 12-7. Example of EMI shielded cabinets. (Courtesy of Cabtron Systems, Inc.)

A typical example is shown in Fig. 12-7. These cabinets are similar in appearance to most other racks or consoles that have not been EMI-hardened. Although both offer shielding, there are important differences. EMI shielded cabinets differ from other cabinets in the following particulars:

1. The rear access door uses a continuous piano hinge for better door-to-frame bonding by equalizing the pressure along the hinge side.
2. The rear door is EMI-gasketed when in the closed position to provide shielding integrity all around.
3. The front frame members are gasketed to secure the cover panels and drawers.
4. Honeycomb ventilation apertures are used.
5. The cabinet frame enclosure is usually seam-welded to provide a continuous homogeneous bond.
6. The cabinet is often made of light-gauge steel, which provides better magnetic shielding at low frequencies than does aluminum.
7. Separate heavy power and signal grounding busses are run vertically up the rear side walls.
8. Large strike-plate holes on the back of drawers and a dagger-pin arrangement on the back of cabinet grounding busses result in lower common-mode impedance problems. In lieu of this, insulated grounding bus jumper cables to drawers may be used.
9. Frame bonding and grounding often follows that shown in Fig. 9-15 (above).

SHIELDED ROOMS AND ENCLOSURES

This section continues with shielded housings but for an increase in size to that of the shielded-room configuration. Here, sizes may range from 2.5-m^3 small rooms to large aircraft hangars.

The shielded room, or shielded enclosure, has been in use for many years for performing electronic measurements where a low electromagnetic ambient is required, or where potentially damaging emissions must be contained. (The term "screen room" is still used in some circles because the early shielded enclosures were made of copper screen.) The use of shielded enclosures has spread to non-measurement applications, such as protecting personnel working near high-power radar sites, containing certain industrial RF emission sources, and protecting sensitive equipment such as biomedical instruments and computers.

The main advantage of the shielded enclosure used for performing EMI measurements is that it provides RF isolation from the outside world. Its use allows meaningful susceptibility measurement to be made, both conducted and radiated, in locations where high ambients exist and thus such testing would not ordinarily be possible. However, there still exists some residual electromagnetic ambient inside a shielded room, since the room attenuates rather than eliminates outside-world emissions. In most situations where the outside environment is not abnormally high, a modern shielded room provides sufficient attenuation to reduce the effects of most ambient emissions to levels below the sensitivity of typical receivers equipped with state-of-the-art antennas. Magnetic field shielding at and below extremely low frequencies (ELF) is the exception. For example, the enclosure walls to 50 or 60 Hz fields offer little attenuation, and input power lines allow easy emission entry. To reduce 50 or 60 Hz magnetic fields and their first few harmonics, it is necessary to use "extended frequency-range shielded enclosures." (This term has been adopted by the EMI community for shielded enclosures to imply significant attenuation to magnetic fields at power-line frequencies.)

Enclosures Types and Size

Some electronic systems are physically too large to employ normal room-size shielded enclosures for testing (2). The size of a shielded enclosure has no theoretical limit, and enclosures have been built to enclose large aircraft and spacecraft. For example, the Titan ICBM was checked out in an enclosure five stories high with a five-story door. Another example was the testing of an entire computer facility for an early model Atlas ICBM, which was done in a large shielded structure such as is shown in Fig. 12-8. Some structures are lined with absorbing material to form an anechoic chamber, as shown in Fig. 12-9. Shielded enclosures are also constructed in mobile configuations

Fig. 12-8. Example of a large shielded enclosure. (*Courtesy of Interference Control Technologies*)

that are roving electronic laboratories (viz., shielded enclosures constructed in a trailer or van).

One of two metals, either steel or copper, is usually employed as the basic material in the shielded enclosure (1). Steel is generally used in the form of galvanized sheet, as shown for the enclosure in Fig. 12-8, whereas copper is used in either sheet form or fine-mesh screening, as shown in Fig. 12-9. Which

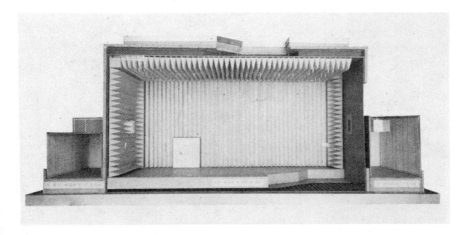

Fig. 12-9. A large shielded anechoic chamber. (*Courtesy of Interference Control Technologies*)

material is to be used depends on weight restrictions, shielding desired, cost, and other variables. For equal cost, steel generally provides comparable performance to that of copper at radio frequencies down to 150 kHz. Below this, the permeability of steel begins to provide improved magnetic-field shielding due to higher absorption.

The first of the modern shielded-enclosure construction methods was developed a number of years ago at the Naval Air Development Center, Johnsville, Pennsylvania. This room was made with two layers of copper screen, separated by a 2.5 cm frame. The room was constructed of several panels, called cells, each measuring 122 cm × 244 cm. Individual panel edges were butted together and were bolted through the wood frame that provided the shape and strength for each panel. One feature of this room was that it could be disassembled, moved, and reassembled at another location without major modification.

The door for this room was well braced, framed with wood, and also covered with copper screen. To the periphery of the door were attached two sets of spring-finger stock, one to provide contact with the inner edge of the door jamb, and the other set to push against the outer edge of the door frame. This second set of spring fingers actually overlapped the door frame opening.

A similar enclosure construction is the cell type that employs only a single layer of either screening or sheet metal. This enclosure is not widely used because its shielding performance is less satisfactory than the double-layered cell-type, while it is almost as costly as that type. Figure 12-10 illustrates the three most common types of demountable shielded enclosure construction.

Fig. 12-10a. RF modular shielded enclosure. (*Courtesy of LectroMagnetics, Inc.*)

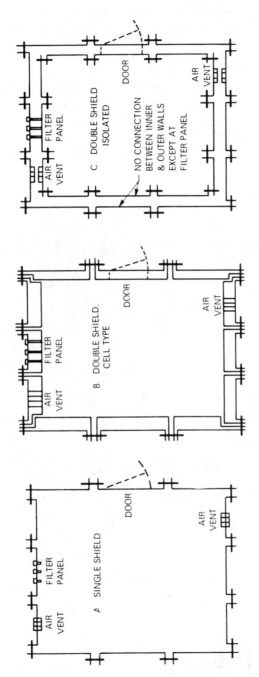

Fig. 12-10b. Batted or clamped-together enclosures. (*Courtesy of Interference Control Technologies*)

There appears to be some controversy over the purported advantages of completely separate layers in the double-shield, isolated-type room as opposed to the cell-type room, both shown in Fig. 12-10. No evidence firmly indicates any inherent advantage in either type of construction where the same total thickness of the same metal is used. Where bolted seams are used in both, the double-wall isolation construction may be more effective because of the second opportunity to clamp seams closed.

Properly made welded seams (see Fig. 12-11) are preferable for single-shield construction because electromagnetic leaks are virtually eliminated,

Fig. 12-11a. All-welded RF shielded enclosure with double RF shielded sliding door. (*Courtesy LectroMagnetics, Inc.*)

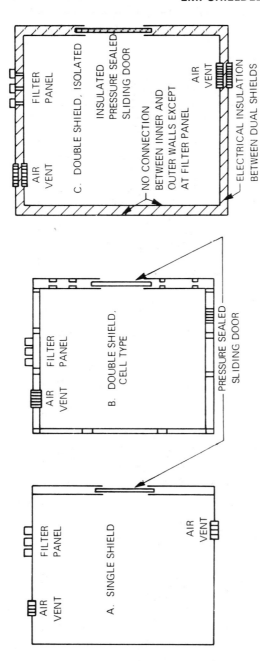

Fig. 12-11b. All seam-welded type enclosures. (*Courtesy of Interference Control Technologies*)

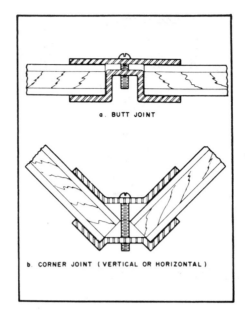

Fig. 12-12. Typical methods of forming sandwich-type panels. (*Courtesy of Interference Control Technologies*)

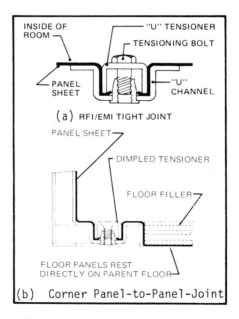

Fig. 12-13. Joiners for solid-sheet-stressed metal enclosures. (*Courtesy of Interference Control Technologies*)

and there are no possibilities of intra-wall resonances. The trick is to seam-weld without developing voids in the weld that may not be seen by the naked eye. This type of construction uses a single sheet of steel on a metal frame-work. The sheet is under some tension because of the manner in which it is welded to the frame, and the frames or panels are welded together. Seam-welded enclosures cannot be moved without literally destroying them in the process. Thus, when planning for an enclosure of the seam-welded type, one should be sure that there will not be a future requirement for demounting it and erecting it elsewhere.

For the double-layered enclosure (i.e., the double-shielded room), a sand-wich panel with two steel sheets bonded to a 19-mm plywood core is often used. The panels are not butted together all the way, but are clamped on each edge by special continuous channels and strapping. The method of joining panels both along the sides of a room and at the corners is shown in Fig. 12-12. Machine screws pull the channel and strap together every few inches.

A different approach to the above sandwich-type enclosure is a single solid wall type in which the shield material is 3 mm rolled steel including U-channels and U-tensioners, as shown in Fig. 12-13. All fittings, bolts, screws, and spline nuts used to fasten the framing members together are plated steel.

Variations are used by several manufacturers of shielded enclosures. Some newer developments in clamp design include a preassembled welded and interlocking three-way corner that eliminates leakage problems. Also incor-porated into the design is the use of a closed threaded insert to eliminate RF leakage and penetration at each of the clamping bolts. Typical features of this design are shown in Fig. 12-14.

Shielded enclosures require periodic maintenance if they are to retain their designed attenuation. The vulnerable areas are the joints and seams of bolted structures, and the door. Fastenings between panels must be kept tight. The enclosure manufacturer usually gives a torque rating on the fasteners. Exclu-sive of doors, no such maintenance is required on all-welded enclosures, since they use no bolts for fastening.

Finger-stock gasketing along the edge of the door must be kept in good condition. If fingers are damaged or broken off, a new section of fingers may be soldered or bonded in position as a replacement. As discussed in the perceding chapter, these fingers provide a good bond between the enclosure and the door by a wiping motion for a short distance along the door frame. To maintain good gasketing action, the door frame must be kept smooth and clean. Some doors use finger stock on the door frame, and the door must be kept clean of oxides and grease. Some manufacturers use recessed finger stock on the door panels to protect them from damage.

Some newer shielded enclosure doors are considerably different in concept. They are stronger, with stronger frames to provide better attenuation to RF

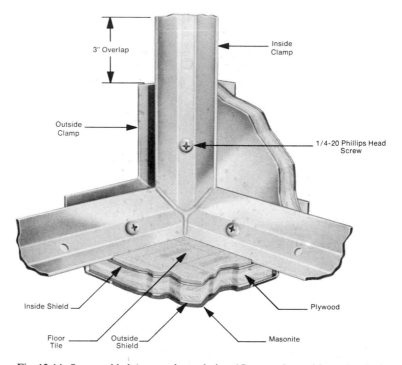

Fig. 12-14. Seam-welded, corner-clamp design. (*Courtesy LectroMagnetics, Inc.*)

emissions around the door. They still use two rows of metallic finger stock or hidden rows to effect a good seal between the door and frame. An example is shown in Fig. 12-15. There are also improvements in the latching arrangements, so that the door may be opened and closed more easily. Doors for all-welded shielded enclosures require uniform clamping or sealing pressures well beyond the standard approach. A very effective and reliable high-performance shielded door is the pneumatic sealed door. This door uses neither finger stock nor RF gaskets. One version employs a rather complex system that makes use of a set of pressurized air bags to force the door edges against a mating flange.

Enclosure Apertures and Penetrations

Part of providing attenuation by using a shielded enclosure is accomplished by assuring that the entire structure provides the required shielding effectiveness, including the wall sections, seams, door, and any apertures. Selecting the proper material and wall thickness to provide a given shielding effectiveness

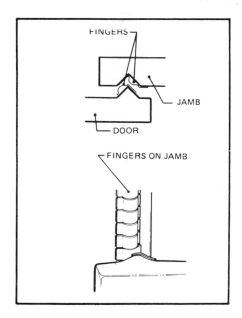

Fig. 12-15. Contact fingers of enclosure door and jamb. (*Courtesy of Interference Control Technologies*)

is of little value, no matter how well designed, if the shielding integrity of the enclosure is compromised by one or more apertures, as suggested in Fig. 12-16.

Other than the access door, the most prominent aperture involves power-service entrance and associated filtering to remove RF noise on the input lines (2). The basic requirement for shielded enclosure filters is that they provide a certain minimum attenuation over the useful frequency range of the room, typically from approximately 14 kHz to 10 GHz. One problem is to determine how much attenuation the filters must offer to complement a particular room design or application, since the filter attenuation requirements are only roughly related to the attenuation of the room.

One approach to the problem is to provide filters with attenuation capability somewhat less than that of the room. For example, if the room offers attenuation of 120 dB over most of the frequency range, filters offering 100 dB should prove concomitantly adequate. If the required enclosure performance exceeds the RF power-line filter capability, the filters may be enclosed in a shielded electrical panel to increase low-frequency magnetic field isolation. Normal installation practice places the filters outside the enclosure, with the line coming through pipe nipples into the room.

The mechanical design of high-attenuation filters is an important factor in their performance. To reduce coupling between the input and output of the separate sections, well-designed compartments are required.

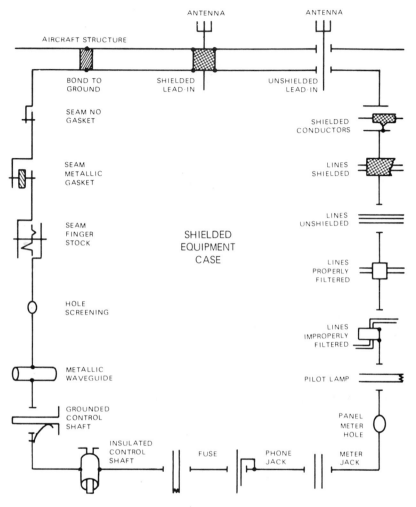

Fig. 12-16. Typical shielded enclosure discontinuities. (*Courtesy of Interference Control Technologies*)

The enclosure design must provide for lighting, heating, and air conditioning. Lighting should be of the incandescent type, since other types usually involve ionization that produces substantial RF noise. Special fluorescent fixtures with built-in filters, shields, and conductive cover windows are available. Although this type of lighting is effective in lowering the electromagnetic ambient within buildings and laboratories, it is not completely suitable for shielded enclosures.

Regarding heating and air conditioning, it is not generally necessary to provide additional facilities, if they are available in the immediate area where the shielded enclosure is to be located. Consequently, forced air ducting is extended from existing ducts to enter the enclosure at the top. To assure adequate forced ventilation, however, a return system is needed. This generally consists of an exhaust fan forcing air to the outside ambient from within the enclosure through a honeycomb vent that protects the shielding integrity of the enclosure.

Additional penetrations of the enclosure walls are often necessary to provide for other services. Gas, water, and compressed air may be furnished through steel or copper piping which acts as a waveguide-beyond-cutoff. If pipe is joined to the enclosure wall in a clean, tight bond, the shielding effectiveness of the room will not be compromised.

The same method is used to bring coaxial lines through the enclosure wall. Special fittings are available that are similar to a threaded pipe nipple, except that suitable coaxial fittings are used. Coaxial cables feeding the fittings from outside the enclosure can reduce the shielding effectiveness of an enclosure by providing a path of entry for high-level signals. This form of EMI develops by penetrating the cable shield, whence it is conducted into the enclosure. For this reason, triax or quadrax cables should be used (see Chapter 6). No cable will stop EMI from entering the enclosure if it exists in the source or load equipment located outside the room.

Enclosure Performance and Checkout

Well designed and installed shielded rooms of the modular clamp-together type discussed in a previous section conform to well-known shielding requirements of MIL-STD-285, USAF Class I Shielding, or NSA Spec. 65-66. The shielding effectiveness of some modular rooms, using two sheets of 24-gauge steel sandwiched on 19-mm plywood, is shown in Fig. 12-17. The performance of other welded rooms is also shown in the figure for different classes of shielded rooms based on the thickness of the shielding steel used. The performance of all-welded rooms tends to follow predictable magnetic field attenuation, while the lesser slope of magnetic shielding for the modular room is the result of a derating effect based on the magnetic seam impedance of the clamping arrangement.

In order not to create the impression that all cell-type shielded enclosures are inferior to seam-welded designs in shielding effectiveness, Fig. 12-18 illustrates the performance of a double, electrically isolated, enclosure. Here, the shielding effectiveness is everywhere superior to that designated "Some Modular Rooms" in Fig. 12-17. The magnetic field attenuation of the double

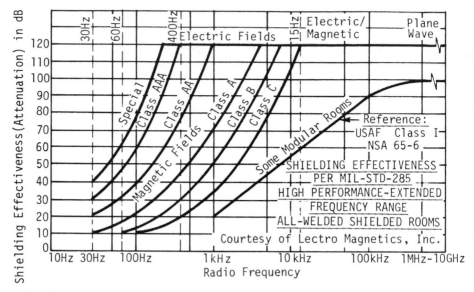

Fig. 12-17. Shielding frequency of seam-welded and some modular rooms. (*Courtesy of Interference Control Technologies*)

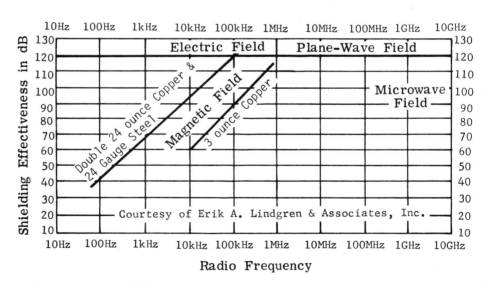

Fig. 12-18. Typical shielding effectiveness of a double electrically isolated shielded enclosure.

Table 12-1. Summary of typical shielded enclosure performance. (Courtesy of Interference Control Technologies.)

Enclosure type	Wall type	Material	Magnetic fields 60 Hz	15 kHz	Electric & plane waves 1 GHz	10 GHz
Double electrically isolated	Screen	Copper		68 dB	120 dB	
		Bronze		40 dB	110 dB	
		Galvanized	50 dB	50 dB		
	Solid	24 ga. steel	15 dB	82 dB	120 dB	90 dB
		Cu & steel	18 dB	93 dB	120 dB	105 dB
Cell-type	Screen	Copper		48 dB	90 dB	
		Bronze				
	Solid	24 ga. steel		68 dB	100 dB	
		Copper				
Single-shield	Screen	24 ga. steel	6 dB		60 dB	
		Bronze				
	Solid	Copper				
		24 ga. steel		48 dB	90 dB	

24-ounce copper–steel design in Fig. 12-18 exceeds that of Classes A, B, and C of Fig. 12-17 up to 70 dB.

It is difficult to make comparisons because the parameters are often different, and measurement details may not be the same. One attempt to present enclosure attenuation on a comparative basis is shown in Table 12-1.

The attenuation of a newly installed shielded enclosure is usually measured to determine whether it performs to specification, which is typically MIL-STD-285. This specification prescribes test frequencies and equipment, as well as antenna separation distances. Below UHF, the testing of an enclosure is performed in the near field, yielding results that may vary widely as a function of distances, antenna types, and frequency. Thus, it is important that the methods indicated in MIL-STD-285, if this is the test specification to be met, be followed as closely as possible so that repeatable results can be obtained.

Testing of an enclosure should be performed periodically to verify that its present attenuation still meets the original specification. In this respect, the shielded enclosure is often considered as an item of equipment in a laboratory inventory, and is placed on a periodic calibration schedule. After the initial checkout, the enclosure should be retested at least once every other year. Interim spot checks may be desirable in conjunction with special interference tests or if degradation of enclosure attenuation is suspected for any reason.

Cost

Cost is always important and should be resolved early when a new shielded enclosure is being considered. Cost and shielding effectiveness are interrelated. If extended frequency performance is not required, the modular room is less expensive in its initial cost for enclosures up to a size of approximately 6 m × 7 m. Operational environments and maintenance must be considered to determine the end cost over a period of years. If performance and/or environments are paramount, the all-welded room becomes less expensive.

One exception to the above involves structures that are significantly larger than 7 m × 9 m, in particular when such structures must pass building codes and other environmental requirements. Here, the all-welded enclosure is generally less expensive regardless of the degree of shielding effectiveness required. In such applications, modular rooms are not self-supporting and require considerable framing, whereas all-welded rooms are in accordance with the Uniform Building Code and are self-supporting.

Where tight budgets exist, the purchaser should not overlook the used market as a source of shielded enclosures of the modular form. The EMI instrument and rental houses are often aware of such sources. Used modular enclosures generally can be purchased for 20 to 50% of the original cost. Prospective purchasers must remember the added cost to disassemble, move, reassemble, and check out the shielding effectiveness of such enclosures.

Shielded Huts and Vehicles

Shielded huts and vehicles are a special case of shielded enclosures. They are characterized by their transportability which provides a substantial amount of flexure in transit. Consequently, modular construction is seldom used, and seam-welding is the most popular variety. Techniques similar to those used in shielded rooms are used to protect shielding integrity at apertures and penetrations. However, owing to structure flexure, shielding effectiveness of the access door becomes the limiting factor in attenuation performance. Consequently, shielded huts and vehicles typically exhibit 20 to 30 dB less shielding effectiveness when compared to their stationary counterparts.

Fig. 12-19. Portable weather tight all-welded shielded enclosure. (*Courtesy LectroMagnetics, Inc.*)

Figure 12-19 shows one form of shielded hut. The air vent flaps are typically backed by gasketed honeycomb. The power supply is generally a remotely located diesel-engine-driven generator. Some versions provide a battery bank of 28 VDC source for quiet operation between periods of battery charging. The hut may be transported by either a flatbed trailer or helicopter. When shielded vehicles are used, an auxiliary power supply, heating, and air conditioning are furnished external to the driver's cab. To minimize flexure, an air-bag suspension system is employed.

SHIELDING OF BUILDINGS

This section shows that most commercial office and industrial buildings offer substantial attenuation for radiation from AM broadcast and lower frequencies, but are fairly transparent to FM and TV broadcast frequencies. The interior of a building, in contrast to rooms located on the perimeter, offers additional natural attenuation, especially at UHF and higher frequencies.

Typical Building Attenuation

Based on measurements performed by Bell Northern Research of Canada, an empirical mathematical model has been developed to describe the natural shielding effectiveness of commercial office and industrial buildings. These buildings are identified as those of steel girder construction having vertical girder separation ranging from 6 to 9 meters and horizontal girder separation ranging from 3 to 5 meters. The floors consist of concrete and reinforcing bars. The outer facade consists of brick, mortar, stone, and/or glass. No known attempt has been made to apply shielding to wall and other surfaces.

Figure 12-20 is a plot of the mathematical model of the above-mentioned building shielding attenuation. Note that there exists a definite and sharp change in slope at about 70 MHz. This corresponds to a half wavelength resonance of the approximately 3-meter floor-to-floor separation for horizontal polarization. Below 70 MHz, the attenuation slope with frequency is 23 dB/decade. Theory indicates this slope should be 20 dB/decade, and the difference

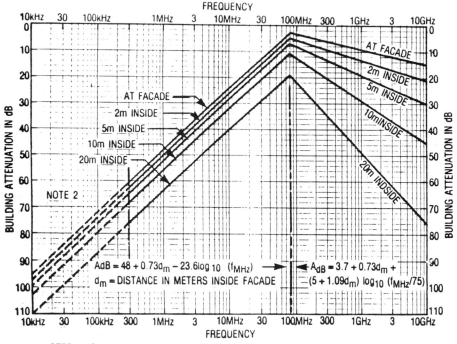

NOTES: 1. Buildings surveyed are business-type, with metal girders.

2. Suggested extrapolation for E field Attenuation—H-field attenuation decreases significantly below 1MHz, to approach 0dB around 10kHz

Fig. 12-20. E-field building attenuation to outside licensed transmissions vs. frequency and distance inside facade. (*Courtesy of Interference Control Technologies*)

is due to measurement variations. Above 70 MHz, the building attenuation increases with frequency due to absorption offered by both the building material and its contents.

Figure 12-20 also shows the results of shielding effectiveness offered by increased distance from the perimeter toward the building interior. Shielding offered by the second, third, and additional rows of girders becomes less effective than that of the original facade girders. This is believed to be due to the perimeter now acting as a Huygens source, and due to scattering and standing wave effects. Well inside the building, its contents serve to absorb the microwave energy. Therefore, a computer installation, for example, would be best shielded from radiation due to outside sources if located in the center or core of the building rather than close to the facade.

Exterior Building Materials

The exterior of most commercial or industrial buildings is either cinder block, concrete, or brick. These materials are nearly transparent to RF energy at and below VHF (300 MHz). Studies conducted at the National Bureau of Standards (NBS) indicate that the median insertion loss to a single layer of a brick wall (brick veneer) at 518 MHz and 1046 MHz was about the same, only 2.5 dB. Plotted on cumulative probability distribution paper, it was shown that for horizontal polarization only 10% of the measured locations exceeded an insertion loss of 5 dB, while the 90th percentile situation corresponded to about 1 dB. This indicates that buildings whose basic construction and surfacing is done with nonmetallic materials are essentially transparent to induced or radiated EMI at or below UHF. (Some small buildings are made of corrugated steel or aluminum, and some houses use finished aluminum siding. This type of construction can provide some shielding effectiveness, depending on if and how the metal members are grounded.)

One obvious solution to improving the RF attenuation of exterior building materials is to use aluminum of steel facing or backing. Screen backs to traditional surface materials and even conductive mastics and plastics have been used. However, they all represent a significant increase in material and installation costs, and a different solution to shielding buildings using basic exterior materials is sought. One such solution involves brick, cinder block, and/or concrete made of a mixture of a reasonably good grade of high-temperature coke.

Coke-Shell Exteriors

Before investigating hybrid building materials (coke-embedded brick, etc.), it is useful to summarize some attenuation studies of coke, per se. Coke achieves its shielding properties by virtue of reflection loss (see Chapter 10) at low frequencies in which the volumetric resistivity approximates 0.1 ohm-cm at DC. At high frequencies, both reflection and transmission (absorption) loss contribute to total attenuation. Thus, the layer thickness of coke chunks is more important at HF. Coke also exhibits considerably less insertion loss at low frequency when it is broken down into small particles or ground into fine dust. Typical values of insertion loss of coke chunks (about 2 cm size) vary from about 60 dB at 100 KHz to 90 dB at 1 GHz per 30 cm thickness for untreated coke, versus about 50 dB at 100 KHz to 70 dB at 1 GHz per 30 cm thickness for coke neutralized with lime slurry. Coke has some sulfur content which produces sulfuric acid. Neutralizing coke with lime tends to deteriorate the shielding effectiveness. Powdered coke dust, on the other hand, offers about 15 dB attenuation per 30 cm at 100 kHz and rises to about 50 dB at 1 GHz.

One approach to the use of coke would be to inject it as a sandwich layer between two surfaces such as wooden siding and brick veneer, or between two slabs of concrete or other surfacing and backing. Based on the above data, a 5-cm sandwich layer of treated coke would be expected to exhibit about 10 to 15 dB insertion loss—hardly an impressive figure.

Another approach would be to make brick, concrete, or cinder block of a specified percentage of coke chunks in such a manner that the compression strength would not be significantly less than that of the normal comparable materials without coke. This tends to result in a relatively low coke content so that the attenuation of the modified materials approximates 10 to 20 dB per 30 cm thickness.

The above data show that coke, while inexpensive and offering good shielding properties, cannot normally be used in most exterior building materials. Where special structures, such as a blockhouse or bunker building, are called for, coke can be gainfully used.

Screen and Conductive Finishes

As discussed in Chapter 10, a number of conductive paints are commercially available. Insertion losses of the order of about 50 dB for frequencies up to 1 GHz are obtainable, where skin depth becomes the limiting factor. However, these paints are very expensive in terms of building finishes, and would require periodic cleaning unless the appearance of oxides is not objectionable. In terms of cost and other factors such as environmental considerations, conductive paints applied to their exteriors do not appear to be a practical solution for enhancing the shielding effectiveness of buildings.

The building paper and vapor barriers that must be used on the outside of building walls present an opportunity to secure a relatively inexpensive RF screen. The screen is attached by staples, tape, or other binding media except at the edges where a 2 cm overlap of the next screen piece is made. While conductive adhesives of flat conductive caulking could be used, this is both expensive and unnecessary. Overlap screening can be secured in the same manner as the basic screen, provided no layer of electrical insulation (e.g., nonconductive tape) is used. Insertion losses of the screen will offer at least 50 dB up to about 1 GHz.

If the above screening technique is applied to the construction of building walls, then the windows become the leaky apertures. A typical 1 m × 2 m window will be fairly opaque to low frequencies, but will become leaky above 20 MHz (0.1 λ). There is no economical way to provide an RF shield to a window without using window screens or conductive glass. Here, the sash would have to be metallized (e.g., by using conductive paint) and would terminate on the building screen discussed in the previous paragraph. As discussed above for shielded enclosures, the objective is to obtain a continuous, conductive surface across the window aperture.

Since all modern buildings in the United States are air-conditioned, window screens are seldom used. They also oxidize and become unsightly after a few years, and they tend to block light. A far better approach is to use conductive glass (see Chapter 11). Although relatively expensive, a light conductive coating can be used. When interior building shielding materials are used (see next section), it is not necessary to use conductive glass on all windows. Only certain floors, sections, or rooms need to be so equipped, where occupants would be expected to use computers, medical instruments, and other devices that could emit or be susceptible to EMI. Conductive glass windows of the type described would provide about 20 to 30 dB of attenuation.

Interior Building Materials

The use of appropriate interior building materials could provide a good way to economically improve building RF attenuation to radiation. Coke aggregate appears not to offer sufficient attenuation per unit thickness, as discussed above. Other materials, however, do offer some promise; they include: (1) conductive mesh in walls and ceilings, (2) metallic wallpaper, (3) special windows and doors, and (4) conductive paint. This section reviews the first three topics; conductive paint was discussed in Chapter 10.

Conductive Meshes in Walls, Floors, and Ceilings.   Table 12-2 shows the magnetic and electric field shielding effectiveness of various types of screen materials. Number 22 copper screen mesh provides approximately 65 dB of attenuation to electric fields and plane waves over a frequency range

Table 12-2. Minimum shielding effectiveness of screen materials to high-impedance electric fields and plane waves. (Courtesy of Interference Control Technologies.)

Screen material	Configuration	Thickness	Attenuation
Steel	Perforated Sheet	1.5 mm	58 dB
Aluminum	3 mm dia, 5 mm ctrs.	1.5 mm	48 dB
Aluminum	6 mm dia, 8 mm ctrs.	.9 mm	35 dB
30 mil. galv. steel wire	# 2 Screening	NA	24 dB
30 mil. galv. steel wire	# 4 Screening	NA	28 dB
20 mil. copper wire	# 12 Screening	NA	50 dB
Copper	# 22 Screening	NA	65 dB
20 mil. aluminum wire	# 16 Screening	NA	55 dB

* Minimum attenuation in dB from 100 Hz to 1 GHz.

of 100 Hz to 1 GHz, as shown in Table 12-2. Effectiveness decreases as screen aperture size increases, and Number 2 galvanized steel screens provide a shielding effectiveness of 24 dB over the same frequency range. Number 40 copper screen provides 32 dB of magnetic field attenuation at 300 kHz, 50 dB at 1 MHz, and 60 dB at and above 20 MHz.

The above screen mesh can be impregnated within, or be sandwiched between, drywall and other internal materials, with a screen overlap for seam bonding. Such screening should be electrically continuous around the shell of the room to be shielded. This necessitates either soldering of screens or deploying conductive epoxy or caulking along corner junctions (see Chapter 11). A bond of solid copper cable or sheet metal strap should be connected from the screen to the reference plane grid mesh for that floor. The cost of installing the screen mesh would be wasted unless it is properly grounded.

Regarding the ceilings and floor of such rooms, both can be covered first with a screen mesh similar to that of the walls and bonded along the corner junctions. The ceilings can then be covered with regular celotex, acoustical tile, or other materials, and the floor covered with the standard flooring materials. An alternative approach is to use the same screen-impregnated drywall for the ceiling, and to immerse to concrete floor in screen in which a screen overlap permits seam bonding at corner junctions.

Except for the doors and windows, a room enclosure shielded using the techniques described above will probably provide about 60 dB of attenuation to electric and plane wave electromagnetic fields over several decades of the frequency spectrum. Although expensive relative to normal construction, it is likely to be considerably less expensive than to subsequently have to install the least expensive shielded enclosure. Further cost reduction will depend upon either: (1) the commercial availability of such screen-impregnated drywall; screen-backed 4-foot by 8-foot (122 cm × 244 cm) sheets of 1/8-inch or 1/4-inch (3 mm or 6 mm) thick embossed and grooved hardboard, or panel-veneer finishes for wallcoverings; or (2) the economical purchase in quantity of special screen-impregnated wallboards.

Metallized Wallpaper.   Metallized-foil wallpaper is used as an economical means for the attenuation of electromagnetic fields. Although such material is presently being used as a shielding technique during building construction in a very limited way, its application warrants further consideration. Such metallized foil may perform effectively if installed within walls where moisture is not a problem, and where such material will not be subjected to abusive wear and tear. Applicability of this material as a construction shielding medium depends on site requirements, its availability, and performance evaluation compared to other shielding methods.

Door and Window Design.   The construction of a shielded structure of the type described in the preceding section is difficult to achieve while maintaining the overall shielding effectiveness of the structure. Doors installed in such a shielded structure must employ impregnated screen construction and provide bonding between the door, the door frame, and the overall structure. Two methods may be used to permit a satisfactory bond between the mating surfaces: EMI gasket material and metal finger stock (see Chapter 11). Metal finger stock is preferred, since shielding effectiveness of gasketing material is dependent upon applied pressure. Aging effects are also important if gasketing material is used for the bond, and piano hinges must be used to maintain constant pressure on the gasket. Metal finger stock requires a minimum of maintenance. The metal fingers should be adhesive-bonded to the door frame after mating surfaces have been cleaned of paint, varnish, grease, and so on. The metallized door frame must be bonded to the building ground reference plane.

Two methods are available for maintaining the shielding effectiveness of a shielded room with windows. One method requires a metal-screened window bonded directly to the window frame; this method is preferred because of the ease of installation and reduced costs. A more aesthetic method is to install a commercially available transparent conductive coated glass that is bonded to the metal window frame by a peripheral gasket. This method, while reducing the amount of light entering the room somewhat, is becoming a popular approach where both RF shielding and optical visibility are needed.

Special Shielded Enclosures

In severe cases of offending sources of EMI in the operation of EMI-susceptible equipment or systems, it may be necessary to install one or more shielded enclosures within a building. In this case, one of two practices is used: (1) if the building is dedicated to a single purpose such as a hospital, research lab, or the like, the enclosure(s) can be planned during the building design phase and installed during construction; or (2) if the building is to be used by many different commercial, industrial, and/or professional tenants, individual users can plan and fund their own shielded enclosure installations as needed.

REFERENCES

1. White, Donald R. J. *A Handbook on Electromagnetic Compatibility*, Volume 3. Gainesville, VA: Don White Consultants, Inc., 1973.
2. White, Donald R. J. *A Handbook on Electromagnetic Compatibility*, Volume 4. Gainesville, VA: Don White Consultants, Inc., 1973.

Chapter 13
COMMUNICATIONS AND WAVE FILTERS

Electrical filters used to control EMI may be divided into two types: communications or wave filters used in intersystem EMI control and power-line filters used in intrasystem EMI control (6). This chapter presents a summary of communication filters and filtering techniques, and the next chapter surveys EMI power-line filters.

Filters are used to control intersystem EMI in one or more of the following ways (1):

1. Selectivity in superheterodyne receivers via the intermediate frequency (IF) band-pass response formed by a combination of a fixed-tuned filter and IF amplifier.
2. Radio frequency (RF) selectivity in superheterodyne receivers via a tunable pre-selector that both suppresses the image response and provides additional out-of-band rejection to strong unwanted emissions.
3. RF selectivity in both tuned-radio-frequency (TRF) and crystal-video receivers, via either tunable pre-selectors or fixed-tuned, band-pass filters to reject out-of-band emissions.
4. Band-rejection or notch filters (5) in receivers to suppress EMI from strong adjacent-channel emissions or for use in wide-band amplifiers.
5. Low-pass (or high-pass) filters to protect the front-end of receivers and other susceptible circuits or equipment from emissions existing above (or below) the base-band of operation.
6. High-power, low-pass filters in the output of transmitters to suppress unwanted harmonic radiations.

With few exceptions, these intersystem filters are all characterized by having equal input and output impedances in the pass band of their operational networks. Typical impedance levels are 50 or 72 ohms, but occasionally 600 ohms (or other impedances) may be used at audio frequency and 300 ohms may be used to VHF and lower UHF. They are also characterized by protecting low-level susceptible circuits, especially receivers, having sensitivities ranging from -150 dBm to about -100 dBm. One exception is item (6) above, in which transmitter outputs may operate with peak-pulse powers from $+20$ dBm to about $+100$ dBm. Here, filters must readily dissipate any absorbed power and have very low insertion losses in the pass band so that the signal will not be attenuated.

Filters are also used in intrasystem EMI control in one or more of the following ways but with a different emphasis (1):

1. RF suppression of unwanted signals that can couple into or exit from AC or DC power mains.
2. RF isolation of common-impedance coupled circuitry, such as several networks fed from common power supplies, via low-pass filters.
3. Conducted broadband noise suppression from power tools, appliances, industrial machinery, office equipment, and other devices that develop transients due to arc discharge at the brush–commutator interface of motors.
4. Conducted broadband noise suppression from transient-producing devices such as fluorescent lamps; electric-ignition systems; industrial controls, relays, and solenoids; and other switching-action devices.
5. Protection of susceptible components, such as transducers, computers, and electro-explosive devices (EEDs).

With some exceptions, intrasystem filter applications are characterized by unequal input and output impedances when the filters are installed in their operational environments. For example, power sources typically have impedances of less than 1 ohm at low frequencies, while their loads often represent higher impedances. Furthermore, both source and load impedances are frequency-dependent. Emphasis for intrasystem filtering is placed upon suppressing the conducted EMI at the source rather than protecting susceptible circuits as intended by item (5) above.

As a result of the above distinction between inter- and intrasystem filters and filtering techniques, it is convenient to discuss the subject in two separate chapters. This chapter discusses communication filters for intersystem applications.

MATHEMATICAL FILTER MODELS

There is no simple way to mathematically model an electrical filter for use in EMI prediction when either (1) measured data are unavailable, or (2) the equivalent circuit or contents of the filter are unknown. This is so because filters may range in rate of attenuation beyond cutoff frequency from 20 dB per decade, corresponding to a simple feed-thru capacitor or series inductor, to 200 dB (or more) per decade for a multistage L-C network. This is illus-

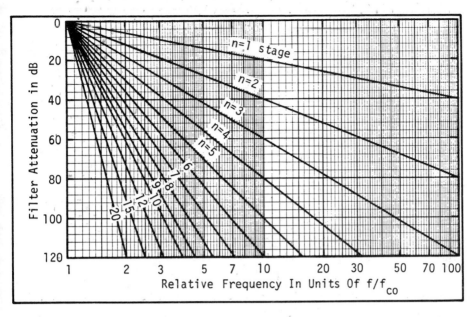

Fig. 13-1. Filter attenuation vs. frequency. (*Courtesy of Interference Control Technologies*)

trated in Fig. 13-1 for $n = 1$ to 20 stages. The abscissa is shown in normalized units of frequency with the 0-dB notch cutoff frequency appearing in the upper left corner.

It is dangerous to use Fig. 13-1 as a general model for predicting EMI suppression of low-pass filters, or the performance of derived high-pass, band-pass, or band-rejection filters discussed later, because (1) attenuations of more than 100 dB are difficult to achieve owing to input–output crosstalk coupling, and (2) the filter may completely degenerate in performance a few decades above cutoff because of parasitic capacitances and inductances. Where open circuitry is used, not involving either connectors or filter shields, it is not uncommon to have direct input–output coupling on the order of 40 to 60 dB, especially in miniature and integrated circuits. Regarding parasitics, unless special precautions are taken in the filter design and fabrication, a filter may offer little to no attenuation at frequencies two or more decades above cutoff.

Therefore, it is important that filter data obtained (measured) under actual installation conditions be used for EMI prediction whenever such data are available. The next best data to use are those obtained (measured) with the filter installed in a circuit configuration that closely resembles the configuration intended, or use nominal (manufacturer) data. Finally, if all else fails, and the only known data are (1) the filter type and the number of stages, (2) the physical size and whether a shield and connectors are used, and (3) the

Table 13.1 - Model of Maximum Average Attenuation of
Electrical Filters Outside Their Pass Bands

Rejection-Band Frequency Range	Shield & Connectors	Shield Only	No Shield/ Connectors
Microminature or IC Filters			
$f_{co} \leq f \leq 10\ f_{co}$	NA	60 dB	50 dB
$10\ f_{co} \leq f \leq 100\ f_{co}$	NA	40 dB	30 dB
$f > 100\ f_{co}$	NA	20 dB	10 dB
Communication Filters (No Special EMI Precautions)			
$f_{co} \leq f \leq 10\ f_{co}$	80 dB	70 dB	60 dB
$10\ f_{co} \leq f \leq 100\ f_{co}$	60 dB	50 dB	40 dB
$f \geq 100\ f_{co}$	40 dB	30 dB	20 dB
Communications Filters (EMI Hardened)			
$f_{co} \leq f \leq 10\ f_{co}$	90 dB	NA	NA
$10\ f_{co} \leq f \leq 100\ f_{co}$	80 dB	NA	NA
$f > 100\ f_{co}$	70 dB	NA	NA
Power-Line Filters $\leq$ 10 Amps (EMI Type)			
$f_{co} \leq f \leq 10\ f_{co}$	80 dB	NA	NA
$10\ f_{co} \leq f \leq 100\ f_{co}$	80 dB	NA	NA
$f > 100\ f_{co}$	70 dB	NA	NA
Power-Line Filters > 10 Amps (EMI Type)			
$f_{co} \leq f \leq 10\ f_{co}$	100 dB	NA	NA
$10\ f_{co} \leq f \leq 100\ f_{co}$	100 dB	NA	NA
$f > 100\ f_{co}$	90 dB	NA	NA

general installation plan, then the maximum performance suggested in Table 13-1 may be used.

In contrast to power-line filters (see next chapter), one distinguishing feature of communications filters (sometimes called wave filters or electrical filters), is that both the source and load impedances are constant over the pass band, or they are at least definable and predictable. This is so because communications filters usually are derived from and terminated by a relatively simple equivalent circuit element. In contrast, the equivalent circuit of a power distribution source at RF, with its myriad parasitic elements and the enormous load variations offered by many devices at RF such as motors, incandescent bulb elements, transformers, and so on, results in complex impedances that are difficult (at best), and usually impossible, to predict. This realistic factor renders power filter design and fabrication to achieve a predictable level of performance quite difficult.

Until about 1950, communication-filter design was still much of an art because of the unwieldy, mostly empirical, "constant K" and "M-derived" techniques that were used (2). The problem involved one of uncertainty in both theoretical design and physical realizability. By the mid-1950s, however, the use of modern network synthesis became widespread, and an entirely new, mathematically based, complex-variable technique of design had evolved (3, 4). This, at least, removed much uncertainty in theoretical design so that concentration could be directed to physical realizability. Further developments in the capability to define filter capacitive, inductive, and resistive-element parasitics encourage still further advances in filter synthesis, with appropriate adjustment of element values to give a still more predictable performance. Finally, with widespread use of digital computers, it became possible to compile an enormous amount of design data accounting for specific source-to-load impedance ratios, ranges of Q factors of both capacitors and inductors, and so on (3, 4).

Prototype Filters

Modern network synthesis provides a means to derive any low-pass, high-pass, band-pass, or band-rejection filter, having any selectivity slope in dB/decade or dB/octave, any cutoff frequency, and any impedance level using a family of low-pass filter prototypes (3). A prototype filter is defined in terms of an angular cutoff frequency, ω_c, of 1 radian, and an impedance level of 1 ohm, as shown in Fig. 13-2, with its dual network. The dual of a network

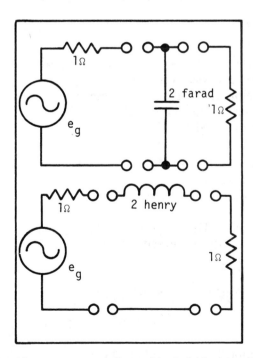

Fig. 13-2. Fundamental low-pass prototype filter and its dual. (*Courtesy of Interference Control Technologies*)

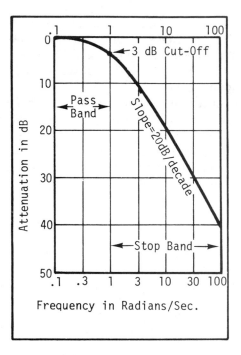

Fig. 13-3. Frequency response of low-pass prototype. (*Courtesy of Interference Control Technologies*)

is obtained by replacing any shunt (or series) capacitance with a series (or shunt) inductance, and vice versa. Dual circuits are electrically equivalent. The spectrum amplitude responses of these prototype, low-pass dual networks is shown in Fig. 13-3, where the 3-dB frequency and 20 dB/decade slope are readily observed.

Bandwidth and Impedance Scaling.　The low-pass filter shown in Fig. 13-2 has no practical value per se. However, it may be made practical by scaling the bandwidth and leveling impedance to any value by using the following rules (3):

1. *Bandwidth Scaling.* Divide all reactive (*L-C*) components by the desired new cutoff angular frequency in radians, $\omega_c = 2\pi f_c$, where f_c is the cutoff frequency in Hz. Thus:

$$L_a = L_b/2\pi f_c \qquad (13\text{-}1)$$

$$C_a = C_b/2\pi f_c \qquad (13\text{-}2)$$

where, the subscripts a = after and b = before scaling.

2. *Impedance Leveling.* Multiply all resisters and inductors by the desired new impedance level of source and load, and divide all capacitors by the new impedance level, Z. Thus:

$$R_a = ZR_b \tag{13-3}$$

$$L_a = ZL_b \tag{13-4}$$

$$C_a = C_b/Z \tag{13-5}$$

3. *Combined Bandwidth and Impedance Scaling.* Both scalings may be merged into a single operation by combining equations (13-1) through (13-5). Thus:

$$R_a = ZR_b \tag{13-6}$$

$$L_a = ZL_b/2\pi f_c \tag{13-7}$$

$$C_a = C_b/Z2\pi f_c \tag{13-8}$$

Figure 13-4 shows each 50-ohm filter system having a response identical to that of Fig. 13-3 except that the cutoff frequency is 1 MHz instead of corresponding to 1 radian/sec.

Response Functions. The preceding filter design was quite simple, since it involved a single reactive element giving 20 dB/decade (6 dB/octave) attenuation in the stop band. Actually, this is not a sufficient rate of attenuation for most practical applications where a small insertion loss (ideally, zero loss) is desired in the pass band, and substantial attenuation may be desired outside the pass band. For these situations, multistage filters are required in which the rate of stop-band attenuation beyond cutoff is $20n$ dB/decade ($6n$ dB/octave), where n is the number of reactive filter elements, or stages, or poles in the mathematical complex plane.

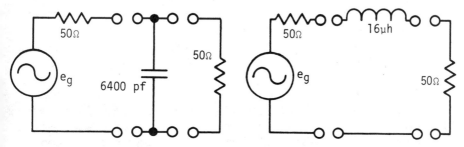

Fig. 13-4. Single-stage, low-level filters (duals) with 1-MHz cut-off and 50-ohm impedance source and load. (*Courtesy of Interference Control Technologies*)

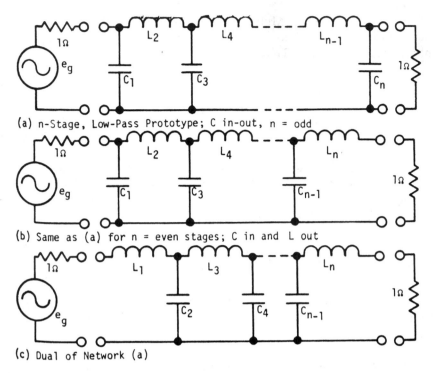

(a) n-Stage, Low-Pass Prototype; C in-out, n = odd

(b) Same as (a) for n = even stages; C in and L out

(c) Dual of Network (a)

Fig. 13-5. Fundamental, *n*-stage, low-pass prototype filters. (*Courtesy of Interference Control Technologies*)

There also exist prototype, low-pass filters similar to the illustration of Fig. 13-2, but having more than one stage. The value of each reactive elements in such filters is determined from modern network synthesis, and may belong to one of the following filter response classes:

1. *Butterworth*, or maximally flat, amplitude response over the pass band. The time domain response, however, has some phase and time delay distortion over its pass band, especially near the cutoff frequency region. Consequently, its transient response exhibits considerable overshoot. Yet, the Butterworth response is one of the most widely used functions.
2. *Tchebycheff*, or equal-ripple, amplitude response over the pass band. This configuration offers a faster rate of attenuation just outside the stop band than the Butterworth response. The price paid for this is the ripple in the pass band, which may be designed to have any value, typically from 0.1 dB to 1 dB. The time delay and phase distortion are greater than for the Butterworth response, although the overshoot of the Tchebycheff response is somewhat less.

3. *Bessel*, or maximally flat, time delay response. This function exhibits extremely small overshoot. Its rise-time, however, is longer than for either the Butterworth or Tchebycheff.
4. *Butterworth–Thompson* response. This configuration exhibits some of the tradeoffs of the fast rise-time and flat amplitude properties of the Butterworth and the low overshoot of the Bessel function.
5. The *elliptic* response, which combines some of the advantages of the Tchebycheff amplitude response with a "suck out" (large attenuation) just outside the stop band similar to that of the old M-derived filters. Its transient response is relatively poor.

FILTER-DESIGN GRAPHS AND TABLES

The above prototype responses are discussed elsewhere in the literature (3). For the purposes of the remaining discussion, only the Butterworth response is presented. The low-pass prototype of any n-stage filter is shown in Fig. 13-5 for either an odd or even number of stages, together with a dual network of the former. Table 13-2 lists the L-C element values of those prototypes for any number of stages from $n = 1$ to $n = 20$. Notice that the element values are symmetrical about the center of the filter, with the first and last stages being equal, and so on.

A way to calculate the number of stages required needs to be developed. The normalized frequency response function for the Butterworth filter was shown earlier, in Fig. 13-1, for $n = 1$ to $n = 20$ stages. The relative cutoff or notch frequency is 1.0, which appears at the left margin of the figure, and the relative stop-band frequency abscissa is extended over two decades. The attenuation (transmission loss) is presented over a 120-dB range. To determine the number of stages required at twice the cutoff frequency ($f/f_{co} = 2$) and to offer an attenuation of 50 dB, for example, Fig. 13-1 shows that $n = 8.3$ is required. Since n must be an integer, the next highest integer ($n = 9$) is chosen.

Illustrative Example: HF Receiver Filter Design

Additional protection is desired at the front end of a sensitive high frequency (HF) receiver, operating within the frequency range of 2 to 30 MHz, from VHF TV transmitters located nearby. An analysis of geographical locations, signal levels, and the frequencies of TV channels 4, 5, 7, and 9, indicates that the emission levels of channel 4 (frequency range of 66–72 MHz) will likely be the most troublesome. Further analysis indicates that at least 30 dB of additional attenuation is required to prevent intermodulation and other spurious responses within the 72-ohm receiver antenna system.

Table 13-2. Element values of Butterworth low-pass filter prototypes. (Courtesy of Interference Control Technologies.) (Use this table when load and source resistance are within 30% of each other, viz. when $0.7 < \bar{R} \le 1.0$.)

n	C_1	L_2	C_3	L_4	C_5	L_6	C_7	L_8	C_9	L_{10}	n
1	2.000										1
2	1.414	1.414									2
3	1.000	2.000	1.000								3
4	0.765	1.848	1.848	0.765							4
5	0.618	1.618	2.000	1.618	0.618						5
6	0.518	1.414	1.932	1.932	1.414	0.518					6
7	0.445	1.247	1.802	2.000	1.802	1.247	0.445				7
8	0.390	1.111	1.663	1.962	1.962	1.663	1.111	0.390			8
9	0.347	1.000	1.532	1.879	2.000	1.879	1.532	1.000	0.347		9
10	0.313	0.908	1.414	1.782	1.975	1.975	1.782	1.414	0.908	0.313	10
11	0.285	0.832	1.319	1.683	1.920	2.000	1.920	1.683	1.319	0.832	11
12	0.261	0.765	1.220	1.591	1.849	1.983	1.983	1.849	1.591	1.220	12
13	0.240	0.707	1.133	1.493	1.768	1.943	2.000	1.943	1.768	1.493	13
14	0.223	0.661	1.066	1.414	1.694	1.889	1.988	1.988	1.889	1.694	14
15	0.209	0.618	1.000	1.338	1.618	1.827	1.956	2.000	1.956	1.827	15
16	0.199	0.581	0.942	1.269	1.545	1.764	1.913	1.990	1.990	1.913	16
17	0.185	0.548	0.892	1.206	1.479	1.699	1.866	1.966	2.000	1.966	17
18	0.174	0.518	0.845	1.147	1.414	1.638	1.813	1.932	1.992	1.992	18
19	0.164	0.491	0.804	1.095	1.354	1.578	1.759	1.891	1.973	2.000	19
20	0.157	0.467	0.765	1.045	1.299	1.521	1.705	1.848	1.945	1.994	20
n	L_1	C_2	L_3	C_4	L_5	C_6	L_7	C_8	L_9	C_{10}	n

n	C_{11}	L_{12}	C_{13}	L_{14}	C_{15}	L_{16}	C_{17}	L_{18}	C_{19}	L_{20}	n
1											1
2											2
3											3
4											4
5											5
6					ALL L's in henrys.						6
7					ALL C's in farads.						7
8											8
9											9
10											10
11	0.285										11
12	0.765	0.261									12
13	1.133	0.707	0.240								13
14	0.414	1.066	0.661	0.223							14
15	1.618	1.338	1.000	0.618	0.209						15
16	1.764	1.545	1.269	0.942	0.581	0.199					16
17	1.866	1.699	1.479	1.206	0.892	0.548	0.185				17
18	1.932	1.813	1.638	1.414	1.147	0.845	0.518	0.174			18
19	1.973	1.891	1.759	1.578	1.354	1.095	0.804	0.491	0.164		19
20	1.994	1.945	1.848	1.705	1.521	1.299	1.045	0.765	0.467	0.157	20
n	L_{11}	C_{12}	L_{13}	C_{14}	L_{15}	C_{16}	L_{17}	C_{18}	L_{19}	C_{20}	n

$$C_1 = C_5 = C_b / Z2\pi f_c$$
$$= 0618/72 \times 2\pi \times 32 \times 10^6 = 43 \text{ pf}$$

$$L_2 = L_4 = ZL_b / 2\pi f_c$$
$$= 72 \times 1.618/2\pi \times 32 \times 10^6 = 0.58 \ \mu h$$

$$C_3 = C_b / Z2\pi f_c$$
$$= 2.00/72 \times 2 \times 32 \times 10^6 = 138 \text{ pf}$$

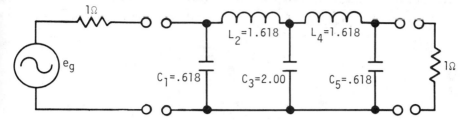

Fig. 13-6. 5-stage, low-pass Butterworth prototype. (*Courtesy of Interference Control Technologies*)

Example Solution. To assure that a cutoff frequency of not less than 30 MHz will exist with a given component tolerance of 5%, a design cutoff frequency of $f_c = 32$ MHz is selected. At the lowest EMI frequency, $f_e = 66$ MHz, against which the HF receiver is to be protected, the normalized (relative) frequency becomes $f = f_e/f_c = 66$ MHz/32 MHz = 2.06. Referring to the filter characteristic illustrated in Fig. 13-1, to achieve at least 30 dB of attenuation at 66 MHz, the number of stages is $n = 5$ (five-pole filter). The normalized, low-pass Butterworth prototype filter is illustrated in Fig. 13-6, and Fig. 13-7 illustrates the final low-pass filter design. For load and source impedance values of $Z = 72$ ohms, the final circuit elements are determined as follows:

$$R_g = R_L = Z \times R_b = 72 \times 1 = 72 \text{ ohms}$$

$$C_1 = C_5 = C_b/Z \times 2\pi f_c$$
$$= 0.618/72 \times 2\pi \times 32 \times 10^6 = 43 \text{ pf}$$

$$L_2 = L_4 = Z \times L_b/2\pi f_c$$
$$= 72 \times 1.618/2\pi \times 32 \times 10^6 = 0.58 \text{ μH}$$

$$C_3 = C_b/Z \times 2\pi f_c$$
$$= 2.00/72 \times 2\pi \times 32 \times 10^6 = 138 \text{ pf}$$

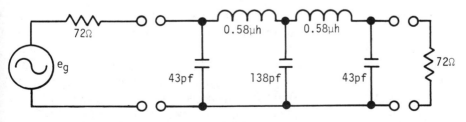

Fig. 13-7. 5-stage, low-pass filter with a 32 MHz cut-off and 72-ohm impedance source and load. (*Courtesy of Interference Control Technologies*)

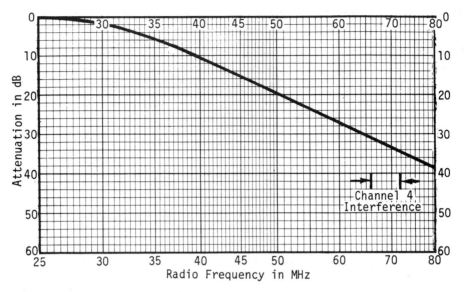

Fig. 13-8. Frequency response of filter shown in Fig. 13-7. (*Courtesy of Interference Control Technologies*)

Figure 13-8 shows the frequency response of the final filter circuit. Note that 31 dB of attenuation exists at a frequency of 66 MHz, which satisfies the 30-dB requirement. The insertion loss is about 2 dB at 30 MHz and is less than 1 dB at frequencies below 28 MHz.

Other Filter Types

The design of high-pass, band-pass, and band-rejection filters may be obtained directly from the low-pass prototype by a change in the frequency variable (2–4).

HIGH-PASS FILTERS

High-pass filters having geometric frequency symmetry can be obtained from low-pass filters (2, 3). By substituting $\omega_{hp} = 1/\omega_{lp}$ in the transfer function of the low-pass prototype, a high-pass prototype is obtained. By using this transformation, the impedance of an inductance, $L\omega_{lp}$, becomes the impedance L/ω_{hp}; the impedance of a capacitance, $1/\omega_{lp}C$, becomes ω_{hp}/C; and the value of the resistance(s) remains unchanged.

This transformation is equivalent to replacing all capacitances and inductances with inductances and capacitances, respectively, with each taking on the value of the reciprocal of the replaced component. Impedance leveling and bandwidth scaling of the new high-pass prototype, which also has an impedance level of 1 ohm and a cutoff angular frequency of 1 radian/sec, respectively, are accomplished in the same manner as in equations (13-6) through (13-8).

Illustrative Example: High-Pass Filter Design

Assume that a high-pass filter is desired having a 600-ohm input/output impedance, Z; a cutoff frequency, f_c, of 1 MHz; an attenuation of 70 dB at 250 kHz (f_1); and no ripple (a maximally flat response) within the pass band above 1 MHz.

Example Solution.   Since the desired attenuation or band-rejection of a high-pass filter lies below the cutoff frequency, the normalized frequency is determined as follows:

$$\bar{\omega}_{hp} = \omega_c/\omega_1 = 2\pi \times 10^6/2\pi \times 250 \times 10^3 = 4.0$$

From Fig. 13-1, the required number of stages for the Butterworth, lowpass prototype is $n = 6$ for a relative (normalized) frequency of 4.0 and no less than 70 dB of attenuation. Arbitrarily selecting a filter configuration with a capacitor input, Table 13-2 indicates that the high-pass prototype element values can be determined as follows. The high-pass elements are indicated by the primes (') and the low-pass by the double-primes ('').

$$C_1' = L_6' = 1/L_1 = 1/C_6 = 1/0.518 = 1.932 \text{ farads or henries}$$

$$L_2' = C_5' = 1/C_2 = 1/L_5 = 1/1.414 = 0.707 \text{ henry or farad}$$

$$C_3' = L_4' = 1/L_3 = 1/C_4 = 1/1.932 = 0.518 \text{ farad for henry}$$

Employing equations (13-6) through (13-8), the final element values can be obtained:

$$C_1'' = C_1'/2\pi Z f_c = 1.932/2\pi \times 600 \times 10^6 = 512 \text{ pf}$$

$$L_2'' = Z L_2/2\pi f_c = 600 \times 0.707/2\pi \times 10^6 = 67.5 \text{ μH}$$

$$C_3'' = C_3/2\pi Z f_c = 0.518/2\pi \times 600 \times 10^6 = 137 \text{ pf}$$

$$L_4'' = Z L_4/2\pi f_c = 600 \times 0.518/2\pi \times 10^6 = 49.4 \text{ μH}$$

$$C_5'' = C_5/2\pi Z f_c = 0.707/2\pi \times 600 \times 10^6 = 188 \text{ pf}$$

$$L_6'' = Z L_6/2\pi f_c = 600 \times 1.932/2\pi \times 10^6 = 184 \text{ μH}$$

The fact that all the final filter element values are not symmetrical from the ends to the center has nothing to do with the fact that this is a high-pass rather than a low-pass filter. It lacks symmetry only because the design has

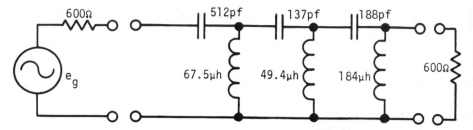

Fig. 13-9. Six-stage Butterworth high-pass filter having 1 MHz cutoff frequency. (*Courtesy of Interference Control Technologies*)

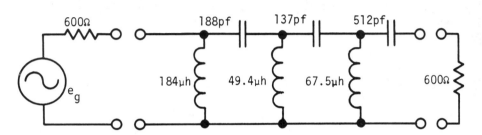

Fig. 13-10. Dual of network shown in Fig. 13-9. (*Courtesy of Interference Control Technologies*)

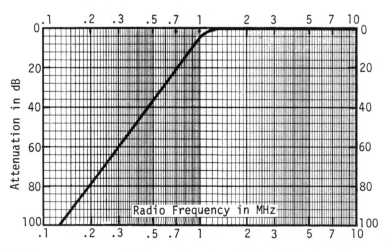

Fig. 13-11. Frequency response of high-pass filters shown in Figs. 13-9 and 13-10. (*Courtesy of Interference Control Technologies*)

an even number of stages rather than an odd number, and the impedance level is other than 1 ohm. The desired high-pass filter circuit is shown in Fig. 13-9, and its dual in Fig. 13-10. The frequency response of both circuits is shown in Fig. 13-11. Formation of the dual network, having an even number of elements, is exactly equal to interchanging the source and termination of the filter or turning the filter end for end. This, in turn, is an illustration of reciprocity. This equivalence of duality and reciprocity does not apply to a network with an odd number of elements.

BAND-PASS AND BAND-REJECTION FILTERS

Like the high-pass filter, the design of band-pass filters may also be directly obtained from the low-pass prototype by a change in the frequency variable of the transfer function. The low-pass prototype has a center frequency (in the parlance of band-pass filters) of 0 rad/sec. In order to make a low-pass to band-pass filter transformation, therefore, the frequency variable, ω, must be replaced by a variable displaying a resonance (pole of the transfer function) at $\omega = \omega_0$ rad/sec instead of at 0 rad/sec. Since LC networks can display this resonant effect, the transformed variable will be of the form:

$$\omega_{bp} = \omega - 1/\omega \tag{13-9}$$

This is equivalent to replacing, in the low-pass prototype, all shunt capacitances (impedances vary with frequency as $1/\omega$) with parallel-tuned circuits, and all series inductances (impedances vary with frequency as ω) with series-tuned circuits.

The frequency at which equation (13-9) is resonant is:

$$\omega - 1/\omega = 0$$

or:

$$\omega^2 = 1; \omega = \pm 1 \text{ rad/sec} \tag{13-10}$$

In order for either the impedance of a series-tuned or the admittance of a parallel-tuned network to be reduced to zero (to give band-pass filter action) at $\omega = \omega_0$ rather than at $\omega = \pm 1$ rad/sec, the frequency variable in equation (13-9) must be normalized to the resonant frequency, ω_0:

$$\omega'_{bp} = \omega_0 \left(\frac{\omega}{\omega_0} - \frac{\omega_0}{\omega} \right) \tag{13-11}$$

where the order of the terms is chosen, as in equation (13-9), to correspond to a negative reactance or susceptance for $\omega < \omega_0$, which is the case for tuned circuits.

The change in variable of the low-pass prototype to yield a band-pass network having a center frequency of ω_0, a bandwidth of ω_c, and hence a loaded Q-factor of $Q_L = \omega_0/\omega_c$, requires bandwidth scaling by dividing the 1-rad/sec cutoff frequency by ω_c:

$$\frac{\omega'_{bp}}{\omega_c} = \frac{\omega_0}{\omega_c}\left(\frac{\omega}{\omega_0} - \frac{\omega_0}{\omega}\right) = Q_L\left(\frac{\omega}{\omega_0} - \frac{\omega_0}{\omega}\right) \tag{13-12}$$

$$\frac{\omega^2 - \omega_0^2}{\omega\omega_c} = \frac{\omega}{\omega_c} - \frac{\omega_0^2}{\omega\omega_c} \tag{13-13}$$

The right-hand expression of equation (13-13) is equivalent to saying that each series inductance in the low-pass prototype (which varies with frequency as ω) can be replaced by:

$$L_s = \frac{L_k}{\omega_c} \tag{13-14}$$

in series with a capacitance (which varies with frequency as $-1/\omega$). This is expressed in the second term of equation (13-13) as:

$$C_s = \frac{1}{L_k(\omega_0^2/\omega_c)} = \frac{\omega_c}{\omega_0^2 L_k}$$

$$= \frac{1}{\omega_0 Q_L L_k} \tag{13-15}$$

Similarly, each shunt capacitance is replaced by a capacitance:

$$C_p = \frac{C_k}{\omega_c} \tag{13-16}$$

in parallel with an inductance, L_p:

$$L_p = \frac{1}{\omega_0 Q_L C_k} \tag{13-17}$$

As a check, it can be shown that the resonant frequency of the above elements is given by:

$$\omega_0^2 = \frac{1}{L_s C_s} = \frac{1}{L_p C_p} \tag{13-17a}$$

Finally, as in the cases of the low-pass and high-pass filters, the impedance level may be changed from 1 ohm to Z ohms by multiplying all resistances and inductances by Z ohms and dividing all capacitances by Z ohms.

$$L'_{sk} \text{ (new)} = \frac{Z \times L_{sk} \text{ (prototype)}}{\omega_c} \tag{13-18}$$

$$C'_{sk} \text{ (new)} = \frac{1}{\omega_0{}^2 L'_{sk}} \tag{13-19}$$

$$C'_{pk} \text{ (new)} = \frac{C_{pk} \text{ (prototype)}}{Z\omega_c} \tag{13-20}$$

$$L'_{pk} \text{ (new)} = \frac{1}{\omega_0{}^2 C'_{pk}} \tag{13-21}$$

Illustrative Example: Band-Pass Filter Design

A 300-ohm, band-pass filter is desired. The center frequency, f_0, is 15 MHz, and the 3-dB bandwidth is 3 MHz, resulting in $Q_L = f_0/f_c = 15/3 = 5$. A roll-off rejection of at least 30 dB is desired at 3 MHz on either side of the 15 MHz center frequency. Assume that a maximally flat (Butterworth) response is desired.

Example Solution. The number of half-bandwidths off the center frequency is $3/(3/2) = 2$, to yield the normalized frequency $\omega_{bp} = 2$. Figure 13-1 indicates that a five-stage Butterworth filter will have the required results. From Table 13-2 (same source and load impedance), the five-element prototype values for a five-stage Butterworth filter may be obtained directly. These values, together with the above ω_0, ω_c, and Z values, are substituted into equations (13-18) through (13-21) as follows:

$$C'_1 = \frac{C_1}{2\pi R f_c} = \frac{0.618}{2\pi \times 300 \times 3 \times 10^6} = 109 \text{ pf}$$

$$L'_1 = \frac{R}{2\pi f_0 Q_L C_1} = \frac{300}{2\pi \times 15 \times 10^6 \times 5 \times 0.618} = 1.0 \text{ μH}$$

$$L'_2 = \frac{R L_2}{2\pi f_c} = \frac{300 \times 1.618}{2\pi \times 3 \times 10^6} = 25.8 \text{ μH}$$

$$C'_2 = C'_4 = \frac{1}{2\pi f_0 Q_L R L_2} = \frac{1}{2\pi \times 15 \times 10^6 \times 5 \times 300 \times 1.618} = 4.4 \text{ pf}$$

$$C'_3 = \frac{C_3}{2\pi R f_0} = \frac{2.00}{2\pi \times 300 \times 3 \times 10^6} = 354 \text{ pf}$$

$$L'_3 = \frac{R}{2\pi f_0 Q_L C_3} = \frac{300}{2\pi \times 15 \times 10^6 \times 5 \times 2.00} = 0.32 \text{ μH}$$

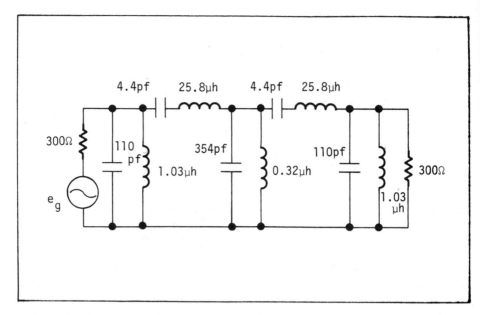

Fig. 13-12. Five-stage, 13 MHz Butterworth band-pass filter. (*Courtesy of Interference Control Technologies*)

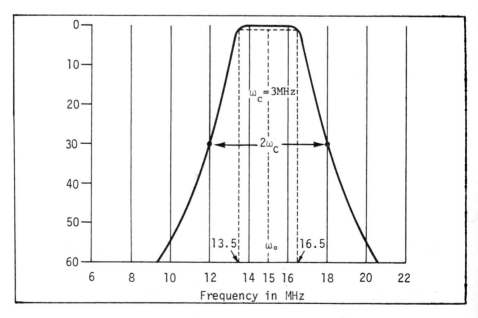

Fig. 13-13. Transmission response of five-stage filter depicted in Fig. 13-12. (*Courtesy of Interference Control Technologies*)

The resulting network is shown in Figs. 13-12 and 13-13.

Like band-pass filters, band-rejection filters may also be derived from the low-pass prototype by a change in the frequency variable of the transfer function.

PHYSICAL REALIZABILITY OF FILTERS

The desired communication filter having been designated on paper, its fabrication, alignment, if applicable (to band-pass and band-rejection filters), performance testing, and installation remain to be accomplished. Fabrication of a filter can be an enormous undertaking, and is beyond the scope of this chapter. In general, it may be stated that (1) lumped-element, LC (or active) filters are constructed below 300 MHz; (2) distributed-element stripline, coaxial, or waveguide filters are fabricated above 1 GHz; and (3) combinations of either lumped-element or distributed-element filters, or hybrids that include both techniques, may be used within the frequency range between about 300 MHz and 1 GHz.

The following discussion summarizes physical realizability of lumped passive element, low- and high-pass filters over (1) the cutoff frequency range from 3 Hz to 300 MHz and (2) the impedance level range from 1 ohm to 50 ohms. For other filter types, and ranges, the references should be consulted.

The following four definitions of LC element physical realizability are established without regard to the current rating of the inductors or the voltage rating of the capacitors, other than to assume that the filter are used for low-level power transmission of less than 1 watt. Thus, some degree of caution obviously must be exercised in employing the following definitions, since certain exceptions may exist.

$$R = \text{Readily realizable} \tag{13-22}$$

$$1 \ \mu H \leq L \leq 1 \ H$$

$$5 \ pf \leq C \leq 1 \ \mu f$$

$$P = \text{Practical} \tag{13-23}$$

$$0.2 \ \mu H \leq L \leq 10 \ H$$

$$2 \ pf \leq C \leq 10 \ \mu f$$

$$M = \text{Marginally practical} \tag{13-24}$$

$$50 \ nH \leq L \leq 100 \ H$$

$$0.5 \ pf \leq C \leq 500 \ \mu f$$

Table 13-3. Degree of physical realizability of four-pass and high-pass Butterworth filters.

Impedance level in ohms	Cut-off frequency, f_c							
	3 Hz–30 Hz	30 Hz–300 Hz	300 Hz–3 kHz	3 kHz–30 kHz	30 kHz–300 kHz	300 kHz–3 MHz	3 MHz–30 MHz	30 MHz–300 MHz
1 Ω–10 Ω	I	M	M	P	R	P	M	I
10 Ω–150 Ω	M	M	M	R	R	R	R	M
150 Ω–2500 Ω	M	P	R	R	R	R	R	R
2500 Ω––50 kΩ	I	M	P	R	R	R	P	I

Considered impractical (I) from a physical-realizability viewpoint are filters that require element values exceeding the bounds of marginal:

$$I = \text{Impractical} \tag{13-25}$$

$$50 \text{ nH} > L > 100 \text{ H}$$

$$0.5 \text{ pf} > C > 500 \text{ } \mu\text{f}$$

Table 13-3 lists the physical-realizability scores of R, P, M, and I for four different driving and terminating impedance loads covering eight decades in the frequency spectrum. Note that: (1) filters are quite realizable at any impedance level for cutoff frequencies in the 3 kHz to 3 MHz frequency range, and (2) an optimum impedance range appears to be about 100 to 1000 ohms. Conversely, physical realizability of low- and high-pass filters is impractical at the four extreme regions shown in Table 13-3.

REFERENCES

1. Crawford, R. A. "High Frequency Quartz Crystal Bandpass Filters." *Electrical Design News*, Volume 7, Number 10, September 1962.
2. Converse, M. E. "Time Domain Filters—Principles and Applications." *Record of the 1975 IEEE Electromagnetic Compatibility Symposium*, Volume 75CH1001-5, October 7–9, 1975, San Antonio, TX.
3. Favors, H. A. "Trade-off Considerations in the Design of Wave Filters for TEMPEST, EMP, and Communications Applications." *Record of the 1975 IEEE Electromagnetic Compatibility Symposium*, Volume 75CH1001-5, October 7–9, 1975, San Antonio, TX.
4. Schiffres, Paul. "A Dissipative Coaxial RFI Filter." *IEEE Transactions on Electromagnetic Compatibility*, Volume EMC-6, January 1964, pp. 55–61.
5. Warren, W. B., Jr, "Tracking Notch Filter for the Reflection of CW Interference." *The Ninth Tri-Service Conference on EMC*, October 15–17, 1963, Chicago IL, pp. 310–319.
6. White, Donald R. J. *A Handbook on Electromagnetic Interference and Compatibility*, Volume 3. Gainesville, VA: Don White Consultants, 1973.

Chapter 14
POWER-LINE EMC FILTERS

Most conducted forms of intrasystem EMI result from various equipment or systems sharing the same source of AC power mains (6). Here, an electrically noisy source may pollute the power distribution wiring by injecting broadband conducted emissions into wires that also feed other potentially susceptible equipments. In this instance, the common-mode noise voltage is generated across the power distribution source and circuit impedance and is therefore "seen" by all equipment connected to the power source. Another mechanism involves common impedance coupling in which two or more circuits are fed from a common regulated or unregulated power supply with an (usually unintentional) impedance element appearing in (i.e., in common with) both circuits. For example, this latter situation often occurs when two circuits share a common return path, such as a ground plane, for the power current, determines the level of induced voltage and the susceptibility of the interference depends upon the level of the path impedance, which, for a given current, determines the level of induced voltage; and the susceptibility of the potential victim circuit. Power mains pollution can also occur if the power lines are electromagnetically contaminated as a result of the coupling of radiated emissions, such as the signals from licensed transmitters, into the lines. The radio frequency (RF) noise is thence conducted into equipment circuits via the power-line conductors.

This chapter covers the topic of AC and DC power lines and the role of power-line filters in decoupling noisy EMI sources from potentially susceptible equipment circuits that must share the same power source. This topic includes the filtering of AC power supplied by electric utilities. Power-line filter considerations, voltage and current ratings, and allowable reactive power are reviewed, along with filter performance and alternative measurement methods. Finally, a survey is made of available EMI power-line filters, including active filters.

UTILITY-SUPPLIED POWER MAINS

Chapter 2 discussed radiation from, and conducted noise existing on, power transmission lines. This section investigates the power distribution system, starting at the utility service poles and ending at the duplex outlet or power breaker distribution panels within a building, structure, or vehicle.

Most AC power is provided by the local electrical utility company to a building or other consumer site. In the United States, low-capacity service, on the order of 500 amperes or less, is generally furnished at a frequency of 60 Hz via single-phase, three-wire systems, with the neutral conductor grounded at the service entrance power panel. Voltage between either of the two-phase lines and neutral is nominally 115 VAC, and the voltage between the two phase lines is 230 VAC. Larger-capacity service to a facility is generally furnished via three-phase, four-wire distribution systems, with the neutral grounded at the input service entrance panel.

Single-Phase Service

Figure 14-1 illustrates two separate electrical power supply service configurations having different capacities. The illustration shows that a single-phase output transformer, with a center-tap neutral wire grounded at the building, is taken from one of the three-phase line pairs of the primary feeder from a nearby power pole or other external distribution configuration. In many modern residential houses and small buildings in urban areas, the power lines are buried underground, and the transformer is located in a protective concrete structure above ground.

The input power lines are connected to the power meter and thence to a circuit breaker and distribution service entrance panel where the neutral terminal is grounded to earth. Power at 115 VAC is then distributed to electrical loads via branch circuits, formed by either phase conductor and the neutral conductor, to the various AC outlet receptacles. Reasonable effort is used to balance the two 115 VAC load distributions. In viewing this arrangement in Fig. 14-1, it is noticed that the primary feed lines, input service lines, and local power distribution system throughout the building act as one extended pickup antenna system with various breaks and appendages. This antenna system picks up radiation from the outside world as well as electrical noise from internally located devices within the building or facility. This appears as common-mode currents that flow in the respective conductors, and voltages that appear between the phase and neutral conductors and a common-mode reference, such as ground.

One of the best ways to demonstrate the magnitude of typical intercepted radiation is to place an RF current probe on one or more of the internal power distribution lines and observe its output voltage on an EMI receiver or other measuring instrument, such as an oscilloscope or spectrum analyzer. Figure 14-2 shows a measured amplitude versus frequency spectrum plotted on an X-Y recorder using automatic test equipment. Sixteen

3 Phase, 3 Wire, 2200 Volt Primary Feeder on Utility Pole

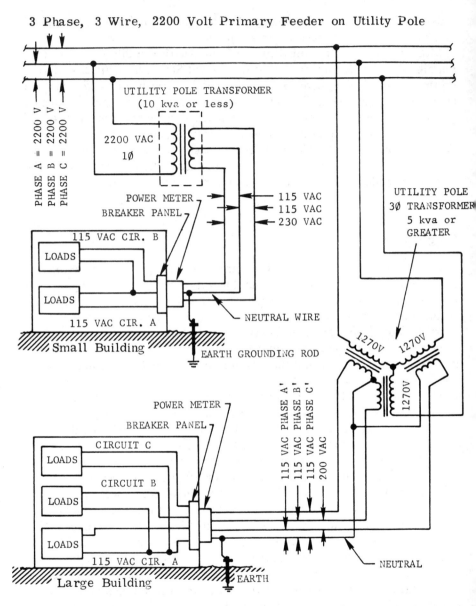

Fig. 14-1. Single- and three-phase utility input and power distribution systems acting as an extended pickup antenna system. (*Courtesy of Interference Control Technologies*)

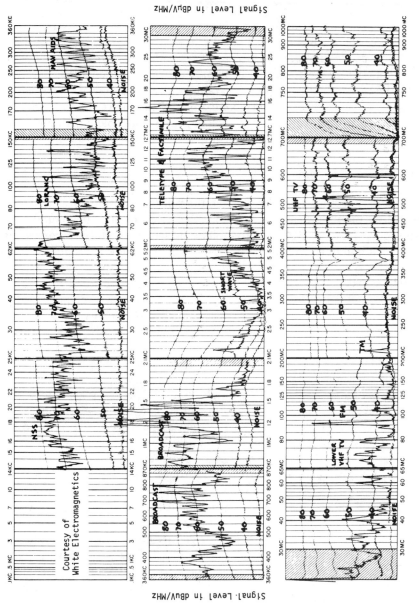

Fig. 14-2. Typical conducted signals and noise appearing on a 115 VAC, 60 Hz power line. (*Courtesy of Interference Control Technologies*)

octaves of the spectrum were examined from 14 kHz to 1 GHz. The power-bus contamination is readily recognized in this *X-Y* plot by observing the AM broadcast, shortwave, HF, TV, FM, and other common-mode current intercepts. The high level of broadband noise, evidenced at frequencies below approximately 1 MHz, results from fluorescent lamp noise emission within the building itself for this particular example. Thus, the power furnished to the building by the electric utility company and the internal branch circuit distribution system act as a system of pickup antennas replete with potential EMI signals that may jam sensitive equipment.

Three-Phase Service

For service capacities typically greater than about 500 amperes at 60 Hz provided by the public utilities in North America, it is not uncommon for the power to be furnished via three-phase, four-wire, delta-wye connected systems to a building or facility. The purpose is to balance the load distributions efficiently on the primary feeders in a given locale, especially where one customer consumes considerably more power than several other users.

The lower portion of Figure 14-1 also illustrates a typical three-phase, four-wire, delta-wye connected service system provided to a facility. The primary voltage reduction transformer exists at a utility pole or other local distribution point as before. Three-phase, four-wire power, however, is now furnished to the facility rather than single-phase, three-wire power as discussed above. The neutral conductor is grounded at the utility input power meter of the facility. The voltage between any single phase and neutral is also 115 VAC, at a frequency of 60 Hz. The voltage measured between any two phases is approximately 200 VAC due to the 120° phase difference between the voltages of any two phase conductors. For facilities requiring higher levels of power, 230-volt or 440-volt service can be provided.

The distribution from a three-phase, four-wire system within a building or facility, corresponding to larger capacity requirements, is somewhat similar to the single-phase, three-wire distribution. From am EMI point of view, the same general problems exist with regard to the primary feeder, input service lines, and internal distribution system acting as an extended pickup antenna complex wherein RF signals can be coupled. Spectrum measurements of amplitude versus frequency, performed on typical three-phase internal line distribution systems within a facility, will likely indicate the presence of common-mode emissions conducted on the line similar to those shown in Fig. 14-2.

From the foregoing, it can be concluded that if susceptible equipment circuits and systems are to perform in such a conducted EMI environment, it is necessary that the conducted emissions picked up by the power distribution system be removed. Accordingly, power-line filters are used at power input terminals of many potentially susceptible equipments. (Additionally, faraday-shielded isolation transformers, EMI-designed motor-alternator sets, and other methods may be employed to remove EMI; see Chapter 15.) Power-line filters are also used in the power lines of EMI emission sources to prevent further pollution of the power mains. These filters and their applications are the topics of the next sections.

POWER-LINE FILTER SPECIFICATIONS

As explained in the preceding section, power lines feeding a given area can act as pickup antennas for signals emitted by AM broadcast, shortwave, HF, FM, TV, communication systems, and radar across the frequency spectrum. Further, these lines can conduct wide-band ignition and overhead fluorescent lamp noise, and virtually any electrical noise that can couple to the input power lines via electric and/or magnetic fields. Since these potentially disturbing EMI conducted noises can cause adverse responses in sensitive equipment, it is paramount to filter them out, preferably before they get to the susceptible circuits. This is accomplished by the use of powerline filters. Such filters are basically of the low-pass type and must pass the DC, 60 Hz, and/or 400 Hz power-mains frequencies with minimum insertion loss (attenuation). They must also provide significant attenuation over a wide frequency range; for example, from a low frequency of 10 kHz up to a high frequency of 1 GHz to 10 GHz, depending upon the EMI frequency bounds of potential equipment susceptibility.

If EMI is to be removed from a power line, it is necessary to determine whether the conducted emissions present are common-mode or differential-mode (3). For power lines, these modes are defined as follows:

Common-mode EMI: Signals that are induced on the black (hot) line, white (neutral) line, and green (safety) line. These induced signals are approximately equal in amplitude and phase. Thus, whenever common-mode noise is present, there exists little to no differential-mode EMI between the phase and neutral lines, but some may exist between phase or neutral and safety wires, depending upon the degree of circuit impedance unbalance. Common-mode noise is developed from the radiation pickup by the power mains cables, as shown on the right in Fig. 14-3. To be effective, then, common-mode filters must have elements placed on all lines. It must be noted that the common-mode reference, or return path, can sometimes be the safety, or green (sometimes green/yellow), wire. Quite often, the common-mode reference is earth ground with common-mode current flowing in the safety wire as well as in the phase and neutral wires. In the latter situation, common-mode voltages appear on the phase, neutral, and safety wires, all determined with respect to the earth ground as the reference. A filter element then must also be injected in the

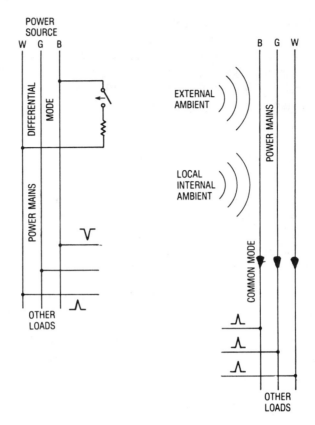

Fig. 14-3. Common and differential mode-coupling on power mains. (*Courtesy of Interference Control Technologies*)

safety wire, or the safety wire must be terminated at the equipment enclosure and not be allowed to enter inside the enclosure (5).

Differential-mode EMI: signals that are induced on and between the phase and neutral wires. These induced signals are approximately equal in amplitude and 180° out of phase. Differential-mode noise is created mostly by other equipment directly connected to the same power mains as the susceptible equipment, as shown in Fig. 14-3 on the left. The situation can be aggravated when changing, or switching, loads are on a circuit and cause switching transients whenever they are actuated on or off. To be effective, filters must be placed to suppress unwanted signals between the phase and neutral wire, with or without the green wire involved.

Figure 14-3 and the above discussion indicate that power-line filters must filter out both common- and differential-mode noise to be effective (3). The classical filters of yesteryear, many of them still in use, were designed to remove differential-mode noise only. When installed, these filters have little or no effect on the common-mode conducted emissions originating from radiation pickup.

To suppress differential mode EMI, a filter would be connected between the black and white wires in Fig. 14-3. Since differential-mode signals are out of phase, the filter will more or less perform satisfactorily. It will not work if only common-mode noise is on the power mains because these signals at the filter inputs are in phase and rise and fall together. Therefore, a special common-mode choke filter is required to suppress common-mode emissions. Since the inputs are in phase, the chokes are actually wound on a toroid and a transformer to permit common-mode signal cancellation. Finally combinations of both filter types, or hybrid filters, are used to suppress both differential- and common-mode noise.

There are usually only a few important specifications in selecting the correct power-line filter to accomplish suppression of EMI conducted emissions. These specifications include:

1. Voltage and current rating at the power mains frequency. This includes the allowable voltage drop across the filter under full-load conditions (e.g., less than 0.2 dB, i.e., 2%, at 30 amps, etc.). This also includes allowable harmonic distortion of the power-line frequency under full-load conditions (e.g., all harmonics above 10 kHz are to be more than 80 dB down from the amplitude of the power mains frequency.)
2. Allowable reactive current at the power mains frequency (e.g., not more than 10% of rated full-load current).
3. Attenuation expressed in dB over the operational frequency range for both pass and stop bands for a defined source and load impedance.
4. Mechanical and other considerations, including size and weight; type of housing and mounting; temperature and vibrations; and shielding protection of the filter housing to electric, magnetic, and plane waves.

Voltage and Current Rating

The voltage rating is the important factor that ensures against a breakdown of the insulation of the internal capacitor(s) used in the filter under maximum supply peak voltage conditions, including undesired transients. In order to get the best EMI suppression (attenuation over the designated frequency spectrum) for a defined filter size or weight, more capacitance per unit volume can be attained with the least voltage rating possible. Thus, to prevent overdesign and increased cost and weight, the voltage rating should accommodate the highest peak power supply voltage expected, and no more. For example, a 115 VAC (RMS rating) filter should be specified for a 115 VAC (RMS rating) line, and not a 250 VAC filter, provided that the line transients, say, are not expected to exceed more than 20% of the supply voltage.

The current rating is important to assure that the internal inductors used in the filter will not saturate or otherwise improperly perform. Inductors are made using toroids that can saturate and produce power-line harmonics if underrated and driven to the saturation point of the magnetic toroid material. Also, as the load current (I) is increased, the voltage drop (IZ) across the filter inductor impedance (Z) will increase. This can result in both (1) poor equivalent voltage regulation of the power mains and filter combination, and (2) transient coupling when different loads at the test specimen terminals are turned on or off. The largest transients are likely to develop when a load is turned off, since many loads appear inductive.

The above considerations are illustrated in Fig. 14-4. The voltage change, V_f, across the generator source impedance filter combination may be, say, about a 5% increase (or 5 volts for a 115 VAC supply) due to source regulation when a significant load is removed (turned off). However, the Q-factor ($Q = 2\pi fL/R$ total) of the inductor–generator–load combination may be on the order of 20 at some frequencies, so that a momentary transient of voltage amplitude, $V = QV_f = 20 \times 5$ or about 100 volts, can surge to the remaining load. The loads, on the other hand, may generate transients of several hundred volts, especially if inductive in nature.

To mitigate the above Q-factor problem, light loads should not be operated when fed by a power-line filter unless a dummy load is also connected. Thus, many filter manufacturers recommend that loads (real, or real plus dummy) should be at least 10% of the maximum filter rating.

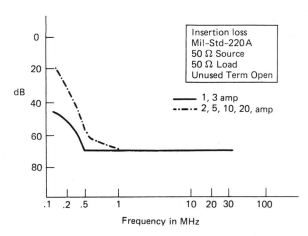

Fig. 14-4. Simplified equivalent circuit of one power source, filter, and loads. (*Courtesy of Interference Control Technologies*)

Allowable Reactive Power

For low-frequency performance of power-line filters, capacitors and inductors are generally used (cf. active filters, below). Among other parameters, the filter attenuation at any out-of-band frequency increases with an increase in either the capacitor or inductor values. Thus, disregarding size or economic considerations, the value of the first capacitor, for example, could be made arbitrarily large. Capacitors of 10-μf values, capable of performing up to or beyond 1 GHz, are not uncommon. At a frequency of 60 Hz, the value of the resulting capacitive reactance, X_c, is:

$$X_c = \frac{1}{2\pi f C} = \frac{1}{6.28 \times 60 \times 10^{-5}} = 265 \text{ ohms} \qquad (14\text{-}1)$$

For a 115 VAC supply, the reactive current, I, is $I = 115/265 = 0.43$ ampere. Should the first capacitor be increased, say, to 250 μf, to substantially lower the cutoff frequency, then the reactive current would be about 10 amperes. This not only enormously increases the shock hazard for line-to-line frame (safety) filtering, but it requires that the source provide the necessary reactive power capacity. Stating this another way, if an auxiliary gasoline-engine–driven generator power supply were used, a wasted capacity of 10 amperes × 115 volts, or more than 1000 volt-amperes, would result.

From the foregoing, it is seen that if the power supply capacity were limited, then the user would want also to specify filters that do not draw more than some maximum allowable reactive current or volt-amperes. If a computation is not made, then a figure such as 10% of rated load might be used for maximum allowable filter reactive power.

Filtering of differential-mode emissions is usually carried out by placing series filtering on each line and parallel decoupling from line to line. The case of each such filter is bonded to the skin of the equipment at the point of power entry. To avoid any shock hazard, filtering is done from line to line rather than from line to safety. Any line-to-safety capacitance may have to be limited to about 0.01 μf to limit the amount of leakage to the safety wire, to comply with safety codes in North America.

Filter Attenuation Performance

Attenuation, as a function of frequency over a prescribed frequency range, is perhaps the most common way of specifying filter spectrum performance; and it is also one of the most abused terms in EMI filters. Filter attenuation refers to the ratio of output voltages, before and after filter insertion, as a function of frequency. Attenuation expressed in decibels, A_{dB}, is derived

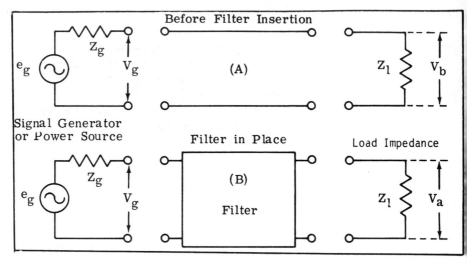

Fig. 14-5. Computing filter attenuation by measuring the voltage before (V_b) and after (V_a) filter insertion (insertion loss method). (*Courtesy of Interference Control Technologies*)

in the following manner (see Fig. 14-5):

$$A_{dB} = 10 \log_{10} \frac{P_b}{P_a} \qquad (14\text{-}2)$$

where:

P_b = power delivered to the load *before* the insertion of the filter
P_a = power delivered to the load *after* the insertion of the filter

Since the *real* load impedance, Z, is assumed to remain unchanged in both the *before* and *after* cases, the respective powers of equation (14-2) may be replaced by their expressions in terms of the respective load voltages and impedances in equation (14-2):

$$P_a = \frac{V_a^2}{Z} \quad \text{and} \quad P_b = \frac{V_b^2}{Z} \qquad (14\text{-}3)$$

Thus:

$$A_{dB} = \log_{10} \frac{V_b^2/Z}{V_a^2/Z} \qquad (14\text{-}4)$$

and:

$$A_{dB} = 10 \log_{10} \left(\frac{V_b}{V_a}\right)^2 = 20 \log_{10} \left(\frac{V_b}{V_a}\right) dB \qquad (14\text{-}5)$$

Note that equation (14-5) is truly *insertion loss* expressed in dB and requires that the measurements at any frequency be made by removing and inserting the filter.

Since it is very time-consuming to make a series of measurements with the filter both in and then out of the installation network at each frequency, a more common method is to permit rapid switching between the two situations, such as shown in Fig. 14-6. This is more fully described in MIL-STD-220A.

For power filters, however, the above attenuation measurement method is invalid because the impedances of both the source and load of real-life installations are significantly different from the 50 ohms used as a convenient coaxial measuring system. Since the filter's frequency response behavior is dependent upon the source and load impedance level, the MIL-STD-220A test method is useful only as a means of comparing filters used in a 50-ohm circuit (1). For example, a 115 VAC power supply mains may provide a 100-ampere service with not more than 5% voltage drop. This corresponds to

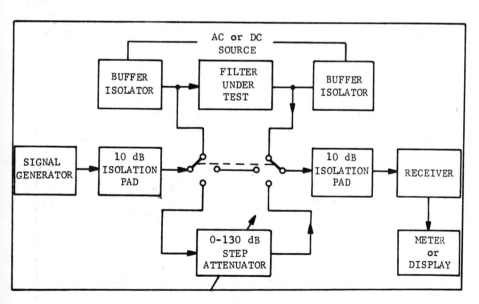

Fig. 14-6. Setup for rapid measurement of filter attenuation (insertion loss) in a 50-ohm measurement system (per MIL-STD-220A). (*Courtesy of Interference Control Technologies*)

an effective impedance source, Z_g, at 60 Hz, of 0.06 ohm, derivable as follows (see Fig. 14-4):

$$V_g = E_g - IZ_g = E_g(1 - 0.05)$$

or:

$$Z_g = \frac{0.05E_g}{I} = \frac{0.05 \times 115 \text{ V}}{100 \text{ amp}} = 0.06 \text{ ohm} \qquad (14\text{-}6)$$

For a 60-ampere load, for example, the termination impedance, Z_L, is:

$$Z_L = \frac{V_L}{I_L} = \frac{115 \text{ V}}{60 \text{ amp}} \cong 2 \text{ ohms} \qquad (14\text{-}7)$$

The above example illustrates one typical low-frequency power source impedance of 60 milliohms and a termination load of 2 ohms at 60 Hz. Now, if a typical single-element filter (e.g., capacitor or inductor) were measured in an impedance system of this amount compared to the 50-ohm system generally used for rating purposes, the results of the spectrum attenuation performance would be significantly different. Figure 14-7 illustrates this, where either a single shunt capacitor ($C = 0.63$ µf) or a single series inductor ($L = 1.6$ mH) results in a cutoff frequency of 10 kHz when measured in a 50-ohm system. Beyond cutoff, the rate of attenuation is 6 dB/octave or

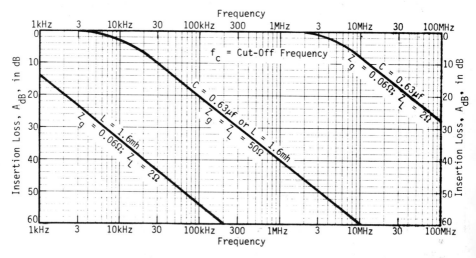

Fig. 14-7. Attenuation (insertion-loss) of a single element filter in a 50 ohm and in a low-impedance source and load system.

Table 14-1. Equivalent number of filter stages for installation impedance differing from design source and load impedances.
(Courtesy of Interference Control Technologies.)

Installation	Impedance	Equivalent Number of Filter Stages	Cutoff Frequency for $n = 1$ Stage	
Source	Load		Capacitor	Inductor
Lower	Lower	n	Increase	Decrease
Lower	Same	$n - 1$	Increase	Same
Lower	Higher	$n - 2$	Increase	Increase
Same	Lower	$n + 1$	Same	Decrease
Same	Same	n	Same	Same
Same	Higher	$n - 1$	Same	Increase
Higher	Lower	$n + 2$	Decrease	Decrease
Higher	Same	$n + 1$	Decrease	Same
Higher	Higher	n	Decrease	Increase

20 dB/decade. Thus, at 10 MHz, for example, the measured attenuation would be 60 dB.

If the above series inductor or shunt capacitor filter is used in a MIL-STD conducted emission test setup, where the HF impedance is artificially stabilized to less than 100 milliohms (by a 10 µf bypass capacitor), entirely different results are obtained. Figure 14-7 shows that the cutoff frequency of the capacitor filter has increased from 10 kHz to 4.3 MHz, and the attenuation thereafter is 52 dB poorer than when measured in a 50-ohm system (per MIL-STD-220A). On the other hand, Fig. 14-7 shows that the cutoff frequency of the inductor filter has decreased to 205 Hz, and the attenuation thereafter is 33 dB greater than when measured in a 50-ohm system. Thus, it is concluded that MIL-STD-220A gives meaningless results regarding attenuation unless the filter is used in a situation whereby the source and load impedances are 50 ohms. Both source and load impedances are pertinent to the filter's performance.

It so happens that the actual performance situation for three or more stages of L-C filtering is not quite so bad as implied above for one-stage elements. Depending upon both source and termination impedance, one or more filter stages may be offered as "sacrificial elements" to establish a pseudo-source and/or pseudo-load impedance. This results in a different number of equivalent filter stages, n, as shown in Table 14-1.

Figures 14-8 through 14-11 illustrate basic filter configurations for application to circuits with different high–low, source–load impedance combinations. All filters shown are of the low-pass type (i.e., they use series inductors

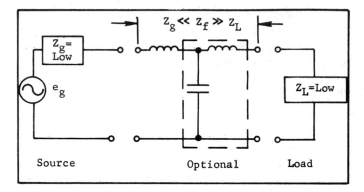

Fig. 14-8. Filter network for low-source and low-load impedances. (*Courtesy of Interference Control Technologies*)

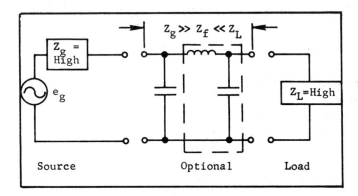

Fig. 14-9. Filter network for high-source and high-load impedance. (*Courtesy of Interference Control Technologies*)

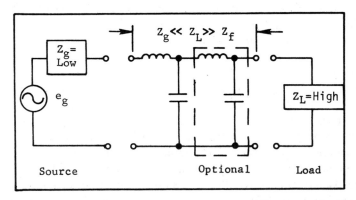

Fig. 14-10. Filter network for low-source and high-load impedances. (*Courtesy of Interference Control Technologies*)

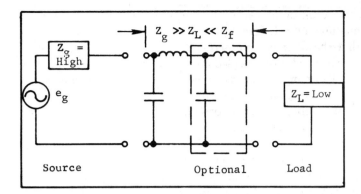

Fig. 14-11. Filter network for high-source and low-load impedances. (*Courtesy of Interference Control Technologies*)

and shunt capacitors). The object is to connect either (1) a filter series inductor to a low-impedance source or (2) a shunt capacitor to a high-impedance source, such that the impedances of source and filter element are about equal at the desired cutoff frequency. Similarly, a series inductor should face a low-impedance load, and a shunt capacitor should face a high-impedance load. This ensures optimum use of filter elements and in part compensates for some source and/or load impedances of typical power mains varying over wide ranges of frequency, starting at about 100 times the power frequencies.

Filter Sizes, Weight, and Mounting

As shown in the next section on EMI power lines, low-pass filters may assume any one of a number of configurations. Their size and weight range can vary up to about six orders of magnitude. The smallest size and weight are about 0.08 cm^3 and 0.5 gram; the largest are about 10^5 cm^3 and 100 kilograms. These exclude filter-pin connectors (see Chapter 7). In general, the filter size and weight will increase with:

- Increase in voltage rating, V volts.
- Increase in current rating, I amps.
- Decrease in internal voltage drop at rated full load current.
- Decrease in cutoff frequency, f_c, in kHz.
- Increase in out-of-band attenuation just above cutoff frequency (i.e., number of stages).

Crosstalk between the filter input and output terminals may be significant (i.e., EMI input-to-output coupling in excess of -60 dB can exist) unless an infinite baffle is used between the two-terminal pair. This coupling path is usually the result of filter input-to-output capacitance. Consequently, whenever a manufacturer determines filter attenuation with frequency, it is understood that the filter is mounted onto a suitable bulkhead that is usually properly grounded. For shielded enclosure filters, where attenuation is rated up to 100 dB or more, the entire assembly is shielded in a permeable case. Thus, after filtering, there is reasonable assurance that there will be no cross-coupling of magnetic, electric, or electromagnetic fields to the filter output leads which are located inside the enclosure.

In addition to the more or less obvious electrical characteristics of a filter, ambient temperature and shock and vibration specifications for the filter must be considered for certain applications unless the filter is to be operated within a known benign thermal and mechanical environment.

AVAILABLE POWER-LINE FILTERS

This section discusses LC, lossy-line, and active filters which are available from a number of manufacturers of power-line filters. The latter two filter types are included because they offer interesting possibilities where passive LC filters run into limitations.

LC Low-Pass Filters

Nearly every EMI power-line filter on the market is of the low-pass type, passing power frequencies from DC to 60 Hz or 400 Hz, and cutting off above these frequencies. There are several manufacturers (the authors are aware of over 20) of power-line filters that are used to filter the power mains into an open area or a shielded enclosure, or to an instrument, equipment, or device.

Figure 14-12 shows a few typical power-line filters available from suppliers. Figure 14-13 illustrates two typical spectrum attenuation plots of available power-line filters used for shielded enclosures.

Lossy-Line Filters

Lossy-line filters are based on one of two principles of operation: dielectric (electric) losses and/or permeable (magnetic) losses. In Fig. 14-14, the dielectric medium intentionally corresponds to one with a high dissipation factor or loss tangent. The equivalent circuit shown in Fig. 14-15 shows that the ferrite-impregnated dielectric conductive losses convert the RF energy into heat. In ordinary transmission lines, these losses are negligible, since low dissipation factor dielectrics are used; but for EMI filter design, the high dissipation factor corresponds to a cutoff frequency of about 10 MHz when used in

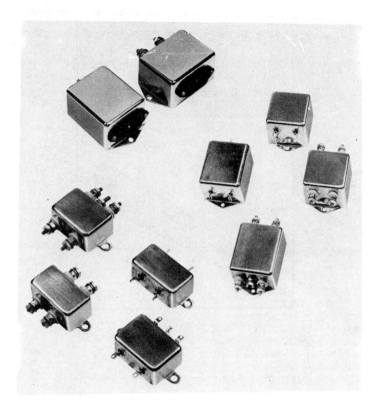

Fig. 14-12. Typical commercially a available power line filters. (*Courtesy LectroMagnetics, Inc.*)

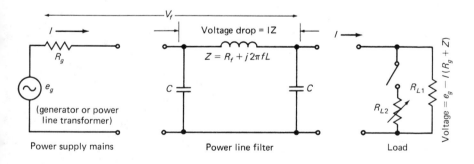

where:

e_g = Power supply voltage

I = Current delivered to load from power supply

R_f = Resistance of filter inductor in ohms

R_{L1} = Fixed load resistance

R_{L2} = Variable load resistance

L = Inductance in henries

C = Capacitance in farads

Z = Impedance in ohms

f = Frequency in Hertz

Fig. 14-13. Typical insertion loss characteristics of power line filters measured per MIL-STD-220A. (*Courtesy of Interference Control Technologies*)

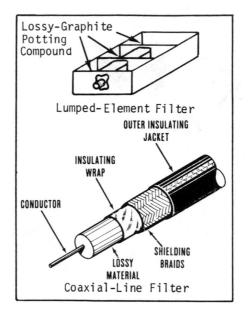

Fig. 14-14. Two examples of lossy-line filters. (*Courtesy of Interference Control Technologies*)

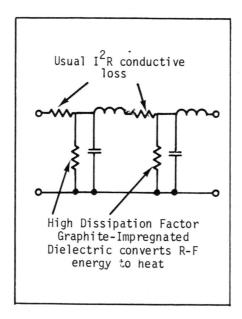

Fig. 14-15. Equivalent circuit of lossy-line filter shown in fig. 14-14. (*Courtesy of Interference Control Technologies*)

a 50-ohm system. Thus, lumped element EMI filters that would fail to perform above, say, 100 MHz, perform well to 10 GHz when the lossy ferrite is used as a potting compound.

A variation of this technique employs a flexible tubing material that may be slipped over an insulated or uninsulated conductor of any standard size. Because a material of lower permeability than the material used for ferrite beads and rods is used in flexible tubing, little attenuation of EMI is offered by the tubing below 10 MHz. On the other hand, no saturation or resonant frequency properties are exhibited, and attenuation above 100 MHz becomes significant.

The principle of operation of the EMI suppressant tubing is similar to that of ferrite beads and rods. Having an equivalent permeability of about 10, the self-inductance of a wire covered with the tubing is increased so that it acts as a one-stage, distributed filter, with a series inductive impedance as shown in Fig. 14-16. However, the material loss is more important than the series impedance because it converts unwanted high-frequency energy into heat. By avoiding the alternating high and low incremental inductance of beads and rods, the tendency to radiate between elements is avoided. The suppressant tubing is also available with a shielded layer of metallized mylar for capacitance shielding (shielding against electric field pickup) at lower frequencies. The tubing exhibits no saturation to any DC, 60 Hz, or 400 Hz

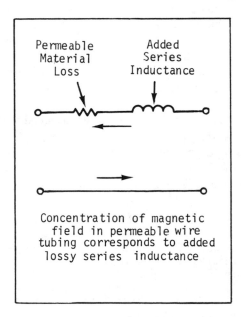

Fig. 14-16. Equivalent circuit of permeable flexible wire tubing. (*Courtesy of Interference Control Technologies*)

power-line current when slipped over power-mains buses for low-pass filter operation.

A combination of the above techniques results in a dissipative coaxial line, ferrite filter. Here, the dielectric loss tangents are very high, the relative permeability is in excess of 10^3, and the relative permittivity is about 10^5 with attenuation on the order of 20 dB at 100 kHz and 100 dB at 10 MHz being achievable. However, the use of this kind of a ferrite filter is limited to applications such as squib initiators (electroexplosive devices, or EEDs) where a low DC resistance between conductors is not objectionable.

Active Filters

The previous section indicated that passive, low-pass, LC filters will become very large as the cutoff frequency is lowered into the audio region. While capacitors can be chosen with relatively low values (e.g., 0.01 µf), the inductor becomes impractically enormous in size and weight. Thus, an inductorless filter is needed for extremely-low-frequency (ELF) applications. An active filter is such a device, in which operational amplifiers transfer the impedance of an RC network to make it appear as an inductor. Active parallel or Twin-Tee tuned audio networks are one example. For power-line applications, however, the main power will have to bypass the active filter transistor if large supply currents are to be processed (i.e., EMI removed).

When the passband power cannot pass through transistors in active filters, then verters and separators may be used. The term "verter" represents impedance converter or inverter with impedance transformations having either positive or negative values. The separator passes no energy except at those frequencies for which the feedback of the active elements is made inoperative by low-power filters.

Active low-pass, power-line filters have been designed with supply current ratings in excess of 100 amperes. In the case of one 24VDC unit furnishing 30 amperes, 40 dB of attenuation existed at a frequency of 1 Hz and 60 dB at 20 Hz. The physical size of an active filter, when compared to a comparable passive LC network, is typically a ratio of 1000 to 1.

REFERENCES

1. Bridges, J. E. "Determination of Filter Performance for Any Arbitrary Source or Load Impedance Based on Experimental Measurements." *Record of the 1975 IEEE Electromagnetic Compatibility Symposium*, Volume 75CH1002-5, October 7–9, 1975, San Antonio, Tx.
2. Cowdell, R. B. "DC to Daylight Filters for DC Lines." *Record of the 1976 International Symposium on Electromagnetic Compatibility*, Volume 76-CH1104-9 EMC, July 13–15, 1976, Washington, DC.

3. Parker, W. H. "How to Specify an EMI Filter." *Record of the 1976 International Symposium on Electromagnetic Compatibility*, Vol 76-CH1104-9 EMC, July 13–15, 1976, Washington, DC.

4. Pontin, G. W., and Thomson, J. M. "Aircraft Interference Suppressors—Passive and Active." *Record of the 1976 International Symposium on Electromagnetic Compatibility*, Volume 76-CH1104-9 EMC, July 13–15, 1976, Washington, DC.

5. Rostek, P. M. "Techniques of Shielding and Filtering Digital Computers for EMI Emissions and Susceptibility." *Record of the 1975 IEEE Electromagnetic Compatibility Symposium*, Volume 75CH1002-5, October 7–9, 1975, San Antonio, TX.

6. White, Donald R. J. *A Handbook on Electromagnetic Interference and Compatibility*, Volume 3. Gainesville, Va: Don White Consultants, 1973.

Chapter 15
POWER-LINE ISOLATION DEVICES

A great deal of EMI comes from electromagnetic pollution of the AC power mains (11, 12). Among the more predominant sources of contamination of primary power are: common-mode noise due to radiated pickup, differential-mode noise due to other users on the same distribution line, dropouts, and interrupts. These primary AC power-line EMI disturbances are discussed in the first section of this chapter.

The next sections of Chapter 15 deal with isolation devices used to reduce or eliminate pollution of the AC power mains. These devices include isolation transformers, faraday-shielded isolation transformers, ferroresonant transformers, EMI-hardened motor-generator or motor-alternator sets, uninterruptible power supplies, and power conditioners.

Typical waveforms and amplitudes of lightning-induced transients on AC power mains are then discussed. Since telephone landlines are affected by lightning strikes in somewhat the same manner, the amplitudes and waveforms of "expected" transients on these lines are also discussed. The final section describes typical protection devices and circuits used to reduce the severity of lightning-induced transients.

POLLUTION OF THE AC POWER MAINS

This section summarizes the major sources of AC power-mains pollution, including generation and distribution effects, multiple users, radiated electromagnetic ambients, and lightning effects.

AC Power Generation and Distribution Effects

Sources that generate noise in power transmission systems were discussed in Chapter 2, and will not be reviewed here. Most, but not all, of the same effects appear on power distribution lines (i.e., those lines that originate at substations and terminate at the user's facilities). Substations also contribute to power-mains pollution. The switching of capacitor banks for power factor correction, phase load balancing, and fault circuit interrupts, all result in the injection of transients into power distribution systems.

EMI Due to Multiple Users

Two kinds of broadband noise are injected onto power mains at the user location: (1) line-voltage shift due to inadequate voltage regulation, and (2) inductive transients due to equipment switching. The former results in differential-mode, broadband noise (see Chapter 14) due to changes in user circuit loads. Heavy inrush transient currents at the time of load connection result in power line voltage surges and drops due to power distribution system source impedance, distributed inductance, and limited line-voltage regulation. During load disconnect, the line voltage also changes, because of the time-changing current and limited regulation.

The second effect, that of inductive transients (LdI/dt), is usually associated with motors, solenoids, relays, and other inductive loads, especially when those loads are switched on or off, thereby causing time-changing currents. The amplitudes of these voltage transients can reach levels that are up to 15 times the nominal line voltage and have rise-times that can be as short as 10 nanoseconds.

Radiated Electromagnetic Ambient Effects

Electromagnetic ambients, whether originating from radiated transmissions (e.g., broadcast) or from man-made broadband noise, result in common-mode noise pickup on the power mains. This form of noise and solutions for its filtering or suppression were discussed in Chapter 14. Both differential- and common-mode noise will exist simultaneously, and both must be suppressed to levels below susceptibility levels of equipment connected to the mains.

Lightning

Lightning can burn out power distribution transformers, trip breakers (because of surges), fell trees onto distribution lines, and so on. Resulting power outages can last for many hours, depending upon the cause and repair time. The lightning stroke itself typically has a rise-time that can be a fraction of a microsecond, and a typical pulse duration of 20 to 30 microseconds. The resulting transient can appear as common-mode spikes on the power mains due to lightning magnetic-field coupling.

Direct strikes result in common- and differential-mode transients. When a cable containing phase, neutral, and ground conductors is struck, a common-mode transient is induced into the cable. This common-mode transient can be converted to differential-mode when the lines are unbalanced (which often happens; see Chapter 5). The operation of power-line lightning arresters also causes common- to differential-mode conversion when the lightning arresters "fire." These arresters are connected from each phase line to ground, and, as they are usually unmatched, they do not conduct at the same instant, resulting in an impulsive potential difference between conductors.

ISOLATION TRANSFORMERS

The isolation transformer is one of the most frequently used devices to reduce noise on the AC power mains. It will suppress low-frequency, common-mode noise, since both sides (hot line and neutral) of the transformer primary rise and fall together. When an unbalance exists between the common-mode noise component on both sides of the primary, then a residual differential-mode component results. A differential-mode noise component can also be coupled directly into a power line. This low-frequency, differential-mode noise component can be magnetically coupled from primary to secondary. Thus, isolation transformers are of marginal value in reducing low-frequency, differential-mode noise.

The primary-to-secondary winding capacitance in isolation transformers is the significant factor in coupling common-mode noise from the primary to the secondary winding at high frequencies. Figure 15-1 shows the equivalent circuit for the flow of common-mode current due to an induced common-mode noise voltage, e_n. The primary (P) terminals are marked A and B, and the secondary (S) terminals are marked C and D. The noise voltage, e_n, is produced from either radiation into a ground-loop area or common-mode impedance cross-coupling, as discussed in Chapter 5. It can be seen in Fig. 15-1 that the undesired primary to secondary winding coupling capacitance is the problem, since its low reactance at high frequencies permits common-

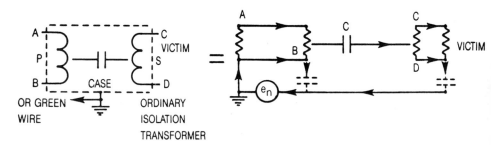

Fig. 15-1. Common-mode noise originating in primary and coupling into secondary. (*Courtesy of Interference Control Technologies*)

mode current induced in the primary to flow into the transformer secondary and into the potential victim load.

To estimate the magnitude of primary-to-secondary coupling of common-mode transients induced into the primary, a typical 1-kW, off-the-shelf, unshielded isolation transformer was selected. A 1-μsec, common-mode, triangular pulse was injected at e_n in Fig. 15-1. The resultant transient that appeared between the transformer secondary and ground corresponded to 5 dB attenuation. Longer pulses, exhibiting more energy at lower frequencies, resulted in considerably greater attenuation.

FARADAY-SHIELDED ISOLATION TRANSFORMERS

One approach used to reduce the primary-to-secondary capacitance in an isolation transformer is to physically increase the distance between primary and secondary windings on a common core. This works to a degree, but results in a larger transformer. It it is desired to reduce the coupling further, the physical separation requirement then becomes too great for practical enclosure sizes. Another approach is then needed, and this is the faraday-shielded isolation transformer, simply called a shielded isolation transformer.

Common-Mode Suppression

Figure 15-2 shows the insertion of a shield between the primary (P) and secondary (S) windings of the isolation transformer. This shield connection is brought out on an insulated terminal mounted on the transformer case. For this illustration, the shield is assumed to divide the primary-to-secondary capacitance into two values, as shown in Fig. 15-2. If the shield is grounded

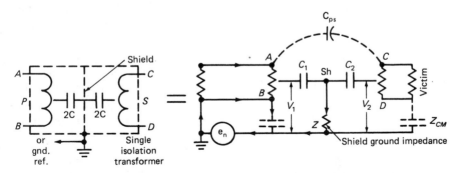

Fig. 15-2. Shielded isolation transformer. Common-mode noise origination in primary and bypassed before secondary. (*Courtesy of Interference Control Technologies*)

to the common-mode noise voltage reference as shown in Fig. 15-2, a primary-to-secondary attenuation, A_{dB}, results as follows:

$$A_{dB} = 20 \log \frac{|V_2|}{|V_1|} \qquad (15\text{-}1)$$

where the logarithm is that of the magnitude of the voltage ratio (V_2/V_1). From Fig. 15-2, this ratio is determined from the following expression, involving the impedances shown:

$$\frac{V_2}{V_1} = \frac{Z(Z_{c2} + Z_{CM})}{Z(Z_{c2} + Z_{CM}) + Z_{c1}(Z + Z_{c2} + Z_{CM})} \frac{Z_{CM}}{Z_{CM} + Z_{c2}} \qquad (15\text{-}2)$$

where:

Z = impedance of the shield ground connection.

Z_{c1}, Z_{c2} = impedance of primary-to-shield and secondary-to-shield coupling capacitances, C_1 and C_2, respectively.

Z_{CM} = effective secondary common-mode impedance to ground.

Z_{ps} = impedance of leakage capacitance, C_{ps}; neglected in deriving equation (15-2).

The common-mode impedance, Z_{CM}, is generally capacitive and can be considered to be relatively large at low frequencies. Equation (15-2) shows the importance of maintaining a low value of the shield grounding impedance, Z, especially at high frequencies. The impedance Z is usually the impedance of a wire or strap used to ground the transformer shield. Its inductive impedance and resistance, the latter due to skin effect, become significant at high frequencies. Equation (15-1) approaches zero as Z approaches zero. Thus, the effectiveness of a high-quality shielded isolation transformer may be significantly degraded unless Z can be reduced to low values at and above VHF.

It should be noted that at high frequencies, the primary-to-secondary leakage capacitance, C_{ps}, shown in Fig. 15-2, represents the capacitance that bypasses the shield. This capacitance is usually quite small in high-quality isolation transformers, but the coupling associated with C_{ps} becomes significant at high frequencies. This capacitance was neglected in deriving equation (15-2).

Differential-Mode Suppression

Both common-mode and differential-mode currents can pass through an unshielded isolation transformer. Figure 15-3 shows differential-mode (i.e.,

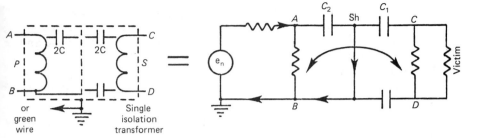

Fig. 15-3. Differential-mode noise origination in primary and bypassed before secondary. (*Courtesy of Interference Control Technologies*)

normal-mode) noise voltage, where the transformer primary currents appear as *push–pull* currents, rather than the *push–push* form for common-mode currents. The added primary-to-secondary shield is shown in the same manner as in Fig. 15-2. The equivalent circuit is shown at the right in Fig. 15-3.

In order to short out the undesired differential-mode voltage appearing across the primary, it is necessary to connect the shield to the primary, preferably at the center tap, if one is available. If a center tap is not available, connect the shield with a low impedance strap to the low or neutral side of the primary. For 50 Hz, 60 Hz, or 400 Hz power-mains frequencies, the value of the capacitive reactance will be high, and the shield will tend to short terminals A and B to reduce the undesired differential-mode voltage.

It is necessary to connect the transformer shield in two different ways to combat the two different types of noise voltage. Common-mode and differential-mode noise cannot be suppressed simultaneously with a single primary-to-secondary shield; a second shield is needed (2).

Figure 15-4 shows a double-shielded isolation transformer and the simultaneous appearance of both common-mode and differential-mode noise. One shield is connected to the primary to reduce differential-mode noise, while

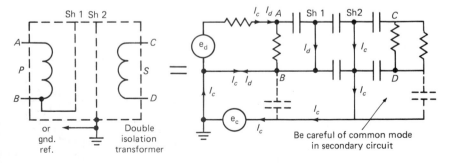

Fig. 15-4. Common - and differential - mode rejection in double shielded isolation transformer. (*Courtesy of Interference Control Technologies*)

the other shield is connected to the common-mode noise reference. The equivalent circuit at the right of Fig. 15-4 shows that it is a composite of Figs. 15-2 and 15-3. Figure 15-4 illustrates a solution to both common-mode and differential-mode noise.

The metal housing of the isolation transformer is also connected to the ground plane or safety wire or, more generally, to the cabinet or rack containing the equipment to be protected. The reasons for this will be explained later. Common-mode noise appearing on the safety or ground wire could be conducted into the victim system unless certain measures are taken (see below).

Triple-Shielded Isolation Transformers

Sometimes the power mains and/or load contains such excessive differential-mode noise that an additional primary-to-secondary shield will give added relief. This is shown at the left in Fig. 15-5, in which the center shield is used for common-mode (CM) noise suppression while the outer shields are used for differential-mode (DM) suppression. Note, however, that the ground safety wire (G) is connected from the supply at the left to the equipment frame. Thus, noise on this line *is directly conducted to the frame*, which is not a desirable practice.

To combat the above situation, a better practice is to disconnect the green safety wire from the isolation transformer housing–equipment frame combination. This is shown in the center part of Fig. 15-5, captioned "BETTER." Any common-mode noise currents on the safety/ground lines will be reduced, since the circuit impedance will have been increased. Some undesired capacitive coupling at high frequencies, however, still exists between the first and the second shield in the ground-safety path. To further reduce the ground-safety common-mode noise currents, the third approach shown in Fig. 15-5, captioned "BEST," is recommended. Here, the polluted safety wire is removed

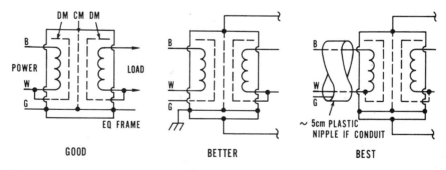

Fig. 15-5. Grounding of triple-shielded isolation transformer. (*Courtesy of Interference Control Technologies*)

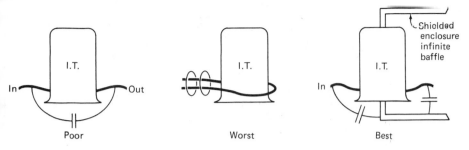

Fig. 15-6. Isolate inputs and outputs of isolation transformers. (*Courtesy of Interference Control Technologies*)

from any direct or indirect coupling to the isolation transformer. The transformer, as in the center figure, is connected to the equipment frame safety ground. If the primary power-mains feed is located in a metal conduit, it too should be electrically broken by using a 5-cm (more or less) plastic nipple.

Isolation Transformer Mounting

For filters (see Chapter 14) and shielded isolation transformers to work properly, they must (ideally) be mounted in an infinite conducting baffle, or a shielded enclosure bulkhead. Figure 15-6 shows at the left the undesired input–output capacitive coupling between transformer primary and secondary when the transformer is sitting on a floor or other structure. The coupling could, of course, even be made higher if the input–output leads were mistakenly laced together as shown in the center sketch. The best approach is to totally isolate input–output leads through a shielded metal bulkhead as shown at the right in Fig. 15-6. Manufacturers of shielded isolation transformers rate their product performance only under this latter condition.

Figure 15-7 shows typical performance curves for single-shielded isolation transformers, which are tested within an infinite baffle. For triple-shielded transformer designs, the rejection is typically 40 to 60 dB better above 100 kHz because of the further decrease in parasitic capacitance (<0.001 pf).

FERRORESONANT TRANSFORMERS (FRTs)

The use of ferroresonant transformers (FRTs) for power-line isolation has increased as a result of the various AC mains disturbances discussed in the first sections of this chapter. They are commonly found incorporated in uninterruptible power supplies (UPSs) and power conditioners, and are sometimes used as stand-alone units incorporated with a separate voltage regulator. The basic operation of ferroresonant transformers will now be discussed, along with their advantages and applications as power-line isolation devices.

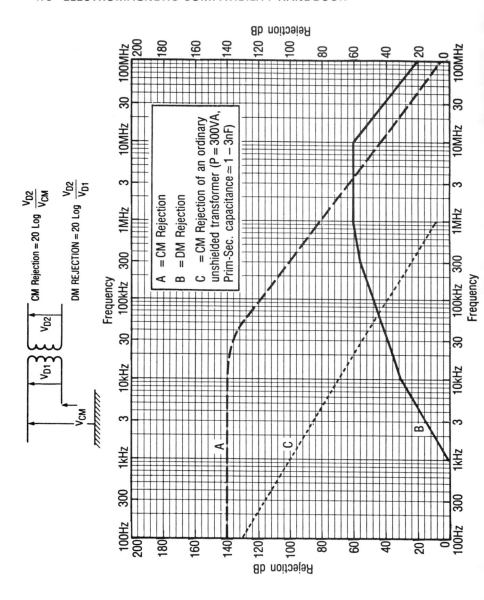

Operation of Ferroresonant Transformers

In contrast to an ordinary power transformer, which operates with its core unsaturated, the ferroresonant transformer is designed to operate in saturation. In ordinary unsaturated transformers, a primary current creates a flux in the core. A change in primary current causes a large change in core flux, (approximately) linearly proportional to the change in primary current. A time-changing flux that flows in the core cuts the turns in the secondary winding, inducing a voltage across the winding, with the induced voltage depending on the turns ratio between the primary and the secondary windings. When a transformer is operated at sufficiently high flux levels (created by a large primary current), the core is in saturation, and a change in the primary current no longer results in a large change in flux. In saturation, if the input current varies (as a result of a variation of input voltage), the output voltage will essentially remain constant. Thus, an ordinary saturated transformer can be used as a regulator, but the input (primary) saturation current is almost as high as the short circuit current.

Ferroresonant transformers operate with the core intentionally driven into saturation. A simplified schematic of an FRT is presented in Fig. 15-8, where the primary winding W_p, the secondary winding W_s, a central air gap, an output capacitance C, and primary and secondary flux paths are shown (5).

The regulating characteristics of the transformer depend on the relationship between the output capacitance C and W_s and the size of the air gap. At low levels of input voltage (E_{in}), the flux flows primarily around the outer legs of the transformer, with some flux flowing through the high reluctance of the center gap, and the output voltage is determined by the turns ratio. At high input voltage levels, the flux increases in the core and across the air gap, and an increase in the inductance of W_s results.

As the reactance of W_s rises, a resonant condition exists when the reactance of W_s equals, but is 180° out of phase with, the reactance of C. The voltage across the secondary rises to a level higher than that which is a result of the turns ratio, which reduces the reluctance of the air gap to a low level. The

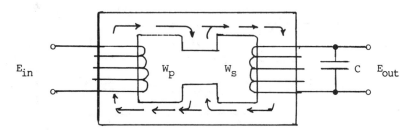

Fig. 15-8. Simplified schematic of a ferroresonant transformer.

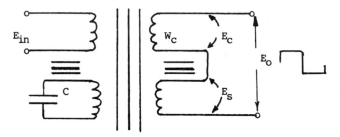

Fig. 15-9. Connection of a compensating winding to offset minor voltage changes.

low effective reluctance of the air gap makes the secondary appear saturated, and a change in input current (due to a change in input voltage), will result in a small change in output voltage.

Additional regulation is provided by a compensating winding that is connected in series opposing the main secondary winding. This connection is shown in Fig. 15-9 (5). The effect of this winding is to provide further compensation for changes in the output voltage. Unfortunately, the output of this type of FRT is a square wave containing high harmonic content that may make it suitable for equipment that is immune to these harmonics, but unsuitable for data processing equipment and other critical loads.

To correct for the square wave output, a neutralizing winding is connected to the secondary in series aiding. The effect of this winding is to add in phase the fundamental of the voltage to the output, but to add out of phase all the harmonics of the voltage. This connection is shown in Fig. 15-10. Thus, the bulk of the harmonics are canceled and, with some additional filtering, the output will be a fairly clean sinusoidal AC voltage (5).

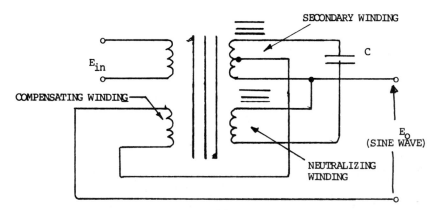

Fig. 15-10. Neutralizing winding added to ferroresonant transformer to produce a sinusoidal output.

An additional benefit of FRTs is their ability to tolerate overloads without damage. Typically, a direct short will cause the output voltage to decrease almost 100%, with an increase in output current approximately 50% of full load. However, an FRT can tolerate this condition almost indefinitely without damage (5).

Although effective against differential-mode variations, a basic ferroresonant transformer is essentially transparent to common-mode noise. The solution is to provide shielding between the windings in the same manner as is done in faraday-shielded isolation transformers, discussed earlier.

MOTOR-GENERATOR SETS

EMI-hardened motor-generator (MG) sets provide another means of power-line isolation from conducted RF emissions. They overcome almost all of the limitations in both the preceding and other types of isolated devices. These devices: (1) are relatively quiet, (2) do not develop toxic fumes, (3) have no fuel handling and storage problems, (4) have a wide range of power-handling capacity and do not require recharging, and (5) are relatively maintenance-free. In addition, they can be obtained to provide 60 Hz to 400 Hz conversion. The MG sets are driven directly from the AC power mains.

MG sets rely on the electromechanical isolation properties of the motor and generator to provide power-mains decoupling. However, it does not follow that any MG set will accomplish this, since some (especially smaller-sized modern units) use a common stator housing to support both the motor and the generator. Here, magnetic-field coupling at low frequencies and parasitic-capacitive coupling at high frequencies can result in a relatively small EMI attenuation over a broad portion of the spectrum. While capacitive coupling can be significantly decreased when a faraday-shield partition is used between the motor and generator, magnetic inductive coupling at low frequencies remains unaffected.

All MG sets properly chosen for EMI power-mains isolation will employ separate and distinct motor and generator housings—often with a longer common shaft to further physically isolate them. To prevent a circulating RF loop, the common shaft may even be conductively isolated by using a nonconductive sleeve bushing or universal joint. (According to discussions that the authors have held with users of MG sets, this latter electrical isolation, via the use of nonconductive mechanical coupling between the motor and generator, is very effective, and often is the difference between satisfactory and unsatisfactory operation.) For brief power interruptions of a few cycles up to a few seconds, the mechanical flywheel effect will smooth out such short-duration voltage variations.

UNINTERRUPTIBLE POWER SUPPLIES

As discussed in the first section of this chapter, blackouts, transients, and dropouts or brief power interrupts are the principal problems associated with the power supplied by the public utilities. Where user operations are critical, as in air traffic control regions, hospital operating rooms, and large computer service bureaus, some form of uninterruptible power is needed for periods of time ranging from several minutes to several hours. Devices called uninterruptible power supplies (UPSs) fill this need.

Most UPS systems operate on the same principle. As shown in Fig. 15-11, the primary 50 Hz, 60 Hz, or 400 Hz power mains feed input power to the UPS on the left. The transformer and rectifier convert this to pulsating DC, which is used to trickle-charge the battery supply. The battery bank directly drives a DC to AC converter that produces the operational AC power fed via the UPS output to drive the loads to be protected. In one type of UPS system, the converter consists of a DC motor driving a brushless AC generator. These are called rotary UPS systems (RUPSs). In older RUPSs, the speed of the DC motor was difficult to control accurately, so that the frequency of the AC output varied. Modern RUPSs, however, control the frequency of the output voltage to ± 0.2 Hz. Most converters use solid state devices, sometimes referred to as *static inverters*, to accomplish the DC to AC conversion. Additionally, a *static transfer switch* connected to the output of the UPS is used to transfer between the output of the UPS and the AC line. In case of UPS failure, the static transfer switch performs an in-phase transfer of the load from the UPS to the AC mains.

For increased protection from long-term blackouts, a motor-generator set is added to the system, to be switched on after a specified period of time. The motor-generator powers the critical load and recharges the batteries. Since batteries exhibit the characteristics of a large capacitor, they can and do act as an excellent sink for power-line transients. To accomplish this, however, it is imperative that the impedance of the feeder line from the batteries to the rectifier be as low as possible. Thus, the cross section of these

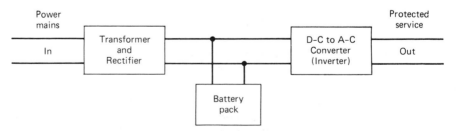

Fig. 15-11. Simplified schematic of a UPS (uninterruptible power supply). (*Courtesy of Interference Control Technologies*)

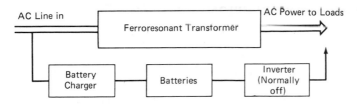

Fig. 15-12. Schematics of FERRUPS™ supply.

lines must correspond to a low inductance, as described in Chapter 9 in the section on bonding. Additional transient and surge suppressors are located at the rectifier to provide added transient suppression.

Another design utilizes ferroresonant transformers in the UPS system. These devices are known as FERRUPS™ (BEST Technology). Unlike the previous design, in which the inverter is on constantly, a FERRUPS uses an inverter only when the AC power is interrupted. This design improves system efficiency. A schematic of a FERRUPS is shown in Fig. 15-12, where the AC is connected to a ferroresonant transformer that provides voltage regulation and filtering to the load during normal operation as well as trickle-charging the battery pack. If the power fails, an inverter, powered by the batteries, is switched *in phase* to power the critical load.

POWER CONDITIONERS

Power conditioners are generally less expensive than UPS systems because they do not provide for long-term battery-powered backup. They are effective against spikes, undervoltages, oscillatory disturbances on the power line, and short-duration power outages. Some power conditioners use a linear amplifier connected to the secondary of an ordinary or isolation transformer in a complex feedback arrangement to cancel the effects of common-mode fluctuations and common-mode noise. A more common (and less expensive) power conditioner uses ferroresonant transformers to compensate for input line voltage fluctuations, transient suppression for reducing the severity of spikes, and shielding (as described for isolation transformers) to suppress common-mode noise.

A power conditioner is not a replacement for a UPS system, as it can only "hold" the output voltage constant for a short time (a couple of milliseconds) after removal of the input line voltage. But, a power conditioner is far less expensive than a UPS in areas where noise, spikes, and line regulation are the main source of trouble, and the user is willing to incur a certain amount of risk from loss of data, operational upset, and so forth resulting from a complete loss of power.

TRANSIENT SUPPRESSION—AMPLITUDES AND WAVEFORMS

The coupling of transients from AC power lines into equipment can affect the operation of the equipment in varying degrees, ranging from temporary upset and loss of data to catastrophic failure and burnout of sensitive circuitry. The effect of the transient on the equipment depends on the magnitude of the transient and the susceptibility of the equipment. The magnitude of the transients on an AC power line is of statistical nature, while the susceptibility of the equipment depends on the inherent "hardness" of the particular piece of equipment and whatever steps the user has taken to increase the hardness (the use of suppression techniques, isolation of sensitive loads from potential transient sources, etc.), as well as utility company–installed protection on the power lines (lightning arresters).

Lightning energy coupling into the telephone system is another source of potentially damaging transients that, when coupled into receivers, modems, PABXs, and so on, can also degrade or destroy components and circuits. Severe lightning-induced transients have been observed to lift copper traces off printed circuit board substrates.

This section presents examples of the "expected" magnitudes of common-mode transients on low-voltage power distribution lines and telephone land-lines, as well as methods to design transient-suppression networks to reduce the magnitude of transients on power and phone lines.

The amplitude of a transient depends on the strength of the source causing the transient (such as the current in a lightning flash), the type of distribution system (underground or above ground), and the electrical parameters of the system. One important quantity is the surge impedance of the line. Viewing a distribution or branch circuit as a transmission line, the surge impedance, Z_0, is equal to the characteristic impedance of the line, or $Z_0 = \sqrt{L/C}$ (8, 10), where L and C are the inductance and capacitance of the transmission line, respectively.

The transmission line approach to surge modeling leads to a Thevenin equivalent circuit of the line, as shown in Fig. 15-13, where V_0 is the open

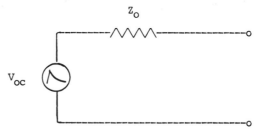

Fig. 15-13. Thevenin equivalent of transmission line. (*Courtesy of Arlington Science Applications, Inc.*)

circuit surge voltage and the available surge current from the source is equal to $I_{sc} = V_{oc}/Z_0$.

The value of the surge impedance Z_0 of a wire or cable above ground is found from equation (15-3)(13):

$$Z_0 = 60 \ln_e \left[\frac{2h}{r} \right] \qquad (15\text{-}3)$$

where:

h = height in meters above ground
r = radius of wire in meters

For multiple conductor groups, the effective radius r' is used in equation (15-3) in place of r. The effective radius is:

$$r' = \frac{ra^{(z-1)}}{z} \qquad (15\text{-}4)$$

where:

a = interwire spacing
z = number of conductors in the cable

From equations (15-3) and (15-4) it is apparent that an increase in the number of conductors per cable will result in a decrease in the surge impedance of the line. Effectively, this means that the surge current will be divided among the wires in the cable, and the surge current for any one wire in a group will therefore be less than it would be for a single wire alone.

Power Line Transients

ANSI/IEEE Standard C62.41-1980 (IEEE Std 587) (1) presents expected surge voltage and current magnitudes as well as the types of waveforms associated with these surges. These values are "deemed to represent the indoor environment" and are to be used to design protective systems in AC power distribution lines of 600 volts and less. This category of distribution voltage is discussed because a large percentage of electronic equipment operates on line voltages of less than 600 volts.

According to the standard, the magnitude of the expected transient depends upon the location of the branch circuit or feeder relative to the service entrance of a building. Three major location categories have been defined, as shown in Fig. 15-14. (Categories A, B, and C correspond to IEC No. 664 Categories II, III, and IV, respectively.)

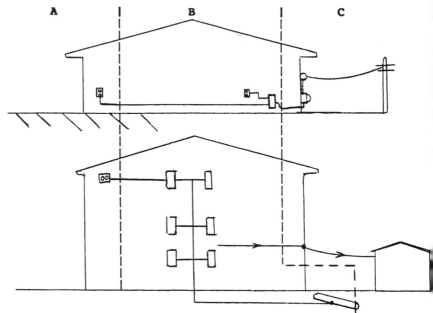

A. Outlets and Long Branches: All outlets at more than 30 ft. (10 m) from Category B with wires #14—10.

B. Major Feeders and Short Branch Circuits: Distribution panel devices; bus and feeder systems in industrial plants; heavy appliance outlets with "short" connections to the service entrance; and lighting systems in commercial buildings.

C. Outside and Service Entrance: Service drop from pole to building entrance; run between meter and distribution panel; overhead line to detached buildings; and underground lines to well pumps.

Fig. 15-14. Location categories.

An impulsive type wave, impinging on a distribution circuit, excites the natural frequencies of the circuit, and, depending on the losses and reactive quantities of the circuit, an oscillatory wave may result. The type of waveform that has been proposed in ANSI/IEEE C62.41-1980 is a "ring wave" with an an exponentially decaying amplitude and a ring frequency believed to be about 100 kHz. The rise-time (defined as the time from 0.1 peak amplitude to 0.9 peak amplitude) of this wave is 0.5 μs (microseconds). This waveform is shown in Fig. 15-15 and represents an open circuit voltage waveform. This is the type of waveform expected at indoor locations, such as Categories B and C in Fig. 15-14.

Outdoor waveforms are generally unidirectional. Figure 15-16 (a and b) shows the open circuit voltage and short circuit current waveforms, respectively, that are expected at outdoor locations such as location C, Fig. 15-14.

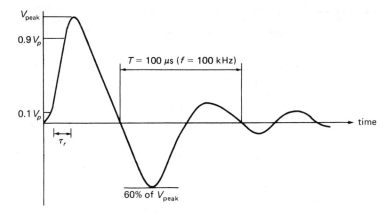

Fig. 15-15. Open circuit voltage waveform per ANSI/IEEE C62.41-1980.

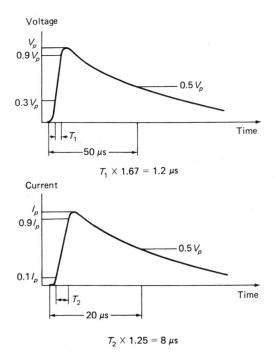

Fig. 15-16. Unidirectional waveshapes (ANSI/IEEE Std 28-1974). (a) open circuit waveform. (b) discharge current waveform. Based upon this standard, testing of transient suppressors is performed.

Table 15-1. Surge voltages and currents deemed to represent the indoor environment.

Location category	Waveform	Medium exposure amplitude	Type of load circuit
A	0.5 μs–100 kHz	6 kV 200 A	High impedance Low impedance
B	1.2 × 50 μs 8 × 20 μs	6 kV 3 kV	High impedance Low impedance
	0.5 μs–100 kHz	6 kV 500 A	High impedance Low impedance

Note: High impedance and low impedance refer to the type of load connected to the circuit, or, in the case of testing, refer to the impedance of the test generator. A low impedance test generator produces a current waveform. A high impedance test generator produces a voltage waveform.

Table 15-1 shows the magnitudes of the transients expected at indoor locations A and B, as well as the associated waveform. The designation of "0.5 μs–100 kHz" for location A refers to the rise-time of the waveform and the ring frequency of the oscillatory wave, the 0.5 us (microseconds) being the rise-time as defined above.

The 6 kV maximum is determined by the "sparkover of clearances" of most types of wiring, such as the air space between terminals on a duplex outlet, which will undergo an arc-discharge type of breakdown at 6 kV, thereby limiting the amplitude of the transient to 6 kV. The surge current is the available short circuit current and thus depends on the surge impedance of the power system as seen from the different locations, typically in the tens of ohms.

The data in Table 15-1 are given for "Medium exposure" systems, which are systems in geographical areas known for high lightning activity and with severe and frequent load switching activity.

The magnitude of transients at location C can be expected to be much higher than at A and B because the limiting effect of the sparkover of clearances is not available. Thus, it is expected that voltage transients of up to 10 kV can occur, with associated surge currents of 10 kA and more. It is recommended that unprotected equipment not be placed in location C unless substantial protection has been applied to the power system.

Telephone Line Transients

Cianos and Pierce have developed a model for predicting the lightning-induced transient amplitude resulting from a direct strike to a telecommunications line. The results are shown in Table 15-2. Column 1 shows the peak stroke current into the shield of the line, and Column 2 is probability of occurrence of that amplitude flash. The peak open circuit voltage and short circuit current are given in Columns 3 and 4, respectively. It is noted that

Table 15-2. Lightning-induced transients on telecommunications line (6).

Lightning stroke, peak current (kA)	Probability of occurrence (%)	Open circuit voltage (volts)	Short circuit current (amps)
175	1	32,200	621
100	5	18,400	355
60	15	11,040	213
20	50	3,680	71

any voltage greater than about 10 kV would probably not be sustained, owing to the breakdown of the insulation about the wire. These data are for a *single wire* with a surge impedance of 50 ohms above ground, with the stroke point being 4.4 kilometers (2.75 miles) from the cable end.

As previously mentioned, the induced current will be divided among multiple conductors in a telephone cable, such that the overall effect on any one wire or wire pair will be less than that indicated in Table 15-2. Table 15-3 shows the results of calculations that were performed to estimate the peak lightning-induced current in a single wire in a cable bundle at various distances from the stroke point. The currents are shown for *each* wire.

The time behavior of the lightning-induced waveforms is defined to be 1.2 × 50 us for an open circuit discharge voltage and 8 × 20 µs for the discharge current. The quantities "1.2" and "8" refer to the rise-time of the waveform in microseconds, and the "50" and "20" refer to the duration of the pulse.

A "severe" current waveshape that may occur on telecommunications lines consists of a "10 × 1000 µs" wave, where the rise-time of the transient is 10 µs, and the pulse duration is 1000 µs.

Table 15-3. Peak lightning-induced currents in telephone cables (6).

Distance to stroke point (km)	Peak Currents (A)			
	At stroke	Number of Conductors		
		1	6 pair	12 pair
4.4	630	355	—	—
2.4	630	637	—	—
1.6	734	799	—	—
0.8	1110	1120	712	453
0.4	1480	1480	852	463

TRANSIENT SUPPRESSION DEVICES AND DESIGN

Protection against the effects of transients requires that the energy in the transient be dissipated or diverted at a voltage that is low enough to prevent unacceptable damage or upset of circuit components and systems. A variety of transient suppressors (sometimes referred to as terminal protection devices, or TPDs) are available, such as gas tubes, metal-oxide varistors (MOVs), and silicon avalanche suppressors.

Transient protection often involves the use of several types of transient suppressors working together to provide the necessary level of protection (3). To choose a suppressor or combination of suppressors, the designer must determine the maximum overvoltage that the equipment can withstand and then pick the suppressor(s) with the proper characteristics to provide that level of protection. The maximum allowed overvoltage depends on whether the designer can risk: (1) no upset, (2) temporary upset (such as a "latch" condition requiring a restart of the system), or (3) damage to the equipment. The "no upset" condition does not mean that there will be no data loss, because unless *extreme* protection is devised, some data will almost certainly be corrupted at the instant of the transient. The type and choice of suppressor depends on the amount of protection that the equipment demands weighed against the cost of the protection. There is always a certain risk that the energy of a transient may exceed the "expected" energy, resulting in the destruction of the transient suppression network and possible damage to the equipment.

Types of Transient Suppressors

There are essentially three types of transient suppressors: spark gaps (including gas tubes), varistors, and silicon avalanche suppressors (SASs). Each of these will now be discussed, along with their relative advantages and disadvantages. A final section will cover the use of hybrid networks, which consist of combinations of the above types of suppressors.

Spark Gaps. These devices were the first types of transient suppressors used by the power and telecommunications industries. The oldest and most commonly used gap is a carbon block spark gap. In their most basic form, spark gaps consist of a line electrode and a ground electrode spaced some distance apart from each other and separated by air (8).

When a transient of sufficient energy encounters the line electrode, the voltage between the two electrodes causes the air between them to undergo breakdown, in the form of an arc across which the transient current flows. The result is that the voltage is clamped to a very low value. Spark gaps are often called lightning arresters, although a more apt designation would be

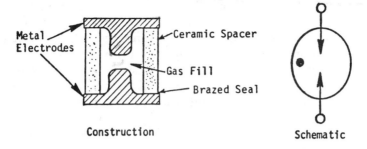

Metal Electrodes — Ceramic Spacer — Gas Fill — Brazed Seal

Construction   Schematic

Fig. 15-17. Typical spark gap. (*Courtesy of Interference Control Technologies*)

lightning diverter because of the diverting action of the device on the transient current. To increase the breakdown voltage of the device, several gaps are placed in series.

Earlier gaps had problems with power follow current; power frequency current would continue to flow across the gap well after the transient had passed. This is a result of the low impedance of the ionized path, which continues to exist across the gap after the overvoltage has passed. The power-line voltage drives a power frequency current through the low impedance across the gap. In order to limit the power follow current, a nonlinear resistance material (such as silicon carbide) is placed in series with the gap. This material has a low resistance under high voltages and a high resistance under low voltages. Gaps such as these are referred to as "valves." The best valves quench the power follow current within about a half-cycle of the power frequency.

Further improvements in gap technology have yielded the modern gas tube, which is a gap situated in an envelope of inert gases under low pressures. These types of gaps have better controlled breakdown characteristics compared to the earlier air gaps. A typical gas tube/spark gap is shown in Fig. 15-17, along with its schematic representation.

One type of gas tube employs three electrodes or terminals, two of which are connected to the current-carryin lines while the third is connected to ground. Figure 15-18 is a representation of this type of gap (9). The benefit

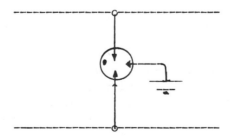

Fig. 15-18. Three terminal spark gap.

of this type of configuration is that this type of device can be effective against both common- and differential-mode transients because of the protection from line to line and from line to ground. Three-electrode gaps can also be effective against unbalanced common-mode transients.

If the amplitudes of the transients on each line are not equal, a differential-mode transient also exists. Whether the initial transient that encounters the gap is a common or differential-mode transient, the chamber of this device will become ionized by that transient. Once the gap is ionized, any subsequent transient energy will be shorted to ground or from line to line through the low impedance of the ionized path.

Another type of gap is the expulsion tube arrester. This device consists of a set of electrodes in a vented chamber that contains a gas-evolving material, usually configured as disks that are placed around the electrodes. The material turns to a gas under the influence of the arc during a discharge, and the gas then expands and blows the arc out of the chamber.

The choice of a gap for a particular application depends upon an estimate of the maximum voltage transient expected and the rate of rise of the transient waveform. Gaps can have a significant voltage overshoot, depending on the transient rise-time. Referred to as the impulse breakdown voltage of the device, the overshoot is often plotted against the rate of rise in V/µs, kV/µs, and MV/µs (EMP-transient). Given the maximum voltage (the impulse break-down voltage) that will be incident on the equipment to be protected, an assessment is made to determine whether this transient level is too high. In the event that the voltage is too high, either a higher-rated or faster-acting gas tube can be used or a hybrid protection circuit devised (see discussion on hybrid circuits below).

Varistors. These devices are made of a metal oxide material (usually a zinc oxide compound) that exhibits highly nonlinear resistance properties. The current through the device is described by the equation (6):

$$I = kV^\alpha \tag{15-5}$$

where k is a constant, V the applied voltage, and α a constant that depends on the material used in the varistor. The constant α is equal to 25 or more for the best varistors. The I-V characteristic of the device roughly resembles that of a pair of paralleled zener diodes of opposite polarity. Thus, the clamping characteristics of these devices are symmetrical, which makes them very useful when the polarity of the transient is unknown, or both polarities are expected. Additional voltage-limiting performance is a result of the high capacitance of these devices, which tends to limit the rate of rise of the transient wave.

The characteristics of these devices make them suitable for a wide range of transient voltage protection. One drawback is the fact that they are always drawing some power from the line, which may or may not be a serious problem. Usually the power amounts to a couple of watts. The impact of the capacitance of a varistor on the impedance of the line must also be assessed. The schematic symbol of a varistor is shown in Fig. 15–19a.

Figure 15-19b shows a number of varistor types and packages. The above varistors range in application from printed circuit board mounting (voltage of 15 VDC) to connection to the AC mains (upwards of 2.8 kV rms).

Figure 15-19c shows surface-mount varistors that can be mounted directly to the substrate or printed circuit board. Lead inductance, which can decrease the effectiveness of the varistor against very fast transients, is significantly reduced in this type of package. The peak operating voltage ratings on these surface-mount GE-MOVs™ is from 5.5 to 369 VDC, with transient current-handling capabilities up to 4500 amperes.

The suitability of a varistor for a particular application depends upon the energy-handling capabilities, voltage rating, and clamping voltage of the device when conducting a certain-amplitude transient current.

A very important consideration when designing varistors is to observe the maximum energy ratings of the varistor. The energy in a wave of a certain shape can be found from equation (15-6):

$$E = V_c(t)I(t)\,\Delta t = KV_cI\tau \tag{15-6}$$

where:

E = energy, in joules, absorbed by varistor.
V_c = clamping voltage of the varistor.
I = current through the varistor at clamping voltage, V_c.
K = energy form factor, which depends on the shape of the wave.
τ = duration of transient pulse, from the peak amplitude to 0.5 times the peak amplitude.

These quantities are found in the previous section on amplitudes and waveforms, and the K factor is found in Fig. 15-20 for different types of waveshapes (6).

The energy found in equation (15-6) using the anticipated waveshape and amplitudes is compared to the varistor energy ratings.

The failure mode of a varistor, if the energy rating is exceeded, is usually a short circuit. Depending on the equipment to which it is connected, the short across the varistor may or may not blow the system fuse. If the system fuse does not blow, the current is mainly limited by the system source impedance. If a large current continues to flow through the varistor package,

Fig. 15-19a. Surface mount varistors. (*Courtesy General Electric Co.*)

Fig. 15-19b. Varistor packages. (*Courtesy of General Electric Co.*)

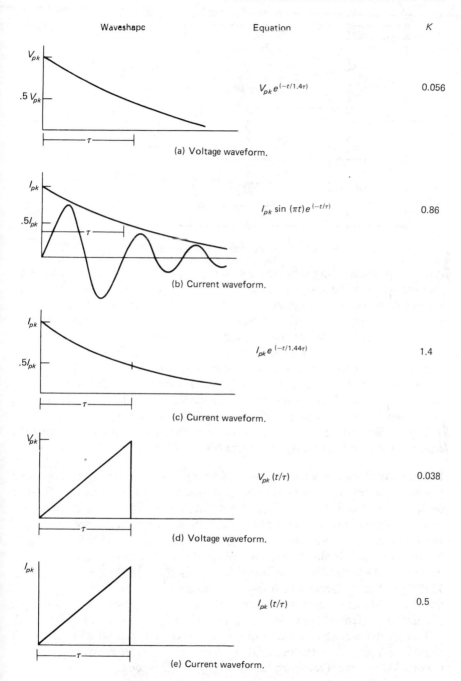

Waveshape	Equation	K
(a) Voltage waveform.	$V_{pk}e^{(-t/1.4\tau)}$	0.056
(b) Current waveform.	$I_{pk}\sin(\pi t)e^{(-t/\tau)}$	0.86
(c) Current waveform.	$I_{pk}e^{(-t/1.44\tau)}$	1.4
(d) Voltage waveform.	$V_{pk}(t/\tau)$	0.038
(e) Current waveform.	$I_{pk}(t/\tau)$	0.5

Fig. 15-20. Energy form factor constants (6).

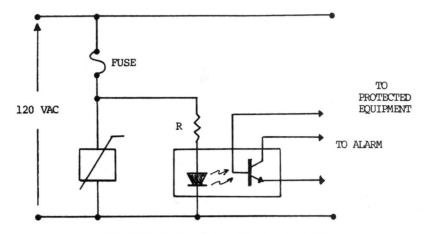

Fig. 15-21. Varistor fault monitoring circuit (6).

it may rupture, ejecting solid and gaseous material. Thus, it is recommended that the leads to the varistor be fused. The size of the fuse should be selected such that the current is limited to less than damage values. Circuit breakers are not recommended because they normally react too slowly to prevent high-fault energy from flowing through the varistor.

A monitoring circuit may be installed in order to determine that the fuse has not blown and to assure that the equipment is still protected. One such monitoring circuit, shown in Fig. 15-21, maintains the base drive to the opto-transistor via the LED if the fuse is in the circuit. The opto-transistor is connected to an appropriate alarm circuit. If the fuse opens up, the LED will turn off, shutting down the base drive of the opto-transistor, turning the transistor off, and sounding the alarm (6).

Silicon Avalanche Suppressors (SASs). These devices are large-area p-n junction devices. The *I-V* curves resemble those of zener diodes. Bipolar devices are available whose *I-V* curves resemble those of back-to-back zeners. The main benefit of these types of devices is the extremely fast response time of the p-n junction; theoretically, they respond in nanoseconds. Their main use is for low-voltage circuit protection, often down line from other forms of protection. One drawback to these devices is the large capacitance (up to 15,000 pf), the impact of which should be assessed (although the capacitance does help to limit the rate of the transient voltage rise). The schematic symbol of a unipolar General Semiconductor Transzorb™ is shown in Fig. 15-22 (7).

Transzorbs are selected according to the reverse stand-off voltage, which should be higher than the peak DC operating voltage or maximum instantaneous AC voltage. Transzorb ratings are generally specified in watts, defined as the peak current through the device times the peak transient voltage (see previous section), and range from 1500 W to 15 kW. Transzorbs dissipate

Fig. 15-22. Transzorb™ schematic symbol.

some current from the line under steady-state conditions, called the reverse leakage current, usually amounting to tens to hundreds of microamperes.

Hybrid Devices

A hybrid device is a combination of two or more transient suppressors. Hybrid devices, or networks, are used to increase the voltage/energy/current handling capability of a suppression network, to increase the response of the network to a transient, and/or to offset one or more undesirable characteristics that may accompany a single device acting alone (3). Figure 15-23 shows various parallel and series combinations of different devices along with the relative merits and drawbacks of the various combinations.

Hybrid networks are sometimes configured in a π-type network; the device with the highest rating is placed on the power or signal line, the lower-rated device on the equipment side of the line, and an isolating impedance on the line between the two transient suppressors. This is shown in Fig. 15-24.

The first transient suppressor, $T1$, clamps the voltage to a relatively low value and dissipates most of the energy. The isolating impedance, Z (an inductor in power-line applications and a small resistor in signal-line applications), limits the current flow and also helps the voltage to build up across $T1$, which aids in the "firing" of $T1$. The second suppressor clamps the transient to an even lower value. $T2$ is usually a fast-acting device, such as a Transzorb™, that clamps the leading edge of the transient waveform. The isolating impedance need not be a discrete impedance, but may represent the impedance of the power line between the primary arrester on a power-line pole and a suppressor installed at the service entrance to a building. Thus, the impedance depends on the length of line and the total inductive reactance at the transient frequencies.

Hybrid devices can be configured using gas tubes, varistors, or silicon avalanche suppressors in a variety of configurations, depending on the application.

CIRCUIT	APPLICATION	ADVANTAGES	DISADVANTAGES
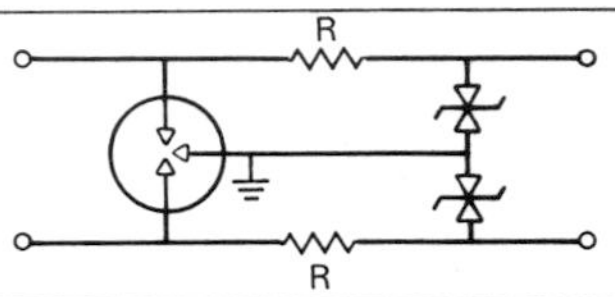	Balanced pair communication lines	High surge current capability low clamp voltage output	Cost, high capacitance, power loss in D-C applications
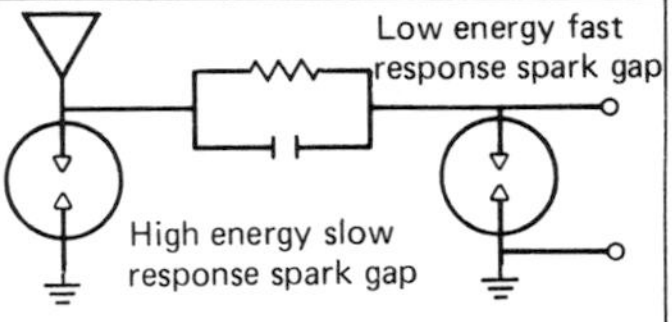 Low energy fast response spark gap / High energy slow response spark gap	High frequency antenna systems	High direct stroke current capability with medium clamping voltage	Insufficiently low clamping voltage for solid state protection
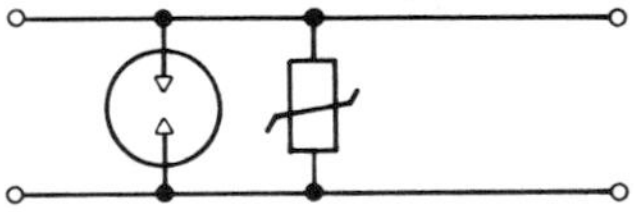	Power circuits	High surge current capability without typical high spark gap impulse over voltage	Cost
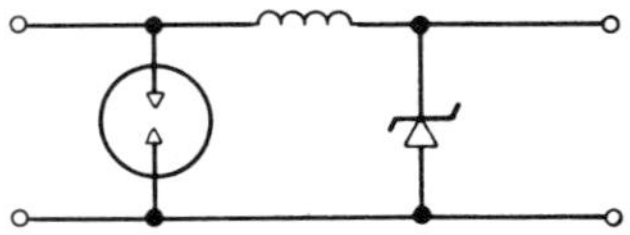	Communication circuits	High surge current capability with low clamp voltage output	Cost, tendency for surges having slow rise times to be concentrated through zener
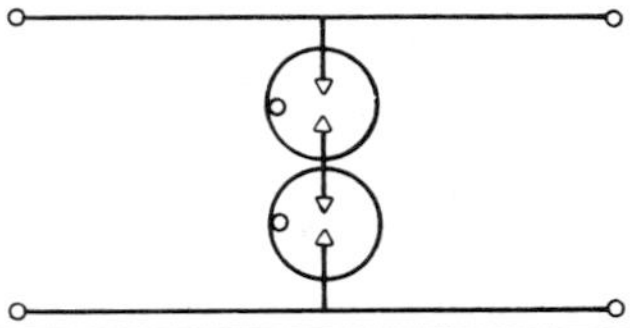	D-C power circuit in 20–40 volt range	Ability to extinguish following surge	Poorer response than single spark gap, cost
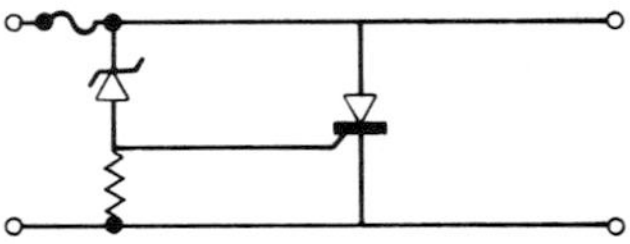	A-C circuit crowbar	Zener stage clamps small energy transients while SCR crowbar opens circuit during larges surges	Cost, SCR stage slow to operate
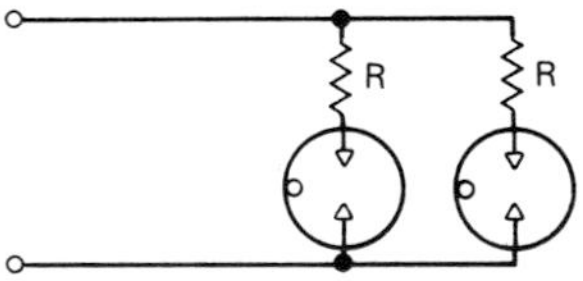	A-C power circuits	Double the surge energy capability of single circuit. Self balancing and lowest discharge voltage	Cost
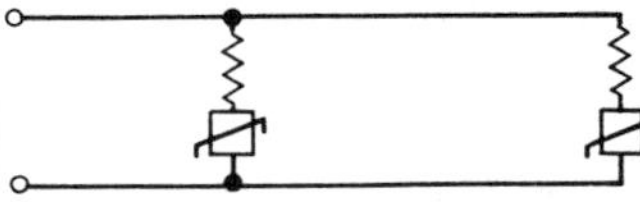	A-C power	Double the surge energy capability of single circuit. Self balancing	Cost

Fig. 15-23. Comparison of hybrids. (*Courtesy of Interference Control Technologies*)

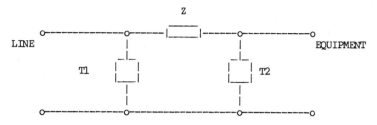

Fig. 15-24. Hybrid device arrangement.

Criteria to specify in the choice of transient suppressors are listed below. Some criteria are characteristic of only certain types of suppressors.

1. *Voltage rating*: maximum rms or peak continuous system operating voltage.
2. *Clamping voltage*: voltage across arrester terminals while conducting a surge current of certain amplitude.
3. *Clamping current*: current that flows through arrester while maintaining a voltage of certain amplitude.
4. *Response time*: the time the device takes to "fire" after application of a transient of a certain rate of rise.
5. *Impulse sparkover voltage, or overshoot*: the peak voltage amplitude that appears across the arrester terminals before the arrester fires (gas tubes).
6. *AC sparkover voltage*: AC power line voltage that will cause arrester to fire.
7. *Power follow current*: power frequency current that flows across the low impedance of an arrester while firing (gas tubes).
8. *Power or energy ratings*: amount of transient power or energy an arrester can dissipate before failing.
9. *Power consumption*: AC or DC power dissipated by device when connected across the line (varistors and SASs).
10. *Failure mode*: a short circuit, an open circuit, or a change in characteristics; caused by exceeding or power rating of device.
11. *Lifetime*: minimum number of transients of a certain amplitude that a device can suppress before failing.
12. *Environmental considerations*: temperature range and maximum altitude.

Installation of Transient Suppressors

To protect equipment against common-mode transients, a suppressor is placed from line to ground. Each line that is carrying a common-mode

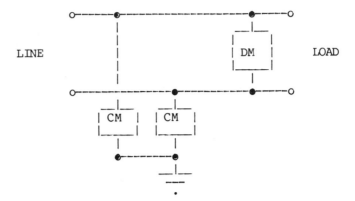

Fig. 15-25. Installation of common (CM) and differential mode (DM) transient suppressors.

transient must have matched transient protection. If this condition is unmet, then a differential-mode voltage will appear on the line that either (1) has no transient protection or (2) has a transient suppressor that fires later or has a higher clamping voltage than the suppressor placed on the other line. It is often necessary to apply differential-mode transient protection because perfectly matched devices are difficult to obtain and because a line may be subject to differential-mode transients. Figure 15-25 shows the installation of transient protection configured for both common mode and differential mode. Three terminal gas tubes may sometimes be used in this situation.

The installation of transient suppressors should be performed with the suppressor leads kept as short as possible. The general installation procedure is shown in Fig. 15-26, where the line leads are brought to the suppressor rather than having long suppressor leads. This configuration minimizes lead inductance and increases the effectiveness of the suppressor against very fast transients.

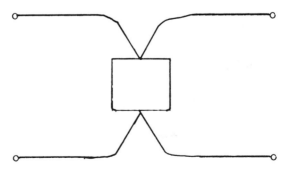

Fig. 15-26. Installation of suppressor to minimize lead inductance.

Keeping the impedance of all suppressor leads as low as possible is especially applicable in the case of grounding conductors. As lightning-induced line transients are generally common-mode and have a very fast rise-time, a significant voltage can build up on the grounding conductor, reducing the effectiveness of the suppressor (i.e., allowing too much transient energy to enter the protected equipment) or resulting in an arcing phenomena called flashover.

Flashover may occur when a conductor of high impedance, such as a poorly designed lightning down conductor or ground conductor, carries a rapidly time-changing current. There are three components of ground impedance: lead inductance, ohmic resistance of the conductor, and the grounding resistance that exists between the ground rod and the common-mode reference or ground. The inductance can be the most significant component of the impedance, and a large voltage, equal to Ldi/dt, can result in the case of a lightning transient. If the conductor passes by a grounded metallic body, such as a water pipe, cable conduit, or metal building framework, an arc may jump from the lightning conductor to the metal body, resulting in radiated and conducted EMI. This situation can be avoided by assuring that the inductance is kept to a minimum by using as short and as large a conductor as possible. Sometimes the size of the conductor necessary to prevent large voltages that may cause flashover is prohibitive. If this is the case, the ground conductor should be well bonded to any nearby metallic objects.

Another component of the grounding impedance is due to the resistance between the ground point and the common-mode return path of the transient. Aspects of grounding and bonding are discussed in Chapters 8 and 9.

REFERENCES

1. ANSI/IEEE C62.41-1980. *IEEE Guide for Surge Voltages in Low Voltage AC Power Circuits.* New York: The Insitute of Electrical and Electronics Engineers, Inc., 1981.
2. ANSI/IEEE 28-1974. *Standard for Surge Arresters for AC Power Circuits.* New York: The Insitute of Electrical and Electronics Engineers, 1974.
3. Bent, Rodney B. "Lightning Protection." *Proceedings of the FAA–Georgia Institute of Technology Workshop on Grounding and Lightning Protection*, May 1977.
4. Bloom, S. D., and Massey, R. P. "Emission Standards and Design Techniques for EMI Control of Multiple DC-DC Converter Systems." *Record of the IEEE 1976 International Symposium of Electromagnetic Compatibility*, Volume 76-CH-1104-9 EMC, July 13–15, 1976, Washington, DC.
5. Deltec Corporation. *AC Power Handbook of Problems and Solutions*, Second Edition. San Diego, CA: Deltec Corporation, 1980.
6. General Electric Company. *Transient Suppression Manual*, Fourth Edition. Auburn, NY: The General Electric Co., 1983.
7. General Semiconductor Industries, Inc. *Transzorb Application Notes.* Tempe, AZ: General Semiconductor Inc., 1983.
8. Greenwood, Allan. *Electrical Transients in Power Systems.* New York: John Wiley and Sons, Inc., 1971.

9. Joslyn Electronics Systems Division. *AC Arrester Application Notes.* Goleta, CA: Joslyn Electronic Systems, 1983.

10. Weeks, Walter L. *Transmission and Distribution of Electrical Energy.* New York: Harper & Row, 1981.

11. White, Donald R. J. *EMI Control Methodology and Procedures.* Gainesville, VA: Interference Control Technologies, 1985.

12. White, Donald R. J. *A Handbook on Electromagnetic Interference and Compatibility,* Gainesville, VA: Don White Consultants, 1973.

13. The International Telegraph and Telephone Consultative Committee (C.C.I.T.T.) *The Protection of Telecommunications Lines and Equipment Against Lightning Discharges,* Geneva: The International Telecommunication Union, 1974.

Chapter 16
EMI CONTROL IN COMPONENTS

Previous chapters have addressed topics such as cabling, grounding, shielding, and filtering that represent EMI problems between sources and victims. This chapter and Chapter 17 will examine the essence of EMC, namely, EMI-control techniques that are applied at component, circuit, and equipment levels.

Fundamentals are stressed. Basic building-block elements of resistors, inductors, capacitors, insulators, conductors, diodes, and logical components are reviewed with regard to their role in EMI problems and control techniques. Several sections on transient-producing devices are included, with discussion of EMI control in transformers, relays, solenoids, motors, and generators. Since these devices also produce significant magnetic fields, flux leakage control provided by shielding is considered. The chapter ends with a discussion of EMI problems and control in fluorescent lamps.

R, L, AND C COMPONENTS

This section discusses potential EMI problems and control techniques that are employed in fabricating and using resistors, inductors, and capacitors. It is shown that these fundamental passive electronic parts in reality do not behave at their stated values (8), especially at high frequencies, because of parasitic inductance and capacitance. Under certain conditions, their performance degrades at frequencies as low as 1 MHz, or even lower. Thus, filters, for example, often do not perform as expected at frequencies equal to ten times the filter cutoff frequency, and amplifiers may exhibit out-of-band prarasitic oscillations and spurious responses. These phenomena result from the nonideal nature of components that manifests itself at high frequencies.

Resistors (6, 8)

The resistors considered here are: carbon composition, deposited carbon-composition film, pyrolytic carbon film, metal film, wirewound, microelectronic, and special purpose. The type of resistor to be used is determined by considerations of resistance, wattage, cost, compactness, precision, distributed capacitance, distributed inductance, life, and internal noise. Composition resistors may be of the pellet or filament type, and are made of finely divided carbon and a binder pressed into a slug with leads embedded in each end. The slug is then enclosed in a phenolic or other case, and the resistor body is molded. In some cases, the resistor is enclosed in a ceramic tube with cement covering both ends. The filament type has the carbon and binder mixture coated on the outer surface of a glass tube with the leads inserted therein. A phenolic tube is then molded around the resistor body.

Resistor Characteristics. Carbon or metal fixed-film resistors are usually made by depositing a controllable thickness of resistive material in a continuous film into a base. The resistor body is then covered with a plastic or epoxy. The geometry of film resistors enhances their high-frequency characteristics when they are used at typical frequencies up to several hundred megahertz.

A microelectronic resistor is a thin layer of silicon on a base or metal placed over a semiconductor. Close spacing increases capacitance and coupling leakage. The small size limits available resistances, and undesired semiconductor junctions may be formed.

Any covering on the resistor body acts as a thermal barrier as well as protection against moisture. Thus, dissipated energy is conducted primarily by the leads. Special metal jackets are made to help heat energy to leave the resistor body. Bifilar winding of a wirewound resistor reduces internal inductance because adjacent turns carry currents in opposite directions. However, adjacent turns may exhibit appreciable shunt capacity, and capacitive currents are likely to have adverse effects on RF applications. The Ayrton-Perry winding is preferred, as each resistor is constructed of two parallel windings in opposite directions; the turns cross each other at points of zero (i.e., minimum) potential difference. A typical Ayrton-Perry resistor wound on a cylindrical spool exhibits 1% of the inductance of a conventional spool-wound power resistor.

A composition resistor can exhibit an AC resistance that is lower than its DC values. This characteristic, known as the Boella effect, is primarily due to the shunting effect of distributed capacitance that results from the large number of conducting particles mixed with the dielectric material. To reduce this effect, resistors with a minimum amount of dielectric material are used to minimize the value of the dielectric constant and associated loss factors. Decreasing the resistor cross section and increasing the resistor length, as in the filament type of resistor, minimizes this problem. Because of the greater amount of dielectric used to construct resistors of higher value, the higher-value resistors exhibit a greater percentage of change in value than the lower-value resistors.

Table 16-1. Ratio of RF resistance to DC resistance of a 1-megohm axial-lead, carbon-composition resistor. (Courtesy of Interference Control Technologies.)

Radio frequency	Manufacturer No. 1			Manufacturer No. 2		
	$\frac{1}{2}$ watt	1 watt	2 watt	$\frac{1}{4}$ watt	$\frac{1}{2}$ watt	1 watt
10 kHz	1.00	1.00	1.00	1.00	1.00	1.00
100 kHz	0.89	0.85	0.75	1.00	1.00	1.00
1 MHz	0.54	0.46	0.37	0.92	0.89	0.90
10 MHz	0.21	0.15	0.12	0.65	0.60	0.67
100 MHz	0.07	0.04	0.04	0.32	0.28	0.36

Skin effect, which occurs at high frequencies, causes current flow to be concentrated at the surface of a device with little current flowing within the rest of the cross section. Because current is not evenly distributed throughout the entire cross section of the conductor, skin effect results in an increased effective resistance at radio frequencies over that of the DC value.

Table 16-1 shows the high-frequency effects of parasitic elements on resistance values at different radio frequencies for some general-purpose, axial-lead, carbon-composition resistors of 1-megohm resistance value. The resistor produced by manufacturer Number 1 is described as a hot molded fixed resistor. The resistance element is a carbon composition film on a glass body. The manufacturer's published frequency characteristics include both inductance and capacitance effects.

Resistor Equivalent Circuits. The equivalent circuit of a resistor depends upon its geometry, manufacturing processes, techniques, and raw

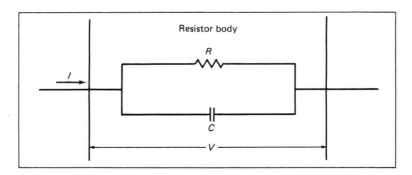

Fig. 16-1. Equivalent circuit of a resistor at low frequencies. (*Courtesy of Interference Control Technologies*)

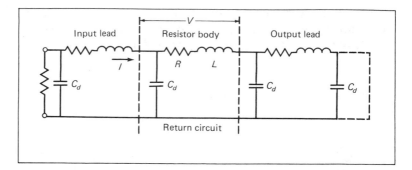

Fig. 16-2. Equivalent of a resistor placed close to the return path. (*Courtesy of Interference Control Technologies*)

materials (6). One such equivalent circuit is shown in Fig. 16-1. The shunt capacitance, C, is on the order of 0.3 pf for a typical 1-watt composition resistor. Figure 6-2 is the equivalent circuit of a resistor located near the return circuit and operating at a frequency where the capacitance of the return circuit is significant and where distributed capacitance, C_d, is low. For composition resistors, the inductance may be negligible.

Wirewound resistors have a relatively large series inductance and distributed capacitance. They are also affected by skin effect, thereby exhibiting an increase in resistance as the frequency increases. The equivalent circuit of a wirewound resistor is shown in Fig. 16-3 (8).

The resistor equivalent circuits illustrated in Figs. 16-1 through 16-3 indicate that the voltage–current relationship for resistors is more complex than the anticipated, or desired, $V = IR$ expression. As the frequency of operation is increased, the parasitic reactive elements cause the resistor impedance to become complex.

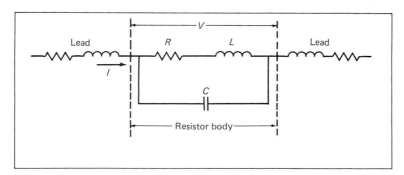

Fig. 16-3. Equivalent circuit of a wire wound resistor. (*Courtesy of Interference Control Technologies*)

Table 16-2. Typical resistor noise for 20 Hz to 20 kHz bandwidth. (Courtesy of Interference Control Technologies.)

Resistor type	μV/V Noise
Metal Film and Wirewound	0.001 to 0.082
Deposited Carbon	0.05 to 0.86
Composition	0.4 to 4.6

All resistors generate thermal and current noise. Table 16-2 shows that metal-film and fixed wirewound resistors generate a lower noise level than other resistor types, although damage or improper manufacturing processes can result in increased noise.

Up to approximately 10 MHz, proper spacing and short leads minimize the effects of self and mutual inductance, while various capacitive effects and dielectric losses are usually negligible.

Between the body ends, the capacitance of a $\frac{1}{2}$-watt resistor is about 0.1 to 0.5 pf, and the inductance of the leads is effectively in series with the capacitance. The inductance and capacitance of helical-form resistors exhibit broadband effects and parallel resonance at certain frequencies. Therefore, these resistors are often limited to power frequency applications.

Strong electromagnetic fields can affect resistors, usually causing a change in resistance due to heating. While composition resistors exhibits only the heating effect, spiral-film and ordinary wirewound resistors can also behave as inductors (magnetic loops) that can couple ambient magnetic field energy into associated circuits.

Capacitors

A capacitance exists between any two physically separated objects. An intentional capacitance, such as a capacitor, is typically constructed of two metal plates or foils separated by a dielectric (4), with the latter selected on the basis of the desired permittivity and frequency range of operation. The capacitance, C, of a parallel plate capacitor is given by the following approximate expression if the fringing electric field can be neglected:

$$C = \frac{A\epsilon}{t} \text{ farads} \tag{16-1}$$

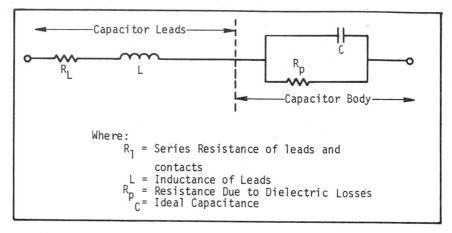

Fig. 16-4. Equivalent circuit of most capacitors over a wide frequency range. (*Courtesy of Interference Control Technologies*)

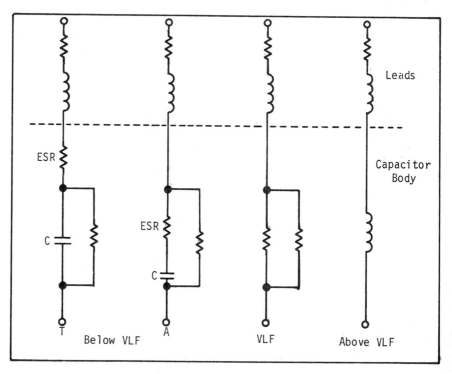

Fig. 16-5. Equivalent circuit of many electrolytic capacitors. (*Courtesy of Interference Control Technologies*)

where

A = common area of the parallel plates in square meters (m^2)

ϵ = permittivity of the dielectric between the plates in farads/m

$\epsilon = \epsilon_r \epsilon_0$

ϵ_r = relative permittivity of the dielectric with respect to free-space (dimensionless)

ϵ_0 = permittivity of free-space (or air) = 8.85×10^{-12} farad/m

t = separation (thickness of the dielectric) between plates in meters

Equation (16-1) applies when the least dimension associated with the area is much greater than the thickness, t. This corresponds to a negligible capacitance due to the fringing field.

The equivalent circuit of most capacitors over a wide frequency range is shown in Fig. 16-4. Figure 16-5 is the equivalent circuit of electrolytic capacitors.

In most capacitors, the lead inductance combines with the capacitance, equivalent series resistance (ESR), and shunt conductance to give an overall complex terminal impedance across the capacitor (1, 5,8) which generally is real and a minimum at the resonant frequency, and increases on either side of this frequency. Above resonance, the capacitor appears as an inductor. Inductance and series resistance limit the rate of change of current during sudden charge or discharge, such as for transients. The series resistance affects the dissipation factor and may cause problems in AC, high-precision, and timing circuits. For most applications, series resistance is considered constant and independent of frequency.

Shunt conductance is caused by current leakage and voltage stress across the dielectric. Current leakage is usually small in solid dielectrics, but may be a problem in both high-precision capacitors and some electrolytic capacitors. Shunt conductance is affected by both the instantaneous and longer-duration application of voltage stress. The effects are energy loss, heating, and change in power factor. Absorption of energy results in reappearance of voltage on the capacitor after it has been discharged. Dielectric absorption causes a voltage stress that is delayed because of the time required to displace charges from the dielectric.

In high voltage circuits, a means of discharging capacitors should be provided to prevent danger to personnel. The self-resonant frequency is determined by many factors, including physical size, dielectric properties, capacitance, lead inductance, and inductance of the plates. Figure 16-6 shows lead-length effects for a 0.5-μf capacitor with 6-mm leads. Note the complete degradation of performance above about 5 MHz. Above this frequency, the

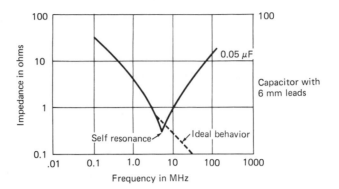

Fig. 16-6. Typical effect of lead length upon capacitor. (*Courtesy of Interference Control Technologies*)

capacitor behaves as an inductor. Were this capacitor to be used as a single-stage filter in a 50-ohm line, the insertion loss would degenerate above 4 MHz, as shown in Fig. 16-7. This may be contrasted with the performance of a 0.05-μf feed-through capacitor in which the associated series inductance is very small. Above 50 MHz, the dielectric and series resistive loss protects the capacitors performance as a filter. Three types of resonances can occur in

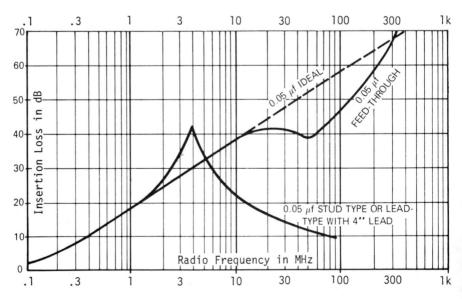

Fig. 16-7. Comparison of filtering performance of feed-through and lead-type capacitors with ideal capacitors. (*Courtesy of Interference Control Technologies*)

disc-type capacitors:

- Low-frequency resonance due to long leads.
- Medium-high frequency resonance when discs are connected in parallel, due to internal leads.
- High-frequency resonance due to resonant cavity effects in high-dielectric capacitors.

The process of selecting a proper capacitor includes, among other considerations, assuring that the dynamic range of operating frequencies anticipated is below the capacitor resonance frequency.

Ceramic dielectric capacitors and filters with ceramic (e.g., barium titanate) dielectrics are useful because of their small size and low weight when compared with capacitors using more conventional materials such as paper or mica. However, some ceramic capacitors are extremely sensitive to temperature, with a significant degradation in equipment performance when they are used as bypass capacitors operating at an ambient temperature of $-40°C$ or at an elevated temperature of $+100°C$. For example, according to one manufacturer's specifications, a nominal 1000-pf ceramic disc, feed-through type capacitor will have a capacity of 330 pf at $-40°C$. While the operating temperature range may be $-55°C$ to $85°C$, the capacity is rated for a temperature of $20 \pm 1°C$. The continuous working voltage rating may be 500 VDC. But the design capacitance may be stated for a 0.5 to 5.0-volt range. With 500 volts applied and a feed-through current of, say, 25 amperes, the device's effective attenuation as a bypass capacitor is reduced by 20 db at $+25°C$. Thus, electrical and environmental operating conditions must be considered when a capacitor or filter incorporating a ceramic dielectric is used.

Tantalum capacitors also offer attractive space and weight savings under some conditions of operation. Tantalum capacitors come in three types: solid, foil, and wet anode. Solid slug types are made by sintering, which forms a spongy slug of metal that has a large effective surface area and is extremely small for its capacity and voltage rating. Feed-through capacitors of the solid tantalum type are effective for frequencies up to 5 GHz.

Foil-type tantalum capacitors can be made in voltage ratings up to 300 volts, compared with about 50 volts for the solid and 125 volts for the wet type. The foil type is limited to audio and low RF applications because of high internal inductance. Sometimes a tantalum-foil capacitor is shunted with a smaller paper or ceramic capacitor to extend to effective bypass frequency range.

The original wet-type tantalum capacitor is no longer used because of the danger of electrolyte leakage. In the newer wet-type construction, the electrolyte is a gel, and the danger of leakage and consequent corrosion is no

Table 16-3. Typical internal inductance of various capacitors. (Courtesy of Interference Control Technologies.)

Capacitor type	Inductance
Porcelain and ceramic fixed capacitors	1.4 nh
Wet-anode tantalum capacitors	25 nh
Solid tantalum capacitors	20 nh
Foil tantalum, tubular case, with leads	50 nh
Foil tantalum, rectangular case, lug terminals	23 nh

longer a factor. The wet-type construction offers the smallest size and the largest capacity of all tantalum types.

The tantalum capacitor has a further advantage: its low-temperature performance is greatly superior to that of the aluminum-electrolytic type. Tantalum capacitors also have a longer shelf life and less current leakage, especially at high temperatures and after long periods of idle time. However, the polarity of the tantalum capacitor must be carefully observed. A tantalum capacitor which has its polarity reversed may quickly burn up.

The total inductance of a capacitor is the sum of the internal (electrode) inductance and the external (lead) inductance. Table 16-3 lists typical values of internal (electrode) inductance for various types of capacitors. The external (lead) inductance, L, can be determined from the self-inductance of a straight round copper wire as explained in the next section. The internal and external inductances may be added to determine the total inductance in series with the capacitance.

Inductors

An inductor is formed by a portion of a conductor and its return, a single conductor coil or, usually a multiturn coil. The inductance of different straight wires as a function of length is shown in Fig. 16-8. An intentional inductor exhibits inductance, resistance, capacitance between turns, and capacitance between turns and ground, shields, and other circuits. Figure 16-9 shows the equivalent circuit of a typical inductor. An air-core inductor may be wound on a nonmagnetic metal or insulator core. Magnetic cores are made with steel or iron alloy in sheet, strip, wire, or powder form.

The distributed capacitance of an inductor acts as an equivalent, lumped shunt capacitance, resulting in parallel resonance at some frequency. Other inductor characteristics are power dissipation or loss, saturation, susceptibility to and generation of stray magnetic fields, and instability of characteristics. High-precision circuitry can lose appreciable precision and stability

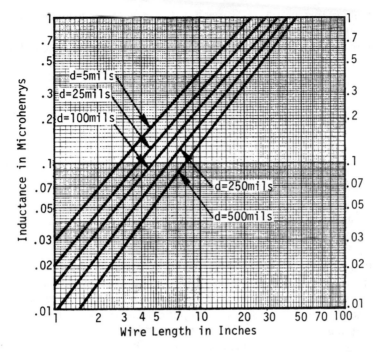

Fig. 16-8. Self-inductance of a straight round wire at high frequencies. (*Courtesy of Interference Control Technologies*)

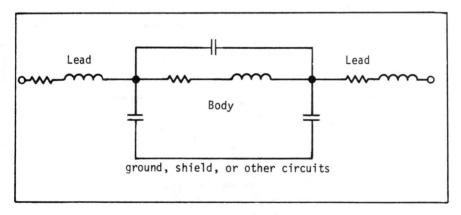

Fig. 16-9. Equivalent circuit of an inductor. (*Courtesy of Interference Control Technologies*)

when temperature and humidity cause dielectric losses from the insulation and inductor supports.

Current flowing through an inductor coil produces a magnetic flux that induces an EMF proportional to the rate of change of flux linkage. The self-inductance is given by:

$$L = \frac{N\phi}{I} \qquad (16\text{-}2)$$

where:

L = self-inductance in henries
N = number of turns, loops, in a winding
ϕ = magnetic flux in webers
I = current in amperes

Mutual inductance (inductive coupling) occurs when magnetic flux lines of one element link with an area of another element. Energy is thereby coupled from one element to the other and can cause interference to other circuits, as suggested in Fig. 16-10. Mutual coupling decreases with distance, and the least coupling occurs when the inductor axes are at right angles to the area of the victim circuit. Inductors made from toroids have considerably reduced external magnetic fields because nearly all the field is inside the magnetic toroid material.

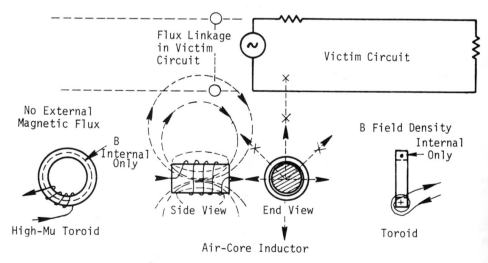

Fig. 16-10. External magnetic fields of an inductor for air and high-permeability cores. (*Courtesy of Interference Control Technologies*)

Inductors are often shielded to keep their electric and magnetic fields within a limited space around the inductor. Electrostatic shields can be made of low-electrical-resistance material such as copper, aluminum, or zinc. These shields decrease the amount of magnetic flux escaping from the vicinity of the inductor because voltages are induced in the shield that set up eddy currents opposing the magnetic fields. A high-permeability material such as Permalloy is used at low frequencies if the flux is undirectional. The use of high-permeability material results in an induced opposing magnetic field outside the shield. High current levels result in strong local magnetic fields that cause saturation of the shield, which reduces the permeability of the shield so that it no longer behaves as a short circuit to the flux.

INDUCTIVE DEVICES

This section discusses inductive devices, most of which produce transients. These devices are mainly characterized by containing one or more relatively high-current, low-voltage sources that produce predominately magnetic fields. As such, these low-impedance devices produce low-impedance fields that are sources of EMI (see Chapter 10). If uncontrolled, they may magnetically couple EMI into other low-impedance suceptible circuits, components, and/ or wiring (see Chapter 6).

The principal inductive devices discussed in this section are inductors, transformers, relays, and solenoids. Motors and generators are discussed in a later section of this chapter. Each discussion emphasizes techniques for controlling EMI. As discussed below, the main techniques available for controlling transient-producing devices involve the use of diodes and filters, and where the magnetic fields are an EMI problem, shielding is used.

Inductors

An earlier section discussed the inductor as a lumped element whose performance degrades at high frequency owing to parasitic capacitance and related effects. Therefore, an inductor may appear as a capacitance above self-resonance. If the inductor is used as a filter element, HF performance may degrade to the point where EMI protection to other circuitry or devices may no longer exist.

An inductor, if not properly constructed or used, may act as a means for coupling undesired EMI into other victim circuits. Figure 16-10 (above) illustrates this, with the magnetic field from the helix coupling into the wiring of another circuit when an air-core inductor is used. When the inductor is wrapped in the shape of a toroid, which has a high permeability (e.g., on the order of 1000), external flux density is available to couple into potential victim circuits. Another advantage of toroids is that their size can be made considerably smaller than air-core inductors because of the greater inductance per unit length achievable with high-permeability materials.

Transformers

Transformer behavior resembles that of inductors, except that two separate EMI source circuits (not the victim circuit) are to be intentionally coupled with a high efficiency. As in the single inductor, external magnetic fields may couple into victim circuits if special precautions are not taken. Transformers and audio chokes are greater EMI offenders because they generally carry considerably more power than RF inductors and associated toroids.

A transformer or choke coil can be enclosed in a high-conductivity metal enclosure, usually called a shading ring. When the coil's alternating magnetic field cuts the shading ring, the induced voltage in the small resistance causes a large flow in the ring. This in turn sets up an alternating field that opposes the coil's original magnetic field. The magnetic field outside the ring is thereby reduced. As the resistance is minimized, the field is decreased. This arrangement can be thought of as a transformer with a shorted secondary.

The shading ring has no effect on DC magnetic fields because the fields do not vary. If a conductive box is used, the sides act as a shading ring, with the top and bottom effectively an infinite number of concentric shading rings.

Figure 16-11 shows the use of a shading ring in a transformer or choke. Television receiver power transformers usually have shading rings to help

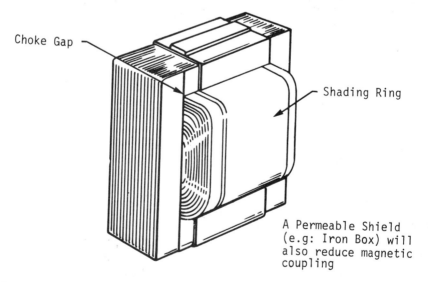

Fig. 16-11. Shading ring in a transformer or choke. (*Courtesy of Interference Control Technologies*)

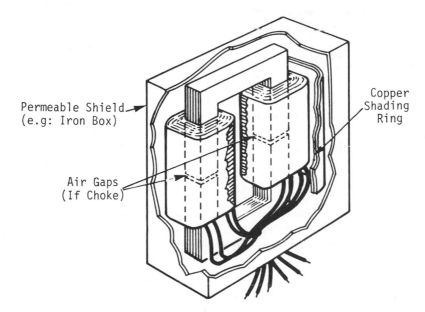

Permeable Shield
(e.g: Iron Box)

Copper
Shading
Ring

Air Gaps
(If Choke)

Fig. 16-12. Magnetically shielded transformer or choke. (*Courtesy of Interference Control Technologies*)

prevent electron-beam deflection in the CRT by the transformer's magnetic field. Chokes are potentially one of the worst interference sources because of the large harmonic content of power supply current, rectifier switching transients, and magnetic-circuit gaps that prevent DC saturation. In the case of this type of choke, the shading ring in Fig. 16-11 must be extended so it covers the gap. The unsymmetric structure of the fields would also require the choke to be enclosed in a high-permeability box of sufficient thickness and distance from the choke to prevent saturation.

Figure 16-12 shows how a semitoroidal choke or transformer is made, minimizing the weight and size. The coils act as shading rings as they are place over a core gap. When the choke is enclosed in a magnetic material housing, external fields at power-mains frequencies can be reduced by approximately 45 dB over a conventional E-lamination transformer.

Another method used to reduce magnetic coupling between transfor
and/or chokes or between transformers and circuit wiring is to mou
at right angles to each other. If the magnetic fields of each are
a plane, then the coupling will be proportional to the cosin
between their planes with a maximum practical decoupli
being achievable. However, this practice is not reliab
generally not confined to a plane, and because com
tioned during design or maintenance.

Switches and Relays

While not inductive devices themselves, switches are used with relays, which are highly inductive. Accordingly, this section reviews switches and relays.

Switches. The function of an electrical switch is to interrupt the flow of current or change the routing of current. An electrical switch causes the impedance of a circuit to change rapidly between a relatively low value and some large value. The rapid change causes a high dI/dt and/or dV/dt in the switched circuit that, in turn, produces steps in current or voltage waveforms capable of causing interference.

When a power circuit is switched by mechanical contacts, high frequencies capable of causing interference are generated during both closing and opening of the contacts. As the contacts close, a step function of voltage is applied to the circuit that excites it, causing oscillations in R-L-C reactive elements. EMI is thereby generated that can be transmitted by radiation and/or conduction. The larger the wattage of the load being switched, the more difficult and expensive it becomes to control the effects of EMI.

Mechanical contacts exhibit the additional problem of contact bounce when closing. Random opening and closing of contacts chops the current, generating HF oscillations and harmonic components. For inductive circuits, abrupt interruptions can lead to high-induced-voltage transients, contact arcing, dielectric breakdown, and associated phenomena. All may cause EMI problems.

The making or breaking of an electrical circuit by a mechanical switch is usually accompanied by the generation of an arc at the switch contacts. Arcing during normal operation of a switch occurs because a highly ionized gas is substituted for a normal metallic circuit as the switch contacts move

The usual DC interruption is caused by making an arc and forcing it into an unstable shape to create instability and eventual collapse. Fast switching minimizes the duration of the arc, although typical inductive DC circuit arcs last longer than 5 msec. Other elements in the circuit may radiate EMI that was conducted to them.

Alternating current switches used to be manufactured so that power was not applied or removed unless the voltage was passing through a zero. This was not totally effective because the non-unity power factor associated with almost any real device resulted in a non-zero current flowing even though the voltage was zero. When AC is interrupted, the duration and magnitude of arcing depend upon the instantaneous voltage at the time of interruption, and the voltage depends on $L di/dt$. Forcing a large current to zero during switching results in a large di/dt, and hence creates an arcing problem. Thus, newer devices are designed so that the current is sensed and the circuit is interrupted when the current passes through zero.

Electrical circuits can be switched by several methods. Arcing is associated with switching by mechanical means such as manual switches, relays, circuit breakers, or thermostats, and generates EMI with a frequency spectrum that extends into the ultraviolet region. Use of solid-state devices may eliminate the arc effect, but solid-state switches exhibit switching transients that may also cause EMI unless special transient-control measures are taken.

In designing equipment containing mechanical switching devices, care must be taken to maintain operational capability of contacts and at the same time suppress EMI generated by switching. Often a gross approach to arc suppression is taken, some standard technique is applied, and a recommended component is used without consideration of all aspects of the problem. The most direct solution, however, may lead to the generation of additional problems. Ramifications such as changes in circuit reliability, cost factor, weight, and system effectiveness must be considered. The system, in fact, may become overdesigned to the point that one loses sight of initial design objectives. Some parameters such as cost factor become serious and intolerable in large systems. For this reason, recognition of the problems of arc suppression is especially important in such systems.

Suppressive networks can be optimized by considering reliability, cost, physical size, weight, and other factors. Special networks can be designed for arc suppression. Such networks may accomplish this by delaying or eliminating any transients that occur during the switching of relay contacts. When the interrupted currents are large, the only effective means of limiting the interference might be to use short leads, to filter, and/or to shield as much of the switching circuit as necessary.

The thyratron-like latching characteristics of the thyristor make it ideal for eliminating interference due to contact opening. Reverse-blocking triode thryistors, commonly called silicon controlled rectifiers (SCRs), and bidirectional triode thyristors, usually referred to as triacs, are the most common types used in switching and power control. Varistors can be used to clamp transients due to switching, though care must be taken to match the energy-handling capability of the device with the anticipated energy in the switching transient.

An SCR can turn off only when the AC current through it naturally reaches zero in the process of reversing polarity, regardless of the load power factor. Because it requires no separate and sophisticated turn-off circuitry, the SCR does not interrupt current abruptly as mechanical contacts do. Instead, it opens the circuit as soon as current reaches zero after the gate drive has been removed. Circuit disturbances are minimal when current is interrupted at this instant. Turn-on logic can be developed from the opposite half-cycle. The turn-on gate control signal can be applied while the SCR is reverse-biased, but it will not begin to conduct until it begins to be forward-biased.

Because an SCR can conduct in only one direction, full-wave operation calls for two SCRs connected in reverse parallel with gate control circuitry arranged accordingly. Alternatively, a triac, which effectively functions like two SCRs in reverse parallel, can be used.

Thyristors are not entirely free of EMI effects, however. At turn-on time, a voltage spike is produced as forward-bias voltage passes through the forward voltage breakover point. At turn-off time, assorted carrier charges produce a current spike until a depletion region is established. Furthermore, a thyristor gate element is susceptible to EMI of high dV/dt. Capacitance charging currents can cause the gate to turn on the thryistor even though the magnitude of the gate voltage does not reach the rated triggering level.

Relays.   A relay is a device that permits one or more circuits to be remotely switched by electrical variations in an independent control circuit. There are several types of relays: electromagnetic, saturable reactor, bimetallic or reed, semiconductor, photosensor, and others. There are many designs, depending upon switching to be performed, power source available, number and type of contacts, and cost.

The most common relay is the electromagnetic solenoid type. EMI problems occur in both the actuator and contact circuits. The electromagnetic solenoid has a large inductance due to the large number of turns and iron mass in the core and armature. When the coil circuit current is interrupted (i.e., deenergized), the collapse of the magnetic field generates a voltage equal to $-L(dI/dt)$. This reverse potential can reach 10 to 20 times the supply voltage in the order of a microsecond, and then decay at a rate determined by the inductance, distributed capacitance, and resistance of the armature winding circuit. The high amplitude voltage surge has a steep wave front that can cause arcing at the point of interruption, along with broadband signals capable of conducting and radiating interference to other circuits. EMI effects of an AC relay are variable because the voltage and current are continually changing in magnitude, producing results according to the momentary state at the time of switching.

Abrupt changes in circuit current will produce waveforms that cause EMI. Ideal contact operation occurs when the contacts go from fully open to fully closed, or vice-versa, without arcing. In actual practice, when contacts close, they bounce and form a closure followed by one or more openings and reclosures. In small relays, a single bounce may occur 10 to 50 μsec after the initial closure; final closure may take 10 or more μsec. In large relays, the bounce may be repeated several times at intervals of a few milliseconds.

Relay arcing can occur when contacts are first opened, continuing until contact spacing is too large to maintain the arc (depending upon the surrounding atmosphere and the applied voltage). The make and break of a contact arc in opening a circuit is proportional to the instantaneous supply voltage, the circuit inductance, and the rate at which the contacts physically separate. EMI may be worse for contact closure than for contact opening. Closed contacts can open or vary in contact resistance because of shock, acceleration, or vibration, causing arcing or at least circuit current changes. If required, contact arc suppression is used.

In digital systems, contact bouncing and erratic opening cause errors in data; so polar or latching relays are often used in such systems. A polar relay, once actuated in a given direction, will remain latched on its internal permanent magnet until its state is reversed by current in the opposite direction.

Transient Suppression Techniques for Relays (8).   The use of a parallel resistance to suppress EMI in a relay circuit is shown in Fig. 16-13 (A). The circuit is not polarity-sensitive and provides relatively good suppression, depending upon the value of resistance used. The effect of relay armature release time (dropout time) is also a function of resistance. The dropout time constant is equal to $(R + R_L)/L$ where R_L and L = relay resistance and inductance, and R = parallel resistance across the relay. The disadvantage of using a single shunt resistor is that it consumes power continuously when the relay coil is energized. Since R effects the dropout time, it will significantly affect the time differential when $R > R_L$.

In Fig. 16-13 (B), the RC EMI suppression circuit limits the transient voltage surge when the relay resistance, R_L, is negligible. The capacitor C is typically chosen to be 0.1 to 1.0 μf with a voltage rating of approximately 20 times the maximum DC input voltage. The use of a capacitor alone will result in a large charging current during relay energizing which may damage the switch contacts or cause transient-noise current surges. Thus, the current-limiting resistor, R, is necessary.

The series–parallel combination of capacitance and resistance shown in Fig. 16-13 (C) is a combination of the above two circuits. Since it offers no particular advantage over either circuit, it is not often used.

Note: A Resistor and Capacitor in Series May Be Used Directly Across Relay Contact(s)

TYPES OF INDUCTIVE SUPPRESSION	VOLTAGE INPUT	RELAY CONTACTS		REMARKS
		CLOSING	DROPOUT	
A 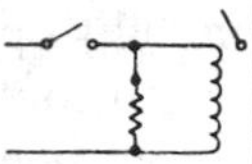 RESISTANCE DAMPING	AC or DC	NO EFFECT	FUNCTION OF RESISTANCE	Increase in power consumption. Resistance should be as low as practicable. Observe power rating E^2/R and heat dissipation.
B 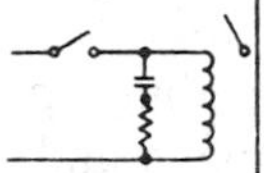 CAPACITANCE SUPPRESSION	DC	SLIGHT EFFECT	SLIGHT EFFECT	Need series resistance of a few ohms. Capacitance value around .01 to 1μf. Capacitance rated 10 times input voltage.
C 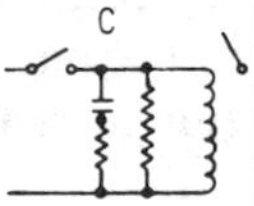 RC SUPPRESSION	DC	SLIGHT EFFECT	FUNCTION OF RESISTANCE	Combination of A and B above.
D 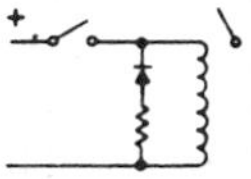 DIODE SUPPRESSION	DC ONLY	NO EFFECT	SLIGHT EFFECT	Polarity critical, diode put in backward or nonconductive direction. PIV should be higher than any transient voltage plus safety factor. Series resistance of a few ohms might be needed to increase inductance life.
E 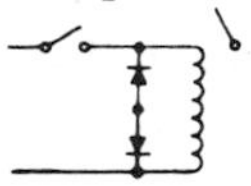 BACK-TO-BACK DIODE SUPPRESSION	AC	NO EFFECT	NO EFFECT	Avalanche voltage should be above input voltage. Power dissipation should be sufficient for transient current. Cost of device is much greater than any of the above divices.

Fig. 16-13. Various techniques for suppressing EMI transients in relays. (*Courtesy of Interference Control Technologies*)

The circuit of Fig. 16-13 (D) results in a polarity-sensitive relay. When the actuating switch is closed, the diode is effectively an open circuit because it is back-biased. Thus, its series resistor is out of the circuit. However, when the switch is opened, the collapsing magnetic field in the relay coil develops a reverse voltage, $V_r = -LdI/dt$, and current flows through the diode. The voltage is limited to the forward diode voltage drop due to the internal resistance on the diode and the voltage drop across resistor R. The peak inverse voltage (PIV) rating of the diode should be higher than the maximum applied input voltage or any transient voltages, and should include a sufficient safety factor. The addition of a series diode (an option) to the shunt-diode suppressed coil circuit provides the added advantage of inadvertent polarity reversal protection by the elimination of excessive shunt diode current.

Most germanium diodes exhibit a low forward resistance and voltage drop that minimize the magnitude of the interference voltage transients. However, silicon diodes are usually used because of cost, current ratings, and high PIV considerations. When R is greater than the effective forward resistance of the diode (typical case), the resistor becomes the primary transient suppressor, and the diode acts as a one-way switch. Thus, the resistor does not consume power when the coil is energized. This circuit provides a small voltage backswing, and there is an increased dropout time: $(R + R_L)/L$. The resistance, R, must be selected to provide the desired voltage backswing. Using a pulldown voltage on the diode will increase the backswing but decrease the dropout time.

The addition of a transistor with the relay coil in the collector circuit increases operating sensitivity and provides significant EMI isolation to transients. High-level switching of the contacts is still retained.

Back-to-back zener diodes, as shown in Fig. 16-13 (E) are effective for both AC and DC circuits. When the actuating switch is opened, the high inverse voltage causes one of the diodes to break down because of the zener effect, and the voltage surge is limited. This method is a compromise between the RC network and the single diode for backswing magnitude and dropout time. Transzorbs can be used as suppressors in place of zeners, as they exhibit I-V characteristics similar to zeners, but can handle more energy.

Other EMI Problems and Control Techniques for Relays.   EMI suppression is especially required for relays located in areas having susceptible circuits or equipment. Power and signal leads must be isolated, twisted, and/or shielded to avoid magnetic and electric coupling (see Chapter 6). Filters should be used on conductors as necessary at points of entry into an enclosure. Signal circuits may have to be shielded and the shields grounded. Where magnetic fields from the relay coil may constitute a potential source of EMI to nearby low-impedance circuits or wiring, the relay should be magnetically shielded.

Solid-state relays now are widely used, only a few years after their introduction; and they are used for the same reasons that transistors and ICs are preferred by designers—because of their small size, reliability, and low power dissipation. Since they have no moving mechanical parts, arcing is inherently prevented, with the result that no EMI is introduced from this source. Thus they have the added advantage of permitting switching in explosive atmospheres.

Many solid-state AC relays are designed so that the relay contacts are closed at the time of zero-axis crossing of the AC supply voltage (called zero-voltage switching). Here the transient is significantly reduced. Additionally, many solid-state AC relays have provisions for turning off at a zero-current crossing of the AC line (called zero-current switching). This eliminates the $-LdI/dt$ arcing problems associated with deenergizing inductive loads.

If the line voltage transients exceeds the maximum contact voltage, the relay contacts will close until the current waveform passes through zero, or for a maximum of one-half an AC cycle; then they will reopen. While such transients will not damage the relay, back-to-back zener diodes can be used across the contacts for suppression. If a dV/dt transient across the contacts exceeds a certain rate, this will cause the contacts to close for one-half AC cycle. Reclosing due to inductive load switching is avoided by an internal RC network across the contacts.

Typical contact resistance (ratios of voltage drop to current through contacts) of a 10-ampere, solid-state relay is about 100 milliohms and rises with a decrease in load current. The off-state leakage is a few milliamps measured at a DC voltage equal to the RMS voltage for which the output is designed. The maximum load current is generally rated at room temperature and smoothly decreases with an increase in temperature until it is completely derated to zero at about 100°C.

INSULATORS AND CONDUCTORS

This section reviews the topics of insulators and conductors as potential sources of EMI and some of the control techniques that are employed to suppress interference. Details regarding both the conductor physics and EMI control methods are covered in Chapter 6.

Insulators

Insulators permit small leakage currents to flow despite their low conductivity. This may lead to voltage surges and changes in current. The insulation around a cable illustrates this phenomenon. When corona occurs, insulation breaks down, and small currents creep along the surface.

Dielectric loss, called loss tangent or power factor, in insulators causes dissipation of power, loss of linearity, and coupling with other circuits. It is especially important in coaxial lines and capacitor construction. This problem is eliminated by selecting proper materials, such as polyethylene or Teflon, for the frequency range of interest.

Surface tracking, a condition in which small currents creep across the insulator, is an important source of EMI. It is caused by surface contamination of the insulation by moisture or solid conductive particles, by chemical degradation of the insulation material, or by momentary overvoltage. When a discharge occurs across a low-impedance circuit, significant energy may be released, and quick catastrophic degradation occurs. If the circuit impedance is high, a slow discharge occurs over a long time interval. The arc changes paths continuously during this interval, extinguishing in one section but igniting in another. These path changes are fast and variable, producing a broad spectrum of interference.

Corona differs from a discharge across insulation because it is highly concentrated at one point, usually shows a visible glow, and has a humming sound, caused by highly ionized air around the conductor. Special cases of corona may develop in gaseous pockets or voids within solid insulation, usually showing no glow. This also causes EMI, as it develops slowly and leads to insulation failure. Internal breakdown of the insulation produces small changes in current as in surface tracking. Small internal paths are overheated, causing chemical changes that results in lower local resistivity and accelerated damage. High-voltage gradients sometimes occur across small mechanical voids in an insulator. Gaseous discharges also occur in the voids, with general current changes leading to degradation.

Some typical insulator problems, causes, and remedial measures are shown in Table 16-4.

Table 16-4. EMI problems, causes, and corrections in insulators. (Courtesy of Interference Control Technologies.)

Problem	Cause	Preventive measure
Surface tracking causing dI/dt changes which produce wide band of frequencies.	Surface contamination, chemical degradation of insulation, momentary overvoltage.	Protection from contamination, use of proper material, proper voltage deisign.
Surface tracking resulting in catastrophic degradation	Low impedance discharge circuit releasing high energy.	Protection from contamination, use of proper material, proper voltage design.
Surface tracking with small but relatively persistent EMI.	High impedance discharge circuit which limits energy to circuit component tolerances.	Protection from contamination, use of proper material, proper voltage design.
Glow discharge which may be visible and audible but causes electrical noise because of dI/dt effects.	Corona with high voltage across one conductor to ground or to another conductor.	Prevention of voids within insulators, protection from contamination, prevention of sharp points at high potentials, low voltage gradient design.

Conductors

A conductor is any material that readily permits an electrical current to flow when subjected to a difference in potential. It is usually desirable to use a low-resistance metal such as copper as the conductor in an electronic circuit. Copper is inexpensive and relatively stable in the ambient temperature range usually encountered. It is easily soldered but will corrode on exposure to the atmosphere. For this reason it is sometimes protected with a plating or coating of tin, silver, or gold.

The selection of a conductor size is generally related to the maximum allowable voltage drop in the conductor or the heating effect by power (I^2R) loss. At RF, skin effect must be considered. Skin effect is the term used to describe an uneven cross-sectional distribution of current density in which there is a concentration of current near the surface or skin of a conductor. This results in a higher conductor resistance at RF. Litzendraht, or litz, wire was designed to reduce skin effect. It is composed of several strands of enamel-coated wire, individually insulated, interwoven, and connected in parallel at each end. Skin effect combined with the effect of flux linkages between a conductor and its return above about 1 GHz makes conduction within a solid conductor very difficult. Accordingly, above 1 GHz (especially above 10 GHz) a transition is often made to waveguide for signal transmission. Waveguide as a conductor is effective up to about 300 GHz.

Skin effect is also reduced by other means. Since a circular-cross-section conductor has the least skin surface per unit of area, it may be advantageous to make the conductor rectangular. Flat, strap, or foil-type conductors have inherently lower AC resistance because of their greater surface per cross-sectional area. Alternatively, removing the center of the conductor and creating a hollow tube will significantly reduce conductor self-inductance. For this reason, tubular conductors are commonly used in RF ground systems and in high-power transmitters.

Parallel wiring systems may be loaded up to a point where the wire fuses or where corona is probable, whereas arcing limits the power in coaxial and waveguide systems. In general, parallel lines exhibit less loss than equivalent coaxial lines. However, coaxial lines have less tendency to radiate. Shielded wire is also relatively more expensive. The attenuation offered by shields is due to energy reflection at the boundary and absorption within as external signals pass through. A typical voltage measurement between an external pickup wire and the central conductor of a coaxial line shows a cross-talk of about 60 dB for the geometry used (see Chapter 6). The use of a double shield may increase shielding effectiveness by about 25 dB for some geometry, and critical circuits may require even more shielding.

Braided sleeving is sometimes used as a bond across a vibration mount because of its flexibility. However, due to its construction, it has a higher inherent inductance than a strap made of solid copper sheet. Braided sleeving also makes a convenient cable shield, but it exhibits a limited effectiveness. For low frequency and especially low-impedance applications, an overall insulating cover should be used so that the braid can be grounded at one point and insulated from ground over the rest of its length. As a high-frequency circuit shield, it can be bare. However, when it is unprotected, the eventual corrosion between intersecting strands will degrade shielding effectiveness.

In many ways Teflon is an ideal insulating material for wires in cables. It offers high insulation with good mechanical protection in extremely thin conductor coatings. This allows a high-density wire within a cable or harness, which in turn may increase circuit-to-circuit coupling within a cable because of closer spacing. Capacitive coupling is a function of the distance between conductors, the length of the conductors, and the dielectric constant of the material between them. Since capacity is proportional to dielectric constant, the capacity will increase by a factor of two when the air between two wires is replaced with Teflon.

Various methods may be used to minimize EMI in conductors:

- Multiple-conductor cables that bundle each conductor with its return conductor carrying current in the opposite direction result in magnetic-field cancellation. The amount of cancellation depends upon the relative equality of the currents and the spacing between conductors.
- When two conductors with similar current levels are twisted together, the field generated by one tends to cancel the field of the other if the currents are opposite in direction. The greater the number of uniform twists, the more effective is the cancellation of magnetic fields (see Chapter 6).
- A conductor surrounded by a return path shield, as in a coaxial cable, theoretically will not have an external magnetic field. Factors degrading this are the lack of solidity of the surrounding conductor and its finite conductivity (see Chapter 6).
- Skin effect at higher frequencies has an increasing effect on electric and magnetic fields. At higher frequencies, hollow conductors should be considered to minimize external fields.
- Magnetic shields can be used to minimize magnetic field penetration. Eddy currents that occur in grounded shields create absorption losses that minimize the external magnetic field. (The absorption losses are discussed in Chapter 10.)
- Bare conductors operating at high-voltage potentials can produce large electrostatic fields at sharp corners or points that ionize the surrounding atmosphere, resulting in broadband, white-noise corona. The minimum bend radius for a high-voltage conductor should be used.
- Conducted interference can be limited by coating conductors with high-permeability material that magnifies skin-effect losses.

- Ferrite beads are tiny cylindrical beads that may be strung on a wire to increase inductance of the wire and thereby cause attenuation of signals due to filter action. Attenuation by inductive reactance and I^2R losses is a function of frequency because there is no DC current in the bead.
- Lossy flexible sleeving made up from ferrite particles in a suitable binder is a continuation of the ferrite bead and rod concept. Above about 10 MHz, conducted EMI is converted to heat, and attenuation becomes significant above about 50 MHz.

CONNECTORS

Connectors, the interfacing and mating devices between cables and harnesses or between cables and other devices or equipment, were discussed at length in Chapter 7. The reader is referred to that chapter for details, but some aspects of the subject are summarized below.

Connnectors can cause EMI by circuit geometric discontinuities when they:

1. Have poor mating contacts, which develop varying contact potentials resulting in broadband noise.
2. Result in unshielded inductive loops and small capacities that interfere with sensitive circuits.
3. Constitute local circuits of inductance and capacitance having natural resonant and anti-resonant frequencies in the pass band of other circuits.
4. Constitute lumped impedance discontinuities (VSWRs) at certain points in the circuit, which cause reflections and standing waves.
5. Provide insertion loss to applied signals.

Ideally, connectors should have:

- Negligible resistance.
- Chemically inert surfaces.
- Resistance to gouging.
- Foolproof alignment, to minimize contact damage.
- Adequate force between contacts.
- Little friction, to minimize increase in resistance with use.
- Contamination-free design.
- Provisions for proper connections, including shielding of backshells.
- Proper dielectric properties.
- Moisture-proofing as required.
- Resistance to degradation due to age, wear, maintenance, and repair.
- Compatibility regardless of varying intersystem contractors.

There should always be proper installation, including a good bond between cable shield(s) and connector shell. Shields should be bonded completely around the periphery of the connector body. All connectors used as conducting paths for EMI should be bonded to the static ground with DC bonding resistance of the order of 1 milliohm or better. Air gaps at the connector–chassis interface should be eliminated by means of woven-mesh EMI gasketing. Other desirable features are protective coverings that extend over the male pins to reduce pin damage, the use of caps on unused connectors, the use of clamps to hold wires steady, contact materials designed for long life and proper pressure, and no loose of faulty contacts that might generate EMI.

Filter pins may be used where interference is in the VHF and UHF range, but most filter pins are not effective below 1 MHz. Use feed-through capacitors or filters mounted in a connector box where conducted interference is below 1 MHz.

SEMICONDUCTORS

This section reviews two types of semiconductors: diodes and logic. Some of the principal EMI emission and susceptibility characteristics of each are discussed.

Inasmuch as solid-state diodes are also semiconductors, they share many of the characteristics of transistors. Diodes act as both sources of EMI and devices that are useful in suppressing EMI. The latter topic is discussed in a previous section on using diodes to suppress transients generated by relays and switches. This section emphasizes the diode as a source of EMI.

Under conditions of forward bias, a solid-state semiconductor stores a certain amount of charge in the form of minority current carriers in the depletion region. If the diode is then reverse-biased, it conducts heavily in the reverse direction until all of the stored charge has been removed. The resulting conditions are presented in Fig. 16-14. The duration, amplitude, and configuration of the recovery-time pulse (also called switching time or period) are a function of the diode characteristics and circuit parameters. These current spikes generate a broad spectrum of conducted transient emissions.

Rectification involves switching from conduction to cutoff repetitively. This causes high dI/dt values dependent upon the input frequency, minority carrier storage in the diode, and circuit characteristics. The interference effect can be minimized by one or more of the following measures:

- Placing a bypass capacitor in parallel with each rectifier diode.
- Placing a resistor in series with each rectifier diode.
- Placing an RF bypass capacitor to ground from one or both sides of each rectifier diode.
- Operating the rectifier diodes well below their rated current capability.

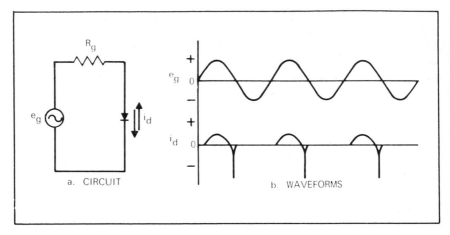

Fig. 16-14. Diode recovery periods and spikes. (*Courtesy of Interference Control Technologies*)

Diodes are also used to switch at a particular voltage level. The switching action results in high values of dV/dt. Diodes are used as limiters or clippers to cut off input waveforms at a certain level. The greater the amount of limiting, the greater will be the number of spurious frequencies that occur due to the steepening waveform. These switching actions can result in EMI interaction with other components, distributed or actual impedances, and signal discontinuities.

Several design considerations minimize switching EMI transients:

- Operate at the lowest possible voltages and currents.
- Anticipate diode-to-diode variation of characteristics.
- Use the lowest possible switching rate or rise-time and amplitude.
- Select diodes with high working and peak inverse voltages.
- Use diodes with a slow recovery time, called soft recovery diodes (inherent with larger current ratings).

RF voltage can change the bias of a diode, resulting in improper switching, distortion, or improper output. All diodes are subject to reverse breakdown if they are exposed to RF voltages greater than their reverse breakdown voltages. Low-power devices (generally those rated 25 mW or less) and small junction devices, such as point-contact diodes, operating in the vicinity of a strong RF field can absorb sufficient radiated energy to be degraded or burned out. Large junction diodes have a large junction capacitance, on the order of 10 to 15 pf that will pass high frequencies. If RF energy, added to the normal energy, exceeds the thermal limit of the device, damage can occur. Therefore, diodes subjected to RF fields should be shielded.

When used as an amplifier, a tunnel diode may couple with related circuitry inductance or capacitance to produce parasitic oscillations, usually about 1 MHz. The oscillations should be suppressed by circuit design methods. All zener diodes generate shot and $1/f$ noise, but the noise level is higher in alloy zeners than in zeners made by a diffusion process. Generally, noise increases with an increase in current, but the noise may occur at some points on the zener curve and not at others (this noise is called spotty). Most commercial zener diodes exhibit noise levels from 1 µV to 1 mV.

LOGIC

This section discusses both EMI emission and susceptibility of different logic families.

When compliance to an EMI specification is one of the design goals, the EMI performance of the different IC logics enters into decisions regarding logic selection. These decisions are also necessary if the susceptibility performance of the logic is a design question.

The choice of the best performing logic from an EMI standpoint will not eliminate packaging and containment problems, but it will make decisions regarding packaging and shielding less critical with respect to achieving compliance.

Table 16-5 shows the results of radiated tests involving six popular logic families. The tests were performed on single chips, and the radiation is proportional to the IAF product where I is the current around the loop, A is

**Table 16-5. Broadband radiated field strength
from six types of digital IC logics.
Radiation level from single 14-pin chip.
(Courtesy of Interference Control Technologies.)**

Logic Family	Clock Rates 1 MHz	3 MHz	10 MHz	30 MHz	Harmonic ※	90 MHz**
CMOS	−48	−36	NA	NA	30	−51
LP-TTL	−27	−17	−6	NA	9	−4
TTL	−25	−15	−4	NA	9	2
LS-TTL	−29	−19	−8	NA	9	−1
S-TTL	−19	−9	2	17	3	16
ECL-10k	−41	−31	−20	−5	3	−6

ASSUMPTIONS: FOUR ON-CHIP DRIVEN GATES, TWO OF WHICH (50%) TOGGLED.
SIX OFF-CHIP DRIVEN GATES, THREE OF WHICH (50%) TOGGLED.

**RADIATION AT 90 MHz FOR INDICATED HARMONIC

the loop area, and F is the frequency of radiation. If the relatively small physical size of each digital circuit is considered, as well as the limited number of integrated circuits used and the relatively limited wiring to serve as effective radiators, the radiated amplitudes generated by all the logics are at a relatively high level.

By examining Table 16-5, we can reach some conclusions concerning logic selection. Since ECL-10k is a high-speed logic, and since it performed comparably to CMOS (2) (a lower-speed logic) but better than the other logics, selection of this logic in many design situations would appear to be a correct engineering decision. The table shows that the Schottky TTL logic gives about 20 dB higher EMI levels than either CMOS or ECL logic. Other TTL logics are 12 to 16 dB noisier than ECL-10k, and 19 to 23 dB noisier than CMOS. One must choose a logic that will meet the operating frequency/speed requirements but is not so fast that radiation problems may result. It would seem that TTL logic should be avoided where EMI noiselessness is critical, but other considerations must be taken into account such as cost and compatibility (a big factor considering the quasi-standard "TTL-compatible"). We can conclude that all IC logics are potential EMI sources, and that the source problem can be minimized by proper logic selection but not eliminated. In general, it is how the logic components are wired together that determines EMC performance.

If the decision to use a certain logic is firm, the options to reduce the radiation depend on reducing that the area in the IAF product. This can be achieved in the following ways:

- Convert from DIPs to flat packs.
- Use low-profile chip carrier sockets.
- Use chip packs soldered into place.

Logic Susceptibility

The reverse of logic noise emission levels is noise susceptibility. It is more often referred to as noise immunity and is the topic of discussion in this section.

The noise margin and noise immunity parameters of gating circuitry used in digital devices are measures of the gate's susceptibility to noise signals. Noise margin is defined as the magnitude in volts of the pulse noise that, when appearing at the input of a digital gate and riding on the worst-case logic level, will cause an undesirable change in the output of the gate. The DC noise margin, as in wide, slow-rise-time noise pulses, is the difference between the worst-case logic voltage level and the worst-case switching threshold voltage of the gate. The transient or AC noise margin, as in fast-rise-time noise pulses, is, in most cases, equal to or greater than the DC noise margin.

Noise immunity is a measure of a gate's immunity to noise generated by neighboring gates, which is determined as follows:

$$\text{Noise immunity} = \frac{\text{Worst-case noise margin}}{\text{Maximum logic voltage swing}} \times 100\% \quad (16\text{-}3)$$

In designing circuits for noise immunity, the inherent noise immunity of the selected logic family should be considered in addition to the usual topics of cost, speed, availability, and compatibility. Noise-immunity comparisons must not be made simply on voltage but also on the basis of the total energy required to trigger a false signal. The amount of voltage developed upon a gate-to-gate signal line is directly related to the output impedance of the device driving the line, the current developed by the noise source, and the input impedance of the logic under consideration. However, this noise voltage must be present long enough to be propagated through the circuit in order for a spurious signal to be generated at its output. Hence, energy becomes the chief criterion.

Radiated noise usually enters logic circuits by capacitive coupling. The noise amplitude of the emission source may reach several thousand volts and be of high impedance. Consequently, the noise appears to be coming from a constant-current source. For this reason, a device with a high-voltage threshold and a relatively high output impedance, such as CMOS, may not have an energy immunity greatly different from one with a relatively low-voltage threshold and a low output impedance, such as TTL. Thus, voltage-noise-immunity specifications published by some manufacturers may not tell the whole story of the important energy considerations.

Another factor is speed of the logic family. Since the noise is capacitively coupled, higher-frequency signals are especially likely to be imposed on signal lines. This means that the slower the response time of the logic, the greater the immunity to noise. This is a result of the large bandwidth necessary for high-speed logic to operate, which results in pickup of many frequencies of EMI. This consideration can often eliminate high-speed logic such as ECL, unless a proper PC board layout is followed.

In essence, the energy–noise immunity of a logic family is a result of its input voltage threshold, the current necessary to force a signal line driven by another gate to that threshold, and the time it takes a gate to respond to the noise pulse, which is a function of the rise-time of the logic.

Part A of Fig. 16-15 shows the calculation of the noise rejection to a pulse for TTL. The manufacturer specifies an AC noise margin and DC noise

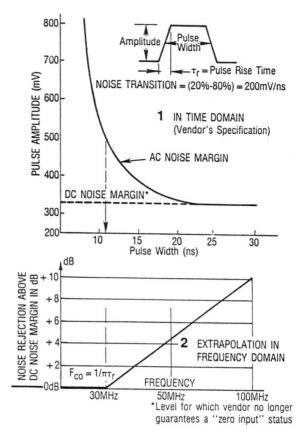

Fig. 16-15. Example of noise rejection of a logic circuit (LS-TTL). (*Courtesy of Interference Control Technologies*)

tolerance (defined as the level at which the manufacturer no longer guarantees "zero change" in the status of the driven gate). The shorter the noise pulse, the larger the amplitude of noise the gate can tolerate before producing an erroneous logic change.

Part B shows the extrapolation to the frequency domain where the cutoff frequency is defined as $F_{co} = (1/\pi\tau_r) \simeq 30$ MHz for TTL. The graph shows the increasing noise immunity (referenced to dB above the DC noise tolerance) of the TTL logic.

Table 16-6 lists the basic characteristics of the logic families and summarizes their DC noise margin or tolerance (7). For CMOS operating at 15 volts and with a rise-time of 50 nsec, the noise immunity is highest. Thus, if family noise immunity were to be the only criterion for choice, CMOS would

Table 16-6. Typical characteristics of various logic families. (Courtesy of Interference Control Technologies.)

Logic Families	Output Voltage Swing	Rise/Fall Time (ns)	Bandwidth (MHz) $1/\pi\tau_r$	Max V_{CC} Voltage Drop (V)	Power Supply Transition Current (mA)	PS Decoupling Capacitor* (pF)	PS Current*** Per Gate Drive (mA)	Input C (pF)	DC Noise** Margin (mV)
Emitter Coupled Logic (ECL-10K)	0.8V	2/2	160	0.2	1	350	1.2	3	100
Emitter Coupled Logic (ECL-100K)		0.75	420						
Transistor-Transistor Logic (TTL)	3V	10	32	0.5	16	2350	1.5	5	400
Low Power TTL (LP-TTL)	3V	20/10	21		8	400	1.6	5	400
Schottky TTL Logic (STTL)	3V	3/2.5	120	0.5	30	1500	4	4	300
Low-Power Schottky TTL (LS-TTL)	3V	10/6	40	0.25	8	3700	2.1	6	300
Complementary Metal Oxide Logic (CMOS) 5V or (15V)	5V (15V)	90/100 (50)	3 (6)		1 (10)		0.2	5	1V (4.5)
High Speed CMOS (5V)	5V	10	32	2	1	150	1	5	1V

* $C = \dfrac{(I \text{ GATE } \& \text{ I FOR 5 DRIVEN GATES}) \cdot \text{RISE TIME}}{0.2 \cdot \text{MAX } V_{CC} \text{ DROP}}$

$0.2 \times$ Max V_{CC} drop is to provide a -14dB safety margin for I/N ratio.

** DC noise margin = difference between V_{OUT} of driving gate and V_{in} required by driven gate to recognize a "1" or "0"

*** Peak instantaneous current that the driving device has to feed into each driven gate.

have a significant advantage. Another advantage of CMOS is its low power consumption. When TTL, with its low cost per gate, ready availability, compatibility, and fair noise immunity, is considered, the final decision sometimes becomes a toss-up.

MOTORS AND GENERATORS

EMI reduction design for electrical machinery is divided into four categories: interference reduction for large motors and generators, for alternators and synchronous motors, for fractional-horse-power machines, and for special-purpose rotating machines. Any rotating machine with sliding contacts should be regarded as a potential source of EMI because the switching and arcing processes of commutation cause rapid current and voltage changes that distribute energy through a wide frequency range. Commutation noise can be eliminated by the use of brushless motors.

Brushes

Brushes and brush leads are the most likely components from which EMI can be radiated or conducted. If a motor or generator is not adequately enclosed, then the brushes and brush leads may require shielding. Provision should be made in the original design of motors or generators for installation of capacitors at the brushes. Brush-generated interference may be reduced by incorporating design features based on certain critical parameters:

- *Brush pressure*: EMI decreases at all frequencies with increasing brush pressure, but increased brush pressure increases the rate of wear. Provision for more frequent brush replacement is often a reasonable compromise for decreased interference.

- *Current density*: EMI decreases with a decrease in current density. As the current density is increased, more heat is generated at the brush surface sliding on the commutator or slip ring, and this heat causes the formation of a thick oxide film on the sliding metal surface. Rapid variation in the sliding contact resistance, resulting from irregularities in this oxide film, causes high-frequency transients that produce interference. To offset the heat increase, a somewhat larger brush surface area than necessary should be designed. Such a design change will reduce heat and losses due to mechanical friction. On the other hand, if too low a current density is used, nonuniform grooves develop on the metal surface of the slip ring or commutator, and frequently increased friction due to the wider brush surface area sets the brushes into a noisy chatter.

- *Brush resistivity*: Brush materials of low resistivity are poor EMI generators and thus are desirable for use in EMI control. One example of such a brush is an electrographitic carbon brush of about 2-millohm specific resistance in machines being used at less than 50 volts. Low-resistance brushes are available with silver-, copper-, or cadmium-impregnated graphite. When it is used with a commutator, the resistance of the brush should match the requirements for good commutation. For use with slip rings, a wide choice of brush material is available because no switching action is involved.

DC Motors and Generators

Of all rotating machinery, DC motors and generators are the most serious offenders in generating EMI because the most common of these machines require commutators for their operation. Commutation is a switching action that is accompanied by interference-producing transients. When a switch is closed in an electrical circuit, the input impedance changes from practically infinity to zero. If the circuit contains inductance and/or capacitance, its voltages and currents cannot return to normal values instantaneously because energy stored in the magnetic field of the inductance (or in the electric field of a capacitance), cannot dissipate instantaneously. Initially, the changing voltages and currents develop steep wave-fronts that decay as a function of time. The bars of a commutator, sliding rapidly past the contacting brushes, produce a switching action that causes extreme variations in impedance, which, in turn, establish the series of voltage transients, or pulses, that cause EMI.

Measures can be taken in designing a generator to minimize the amount of EMI produced by commutator action. Reduction of commutation transients requires the use of design techniques to provide a smooth transition from one value of impedance to another within each armature coil. Interference produced as a result of commutation is reduced by six design techniques: interpoles; compensating windings; increased number of armature coils and commutator bars; laminated brushes; commutator plating; and the use of solid-state commutation.

Interpoles. A good way to improve commutation is by adding interpole windings. Interpoles counterbalance the self-inductance of the armature coils during the commutation period, and also reduce the induced voltage in the armature coils resulting from the coil-cutting fringing-flux during the commutation period. The use of properly designed interpoles produces a rapid change in the armature-coil current at the beginning of the commutating period, reducing the steepness of the transient at the end of the period.

Compensating Windings. To a lesser degree, compensating windings produce the same effect as interpoles, and, in addition, they help to prevent field distortion. They also assist in reducing cross flux produced by armature coils. The use of interpoles and compensating windings lessens critical brush positioning requirements with respect to the commutator, and provides EMF in the coils under commutation that oppose the EMF of self- and mutual induction in these coils.

Increasing the Number of Coils on the Armature. This technique (which thus increases the number of commutator segments, or bars) reduces interference by reducing the current broken per bar and the reactance voltage per coil.

Laminated Brushes. Good commutation can be achieved over a fairly wide range of brush positions, relative to the magnetic neutral, so that brush positioning becomes less critical and less dependent upon armature current. The design of laminated brushes should include two or, at most, three laminations (see Fig. 16-16). The following criteria should be incorporated in the design:

1. The thickness of the leading-edge lamination of a two-lamination brush should be about 90% of the total thickness, and its resistivity should be as high as allowable for heat dissipation.

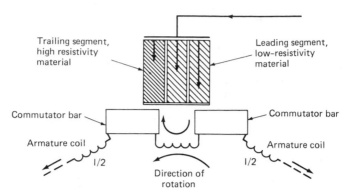

Fig. 16-16. Commutation of an armature coil by laminated brushes. (*Courtesy of Interference Control Technologies*)

2. The resistivity of the trailing-edge lamination should be about 15 times that of the leading edge; this lamination should be thick enough to preclude mechanical weakness.
3. A thermosetting cement of 6 mil thickness should be sufficient to provide electrical insulation between the sections. A cement that will preclude the formation of a smear of conducting particles from brush wear on the rubbing edge should be used; it should have a wear rate equal to that of the brush.
4. A brush with varying resistance characteristics from the leading edge to the trailing edge can be manufactured without the use of insulating separators and will act somewhat like a laminated brush.

Commutator Plating. A copper commutator, after several hours in contact with a carbon or graphite brush, develops a layer of copper oxide mixed with carbon particles (from brush wear). This copper oxide film introduces unidirectional electrical properties (polarity effects) as in a copper oxide rectifier. The oxide layer has a nonlinear resistance of higher value at the brush used as a cathode than at the one used as an anode. The cathode brush, as a result, passes current in discontinuous, high-current-density surges that cause EMI. Approximately ten times as much interference may result from the cathode brush as from the anode brush. Plating the copper commutator with chromium to a thickness of about 1 mil will reduce the EMI level from a cathode brush to that of a relatively quiet anode. No adverse effects will result from the platings; the hard chromium surface prevents threading and grooving of the commutator. The wear rate and sliding friction of many brush materials on chromium are of the same order of magnitude as those for copper.

Solid-State Brushless Commutation. Increased attention in machine design has been given to the placement of rectifier diodes in series with the field windings that perform the commutation necessary with DC motors. The current is then brought out via slip rings positioned on the end of the shaft. Although diodes are a source of EMI (discussed earlier in this chapter), the noise is many orders of magnitude smaller than brush noise. Also, maintenance is decreased in this type of motor, owing to elimination of brush replacement.

Other Design Features That Improve Commutation.   Other techniques also reduce EMI. The most effective and economical one is the installation of capacitors at the brushes. In generators, for example, installing capacitors (e.g., about 1.0 µf) at the brushes applies the remedy as close to the interference source as possible. The interference generated by the commutator and the brushes will be bypassed to the generator housing. The lead from the brush to the capacitor should be as short as possible, and the capacitor should be bonded to the generator housing to provide a low impedance path to ground for EMI currents.

Because of the combined interference-generating characteristics of the commutator and the brushes in a DC generator, an additional capacitor is installed at the output (armature) terminal. The preferred installation is a feed-through capacitor through the generator housing. The alternate installation is a 0.1-µf bypass capacitor, mounted externally, to maintain electrical contact with the generator housing and minimize the lead length between the terminal and the capacitor. Figure 16-17 illustrates the mounting of a bypass capacitor at the armature terminal.

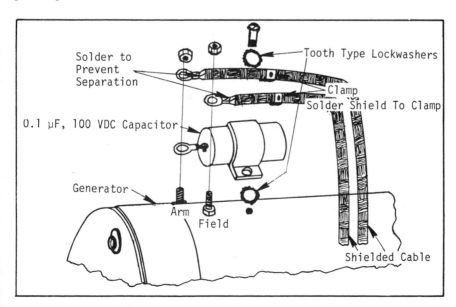

Fig. 16-17. Installation of an EMI bypass capacitor on a generator armature terminal. (*Courtesy of Interference Control Technologies*)

Shielding

Overall shielding is necessary to prevent the radiation of interference from within the generator. This shielding is afforded by the generator housing, which should be designed to provide maximum shielding effectiveness. Ventilation openings should be screened to prevent radiation of interference into space. No matter how perfectly a generator shield is designed, the shaft provides an exit path for interference because it must penetrate the shield. EMI should be bypassed directly to the generator housing by grounding the shaft through a brush riding on a special grounding slip ring (or riding directly on the shaft).

The last shielding consideration for a generator housing is to ensure good contacts and low-impedance paths between the three sections of the generator—the two end plates and the main housing. This is accomplished by the bonding and shielding practices discussed in earlier chapters.

Alternators and Synchronous Motors

Alternators and synchronous motors are very similar to DC generators and motors except that they supply or use AC, and therefore have slip rings instead of commutators. Commutator interference is absent in these machines, but there is EMI from the brushes and from the generation of harmonics. Brush interference is lessened because most alternators and synchronous motors have stationary armature and rotating fields; heavy power currents need not be supplied to the rotor. Only the much smaller field currents have to be supplied through the brushes. Because commutation is not involved in the section of brushes, a much wider choice in brush pressure, size, and material is permitted.

In AC generators, the generation of harmonics and the resonant conditions that create interference can be minimized. Production of as pure a sine wave as possible (an important consideration in the design of alternators) is especially important when EMI reduction techniques are considered. A comparatively small harmonic content may be quite tolerable from all points of view except that of EMI. In the reduction of harmonics, special attention must be given to certain factors, as explained in the following paragraphs.

1. *Flux distribution*: The most important factor determining the waveform of the generated voltage is the distribution of the magnetic flux around the periphery of the armature. Sinusoidal distribution, which produces the least amount of interference, may be achieved by chamfering the pole tips or skewing the pole faces.

2. *Symmetry*: In a perfectly symmetrical machine, all even harmonics disappear; therefore, special care must be exercised to construct identical pole pieces, to make the yoke and armature perfectly symmetrical, to produce a uniform winding on the armature, and to avoid all other irregularities.

3. *External connections.* In a three-phase alternator, the third harmonic and its multiples disappear at the terminals except when the machine is wye-connected and has its neutral grounded. In this case third harmonics are present in the voltage between any phase and neutral. This connection should be avoided, or, if it must be used, special attention should be given to the prevention of the third harmonic and its multiples.

4. *Distribution factor.* The distribution factor should be chosen to eliminate the lowest harmonic not eliminated by any of the devices mentioned in (2) or (3).

5. *Tooth ripples.* The generation of tooth ripples is greatly decreased by skewing, through one slot pitch, either the pole shoes or the armature slots. Tooth ripples may be eliminated altogether by making the number of armature-slots per pole-pair an odd number. The chord factors for the harmonics that are contained in the tooth ripples are then reduced to zero. Slip ring and brush materials should be such that interference is minimized. The design considerations applied to brushes and commutator surface materials in DC machines apply equally well here. The effects of brush bounce, due to vibration or irregularities of armature motion, can be minimized by the use of two or more brushes per slip ring.

Shielding of the alternator is incorporated in the design of its housing. Low-impedance paths between sections of the housing, provisions for bonding, and screening of all ventilating louvres are necessary if the overall interference-reduction design is to be effective. As in DC generators, no matter how perfect the shield, a means of escape from the shield for the interference currents is provided by the alternator shaft which penetrates the shield. The same procedures for shielding DC generators therefore apply to alternators.

The problems of interference suppression for alternators also apply to synchronous motors because they have the same basic components as alternators. A synchronous motor will operate as an alternator, and vice versa. An induction motor should be used instead of a synchronous motor whenever possible because of the lower EMI generated by induction motors.

The primary source of interference within a single-phase induction motor is the starting device. The starting winding is in series with a switch (or capacitor and switch) that is closed when power is off. When the motor reaches approximately 80% of its rated speed, the switch is opened (either by centrifugal force or by a solenoid coil), and a single pulse of interference is generated. This switch should be placed in a shielded housing, and the leads leaving the housing should be filtered.

Portable Fractional-Horsepower Machines

Portable fractional-horsepower machines include such equipment as portable electric drills and saws. Power is furnished by high-speed, lightweight, AC/DC

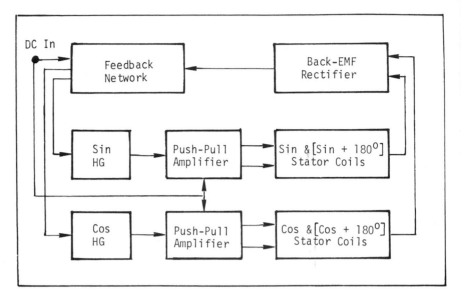

Fig. 16-18. Block diagram of the brushless D-C motor. (*Courtesy of Interference Control Technologies*)

or AC electric motors. Such equipment, using AC/DC (universal motors), is a major source of EMI because commutation is essential in its operation. As in DC motors, an effective, economical method of designing for reduced commutator-brush interference is to install capacitors at the brushes. In some portable AC/DC machines, restrictions of size and shape prevent the installation of capacitors at the brushes, and it is more feasible and economical to mount the capacitors in other parts of the equipment. Installing capacitors at the line side of the switch bypasses interference to the unit housing at the last point of exit to the power lines and prevents the interference from coupling back into an interference-free lead, and from being conducted into the power lines. If the mechanical design of the unit prevents the installation of capacitors on the line side of the switch, it is permissible to install them on the motor side. Shielding may be used to ensure that no interference couples back into the leads before they leave the unit.

Brushless DC Motors

As mentioned before, a new concept in miniature DC motors has been introduced, namely, brushless DC motors featuring solid-state commutation and the associated reduction of conducted and radiated EMI. The technique involves sensing the exact position of the rotor in relation to the stator by using two orthogonally located Hall-effect generators. If the rotor is constructed of a bipolar cylindrical magnet, the induced output in the Hall generators will be in phase quadrature as the rotor rotates.

Figure 16-18 shows a block diagram of the brushless DC motor. The DC source drives the Hall generators (HG) through the feedback and supplies power to the amplifiers. The sensed rotor position signals from the sin HG and cos HG are each amplified by a push–pull amplifier creating four sine waves in quadrature that drive the coil windings of the stator. The amplifier outputs are rectified and applied as back EMFs to the feedback network, which sets the value of the reference voltage applied to the HGs through another set of amplifiers to close the loop. Load, temperature, speed, and voltage compensation is automatically achieved in the process.

LIGHTS

This section describes EMI causes and control in incandescent, fluorescent, and gas lamps.

Incandescent Lamps

An incandescent lamp, once energized, is a fairly stable emitter of infrared and optical energy. Because of its relatively low temperature, comparatively little RF energy is emitted. Consequently, incandescent lamps generally do not create EMI problems. On rare occasions an incandescent lamp will develop a faulty filament that opens and closes, resulting in transient surges. The simplest EMI-control solution is to replace the lamp.

Fluorescent Lamps

Emissions from typical fluorescent lamps were discussed in Chapter 2. It was shown that radiated emission levels are significant up to about 10 MHz, and some types exhibit emissions up to 100 MHz and above. RF radiation from fluorescent lamps comes about as a result of a column of gas being ionized and extinguished 120 times a second for a 60 Hz power-mains supply. Thus, these transient surges result in broadband radiation from the bulb as well as conducted emissions back onto the power lines.

EMI is difficult to control economically in fluorescent lamps mounted in their fixtures. One technique is to shield the light-emitting areas from the lamp-mount fixture with either conductive glass (this is expensive) or wire screen. The AC lines feeding the fixture are also filtered to keep the 120 Hz conducted transients down to controllable levels.

Gas Lamps

EMI emissions from gas lamps are somewhat similar to those of fluorescent lamps. The principal difference is that gas lamps are energized from high-voltage transformers (typically furnishing 10 kV to the lamp), and the gas column remains ionized throughout the AC cycle. Thus, a steady, nontransient radiation takes place.

While EMI control of conducted and radiated emissions from gas lamps could employ the same techniques used for fluorescent lamps, this is rarely done. Most gas lamps, such as neon signs, are too big and cumbersome to shield. Although filtering of the AC power mains may help mitigate some resulting EMI situations, little is done to contain EMI from gas lamps.

REFERENCES

1. Air Force Systems Command. Design Handbook for Electromagnetic Compatibility, 1–4, March, 1984.
2. Bragg, N. L. "ECL for the TTL Designer." *Electronic Engineering*, October 1980.
3. Clark, Charles M., et al. "Electromagnetic Compatibility Principles and Practices." NASA NHB, 5320.3
4. Ginsberg, Gerald L. *A User's Guide to Selecting Electronic Components*. New York: John Wiley & Sons, 1981.
5. Naval Air Systems Command. *Electromagnetic Compatibility Manual*. NAVAIR 5335, 1972.
6. Wellard, Charles L. *Resistance and Resistors*. New York: McGraw-Hill Book Co., 1960.
7. White, Donald R. J. *EMI Control in the Design of Printed Circuit Boards and Backplanes*. Gainesville, VA: Don White Consultants, 1982.
8. White, Donald R. J. *A Handbook on EMI Control Methods and Techniques*. Volume 3, Gainesville, VA: Don White Consultants, Inc., 1973.

Chapter 17
EMI CONTROL IN ANALOG
AND DIGITAL CIRCUITS

The previous chapter emphasized identifying EMI problems and applying control techniques at the component level. It covered resistors, capacitors, inductors, diodes, transistors, transformers, switches, relays, motors, and generators. This chapter continues with EMI problem identification and control techniques, but at the circuit level. Included in this chapter are power supplies, electronic circuits, micro-circuits, amplifiers, SCR devices, and digital circuits.

The higher the echelon, the greater the ensemble of components and devices, the more difficult the EMI problems and the control techniques become. The last chapter illustrated many examples at the component level. It is essential that EMI suppression of noise sources, modes of coupling, and hardening to susceptibility, as applicable, be carried out successfully at the component level, since EMI problems at the next higher level will be multiplied. Correspondingly, failure to adequately identify EMI problems and effect their solutions at this level will render good EMI performance at still higher levels more difficult. These higher levels are the system and system ensemble levels. They include buildings, ships, aircraft, spacecraft, automobile, networks, and so on (2, 10, 11, 12). Intersystem EMI at the level of communications-electronics systems are treated in Volumes 3 and 5 of the DWCI EMC handbook series.

POWER SUPPLIES

EMI problems developing from power supplies originate from one or more of the following:
- Poor voltage regulation, especially under transient loads.
- AC fundamental and harmonics conducted to susceptible circuits.
- Transients and/or switching frequencies conducted to susceptible circuits.
- Fundamental, harmonics, and/or transients radiated or conducted to susceptible circuits.

545

Chapter 16 discussed the last three tropics when the power source is either a DC or an AC generator. EMI control by filtering of the source (Chapter 14) and shielding (Chapters 11 and 12) represent other approaches. While related to filtering, this section discusses EMI control when the power source internal impedance becomes a significant percentage of the load impedance of the equipment using the power source, that is, under heavy load conditions. Called poor voltage regulation, due to common source impedance, this problem can become particularly severe to common users of the power source under transient on–off load conditions. Of related concern in printed circuit board design is as small an internal impedance of the power distribution system as possible. Common source impedance problems cause intra-circuit EMI leading to false triggering, latch-up, and other problems for the logic designer.

Regulated Power Supplies

EMI problems resulting from common source impedance are one of the most prevalent sources of conducted interference. The output voltage of all sources of electric power varies with load and other factors; so the source should exhibit a zero internal impedance. Commercial power system techniques to hold the supply in the power plant and on the distribution system supply voltage within limits make use of tap-changing transformers, load switching, and power factor correction techniques. The line voltage variations result in entirely unsatisfactory supply voltages for communication equipment, industrial controls, and test equipment for research and development laboratories. Often, the end result of these switching and distribution techniques is to introduce transients on the line, thus adding to the total problem.

To provide closer regulation for sensitive equipment, four basic types of units or circuits are employed either directly, jointly, or indirectly via close control of other circuit components (4). These units, shown in Fig. 17-1, are: reference voltage units, magnetic voltage-control units, series-impedance control, and shunt-impedance control. They are also referred to as linear power supplies, as opposed to the various types of switching or nonlinear power supplies. The reference units are typically the zener diode. A simple regulator can be constructed by connecting a zener in series with a suitable resistor across a source of DC. The zener will maintain a constant reference voltage. For very small loads these units alone will often serve to produce satisfactory DC voltage regulation. For larger current outputs, they are bridged across the load circuit to serve as a standard or reference, as shown in Fig. 17-1d. When they are used in this manner, the deviations of the output line voltage appear across the resistor R_x and thus shows the change or error. This error voltage is applied to control units to correct for the deviation, as

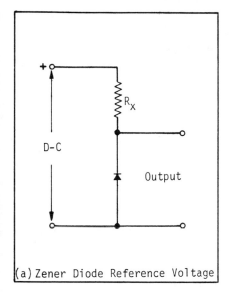

(a) Zener Diode Reference Voltage

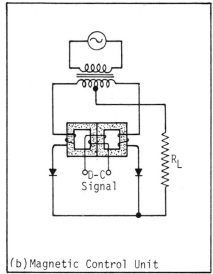

(b) Magnetic Control Unit

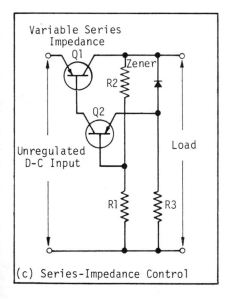

(c) Series-Impedance Control

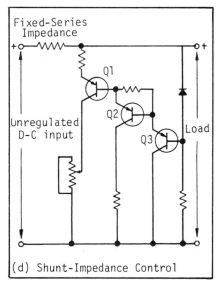

(d) Shunt-Impedance Control

Fig. 17-1. Some basic types of voltage regulators. (*Courtesy of Interference Control Technologies*)

shown in Fig. 17-1b. Zener diodes are available in voltage ratings from 2 to 1500 volts.

Magnetic voltage control units are of three types: constant-voltage transformers, voltage regulator units, and magnetic amplifiers. The constant-voltage transformer maintains a nearly constant secondary voltage under varying load conditions. This transformer may be used as a separate unit for regulating voltage, or it may be used in conjunction with DC rectifiers that have additional control circuits for fine regulation. The magnetic control units are usually a part of the complete AC/DC power supply unit. Both the constant voltage control units and the magnetic control units use similar principles for voltage control.

The magnetic amplifier principle is often used to control the magnitude of the DC voltage and current during the rectification process. One form of the magnetic amplifier as applied to single-phase full-wave rectification is shown in Fig. 17-1b. The amplifier consists of two rectangular iron cores placed adjacently. These cores are constructed of special iron of high permeability having square hysteresis loops and a cross section that assures saturation for normal operation of the equipment. The rectified current from the outer ends of the center-tapped transformer is led through series coils on the outer legs of the iron cores. These coils tend to produce a high reactance to the flow of DC pulses of power. A third coil consisting of many runs of fine wire is placed to surround the inside legs of both magnetic circuits. A variable DC signal control current is applied to this third coil.

If the DC signal control current has a magnitude sufficient to saturate the magnetic cores, the flux in these cores will be nearly unaffected by the rectified current pulses in the series coils. Accordingly, under this condition there is almost no reactance drop, and the only restriction on the rectified current will be the ohmic resistance of the series coil. For a DC signal current value of zero, the rectified pulses of current in the series coils will result in a very high reactance. For DC signal values between the two limiting conditions, the resulting signal flux will follow the magnetization curve for the iron cores. The combination of the ampere-turns of the DC signal coil and the series rectifying coils will produce a varying degree of reactance in the series circuits. Thus, a very small DC signal current in the control winding is very effective in controlling the rectified voltage and current on the load R_L.

The DC output voltage of a rectifier may be controlled by inserting a variable series impedance in the line, as suggested in Fig. 17-1c. This series impedance may be a transistor, as shown in the figure, or it may be a magnetic amplifier. In this and all other voltage regulating systems, the input or rectified voltage is always higher than the desired output to permit a sufficient latitude for control. Transistor $Q1$ in Fig. 17-1c is of the power type and may be replaced by two or more transistors in series if the voltage deviations and the heat energy to be absorbed in the series section are sufficiently large. Transistor $Q2$ is the amplifier for the voltage sensing unit. The sensing unit consists of a zener diode in series with resistor $R3$. This resistor is necessary to assure that the zener diode holds the potential at the emitter of $Q2$ constant. Changes in the load potential control the base current and potential of amplifier $Q2$, which, in turn, controls the base current of the series impedance $Q1$. Integrated circuits are widely available that employ the entire circuit of Fig. 17-1c (and, of course, much more sophisticated circuits) in a single package, and only a few additional discrete components are needed to construct an inexpensive, fully adjustable linear power supply.

In the shunt- or parallel-impedance type of voltage regulator, the voltage drop across the series impedance is maintained constant under load variations. This is accomplished in Fig. 17-1d by varying the transistor shunt impedance in such a manner that when it is combined with the load impedance, both appear as a constant impedance.

For equivalent loop gains, the series-impedance regulator tends to perform better under input line voltage variations than the shunt type, whereas the shunt-impedance regulator performs better under output load variations. Thus, to provide the best regulation under both conditions, it is not uncommon for a well-regulated supply to provide a combination of series–shunt regulation with equivalent source impedances of 5 milliohms, a 0.01% regulation for a 10% input voltage change, a 0.2% regulation for a no-load to full-load change, and 2 mV rms ripple.

The design of reliable DC power supplies presents a number of problems. A major problem arises from the transistor's intolerance to overvoltage. Overvoltages may arise from transients in the load, or they may result from switching and other changes on the input side (5, 7). Instantaneous voltage overshoots or spikes can completely destroy a whole bank of transistors in a matter of microseconds. Conventional fuses and circuit breakers are too slow in operation to offer protection to transistors for transient voltages although they may offer some protection against overheating with heavy overloads or short circuits.

One protection for transistors against voltage spikes is semiconductor devices that are designed for a rated reverse breakdown voltage and can handle relatively high currents. These devices are called Transzorbs™ (trademark of General Semiconductor Industries) and varistors (See Chapter 15). Transzorbs can be obtained with unipolar or bipolar properties. Essentially, a Transzorb behaves like a zener diode, capable of fast response, but with a higher energy-handling capability than an ordinary zener. A bipolar Transzorb is two large-area p-n junction diodes in a back-to-back configuration. A varistor is constructed of a metal oxide compound, most often zinc oxide, and electrically resembles a parallel combination of two diodes with opposite polarity. Such protection devices are placed in parallel with transistors or other circuits to serve as overvoltage transient protectors (1, 3, 15) or across the terminals of inductors to clamp large $L(di/dt)$ voltages created during transient on and off conditions.

A second problem in the design of a DC power supply using transistors is that their sensitivity to temperature affects their operating characteristics and their life. Thus, protection against overloads and overheating is essential. The best solution lies in the use of inherent self-protecting regulation which reduces the output current to near zero on overloads and short circuits. This protection may be attained through the sensing control on a magnetic amplifier or through the transistor regulating circuit.

EMI IN POWER SUPPLIES

This section discusses power supply susceptibility and emissions of power supplies. The typical power supply is surrounded by and contributes to EMI. The amount of EMI that can couple into the power supply and cause problems determines the susceptibility of the power supply and to a large degree the susceptibility of the equipment the power supply serves (6, 16). The power supply can be, and often is, a large source of EMI. Switch mode power supplies generate high frequency radiated and conducted EMI that can result in self-jamming or be a source of interference to other equipment. The subject of emissions from switching power supplies is gaining a lot of attention due to a recent change in West German emissions regulations, dictating that power supplies must meet emissions limits as stand-alone equipment before they can be marketed for use in West Germany. (See Chapter 20).

Radiated EMI Susceptibility Coupling Paths

The coupling of radiated high frequency EMI into systems via the power source usually occurs by coupling into the AC power transmission cables (14). The induced high frequency, common-mode (CM) [and, occasionally, differential-mode (DM)] voltages and currents are then manifested as conducted signals into the equipment power mains through the AC (or DC) power receptacles, and thence coupled into system signal circuits. The coupling paths between the power mains and signal circuits can be formed as a result of common impedance coupling, internal reradiation, and cable-to-cable coupling (crosstalk), and/or via paths due to stray capacitances. Power-mains coupling is discussed in some detail in Chapter 14.

Reduction of Radiated EMI Susceptibility Coupling

Techniques to reduce the effect on power supplies of the coupling of radiated EMI into power mains include the use of:

1. Hybrid power-line filters to suppress both CM and DM coupling.
2. Isolation power transformers, including transformers with an electrostatic shield between primary and secondary.
3. Shielding.
4. Proper circuit layouts to minimize loop areas.
5. Power-line conditioners.
6. Uninterruptible power supplies (UPSs).
7. EMI-hardened motor-generator sets.

For the designer, any number of these options can be considered to prevent the occurence of EMI. The most cost-effective EMI fix is the fix that is implemented in the design stage. For the user of the power supply, a redesign of an off-the-shelf power supply to reduce the loop areas is not always practical; thus, shielding and filtering are more realizable EMI fixes.

Power Supplies and Conducted EMI

Conducted EMI may enter equipment from outside sources via the equipment power mains, or it may be generated internally by sources such as switching power supplies. The usual square wave generated by switchers is rich in high frequency harmonics that can travel via paths formed by stray capacitance. Another source of internally generated noise is diode recovery noise associated with the sudden polarity reversal of rectifier diodes on the output of a power supply. Filtering the output can be accomplished by inserting high frequency capacitors in parallel with the diodes to shunt the noise away from the output; or soft recovery diodes are available that create less of this type of noise.

One type of path that can have a significant effect on the amount of noise conducted onto or from the AC mains is the large interwinding and leakage capacitance that exists between the primary and secondary of power transformers. Common- and differential-mode noise currents, especially at high frequencies, will couple across the capacitance of the transformer, and this can result in a susceptibility and/or an emissions problem. An isolation transformer with grounded electrostatic (faraday) shield between the primary and secondary will reduce the amount of this capacitance, thus decreasing the amount of coupled noise. The EMI current divides between the primary to secondary (leakage) capacitance and the primary to shield capacitance (see Chapter 15). The significance of a low-impedance ground lead from the faraday shield to ground cannot be overstressed. A large impedance associated with the ground lead will increase the effective impedance from the primary to the shield and can nullify much of the benefit obtained by installing such a transformer.

Emission of EMI by Switching Power Supplies

This topic is most significant in the design of switching-type power supplies. Since the efficiency of switching power supplies is much higher than that of linear power supplies, this type of power supply is gaining favor among electronic systems designers and manufacturers. The switching frequencies are now extending from tens to hundreds of kilohertz, (6) with the result that radiation from circuits carrying switched current is becoming more of a problem. Other problems exist where the high frequency emissions capacitively couple onto power cords and subsequently radiate.

Radiated EMI

For emissions radiated directly from the power supply module, the level of radiated emissions can be addressed by a radiating loop model using the IAF product where I is the uniform amplitude of the current around the loop, A is the loop area, and F is the frequency of the current (9). In determining radiated emissions, the frequency F is often a harmonic of the switching frequency (see Chapter 6 for a development of the radiation emitted by a loop). The radiated electric and magnetic fields are proportional to the loop IAF product, provided loop dimensions are small compared to the wavelength of the frequency component in question, and the observation distance is in the far field. This approach can be used in the analysis and design of switching power supplies where the amplitude I is usually determined by the system power requirements, and the frequency F is determined by weight and filtering considerations and desired system efficiency.

A major EMC objective is to minimize the switching current loop area, A, by proper circuit layout, consistent with other system design considerations. It should be noted that keeping loop areas as small as possible provides the double benefit of reducing both emissions and susceptibility. One example of a loop in a switching power supply is shown in Fig. 17-2. The radiating loop is through (1) the transformer primary to (2) the collector of the switching transistor, on through (3) the collector to heat sink capacitance, (4) the heat sink, and (5) the ground path to the power supply input, then back to the transformer.

One type of fix is to install a faraday shield between the collector of the switching transistor and the heat sink, thus reducing the capacitance C and returning the currents to the switching circuit rather than into the ground or chassis. There are commercially available heat sinks that have such a shield already built in, such as Sil-Pads (trademark of Bergquist). Another type of fix is to isolate the heat sink from the ground and connect it to the emitter of the switching transistor. In this situation, the heat sink acts as a faraday shield between the collector and the grounded case.

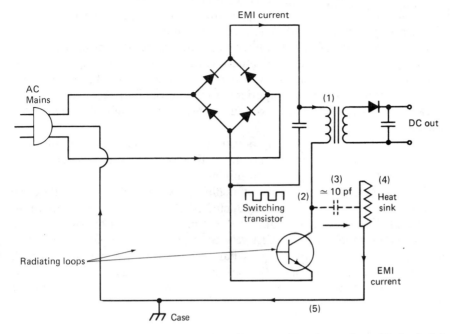

Fig. 17-2. Switching power supply radiating loop. (*Courtesy of Interference Control Technologies*)

The total emissions of a switching power supply can be reduced by reducing the maximum duty cycle of a power supply, thus decreasing the average amount of current flowing through a radiating loop. The reduction in duty cycle for a given power output means that the power supply will have to be larger and therefore costlier. The cost of this solution should be weighed against other EMI-preventive measures.

Radiated emissions from the power supply can also be suppressed by shielding techniques, and overall coupling to other system components reduced by properly locating the power supply with respect to other components, such as printed circuit boards, within the equipment rack or enclosure.

Conducted EMI

Along with the diode recovery noise, mentioned above, switching power supplies generate fast rise-time waveforms leading to high frequency currents that can couple via stray capacitances to the power mains, equipment chassis, and cabling. The usual end result of this coupling is common-mode noise coupling into the mains or into the equipment that the supply serves. The parasitic capacitances associated with ordinary power transformers readily allow this energy to couple onto the mains. Thus, the use of electrostatically shielded isolation transformers reduces this coupling.

The use of shunt electrolytic filter capacitors is usually insufficient to reduce high frequency conducted interference because of the high series inductance of such capacitors. A useful way of reducing this interference employs ferrite beads, placed in series with the output of the rectifier bridge of a switcher, to present a high output impedance to high frequency EMI currents.

The emissions from switching power supplies can also result in an intra-system EMI situation. Proper layout, with respect to other, possibly susceptible equipment is necessary to minimize self-interference.

One beneficial aspect of using switching power supplies is that they are able, to some extent, to prevent conducted EMI from flowing into the equipment that they serve. This is due to the "tuned" nature of the output stage of the power supply, which effectively constitutes a band-pass filter. Thus, emissions that are not in the pass band of the output stage are rejected, and the flow of EMI through the power supply is reduced. This is in contrast to linear supplies which are readily transparent to a large range of EMI frequencies.

EMI Filters

The function of the EMI filter, as it applies to power supplies, is to suppress high frequency noise components from entering or leaving via the AC/DC power cord while allowing power frequency currents to pass unattenuated. EMI filters for power lines are always of the low pass type, and a variety of configurations exist that accomplish the filtering with varying degrees of performance.

The EMI filter has the dual effect of suppressing incoming emissions and also suppressing conducted emissions emanating from within the equipment to which the filter is attached. These emissions are often subject to various military and governmental limits such as MIL-STD 461B, FCC *Rules and Regulations* Parts 15 and 18, and VDE 0871 and 0875. EMI filters are covered in detail in Chapter 14.

SCR POWER CONTROL

The need to replace the inefficient rheostat and the space-wasting variable transformer in power control led to the implementation of semiconductor-based thyristors and silicon controlled rectifiers (SCRs). Thyristors in general and the SCR in particular are now widely used in power control, a familiar example being an incandescent light dimmer. They are smaller and lighter than equivalent variable transformers and result in less heat dissipation and power loss than rheostats, but produce high levels of EMI. The EMI is a result of the control device which switches the power wave on or off rather abruptly, causing harmonic generation and conducted EMI (18). This section discusses prevention of EMI in SCRs.

The AC power control case is reviewed here. Generation of interference in phase control applications is analyzed using Fourier techniques, and measured EMI data are presented. A technique designed to reduce the interference, zero crossover switching, is then analyzed, and measured levels are compared. Techniques of reducing trigger circuit susceptibility are reviewed, and the application of zero crossover techniques in proportional control is discussed.

SCR Phase-Control Switching

In phase control switching in AC power control, SCRs are triggered into conduction at a specified phase angle during each half cycle, as shown in Figure 17-3. Back-to-back SCRs, or triacs, may be used to attain full wave control. Alternately, an SCR or a rectifier bridge and SCR may be used to provide a DC waveform, but the analyses are similar.

A square wave truncated sinusoid, shown in Fig. 17-4, is reviewed.

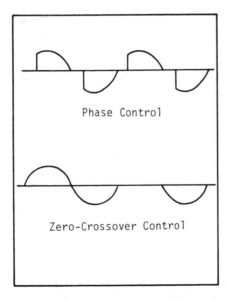

Phase Control

Zero-Crossover Control

Fig. 17-3. Load voltage wave forms. (*Courtesy of Interference Control Technologies*)

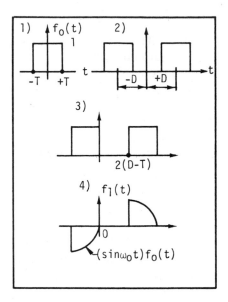

Fig. 17-4. Square wave truncated sinusoid. (*Courtesy of Interference Control Technologies*)

The basic square wave, with the Fourier transform, is:

$$2 \frac{\sin \omega T}{\omega} \tag{17-1}$$

It is shifted to the right and left by an amount D and added to yield:

$$\frac{2 \sin \omega T}{\omega} e^{-j\omega D} + \frac{2 \sin \omega T}{\omega} e^{+j\omega D} = \frac{4 \sin \omega T}{\omega} \cos \omega D \tag{17-2}$$

and then shifted to the right by an amount $(D - T)$ to give:

$$\frac{4 \sin \omega T}{\omega} \cos \omega D e^{-j(D-T)} \tag{17-3}$$

The square waves are then multiplied by the basic power sinusoid:

$$-j \frac{2 \sin (\omega - \omega_0)T}{(\omega - \omega_0)} \cos [(\omega - \omega_0)D] e^{-j(D-T)(\omega - \omega_0)}$$

$$- \frac{2 \sin (\omega + \omega_0)T}{(\omega + \omega_0)} \cos [(\omega + \omega_0)D] e^{-j(D-T)(\omega - \omega_0)} \tag{17-4}$$

The magnitude of the harmonic coefficients is:

$$\frac{2}{T_0} |F(j\omega_0 n)|, \; n = 1, 2, 3, \ldots \tag{17-5}$$

A phase shift of $2\omega_0 (D - T)$ is introduced between the two terms of expression (17-4) by the exponentials, and the phase angle of firing thereby affects the magnitude of the sum. For the case of $D = 2T = T_0/4$, the half power condition, the components are:

$$\frac{1}{\pi} [F(j\omega_0)] = 1 \text{ at } 3\omega_0, \; 1/3 \text{ at } 5\omega_0, \; 1/3 \text{ at } 7\omega_0,$$

$$1/5 \text{ at } 9\omega_0, \; 1/5 \text{ at } 11\omega_0, \ldots \tag{17-6}$$

The amplitude of a square wave truncated sinusoid, in general, falls off at a rate proportional to $1/\omega$, or at 20 dB/decade. This is the rate of decrease of the peaks of the basic $(\sin x)/x$ form.

The turn-on of a real SCR may better be modeled as an exponential than as the abrupt discontinuity in a square wave. An exponential pulse train, shown in Fig. 17-5, is used to truncate the basic power sinusoid. The tail of the pulse distorts the turn-off slightly, but permits a simplified expression.

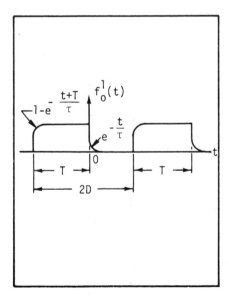

Fig. 17-5. Exponential pulse train. (*Courtesy of Interference Control Technologies*)

For $t \ll T$, this retains the main features of the square wave truncated sinusoid, but introduces a new break frequency at $f \cong (1/2T)$, where T is the time constant of the SCR turn-on process. Typical 10% to 90% rise-times (t_r) for low-current SCRs are on the order of 1 μsec for switching 100 volts and, for a given SCR, are inversely related to voltage switched. The time constant T is $(1/2.2) \, t_r$.

For high-current SCRs, the rate of rise of current must be held to values typically in the tens of amperes per microsecond to avoid device damage. This is accomplished by inductive or current-limiting components in the associated circuitry, and provides a circuitry-limited rather than device-limited rate of rise.

SCR Zero Crossover Switching

Fourier analysis of a zero crossover switched sine wave proceeds in the same manner as that for the square wave truncated sinusoid. Expression (17-4) is valid for this case if:

$$2T = T_0, \quad 2D = \frac{(K-1)}{\omega} T_0 \tag{17-7}$$

where two out of every k half cycles are allowed to reach the load, as shown in Fig. 17-6.

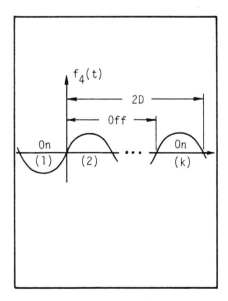

Fig. 17-6. Zero-crossover analysis. (*Courtesy of Interference Control Technologies*)

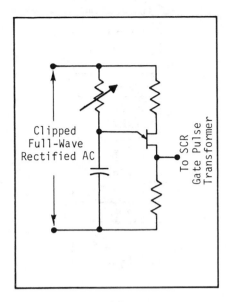

Fig. 17-7. Unijunction transistor oscillator trigger circuit for phase control. (*Courtesy of Inter-ference Control Technologies*)

This allows calculation of the one-of-i and two-of-i half cycles passed con-dition where $i = 1, 2, 3, \ldots$. Since the phase shift terms now are integer multiples of the power frequency, the two terms add to produce a sequence proportional to:

$$\frac{1}{n-1} - \frac{1}{n+1} = \frac{2}{n^2 - 1} \tag{17-8}$$

in which the smoothed envelope of the harmonics decreases in amplitude at a rate proportional to $1/\omega^2$ or at 40 dB/decade.

In this case n is set equal to harmonics of $2\omega_0/k$, and n is not necessarily an integer. A half wave rectified sine wave with a period of 2 seconds and amplitude equal to one is:

$$\frac{1}{\pi}\left(1 + \frac{\pi}{2}\sin t + 2/3 \sin 2t - 2/15 \sin 4t \ldots\right) \tag{17-9}$$

Figure 17-7 shows the familiar unijunction transistor oscillator, phase control technique. Varying the resistance varies the phase of the output waveform to the SCR gate pulse transformer. Figure 17-8 is a simplified block diagram of the zero crossing circuit. The disabling means controls the percentage of half cycles absent in driving the trigger circuit. Both circuits

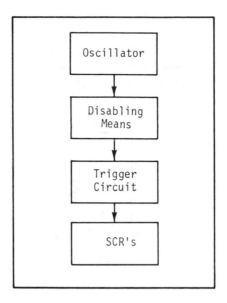

Fig. 17-8. Block diagram of zero crossover circuit. (*Courtesy of Interference Control Technologies*)

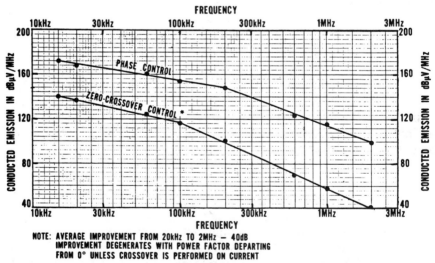

Fig. 17-9. Comparison of conducted emission in a 400 Hz, 115 VAC SCR-controlled 1.2 kW load. (*Courtesy of Interference Control Technologies*)

are powered by the same AC line voltage as that furnished to the SCR and load.

Figure 17-9 shows a comparison from test results of conducted emissions on a 400 Hz, 115 VAC power mains in which a 1.2 kW resistive load is regulated by (1) phase controlling the triggering of SCRs and (2) zero axis control. Since phase control techniques interrupt the current waveform, this method leads to a higher generation of conducted transients as compared to zero axis control, as is noted in Fig. 17-9.

The slope of emission falloff versus frequency below 100 kHz by phase control is about 20 dB/decade, as previously predicted. On the other hand, for zero crossover control, the slope is 30 dB/decade, an amount not simply reconciled compared to the theory of 40 dB/decade. Some of this may be due to (1) measurement error, (2) the true phase angle of the load, and (3) the changing impedance of the source with frequency.

Figure 17-9 could be misleading relative to the amount of EMI reduction because it is measuring broadband noise in voltage units of $dB\mu V/MHz$. For true resistive loads, the voltage and current are in phase, but for inductive loads, such as motors, they are not. Thus, for low impedance devices (less than 377 ohms; see Chapter 10), interference may result primarily by current induction or magnetically coupling means. Accordingly, most the zero crossovers sense the current, not the voltage.

SCR Filtering

In phase control applications, filtering is often an integral part of SCR design. Device dI/dt ratings for high frequency devices are given for the SCR shunted by an RC snubber circuit. Protection of the device from false triggering due to dV/dt or rapid rise of blocking voltage is required. The most important filtering function however, is, control of current rise-time during device turn-on. This limits conducted and radiated electromagnetic interference and protects the device from dI/dt failure.

SCR manufacturers recommend two principal methods of filtering. A single inductor in series with the load SCR combination may be used with a parallel capacitance if an RF ground is available to terminate the capacitance. Otherwise, the unbalanced filter creates RF currents in distribution wiring capacity. Then a two-inductor balanced filter can be used. It is often advantageous to use a split inductor filter with limited inductance and high saturation current levels.

Trigger Circuit Susceptibility

Care must be taken to avoid SCR trigger circuit susceptibility to either line voltage EMI or interference in the gate circuit of the controlled SCRs. Good design practice in regulated supplies for the trigger circuit, together with RF filtering, minimizes the effects of power-line fluctuations and transients. Protection of the gate circuit is more difficult because of the effects of suppression devices that may adversely affect timing or firing characteristics. For example, in a pulse transformer coupled trigger circuit, the HF components of the trigger pulse waveform carry a significant portion of the pulse energy. Rapid application of trigger voltage is desirable to reduce switching stresses on the device. Thus, filtering is not a practical answer. A diode bridge prevents any reverse interaction without reducing trigger effectiveness.

The zero crossover technique is appropriate in many applications requiring proportional control. Here, the term proportional control is used in the general sense to describe any requirement for application of continuous or nearly continuous increments of power to a load, rather than in the special sense of proportional feedback, although the latter is included. Typical applications are: light dimming, heater control for industrial processes, aircraft anti-icing or comfort heating, motor speed control, and control of electrochemical processes. When the needs for low weight and low power loss coincide with an environment that is sensitive to EMI, the zero crossover provides a viable design alternative.

One serious limitation exists. The load must exhibit sufficient damping to avoid intolerable ripple in the output function. In light dimming this requirement can be met with 400 Hz power but not with 60 Hz power. In heating applications, the load thermal inertia is usually sufficient to smooth over variations in average input power over a period of several half cycles of the power waveform. Motor mechanical load systems represent a more critical case, and again a 400 Hz power is preferable, although control is feasible in many 60 Hz applications.

CIRCUITS AND INTERSTAGES

This section reviews EMI problems and control in circuits, amplifiers, amplifier interstages, and micro-circuits. Emphasis is placed on built-in decoupling techniques.

Amplifier Decoupling

External radiation from any source is a function of the radiating circuit area, whether it is an electric field (voltage) or magnetic field (current) source. It is therefore necessary and prudent to decouple amplifiers and other networks stage by stage. Decoupling methods discussed herein emphasize isolation of potential interference voltages and currents.

For high current stages such as a power output stage, as shown in Fig. 17-10, the collector current should be supplied from a low impedance source.

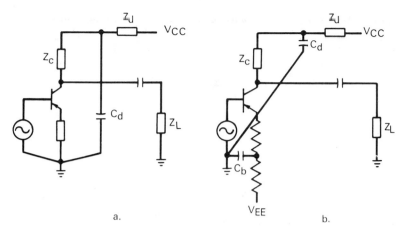

Fig. 17-10. Output stage decoupling. (*Courtesy of Interference Control Technologies*)

Signal currents that flow through Z_c are supplied by the charge on decoupling capacitor C_d and power supply V_{cc}, which is in series with Z_d. If signal current flows through the source impedance of V_{cc}, there will be a voltage drop in both the source and the interconnecting wiring. This voltage will then affect all other circuits connected to that supply, and objectionable interaction may result. In addition, the current through the wiring may induce voltages in other wires (i.e., cause crosstalk). Therefore, a low impedance path, C_d, is provided. A series impedance, Z_d, is provided to raise the impedance of the path and make the power supply a constant current source. The impedance of C_d and Z_d will force signal currents to flow in the power supply below certain frequencies. The value of C_d will force signal currents to flow in the power supply below certain frequencies. The values of C_d and Z_d should be selected so that this frequency is sufficiently low that signal currents do not appear in the power supply and wiring. The self-resonant frequency of C_d must be higher than the highest expected frequency (or harmonic) in the power stage. Capacitor C_d is often a combination of electrolytic and high frequency tantalum or ceramic capacitors.

The signal current that flows through Z_L is the desired output. This current should also be returned to the emitter with a minimum of disturbance to other circuits. If the emitter and the load are both connected to the chassis or printed circuit board ground, the return current can provide a voltage drop in the ground bus impedance that might interfere with other circuits (i.e., common impedance coupling). A loop is also created that can radiate EMI or act as a transformer to couple into other loops. For low frequency circuits, where the signal system is grounded at only one point, the best return path is a wire twisted with the outgoing wire to Z_L.

At high frequencies, the parasitic capacitance of the lead connected to Z_L will cause currents to flow to the chassis along its length and to adjacent wires. This current, too, must return to the emitter. The solution is to provide a shield around the cable to establish a controlled capacitance. The return current for the cable capacitance can then be returned to the emitter by connecting the shield to the emitter. The return current for the load can also be passed through the shield, particularly if either the load or the emitter can be floating. If both emitter and load are grounded, a small percentage of the return current will pass through interchassis grounds. Twisted pairs inside shields and transformer coupling decrease this coupling problem.

Figure 17-10b shows the emitter bypass method used with two-supply biasing. With the connection shown, the signal current flows through the emitter bypass capacitor, C_b. While it may seem preferable to connect C_d to the emitter side of C_b, so that the collector current does not pass through the emitter bypass capacitor, that connection allows disturbances on the power supplies to be coupled into the base emitter signal loop.

Interference involving tuned circuit output is minimized by connecting the resonating capacitance across the tuning coil as shown in Fig. 17-11a, rather than between collector and ground as shown in Fig. 17-11b. With the capacitor across the coil, current through the decoupling capacitor at resonance is $e_0/Q\omega L$, where Q is the quality factor of the tank circuit, L is the impedance of one arm of the tank at resonance, and e_0 is the output voltage. If the capacitor is connected to ground as shown in Fig. 17-11b, C_d becomes part of the series resonant circuit, and current through it is e_0/L, which is higher by a factor Q, which may approach 100. Thus, for easier decoupling, the former connection is preferred. In many cases, most of the current passes through the distributed capacitance to the chassis, in which cases C_d must handle the full tank current.

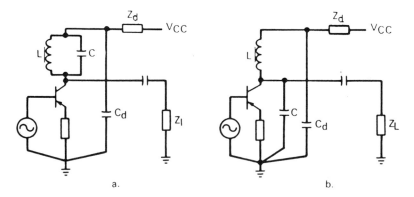

Fig. 17-11. Tuned-output stage decoupling. (*Courtesy of Interference Control Technologies*)

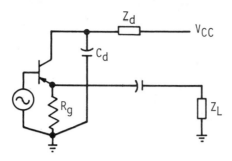

Fig. 17-12. Emitter-follower decoupling. (*Courtesy of Interference Control Technologies*)

If the amplifier has signals at a frequency such as $\omega = 1/\sqrt{L(C + C_d)}$, then series resonance of Z, C, and C_d as a pi network can give an objectional amount of voltage across the decoupling capacitor. The amount of voltage depends on Z_L and Z_d. To approximate the effect, consider the impedance of both to be infinite; then the current through C_d is the short circuit output current of the transistor. Finite values of Z_c and Z_d reduce this current appreciably.

Emitter follower stages should be provided with a collector decoupling network to return the collector signal current to the emitter without flowing through the collector supply. Figure 17-12 shows the collector bypassed to the emitter ground. However, if $Z_L \ll R_g$, which is a typical case when a separate emitter supply is used, current through the chassis can be reduced by returning C_d to the point at which the load current is returned to the chassis.

Interstage coupling of a pair of transistors is shown in Fig. 17-13a. The second transistor is represented by its input impedance, Z_i, and its base biasing resistors are R_1 and R_2. The function of this stage is to amplify the input signal represented by e_g and supply maximum current in Z_i and minimum current to the impedances in common with other circuits. At the same time, disturbances on the supplies or in the chassis impedances should supply minimum current to Z_i.

The emitter is shown bypassed to the input signal ground to return the base current signal directly to the driving source without going through the chassis impedance. The ground point of C_d has conflicting requirements. In the connection shown, all of the transistor current flows through the chassis so that there may be common impedance coupling to other stages. If C_d is connected back to the first stage ground, this chassis current is reduced by the amount of current that flows through Z_c and R_1. If a current exists in the chassis due to some other source, however, the configuration of Fig. 17-13a minimizes the amplification of this undesired current by the following stages. This is demonstrated in equivalent circuits shown in Fig. 17-13 (b and c). The transistor has been replaced by its ouput impedance, r_o, and e_g and Z_g

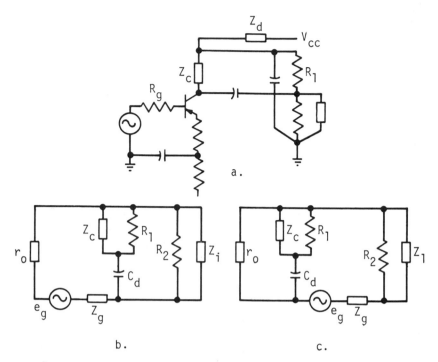

Fig. 17-13. Interstage decoupling. (*Courtesy of Interference Control Technologies*)

represent the interference source due to the chassis currents. Any current that passes through Z_i must pass through r_0, which is usually a high impedance. However, in Fig. 17-13c, which represents the circuit with C_d returned to the emitter ground, the current through Z_i, and flow through R_1 and Z_c, which are usually much lower impedances than r_0.

When flip-flops change state, the current required from the supply voltages changes momentarily. This pulse is rich in high frequency components and can couple into wires adjacent to those carrying the supply current. These transients should be kept on the flip-flop board with a decoupling network consisting of shunt C and series R or L. When series L is used, the filter ringing possibilities must be examined. Where possible, the circuit and board configurations should include both elements of the flip-flop so that the local current transfer between these elements necessitates a minimum of energy storage for the small fraction of time that both elements are simultaneously on or off.

Some computer circuits use a reference voltage to bias clamping diodes to obtain constant-level pulses. The resulting pulse currents in the reference supply and wiring are a source of interfering signals. It is recognized that a series impedance for decoupling can spoil the reference level. However, a capacitor to ground, suitable for the frequency requirements, on each board will usually decrease the amount of high frequency current in the reference supply and wiring. In this case, the power supply source impedance at the point of decoupling must be considered in selecting the size of the decoupling capacitor.

An Audio System Example

An example of the application of EMC techniques to the functional circuits of an equipment such as the power converter and audio amplifier is shown in Fig. 17-14.

Specifications for the audio amplifier are:

1. Supply voltage: 28 volts.
2. Switchable audio frequency inputs for:
   - A long line connected to a remote audio system.
   - A low-level microphone.

3. Audio frequency outputs for:
   - Supplying 5 watts to remote speakers.
   - Providing a tape recorder output.

4. Quality reproduction of voice.
5. Applicable military specifications for conducted and radiation interference over the frequency range of 15 kHz to 400 MHz.
6. That the input power, audio input, and recorder may not be grounded in order to achieve EMI control.

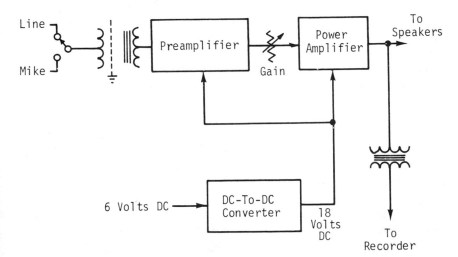

Fig. 17-14. EMC measures in an audio amplifier and power supply. (*Courtesy of Interference Control Technologies*)

The functional circuitry shown in Fig. 17-14 consists of an audio pre-amplifier driving an audio power amplifier, with both powered by a DC-to-DC converter.

A well-shielded input transformer is used to open any ground loops to the distant audio system or microphone. High gain must be used in the pre-amplifier, since the typical microphone has a low power output. Therefore, the input circuitry, including the input transformer, is very susceptible to EMI. An electrostatic shield is used between the primary and secondary winding of the transformer to prevent capacitive coupling of common-mode EMI voltage on the input lines into the preamplifier. Common-mode voltages are those that have the same instantaneous polarity on both conductors. To ensure capacitive and magnetic balance of the transmission line to the distant audio system or to the microphone, these lines should be shielded twisted pairs. The input transformer should be enclosed in magnetic shields and should preferably be of semi-toroidal construction. It may be desirable to place this transformer in a simple sheet iron enclosure or provide a shading ring to prevent it from picking up power supply magnetic interference. The gain control should follow the preamplifier rather than precede it. Otherwise the noise contribution and interference pickup of the control and its wiring will be an appreciable portion of the low-level signal input.

Circuit connections to common buses and circuit decoupling from the common power supply should be designed to minimize common impedance coupling of EMI and to eliminate amplifier instability due to feedback paths. To simplify this, each amplifier stage should be designed for the minimum permissible audio bandwidth that will result in the specified overall amplifier bandwidth.

The DC-to-DC converter portion of Fig. 17-14 is shown in Fig. 17-15. It has a semiconductor chopper that supplies a 6-volt, 20-kHz or higher switching rate, square wave voltage to the power transformer. Switching in the chopper occurs at points of saturation of the power transformer core. The secondary of this transformer feeds a semiconductor bridge rectifier, which, in turn, feeds the output filter that provides the 18 VDC used to power the preamplifier and power amplifier.

The chopper, the transformer, and the rectifier cause switching transients with rise-times on the order of 100 nsec. Because the transformer core saturates, it can radiate a strong magnetic field. To meet EMC requirements, escape of this interference from the power supply shield box must be prevented. This is accomplished by providing a completely enclosed shield box of high-permeable magnetic material.

C_1, C_2, and L form a pi filter on each side of the primary power input line to prevent the escape of conducted EMI. Filters of this type present a reactance to the connected circuits in the stop band and, therefore, reflect EMI. Unless they are carefully made and grounded, they can also have reso-

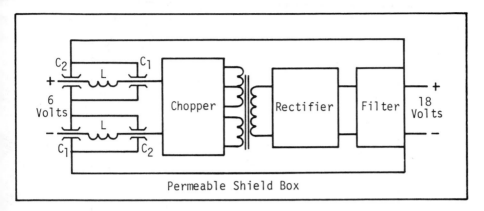

Fig. 17-15. EMC measures for a DC-to-DC power supply. (*Courtesy of Interference Control Technologies*)

nances that permit the transmission of EMI at one or more resonant frequencies. A type of filter that absorbs interference usually consists of a length of conductor embedded in a lossy magnetic material. Ideally, there would be a perfect impedance match to the connected circuits at interference frequencies so that all EMI would be absorbed, and there would be infinite attenuation for interference and no attenuation for the desired frequency pass band.

Similar interference filters should be used on the power supply output leads. Note that separate shield cans are provided on the filter inductors to prevent either pickup or radiation of EMI.

To reduce the magnetic field radiated from a power transformer in this application, the power transformer is of toroidal construction and encased in its own magnetic shield. Only modest electrostatic shielding is necessary for the audio-amplifier circuits.

The power supply transformer also fulfills the requirement that the primary source not be grounded, in order to prevent ground loops through the power system.

EMC IN THE DESIGN OF PRINTED CIRCUIT BOARDS

Proper attention to EMC considerations in the PC board design stage will reduce the occurrence of EMI when the circuit is in the prototype or production stage. The old adage about an ounce of prevention certainly holds true. The types of EMI problems in PC boards include common impedance coupling, crosstalk, radiation generation from wires carrying high frequency currents, and noise pickup via radiation loops formed by interconnecting wiring and PC traces. The higher-speed logic families are especially vulnerable to these types of problems, and for a number of reasons: common

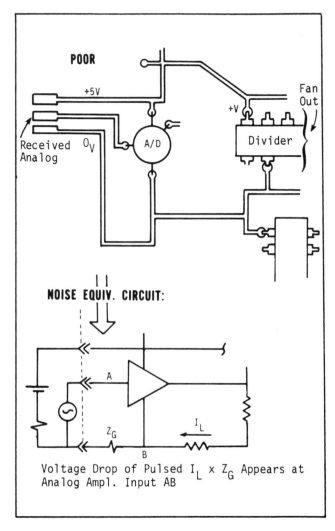

Fig. 17-16. Poor common-mode noise control. (*Courtesy of Interference Control Technologies*)

impedance coupling occurs more frequently because the impedance of supply and ground wires/traces increases with increasing frequency; crosstalk occurs more efficiently because the higher-frequency signals more effectively couple across parasitic capacitances in wiring and traces; and the high-frequency digital signals radiate more effectively because signal loop dimensions become comparable to the wavelength of the clock frequency and its harmonics

(10, 17). Another problem inherent with high frequency design is the problem of impedance mismatch causing reflections on signal lines.

Common Impedance Coupling in PC Boards

A severe common impedance coupling problem can arise when analog and digital signals are mixed on the same PC card, especially if the analog return, the logic return, and the power return are all the same trace. The return currents flow through the finite ground impedance and generate a common-mode voltage that may be higher than the noise sensitivity of analog and digital circuitry connected to the common return. This is shown in Fig. 17-16a for an analog amplifier that shares the same return as a digital gate.

The result of such an arrangement can be the generation of a noise voltage along the finite impedance, Z_G, between the common return and the ground connection of the amplifier. This is shown as noise voltage V_N in the equivalent circuit of Fig. 17-16b.

If the noise voltage is larger than the analog sensitivity, which is usually in the microvolt range, then degradation will result.

A better distribution scheme is suggested in Fig. 17-17. Here, the analog and digital circuits have separate return paths, and the noise voltage does not appear across the amplifier input. Also, notice that the power and return traces have been widened considerably, with the effect that the impedance will decrease, and any noise voltage drops will be less.

The impedance of the printed circuit trace is an important consideration because it is so often the source of the common impedance coupling problem. Table 17-1 presents values of impedance as a function of frequency for various trace widths, thicknesses, and lengths.

To illustrate a typical common impedance coupling problem, consider the following (17). High-frequency gate transitions currents tend to load down power busses that have too high an internal impedance at the transition frequency. If the voltage drop occurring on a bus is higher than the noise immunity of the logic, then an EMI problem exists. Schottky TTL, for example, has a rise-time of 3 nsec and a gate transition current of 30 mA (fanout $= 5$). The 3-nsec rise-time results in significant frequency components upward of 100 MHz ($f_2 = 1/(\pi\tau_r)$). If a power supply and return trace are 3-mm traces 100 mm long (impedance $\simeq 59$ ohms, from Table 17-1), a common-mode voltage equal to $I \times (2/\pi) \times Z = 0.3 \times (2/\pi) \times 59 \simeq 1.1$ volts develops on the bus. Since the noise immunity for STTL is 300 mV, the logic will not work. The solutions are: (1) shorten the trace length, (2) decouple the power distribution system, (3) use a lower impedance distribution, or (4) try a combination of the above.

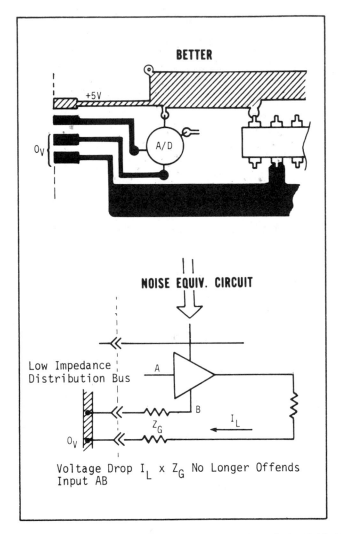

Fig. 17-17. Improvement upon the common-mode noise problem of Fig. 17-16. (*Courtesy of Interference Control Technologies*)

Several options will now be presented that will be useful in the design of circuit boards to reduce EMI. First, the design of single-layer boards will be covered. The rest of the section will then focus on other design options that may be useful when the limitations of using a single-layer board are too great. These include the use of multilayer, multiwire, and press-fit boards. Each of these types will be investigated, and the relative merits and disadvantages presented.

Table 17-1. Impedance of Printed Circuit Board Wiring

FREQ.	w=1mm, t=0.03mm				w=3mm, t=0.03mm			w=10mm, t=0.03mm		
	ℓ=10mm	ℓ=30mm	ℓ=100mm	ℓ=300mm	ℓ=30mm	ℓ=100mm	ℓ=300mm	ℓ=30mm	ℓ=100mm	ℓ=300mm
10Hz	5.74m	17.2m	57.4m	172m	5.74m	19.1m	57.4m	1.72m	5.74m	17.2m
20Hz	5.74m	17.2m	57.4m	172m	5.74m	19.1m	57.4m	1.72m	5.74m	17.2m
30Hz	5.74m	17.2m	57.4m	172m	5.74m	19.1m	57.4m	1.72m	5.74m	17.2m
50Hz	5.74m	17.2m	57.4m	172m	5.74m	19.1m	57.4m	1.72m	5.74m	17.2m
70Hz	5.74m	17.2m	57.4m	172m	5.74m	19.1m	57.4m	1.72m	5.74m	17.2m
100Hz	5.74m	17.2m	57.4m	172m	5.74m	19.1m	57.4m	1.72m	5.74m	17.2m
200Hz	5.74m	17.2m	57.4m	172m	5.74m	19.1m	57.4m	1.72m	5.74m	17.2m
300Hz	5.74m	17.2m	57.4m	172m	5.74m	19.1m	57.4m	1.72m	5.74m	17.2m
500Hz	5.74m	17.2m	57.4m	172m	5.74m	19.1m	57.4m	1.72m	5.75m	17.2m
700Hz	5.74m	17.2m	57.4m	172m	5.74m	19.1m	57.4m	1.72m	5.75m	17.2m
1kHz	5.74m	17.2m	57.4m	172m	5.74m	19.1m	57.5m	1.72m	5.76m	17.3m
2kHz	5.74m	17.2m	57.4m	172m	5.75m	19.1m	57.6m	1.73m	5.81m	17.5m
3kHz	5.74m	17.2m	57.5m	172m	5.76m	19.2m	57.8m	1.74m	5.89m	18.0m
5kHz	5.75m	17.2m	57.5m	172m	5.78m	19.3m	58.4m	1.77m	6.15m	19.2m
7kHz	5.75m	17.2m	57.6m	173m	5.82m	19.5m	59.4m	1.83m	6.52m	21.0m
10kHz	5.76m	17.3m	57.9m	174m	5.89m	20.0m	61.4m	1.93m	7.23m	24.4m
20kHz	5.81m	17.5m	59.2m	180m	6.32m	22.4m	72.1m	2.45m	10.5m	38.6m
30kHz	5.89m	17.9m	61.4m	189m	6.97m	26.0m	87.1m	3.14m	14.4m	54.7m
50kHz	6.14m	19.2m	67.9m	215m	8.14m	35.1m	123m	4.71m	22.7m	88.3m
70kHz	6.51m	21.0m	76.6m	250m	10.8m	45.5m	163m	6.37m	31.3m	122m
100kHz	7.21m	24.3m	92.5m	311m	14.3m	62.0m	225m	8.93m	44.4m	174m
200kHz	10.4m	38.5m	155m	545m	26.9m	119m	440m	17.6m	88.2m	346m
300kHz	14.3m	54.4m	224m	795m	39.9m	177m	657m	26.3m	132m	519m
500kHz	22.5m	87.8m	367m	1.30Ω	66.1m	295m	1.09Ω	43.8m	220m	866m
700kHz	31.1m	121m	510m	1.82Ω	92.4m	413m	1.52Ω	61.4m	308m	1.21Ω
1MHz	44.0m	173m	727m	2.59Ω	131m	590m	2.18Ω	87.7m	440m	1.73Ω
2MHz	87.5m	344m	1.45Ω	5.18Ω	263m	1.17Ω	4.36Ω	175m	880m	3.46Ω
3MHz	131m	516m	2.17Ω	7.76Ω	395m	1.76Ω	6.54Ω	263m	1.32Ω	5.19Ω
5MHz	218m	861m	3.62Ω	12.9Ω	659m	2.94Ω	10.9Ω	438m	2.20Ω	8.66Ω
7MHz	305m	1.20Ω	5.07Ω	18.1Ω	922m	4.12Ω	15.2Ω	613m	3.08Ω	12.1Ω
10MHz	437m	1.72Ω	7.25Ω	25.8Ω	1.31Ω	5.89Ω	21.8Ω	876m	4.40Ω	17.3Ω
20MHz	874m	3.44Ω	14.5Ω	51.7Ω	2.63Ω	11.7Ω	43.6Ω	1.75Ω	8.80Ω	34.6Ω
30MHz	1.31Ω	5.16Ω	21.7Ω	77.6Ω	3.95Ω	17.6Ω	65.4Ω	2.63Ω	13.2Ω	51.9Ω
50MHz	2.18Ω	8.61Ω	36.2Ω	129Ω	6.59Ω	29.4Ω	109Ω	4.38Ω	22.0Ω	86.6Ω
70MHz	3.05Ω	12.0Ω	50.7Ω	181Ω	9.22Ω	41.2Ω	152Ω	6.13Ω	30.8Ω	121Ω
100MHz	4.37Ω	17.2Ω	72.5Ω	258Ω	13.1Ω	58.9Ω	218Ω	8.76Ω	44.0Ω	173Ω
200MHz	8.74Ω	34.4Ω	145Ω	517Ω	26.3Ω	117Ω	436Ω	17.5Ω	88.0Ω	346Ω
300MHz	13.1Ω	51.6Ω	217Ω	776Ω	39.5Ω	176Ω	654Ω	26.3Ω	132Ω	519Ω
500MHz	21.8Ω	86.1Ω	362Ω	1.29kΩ	65.9Ω	294Ω	1.09kΩ	43.8Ω	220Ω	866Ω
700MHz	30.5Ω	120Ω	507Ω	1.81kΩ	92.2Ω	412Ω	1.52kΩ	61.3Ω	308Ω	1.21kΩ
1GHz	43.7Ω	172Ω	725Ω	2.58kΩ	131Ω	589Ω	2.18kΩ	87.6Ω	440Ω	1.73kΩ
2GHz	87.4Ω	344Ω	1.45kΩ	5.17kΩ	263Ω	1.17kΩ	4.36kΩ	175Ω	880Ω	3.46kΩ
3GHz	131Ω	516Ω	2.17kΩ	7.76kΩ	395Ω	1.76kΩ	6.54kΩ	263Ω	1.32kΩ	5.19kΩ
5GHz	218Ω	861Ω	3.62kΩ	12.9kΩ	659Ω	2.94kΩ	10.9kΩ	438Ω	2.20kΩ	8.66kΩ
7GHz	305Ω	1.20kΩ	5.07kΩ	18.1kΩ	922Ω	4.12kΩ	15.2kΩ	613Ω	3.08kΩ	12.1kΩ
10GHz	437Ω	1.72kΩ	7.25kΩ	25.8kΩ	1.31kΩ	5.89kΩ	21.8kΩ	876Ω	4.40kΩ	17.3kΩ

* Wiring dimensions are width x thickness in mm

ℓ = wiring length in mm
m = milliohms
µ = microhms
Ω = ohms

☐ Non-Valid Region for which ℓ ≥ λ/4

Note: Computations appearing in this table are based on the assumption that ℓ >> W. Any interpolations should con sider this assumption.

EMC Design in Single-Layer Printed Circuit Boards

This section will concentrate on proper layout procedures and trace design to reduce EMI problems such as common impedance coupling via the power supply and radiated emissions, as well as impedance control of traces to reduce mismatch. The designs are applicable for one- or two-sided boards.

Logic Layout. In a printed circuit board design program, this is probably the first and relatively easiest aspect of the design to control. The design recommendations can be summarized in a few lines:

- Separate low-level analog and digital circuitry as much as possible (avoid the common impedance coupling problem above).
- Use different areas for high-, medium-, and low-speed logic.

Figure 17-18 shows the optimum board layout for reducing intracard crosstalk, common impedance coupling, and radiated emissions and susceptibility. This type of separation results in short traces that carry high frequency currents, the main contributor to common impedance coupling, crosstalk, and radiation.

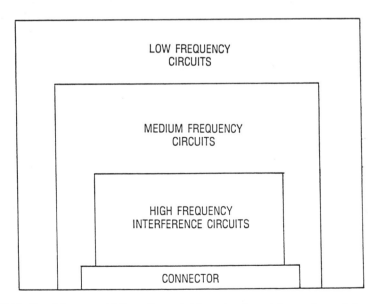

Fig. 17-18. Board layout guidelines. Suggested layout reduces intra-card crosstalk, common impedance problem and radiated EMI (emission and suspectifility. (*Courtesy of Interference Control Technologies*)

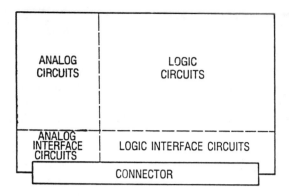

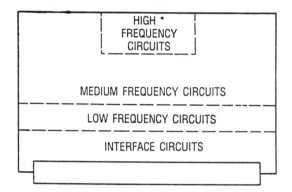

Fig. 17-19. Board layout guidelines. (*Courtesy of Interference Control Technologies*)

Where there are analog circuits on the board, the recommended circuit layout is as shown in Fig. 17-19. The analog circuits are separate from the digital circuits. The same arrangement of logic circuits (i.e., high-speed logic close to the edge) is recommended.

Once the logic is laid out, the trace layout is designed, usually with some computer aided design program.

Printed Circuit Board Trace and Interconnect Design.   Simply stated, one general rule of trace design should be followed: etch away as little copper as possible. This is especially applicable when the power distribution system is being laid out, and less true of signal traces where space is a problem and when the characteristic impedance of the trace must be controlled to prevent impedance mismatch. These general rules apply for trace design and layout (17):

- Route dedicated 0-volt traces and $V_{cc} \geq 1$ mm wide.
- Route power and return traces as closely as possible. The preferred method is to route the power trace on one side of the board and the return trace on the opposite side. This results in low impedance power distribution.
- Landfill open areas with a 0-volt plane. The board should be as optically opaque as possible. This practice also saves etchant.
- Dedicate a 0-volt return for analog circuits.
- Check for crosstalk on long parallel runs. If necessary, increase trace spacing or add a 0-volt guard trace between runs.
- Consider using raised power distribution for high-speed logic.
- Allocate at least 10% of spaced connector pins for PCB 0-volt return traces.
- Dedicate 0-volt traces and connect pins to buffer circuits.

Figure 17-20 shows two ways to design the power and return traces on PC boards. The first shows a bad layout that gives high inductance and a high possibility of crosstalk. Notice the large loops formed by the traces. The second method gives reduced impedance of the power distribution system and results in a decrease in loop areas.

Table 17-1 (above) shows the characteristic impedance of different trace conductor pairs as a function of geometry, viewing the two traces as transmission lines. The first configuration in the table is realizable when the power and return traces are positioned on opposite sides of the board, or in cases of raised power bus distribution (discussed below). The second configuration shows the characteristic impedance of a conductor over a ground plane (as in multilayer PC boards). The third column shows the impedance of side-by-side traces, an arrangement that gives the highest value of impedance.

For example, if Schottky TTL is used, which has a rise-time of 3 nsec, and a gate switch current of 30 mA, the impedance of the power bus must be $Z <$ (Noise immunity level in volts/.03 A) = 300 mV/.03 A = 10 ohms. As a safety margin, the impedance should be, say, less than 8 ohms. The use of the third configuration shown in the table is not very feasible; thus either of the first two configurations should be used.

Raised Power Bus Distribution. This arrangement increases the high frequency performance of the power supply as isolated supply and return systems eventually degrade at high frequencies. The concept is to provide a supply and return trace packaged together that provides the necessary low impedance over a wide frequency range (13).

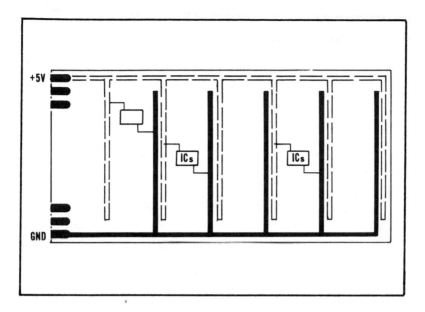

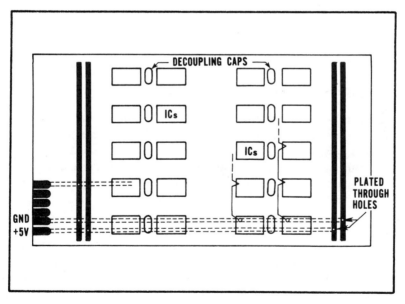

Fig. 17-20. A recommended layout to reduce power distribution and logic return impedances and crosstalk. (*Courtesy of Interference Control Technologies*)

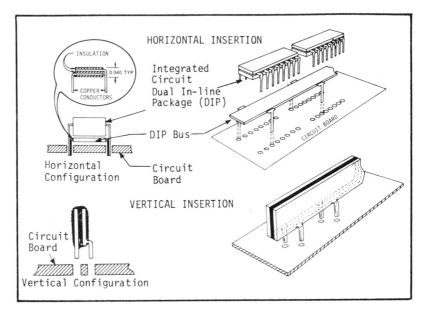

Fig. 17-21. Low-impedance power distribution bus.

Figure 17-21 shows three of the most popular methods of implementing the raised power bus distribution scheme (17).

Power Supply Decoupling

Proper power supply decoupling provides several important benefits in EMI control in printed circuit boards. An increase in the capacitance of the power distribution system will decrease the overall impedance of the distribution system, resulting in less power supply common impedance coupling. Additionally, a decoupling capacitor placed between the supply and 0-volt supply pins on an integrated circuit decreases the loop area of the supply and return circuit for that chip. This is illustrated in Fig. 17-22, which shows the large loop associated with the power distribution and the smaller loop associated with the decoupling capacitor and integrated circuit (IC) (19).

Since the amount of radiation from a current-carrying loop is proportional to the IAF product (current × loop Area × frequency), improper decoupling can significantly affect the EMI performance of a circuit layout, often resulting in a 10 or 20 dB increase in radiated emissions compared to a board that is well decoupled (19).

A significant factor is the impedance of the decoupling loop, Z_L. The ideal capacitor should have low loss and be effective through 200 MHz, though lead inductance can render some types of capacitors useless at a self-resonant

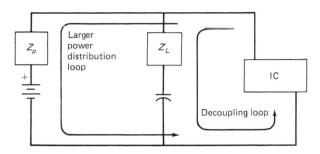

Fig. 17-22. Power distribution model.

frequency of about 10 MHz. The most common types of capacitors are Z5U grade barium titanate ceramic capacitors, which have high dielectric constants and relatively good loss characteristics, depending on the packaging and formulation, from 1 to 20 MHz, above which they become lossy and the capacitance begins to fall. Generally, the Z5U ceramic capacitors are effective in the 1 to 50 MHz range if the lead lengths are very small.

Other dielectrics, such as strontium titanate, NPO, and some polymers, have good high-frequency characteristics, but relatively small dielectric constants, making them unsuitable for low frequency decoupling (DC to 10 MHz). Decoupling then, is a tradeoff between high-capacitance, low-frequency decoupling and low-loss, stable-capacitance, high-frequency decoupling (19).

These factors must be weighed for the specific application, but in general these guidelines for decoupling apply:

- Decouple V_{cc} at the connector with tantalum capacitor in parallel with a 0.01-µf Hf ceramic capacitor or monolithic capacitor (0.001 µf for high-speed logic).
- Decouple V_{cc} for every two DIPs with an HF ceramic disk capacitor. (See below for capacitor value.)
   The three-stage decoupling provides better filtering but with an increased cost and reliability compromise. Table 17-2 provides the decoupling capacitor value for some popular logic families.

The capacitor leads should be as short as possible to reduce the inductance of the lead wires. For a combined 2-cm trace length and 10 nH of lead inductance, the combined inductance is about 36 nH, which will induce a voltage for TTL of $L di/dt = 36 \times 10^{-9} \times 0.03A/3 \times 10^{-9} = 360$ mV. The need to keep the capacitor leads as short as possible is another reason for running the power and return traces close together.

Table 17-2. Decoupling capacitors for some popular logic families

Logic family	Current requirements, mA		$dV = 20\%$ of Nil* mV	$dt =$ rise-time, nsec	$C = \Delta I/(dV/dt)$, pf
	Gate switch	Gate drive			
CMOS	1	1	200	50	500
TTL	16	8	80	10	3000
STTL	30	20	60	3	2500
LSTTL	8	11	60	8	2500
ECL-10K	1	6	20	2	700

* NIL is noise immunity level. Gate fanout is 5, and only one drive gate per chip is assumed to be switched. NIL is spread over five sources: power supply sag, power distribution radiation pickup, common impedance coupling, crosstalk, and reflections due to mismatch.

Terminations.   Correct line termination becomes important when length of line becomes greater than about one-sixth of the wavelength of the frequency of the logic. Stated another way, when the two-way delay time, $2T$, of a length of line, L, exceeds the rise-time of the digital pulse, τ_r, correct line termination is necessary to reduce reflections on the line. The two-way delay is calculated by:

$$2T = \frac{L}{c} \text{ sec} \qquad (17\text{-}10)$$

where L is the length of line in meters and c the velocity of light in free space $= 3 \times 10^8$ m/sec. For propagation along a line in a dielectric, the velocity of propagation is reduced by $\sqrt{\varepsilon_r}$, the relative permittivity of the dielectric. Thus, the speed of propagation in a dielectric is:

$$c_\varepsilon = \frac{c}{\sqrt{\varepsilon_r}} \qquad (17\text{-}11)$$

Hence, the lines on the board should be terminated when the length of the line is related to the rise-time according to equation (17-12):

$$l_{cm} \geq \frac{10\tau_{rns}}{\sqrt{\varepsilon_r}} \qquad (17\text{-}12)$$

This is for a fanout of one. Since the capacitance of each gate increases the line delay, a higher fanout decreases the length of the unterminated line. For a fanout of greater than one, equation (17-13) should be used to determine the maximum unterminated line:

$$l_{cm} \geq \frac{10\tau_{rns}}{\sqrt{\varepsilon_r}\sqrt{F}} \qquad (17\text{-}13)$$

where F is the fanout of the logic.

This capacitive loading has less effect on lines with low characteristic impedance (Z_0). Table 17-3 shows the maximum line length that should be used for unterminated lines with different values of Z_0 and loading (17).

The value of Z_0 for lines on a wire-wrapped board is about 120 ohms. However, this can vary from 100 to 180, depending on distance from the ground plane, proximity of adjacent wires, and configuration of the ground grid. The value of Z_0 for a line on a PC board is much more closely controlled. The impedance depends on board thickness, dielectric, and line width.

ECL-10K. Because of the popularity and high speed of emitter coupled logic (ECL) systems, the remaining discussion on termination will cover aspects of ECL terminations. This is not to imply that the concepts apply only to ECL systems; they apply to many other logic families as well. Unlike some earlier ECL families, ECL-10K does not necessarily require terminated lines. However, the ability to drive terminated lines allows a reduction of line noise.

The ability to terminate the signal lines of ECL-10K circuits provides several design benefits. Terminations can be used to eliminate line reflections,

Table 17-3. Maximum unterminated line lengths vs. characteristic impedance and fanout. (Courtesy of Interference Control Technologies.)

Characteristic impedance in ohms	Maximum line length in centimeters			
	Fanout = 1	Fanout = 2	Fanout = 4	Fanout = 8
Microstrip				
50	21.1	19.1	17.0	14.5
68	17.8	15.7	12.7	10.2
75	17.5	15.0	11.7	9.1
90	16.5	13.7	9.9	7.6
100	16.3	13.0	9.1	6.6
Backplane				
100	16.8	13.7	9.7	7.1
140	15.0	10.9	7.1	4.8
180	13.2	9.1	5.3	3.3

or ringing, completely when line lengths exceed the values in Table 17-3. In addition, terminations can reduce crosstalk between parallel signal lines.

Line reflections occur when the impedance of the load at the receiving end of the line (Z_L) differs from the characteristic impedance of the line (Z_0). The value of the reflected voltage (V_r) depends on the impedance mismatch, as follows:

$$V_r = V_{line} \frac{(Z_L - Z_0)}{(Z_L + Z_0)} \tag{17-14}$$

For any value of Z_L close to Z_0, the reflection in equation (17-14) drops to a negligible value; for $Z_L = Z_0$, $V_r = 0$. Also the polarity of the reflection is the same as the driving signal when $Z_L > Z_0$.

In order to reduce mismatch, a terminating resistor is placed on the line. Both series and parallel terminations can be used in ECL-10K systems (Fig. 17-23). In a series termination a resistor is placed in series with the signal line at the driving end. For parallel terminations, a load resistor is used between the line and V_{EE} (or a V_{rr} voltage) at the end of the lines.

Controlling Crosstalk.   Crosstalk is the coupling of a pulse on one signal line to an adjacent signal line. ECL-10K has a linear input impedance, low output impedance, and capability for signal-line termination. These performance features permit crosstalk on ECL-10K circuits to be analyzed and minimized, with the result that ECL-10K can be a relatively quiet, highspeed family.

The simplest way to minimize crosstalk is to run adjacent signal lines at right angles to each other. Where this is not feasible, a general guideline is to space adjacent lines as far apart as possible. (This is discussed further in Chapter 6.)

Cross-coupling results from the mutual inductance and capacitance between send and receive lines. The signals coupled into the receiving line can be either forward or backward. Forward crosstalk, V_f, consists of a pulse on the victim wire having a width equal to the rise-time of the signal on the sending line and an amplitude that depends on the length and proximity of parallel lines. Pulse voltage, V_f, has a polarity opposite to V_s, the sending signal.

Forward crosstalk is not usually significant in ECL-10K systems, because of the small $\Delta V/\Delta T$ on the signal edge. The value of $\Delta V/\Delta T$ in ECL-10K is about the same as that for TTL and less than a quarter the value for Schottky-TTL.

The backward crosstalk pulse, V_b, has a width that is equal to twice the propagation delay of the sending line. The amplitude depends on the line spacing only and is independent of the length of the parallel lines or signal

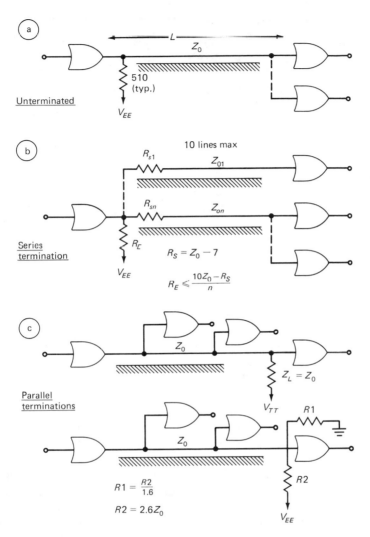

Fig. 17-23. Series of parallel terminations on ECL-10K circuits eliminate noise problem caused by mismatches on unterminated lines (a). in the series configuration (b). each load at the receiving end reduces the noise margin by 0.16 (R_s) mv. In the parallel configuration (c). up to 16 distributed loads can be accommodated. Each load adds about 100 psec of delay. (Reference 8.)

rise-time. The rise- and fall-times of V_b equal the rise- and fall-times, respectively, of V_s.

There exist a number of waveforms of the backward crosstalk signal for a variety of terminations on the receiving line. Since V_b always travels in a direction opposite to V_s, a proper termination completely absorbs any coupled V_b.

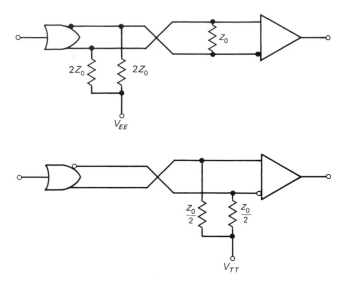

Fig. 17-24. Twisted-pair lines should be terminated in one of two ways. Up to 15 m of line can be driven at greater than 100 MHz. (Reference 8.)

Series-terminated lines generate less crossstalk than parallel-terminated ones. The reason is that only half the logic swing is sent down a series-terminated line. Figure 17-24 illustrates the termination of the twisted-wire pairs.

Board Cable Interconnections. Generally, coaxial or twisted-wire pairs are used for ECL interconnections between boards. Coax should have a characteristic impedance in the range of 50 to 100 ohms, and each end of the transmission line return should be grounded. Twisted pair, AWG 24 to 28, twisted at 100 turns per meter yields a satisfactory Z_0 of about 110 ohms.

If ribbon cable is used, select types having low attenuation at the maximum operating frequency. The maximum recommended attenuation for ECL-10K is 2.5 dB, which may limit the cable length to 5 meters or less. Alternate leads in the ribbon cable should be grounded to maintain the transmission line effect.

At connectors, use multiple ground pins, equally spaced, to reduce the connection impedance to negligible amounts.

Very long lines of 6 meters or more can be driven differentially to maintain noise immunity at high frequencies. The complementary outputs of gates can be used for differential drive, and a 1011415 ECL circuit can be employed as a line receiver.

Fifteen meters of twisted pair can be driven differentially at frequencies over 100 MHz. Differentially driven lines should be terminated as shown in Fig. 17-24, and the fanout should be limited to 4. The twisted pair lines should be shielded when the common-mode noise on the pair exceeds the supply voltage.

Multilayer Printed Circuit Boards

For high-speed logic design, the use of a single-layer board does not fulfill the requirements for electromagnetic compatibility. For these cases the use of multilayer boards should be investigated. The most common type of multilayer board is shown in Fig. 17-25. This figure shows conceptually how the layers are constructed and the functional designation of each layer (17).

The power and return busses depicted are formed of unetched one-ounce copper foil boards. Thus, the power distribution system forms large ground planes, resulting in extremely low distribution source impedance. The benefit of multilayer boards over single-layer boards is that the former are more immune to common impedance coupling, offer shielding (depending on layout), and are better for multiple-level supply voltages. A drawback to mutlilayer construction is that they are more difficult and expensive to construct and repair. The counter to that is that the use of complicated single-layer designs leads to repeatability problems and difficulty in quality control in mass production.

The multilayer board is one or more PC boards sandwiched together, as in Fig. 17-25, with interconnects between the boards made via plated-through holes. In Fig. 17-25, the top board contains the circuit components. The second layer is a solid ground plane except for various through holes that connect the top layer with V_{cc}, layer 3, and the second interconnect board, layer 4. In essence, the traces on the top layer form microstrip lines with the ground plane of layer 2 such that impedances may be tightly controlled. The

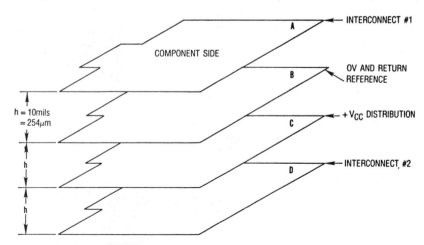

Fig. 17-25. Digital multi-layer printed circuit board. (Typical commercial practice). (*Courtesy of Interference Control Technologies*)

0-volt return layer and the V_{cc} distribution layer form a very low impedance power distribution system, owing to the large capacitance between the planes and the low inductance and low resistance of the copper foil. They also provide shielding from radiated pickup for layers 1 and 4.

Many more layers can be configured, and there are usually $n + 1$ layers for an n-layer board. If two interconnecting boards are to be placed atop one another, the respective interconnect lines are routed 90° to each other. This reduces crosstalk coupling between interconnect boards. High-speed logic circuits for military applications have boards that are usually configured with interconnect boards shielded from one another with a 0-volt return board or V_{cc} board. This also greatly increases impedance control.

Press-Fit Multilayer. This is a variation on multilayer boards. Basically, press-fit is a stacked arrangement of PC boards sandwiched together with press-fit contacts. This is shown in Fig. 17-26.

The benefits of press-fit over multilayer boards are: plated-through holes of press-fit designs are constructed on a layer-by-layer basis compared with traditional multilayer boards which have a hole going through all the PC boards in the stack; the uniform spacing of press-fit allows for more uniform impedance control; hybrid systems can be built with press-fit, where additional circuitry can be stacked atop the existing boards; any board can be

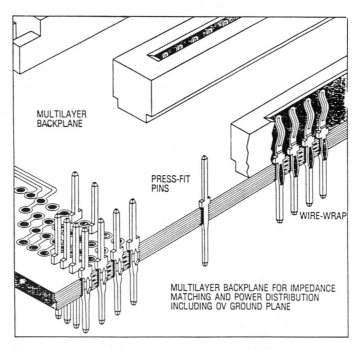

Fig. 17-26. Edge elevation of press-fit board. (*Courtesy of Interference Control Technologies*).

changed up to assembly time because each board is discrete; contacts can be removed with press-fit, thus aiding in repairs or changes. Also, properly done artwork can be interchangeable with multiwire and pressfit, so that either option exists.

Multiwire Boards

An option to multilayer boards is a multiwire board (Multiwire Division, Kolmorgen Corp., 31 Sea Cliff Ave., Glen Cove, NY 11542). These circuit boards are constructed of a customized pattern of AWG #34 wires placed by a high-speed machine on an adhesive-coated substrate, as shown in Figure 17-27.

After the wires have been laid down, the board is covered by an epoxy fiberglass sheet. The polymide insulation is rated at 2000 volts breakdown which makes it possible to pass the wires as close as .07 mm without shorting out. Thus, many layers of wire can be built up, and with a 90° crossover capacitance of less than 1 pf, crosstalk is negligible. The uniform wire width and spacing allow for close control of impedance, nominally 55 ohms, which is necessary in high-speed logic. The 55-ohm impedance requires the use of traditional decoupling capacitors which are reduced by multilayer boards.

Power Plane Distribution

In the above discussion, copper foil planes were used for power distribution. The impedance of the copper plane determines the potential for common impedance coupling via the power distribution system. As will be demonstrated, the common impedance voltage drop can be reduced to levels well below circuit sensitivity using power plane distribution.

The characteristic impedance of a pair of parallel metal planes can be determined by transmission line theory. Two such planes, separated by a dielectric of thickness h that has a relative permittivity of ε_r have an impedance equal to:

$$Z_0 = \frac{120\pi}{\sqrt{\varepsilon_r}} (h/d) \tag{17-15}$$

where d is the length of the smaller of the two sides of a two-dimensional plane.

Table 17-4 shows the impedance of two planes for several values of ε_r and the ratio h/d (17).

For power plane geometries of $h/d < 0.005$ and $\varepsilon_r > 3$, it is possible to have power distribution systems of impedance less than 1 ohm. Thus, 20 or 30 mA of switching current would produce tens of millivolts and less drop in the power distribution system, much less than the noise immunity levels of logic.

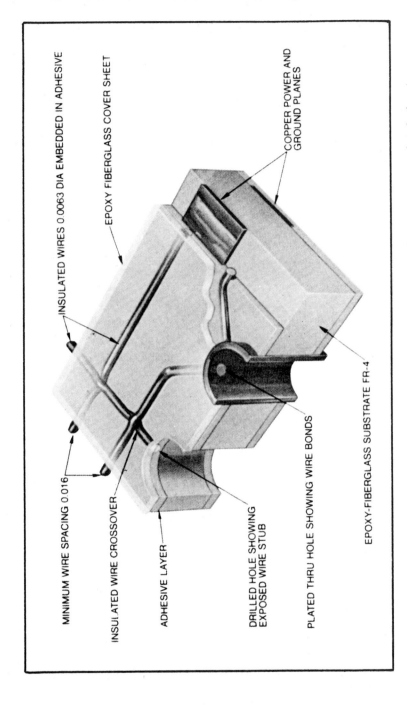

INSULATED WIRES 0.0063 DIA EMBEDDED IN ADHESIVE

EPOXY FIBERGLASS COVER SHEET

COPPER POWER AND GROUND PLANES

MINIMUM WIRE SPACING 0.016

INSULATED WIRE CROSSOVER

ADHESIVE LAYER

DRILLED HOLE SHOWING EXPOSED WIRE STUB

PLATED THRU HOLE SHOWING WIRE BONDS

EPOXY-FIBERGLASS SUBSTRATE FR-4

Fig. 17-27. Expanded view of typical multiwire board realizability. (*Courtesy of Interference Control Technologies*)

Table 17-4. Characteristic Impedances In Ohms Of PCB Supply-Return Planes

Dielectric ε_r	h/d Ratios of Parallel Plane												
	.0005	.0007	.001	.002	.003	.005	.007	.010	.020	.030	.050	.070	.100
1.0	.19	.26	.38	.75	1.13	1.88	2.64	3.77	7.54	11.3	18.8	26.4	37.7
2.0	.13	.19	.27	.53	.80	1.33	1.86	2.67	5.33	8.00	13.3	18.6	26.7
Glass-cloth*	.12	.17	.24	.49	.73	1.22	1.70	2.43	4.87	7.30	12.2	17.0	24.3
3.0	.11	.15	.22	.44	.65	1.09	1.53	2.18	4.35	6.53	10.9	15.3	21.8
4.0	.09	.13	.19	.38	.57	.94	1.32	1.88	3.77	5.65	9.42	13.2	18.8
5.0	.08	.12	.17	.34	.51	.84	1.18	1.69	3.37	5.06	8.43	11.8	16.9
7.0	.08	.11	.15	.31	.46	.77	1.08	1.54	3.08	4.62	7.70	10.8	15.4
7.0	.07	.10	.14	.29	.43	.71	1.00	1.42	2.85	4.27	7.12	10.0	14.2
8.0	.07	.09	.13	.27	.40	.67	.93	1.33	2.67	4.00	6.66	9.33	13.3
9.0	.06	.09	.13	.25	.38	.63	.88	1.26	2.51	3.77	6.28	8.80	12.6
10.0	.06	.08	.12	.24	.36	.60	.83	1.19	2.38	3.58	5.96	8.35	11.5

* Glasscloth = (Woven or non-woven) Teflon, PTFE Laminate, $\varepsilon_r \approx 2.4$.

Wire Wrap Board Design

This section briefly covers some aspects of wire wrap because of its usefulness in prototype design and in small production runs of custom boards. The focus will be on general design guidelines used to reduce crosstalk and board radiation.

As the inductance of round wires is higher than that of flat conductors, such as printed circuit traces, the use of impedance control in wirewrap construction is of great importance. This is especially true in high-speed logic applications. A necessary step in wirewrap design is to establish some type of ground plane, or in this case ground grid, such that some control over impedance is possible.

The first step is, of course, that of mounting the discrete components (resistors, capacitors, diodes, etc.) and DIPs onto the board. For decoupling, a high-frequency ceramic capacitor of 1000 pf is used for every one or two DIPs. The next step is to wire all "$Z1$" wires onto the pins. A $Z1$ wire is one whose length exceeds one-half the diagonal measurement, D, of the board. As the $Z1$ wires have the largest inductance, they are mounted close to the ground plane. As in all interconnects, they are routed point to point, and not in an X-Y fashion. This decreases the possibility of crosstalk by reducing possible parallel wire runs.

After the $Z1$ wires are mounted, an X-Y ground grid is constructed on the board atop the $Z1$ wires. This provides some shielding and decreases the

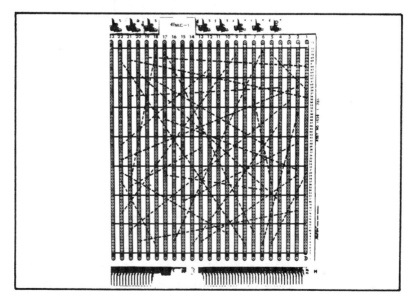

Fig. 17-28. Typical completed wirewrap board. (*Courtesy of Interference Control Technologies*)

impedance of the $Z1$ and $Z2$ wires (defined as wire having length $< D/2$) The $Z2$ wires are mounted next, also in a point-to-point fashion. A completed wirewrap board is shown in Fig. 17-28.

REFERENCES

1. Clark, O. M. "Devices and Methods for EMP Transient Suppression." *Record of the 1975 IEEE Electromagnetic Compatibility Symposium*, Volume 75CH1002-5, October 7–9, 1975, San Antonio, TX.
2. Evans, R. H. "Magnetic and EMC Control of a Deep-Space Satellite Transponder." *Record of the 1975 IEEE Electromagnetic Compatibility Symposium*, Volume 75CH1002-5, October 7-9, 1975, San Antonio, TX.
3. The General Electric Company. *Transient Voltage Suppression Manual*, Fourth Edition. Auburn, NY: The General Electric Company, Semiconductor Products Department, 1983.
4. Gottlieb, Irving M. *Regulated Power Supplies*, Third Edition. Indianapolis, IN: Howard W. Sams & Co., 1981.
5. Greenwood, Allan. *Electrical Transients in Power Systems*. New York: Wiley-Interscience, John Wiley Sons, 1971.
6. Hnatek, Eugene R. *Design of Solid-State Power Supplies*, Second Edition. New York: Van Nostrand Reinhold Company, 1981.
7. Institute of Electrical and Electronics Engineers. ANSI/IEEE C62.41-1981. New York: Institute of Electrical and Electronics Engineers.
8. Jaeger, Robert. "ECL 10,000 Interconnects Economically." *Electronic Design*, September 27, 1974.
9. Mardiguian, Michel. *How to Control Electrical Noise*. Gainesville, VA: Don White Consultants, Inc., 1982.
10. Mardiguian, Michel, *Interference Control in Computers and Microprocessor-Based Equipment.* Gainesville, VA: Don White Consultants, 1984.
11. Mitchell, J. C., et al. "Electromagnetic Compatibility of Cardiac Pacemakers." *IEEE International Symposium on Electromagnetic Compatibility Record*, July 18–20, 1972, pp. 5–10.
12. Naval Air Systems Command. *Electromagnetic Compatibility Manual*. NAVAIR 5335, 1972, Chapter 15.
13. Rogers Corporation. "Bus Bars as Power Distribution Systems," Application Notes.
14. Skomal, Edward N. *Man-Made Radio Noise*. New York: Van Nostrand Reinhold Company, 1978.
15. Van Keuren, E. "Effects of EMP Induced Transients on Integrated Circuits." *Record of the 1975 IEEE Electromagnetic Compatibility Symposium*, Volume 75CH1002-5, October 7–9, 1975, San Antonio, TX.
16. Violette, J. L. Norman, and Violette, Michael F. "Electromagnetic Interference (EMI) in Power Sources." *Proceedings of the Power Sources Conference*, November 27–29, 1984, Boston, MA.
17. White, Donald R. J. *EMI Control in the Design of Printed Circuits and Backplanes*. Gainesville, VA: Don White Consultants. Inc.,
18. White, Donald R. J. *EMI Control Methodology and Procedures Third Edition, Second Printing*. Gainesville, VA: Don White Consultants, Inc., 1982.
19. Johnston, Joseph E. "The Role of Integrated Circuit Decoupling in Electromagnetic Compatibility," *EMC Technology*, Volume 2, Number 4, Gainesville, VA: Don White Consultants, Inc., 1983.

Chapter 18
A SURVEY OF EMI MEASUREMENTS
AND TERMINOLOGY

This chapter is a survey of methods and nomenclature used in measurements of electromagnetic interference (EMI) emission and susceptibility characteristics of equipment and subsystems. The survey is divided into three categories of EMI measurements:

1. *Equipment and subsystem level tests*: The lowest level of EMI tests emphasizes components, equipment, and subsystems, as in MIL-STD-462, MIL-STD-449C, MIL-STD-704C, CISPR, FCC Rules 15 and 18, SAE J551C, and NACSEM 5100 (1). In addition to these EMI test items, this level covers tests on electrical filters used for EMC applications and on shielded enclosures to determine shielding effectiveness.
2. *System- and vehicle-level tests*: The intermediate level of EMI testing involves confirming that EMC exists at the system and vehicle level. Example are MIL-D-6015D(9) and MIL-STD-1541(14). These tests basically involve exercising the system in typical intentional modes of operation and recording whether any malfunction or degradation exists for any receptors or victims due to EMI.
3. *Operational electromagnetic ambient tests of systems*: The highest level of EMI tests involves immersing the system or vehicle in the typical electromagnetic environment (EME) expected for the test item and determining whether there is any undesirable EMI interaction. Specifically, does the EME cause malfunction or degradation to the performance of the test item, or does the item's radiation cause EMI to one or more receptors in the environment?

In general, the higher the level of testing, the more difficult and expensive the tests (and EMI control) become. Therefore, to assist in predicting the outcome, to plan early for EMI control, to help in the highest level of tests, or simply to ascertain the EME for other purposes, this chapter also surveys the need for EMI ambient surveying. The remaining sections discuss testing terminology. Among the topics covered are antennas, current probes, receivers, broad- and narrowband emission, and near and far fields.

THE BASIS FOR EMI-LEVEL TESTING

There exist three levels of EMI testing (see above), one or more of which may apply to any test item:

1. Lowest level: Component, equipment, and subsystem testing.
2. Intermediate level: system and vehicle testing.
3. Highest level: EME interaction with the test item.

Low-Level EMI Testing

The lowest level of EMI testing is that of MIL-STD-462. MIL-STD-461B establishes the specification limits, and MIL-STD-462 outlines the test procedures. The testing applies either to individual black boxes or equipment or to an ensemble of black boxes including their cables. The object of the testing is to ensure that radiations from a piece of equipment and/or input/output wiring do not penetrate another piece to cause EMI when the two are brought together. Similarly, both conducted emission and susceptibility levels of two or more pieces of equipment will correspond to a compatible situation (will not cause EMI) when either interconnected or sharing a common power source.

If an equipment has an antenna terminal, separate tests involving antenna conducted measurement are designed to assure that either (1) emission levels to the outside world (see receiver in Fig. 18-1 emitting local oscillator energy) are below likely interfering levels, or (2) susceptibility levels (intermodulation, out-of-band rejection, and cross-modulation) due to outside world emissions are above these emissions. Thus, the lowest-level EMI tests are intended to give a degree of assurance that at the second level of operation (the system or entire vehicle level) there will be relatively few, if any, EMI problems.

Intermediate-Level Testing

Figure 18-1 indicates (see dashed line) that all housekeeping, avionics, and weapons systems, special payloads, and so on, must operate in an EMC manner. It does not follow that no EMI will exist just because MIL-STD-461B specification levels were met, since the number of variables is too great to guarantee EMC at this level. The number of EMI problems exhibited at this level, however, is likely to be significantly lessened by compliance with MIL-STD-461B.

Basically, the intermediate level of testing, called simply electromagnetic compatibility testing or MIL-E-6051D for avionics systems, involves testing

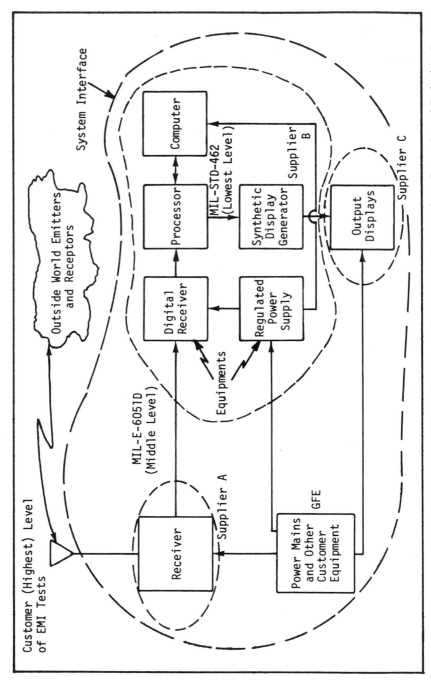

Fig. 18-1. System diagram showing level of EMI specification compliance testing. (*Courtesy of Interference Control Technologies*)

for EMI degradation or malfunction due to self-jamming. Tests are per-formed as the entire vehicular system is exercised in a manner corresponding to the intended modes of operation. If there is no EMI degradation, then EMC exists, and no anticipated EMI problems due to self-jamming are expected in real-life operations. However, this is rarely the case, so that electromagnetic system compatibility testing almost always involves identify-ing culprit–victim situations during or after tests in order to permit redesign or corrective measures to eliminate EMI.

High-Level Testing

With both the lowest and the intermediate levels of EMI compliance testing satisfied, one might expect that the highest level, involving interaction of the system test item with the outside world, would be insignificant (see Fig. 18-1). If the test item were a Voyager probe operating in a nearly sterile ambient electromagnetic environment, perhaps that would be so. However, nearly all test items operate within the earth terrestrial EME, which is highly polluted with emissions (see Chapter 2). For example, consider a warship, such as a guided missile frigate, in which up to 50 antennas representing individual systems are squeezed into a small area. Here a single system can readily be rendered ineffective when immersed in the EME resulting from all other sys-tems. Even if a complete ship is made electromagnetically compatible, EMC becomes a problem when a task force of several ships is assembled, and added to this may be enemy-generated EME.

Another example of EME impact is chosen from the civil sector, as illus-trated in Fig. 18-2. This is an automated rapid transit system serving a metro-politan area. The supersystem includes one or more automated computer-control centrals, track guideway controllers and data communications, and a number of vehicles. The vehicles pick up command and control signals from the guideway and decode them into vehicle speed and acceleration data, track routing and switching, and other commands. The vehicle reports back message receipt confirmation, periodic health status monitoring, and other data. The vehicle also contains its own subsystem propulsion controller, levitation and guidance, housekeeping, and the like.

There are a number of possible intrasystem EMI problems, such as inter-action between thyristors from the propulsion system to levitation and guid-ance sensors, induced transients noise from the vehicle rail shoes into the track guideway telecommunications, and internal substation emissions to the central computer. Whatever the degree of intrasystem EMI control achieved, a whole new set of problems may develop when the rapid-transit system is immersed in the outside-world EME. For example, automobile en-gine ignition noise or land-mobile radiation form an adjoining auto freeway may couple into the guideway telecommunications; out-of-band emissions

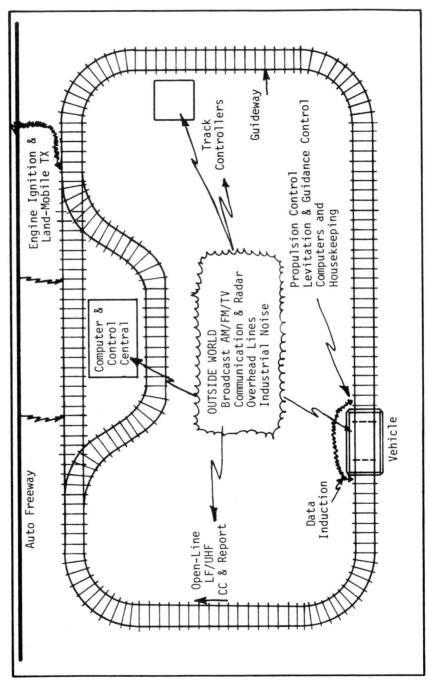

Fig. 18-2. Some principal emitters and receptors in Metro Automatic Rapid Transit System. (*Courtesy of Interference Control Technologies*)

from a nearby 5 MW UHF TV transmitter may jam the communications; a nearby airport radar radiation may couple into the computer control; or an overhead transmission line may couple broadband noise into the vehicle sensor system when it is passing underneath. Whatever the situation, the highest level of testing may reveal entirely new problems when performed in a nonsterile electromagnetic environment.

COMPONENT AND EQUIPMENT-LEVEL TESTING

This section discusses three aspects of EMI testing at the lowest level: (1) early-stage, "quick look" development testing, (2) subsystem and lowest-level EMI compliance testing, and (3) filter and shielded enclosure testing.

Quick Look Development Testing

The basic purpose behind EMI testing at other than the highest level is to demonstrate EMI compliance with specification limits. If the EMI testing is performed late in the development, there may be little if any time remaining to correct out-of-tolerance performance. Hence early quick look EMI tests are suggested so that major EMI culprits or victims can be flagged in time for cost-effective corrective measures to be taken. This introduces the idea of early tests during product or system development. These EMI tests are called developmental tests or simply dev tests.

Dev testing is often performed on an equipment in the brassboard stage—somewhere between the original breadboard and the final packaging. Quick looks are readily made (a) of both conducted and radiated emissions using a calibrated spectrum analyzer, or (b) of susceptibility using a levelized modulated sweeper or signal generator. EMI testing at this stage can be incidental the other design tests while the specimen is on the test bench. The cost of such testing and necessary subsequent fixes runs an order of magnitude or more below the cost of later-stage formal EMI testing and retrofit. They are performed with little interruption to design tests and permit early determination and economical fixes of any flagrant EMI problems.

Lowest Level Specification Compliance Testing

EMI testing here is to contractually demonstrate compliance with EMI specifications. Some typical EMI specifications may include applicable parts of MIL-STD-462, MIL-STD-461, MIL-STD-449D, MIL-STD-704B, FCC *Rules and Regulations* Parts 15 and 18, SAE J551C, NACSEM 5100, and others. A summary of each follows.

MIL-STD-462. MIL-STD-462, "Electromagnetic Interference Characteristics, Measurement of," was issued by the Department of Defense, July 31, 1967. This is a companion tri-service EMI specification to MIL-STD-461A and MIL-STD-463C. The purpose of the standard is to provide measurement techniques for performing the EMI testing to determine emission and susceptibility characteristics of electrical, electronic, and electromechanical equipment, as required by MIL-STD-461B. Test setups are given in block diagram form for each emission and susceptibility test, with some explanations for conducting each measurement. Specification limits that apply to each test in MIL-STD-462 are given in MIL-STD-461B.

The measurement procedures set forth in MIL-STD-462 are brief and leave much to be interpreted by anyone other than a knowledgeable EMI test engineer or technician.

MIL-STD-449C. MIL-STD-449C was issued March 1, 1965 and supersedes MIL-STD-449B of July 29, 1963. This standard is entitled "Radio Frequency Spectrum Characteristics, Measurement of." A supplement to this standard on formats for reporting radar and communications spectrum signatures was also issued on March 1, 1965. MIL-STD-449C was approved by the Department of Defense and is mandatory for use by all tri-service agencies within DoD. The Naval Electronic Systems Command, Washington, D.C. 20360 is the custodian for this standard.

The successful operation of most military systems depends upon the use of information received through electromagnetic radiations. Such operation is degraded if the equipment is interfered with by the presence of the numerous radiations that exist in normal electromagnetic environments. In order to predict the mutual interference effects of electronic equipment, both in existence and planned, more quantitative information is needed than now exists. The entire transmitter emission spectrum needs to be known, as well as the susceptibility of receivers to the various frequencies, power levels, and modulations that may occur in their operational environments.

MIL-STD-449C establishes uniform measurement techniques that are applicable to the determination of the spectral characteristics of radio frequency transmitters and receivers. The ultimate goal is to ensure the electromagnetic compatibility of present and future systems by providing the necessary data for predicting interference situations during design and development stages.

MIL-STD-704C. MIL-STD-704C was issued December 30, 1977 and superseded MIL-STD-704B of November 17, 1975. This military standard is entitled "Aircraft Electric Power Characteristics." The standard, which is mandatory for use by all DoD agencies, is in force as of 1975 and has not been superseded by the MIL-STD-460 family series or other standards.

MIL-STD-704C delineates (1) the characteristics of electrical power supplied to airborne equipment at the equipment terminals and (2) the requirements for utilization of such electrical power by the airborne equipment. The purpose of the standard is to foster electromagnetic compatibility between aircraft systems and between ground-support electric systems and airborne-utilization equipment. It is intended to do this to the extent of confining the aircraft and ground-support electric power characteristics within definitive specification limits and restricting the requirements imposed on the electric power by the airborne-utilization equipment.

FCC Rules and Regulations Parts 15 and 18 (15, 16). This section summarizes some conducted and radiated test compliance required under the Federal Communication Commission's *Rules and Regulations*, Part 15, Low-Level RF Devices; and Part 18, Industrial, Scientific and Medical (ISM) Equipment. Of particular concern is FCC Part 15, Subpart J (FCC 15J), which deals with conducted and radiated emissions from computing devices. Details of FCC 15J are covered in Chapter 20

FCC Part 15, RF Devices. Section 301 of the Communications Act of 1934, as amended, provides for the control by the federal government over all the channels of interstate and foreign radio communications, and further provides, in part, that no person shall use or operate apparatus for the transmission of energy, communications, or signals by radio when the effects of such operation extend beyond state lines or cause interference with the transmission or reception of energy, communications, or signals of any interstate or foreign character by radio, except under and in accordance with the Communications Act and a license granted under the provisions of the Act.

Restricted and incidental radiators emit RF energy on frequencies within the radio spectrum and constitute a harmful interference to authorized radio communication services operating upon the channels of interstate and foreign communication unless precautions are taken that will prevent the creation of any substantial amount of such interference.

RF devices covered under Part 15 include broadcast receivers, low-level telemetry, garage door openers, radio-control toys, carrier-current systems, incidental radiating devices (electrical appliances, power tools, etc.), field disturbance sensors, auditory training devices, Class I-TV devices, and data processing equipment. Part 15 now includes the onerous Subpart J (FCC 15J), which specifies conducting and radiating limits on "Any electronic device or system that generates and uses timing signals or pulses at a rate in excess of 10,000 pulses (cycles) per second and uses digital techniques . . . ." Within this classification are two categories: Class A, which pertains to equipment sold into a commercial, industrial, or business environment; and Class B, which pertains to equipment targeted for use in residential markets. With the advent of the microprocessor in everything from appliances to toys, this includes just about any electronic device on the market. Compliance with the terms of this regulation was mandatory for all devices as of October 1983.

As applicable, each such device must be type-approved or certified and is subject to Public Law 90-379 regarding sale and distribution.

FCC Part 18 ISM Equipment. Examples of the types of equipment that fall into this category are medical diathermy equipment, industrial heating equipment (such as RF furnaces), and RF stabilized arc welders. Threats from this type of equipment to radio communications channels and other users of the radio spectrum can be very grave unless proper precautions are taken that will minimize the amount of radio frequency interference emanating from such equipment.

CISPR Specifications.   The acronym CISPR stands for Comité International Special des Perturbations Radioelectriques (International Special Committee on Radio Interference). CISPR is organized under the general framework of the International Electrotechnical Commission (IEC), which is affiliated with the International Organization for Standardization (ISO). The purpose of CISPR is to permit satisfactory reception of radiotelecommunication services and to facilitate international trade. Countries currently participating in CISPR activities include:

- Austria
- Belgium
- Brazil
- Canada
- Czechoslovakia
- Denmark
- Finlad
- Germany—FDR
- Germany—GDR
- Hungary
- Israel
- Italy
- Japan
- Netherlands
- Norway
- Poland
- Rumania
- South Africa
- Spain
- Sweden

- Switzerland
- U.K.
- United States
- U.S.S.R.
- Yugoslavia

There are 20 CISPR Publications dealing with the subjects of EMI-measuring instrumentation, test recommendations and study questions, CISPR and national specification limits, and rules and regulations. Equipments and devices subject to CISPR radiated and/or conducted EMI emissions control include ignition systems, ISM equipment, electrical appliances, radio and TV broadcast receivers, high-voltage power lines, and electric vehicles

CISPR limits on radio interference (Publication #9) are adopted directly or in modified form by most of the above member nations as their national limits. Similarly, test instrumentation (e.g., receiver bandwidth, synchronous tuning, overload protection, quasi-peak detector, etc.) and test methods and procedures of most of the member nations follow the CISPR publications. (CISPR test instrumentation and methods and procedures are the subject of Chapter 20.)

SAE Standard J551C and 1113A. The Society of Automotive Engineers issued Standard J551C in 1974, superseding earlier Standard J551B. This standard is entitled "Measurement of Electromagnetic Radiation from Motor Vehicles (20 MHz–1000MHz)." J551C provides uniform test procedures and recommends limits for the assistance of engineers in the measurement of electromagnetic radiation from motor vehicle ignition systems to ensure that their operation does not seriously interfere with RF communications or other electronic devices. The standard covers the measurement of radiation from all motor vehicle sources, including auxiliary engines with the exception of short-duty-cycle equipment such as used for starting motors, window regulators, and turn signals over the frequency range from 20 MHz to 1 GHz.

SAE J551C describes the measurement setup and orientation of the measuring antenna with regard to the test vehicle position. The topics of accuracy, impulse bandwidth, antenna height, distance and polarization, test area requirements including ambient noise, and test procedures are discussed. SAE 1113A deals with the susceptibility of automotive electronics to EMI.

NACSEM 5100. FED-STD-222 was issued in 1965, superseding NAG-1A. FED-STD-222 was entitled "Radiation Standard for Communication and Other Information Processing Equipment." It has been superseded by NACSEM's 5100 series published in 1970 to 1973. NACSEM is an acronym for National COMSEC/EMSEC Information Memoranda. These standards are classified and required for use by all U.S. federal agencies concerned with either conducted or radiated emissions where classified information may be involved. These standards specify test procedures for identifying the emanation characteristics of individual equipment in the laboratory environment, and signal limits for the emanations to provide an acceptable degree of protection from possibly compromising the security of the emanations.

Filter and Shielded Enclosure Testing

The MIL-STD-220 family covers methods to be used by Department of Defense (DoD) agencies in conducting measurements of filter performance. It is entitled "Methods of Insertion Loss Measurement." The first of this series was issued on June 25, 1952 and was subsequently superseded on December 25, 1959. The MIL-STD-220 series presents a method of measuring, in a 50-ohm system, the insertion loss of feed-through suppression capacitors and single- and multiple-circuit RF filters. The standard describes basic test setups employing signal generators and tuned receivers for measuring the insertion loss of the test specimen up to 130 dB.

MIL-STD-285, Shielded Enclosure Tests. MIL-STD-285 was issued June 25, 1956, and superseded MIL-A-1823 (Ships) of August 1, 1954. The standard is entitled "Attenuation Measurements for Enclosures, Electromagnetic Shielding for Electronic Test Purposes, Method of." MIL-STD-285 was issued as a Department of Defense standard and is mandatory for use by the tri-service agencies.

MIL-STD-285 covers methods of measuring the attenuation characteristics of electromagnetic shielding enclosures used for electronic test purposes over the frequency range from 100 kHz to 10 GHz. The techniques employed have since been extended down to power-line frequencies and for certain applications up to 26 GHz. Basically the standard gives test procedures for obtaining the shielding effectiveness versus frequency for various barriers to low impedance magnetic fields, high impedance electric fields, and plane waves. See the appendix to Chapter 10 for additional information on testing for compliance to MIL-STD-285.

SYSTEM- AND VEHICLE-LEVEL TESTING

The preceding section discussed EMI testing at the lowest level: component and equipment-level testing. After specification limits for this level of testing have been established, measurements have proved that EMI limit compliance exists, the next level of testing may be undertaken, the system- and vehicle-level tests (3). However, there is no guarantee that EMC will exist at this middle stage because:

1. Lowest-level test limits may not have been properly chosen.
2. Co-located cables and equipment may be closer together than the test item-to-antenna distance originally provided for in the radiated test specifications.
3. Equipment sharing a common power supply will be fed and loaded by a different configuration, source, return impedance, and grounding scheme from that of the tests.
4. Measurement errors will exist in lowest-level tests.

Despite some of these possible causes of EMI at the system level, lowest-level testing compliance will certainly reduce the number and severity of EMI problems at this point.

This section surveys EMC tests that may be called "System Electromagnetic Compatibility Tests"; or in one DoD version, they are known as MIL-E-6051D tests (see Fig. 18-1, above).

This middle level of testing is performed in a sterile electromagnetic environment, and only the EMI interaction of components, equipment, and subsystems is measured on the specimen system or vehicle. Here the test may involve any of the following:

- Aircraft
- Automobile, truck, or bus
- Building
- Combat tank
- Computer and peripherals
- Industrial control system
- Missile or rocket
- Navigation system

- Radar systems
- Rapid transit system
- Reconnaissance system
- Ship
- Surveillance system
- Submarine
- Telecommunications system
- Telemetry system

Note that each system operates as an entity, more or less independent of the outside world, and each has one or more specific purposes or missions. Some systems, such as vehicles, are themselves an ensemble of systems. One example is an aircraft that is essentially self-contained and contains complete power mains, communications, navigation, radar, housekeeping, and payload systems (sometimes called subsystems). Whether the item under test is a system or a collection of systems, this section surveys tests at this level.

System Electromagnetic Compatibility Tests

System-level EMC testing is performed with the system "going through its paces" and with all potentially susceptible receptors being monitored for indications of interference-induced upset. The tests, in other words, are used to determine if any intrasystem jamming is occurring, the culprits involved, and a measure of the amount of EMI hardening required to attain EMC within the overall system.

Malfunction Indication. In an intrasystem EMI test, the system is used to determine its own self-degradation as it is operated as it would be during a typical mission. It is necessary to determine what constitutes a malfunction and how it is measured or calibrated. The answer lies in what constitutes the performance specifications of the system and its associated subsystems; for if the system is performing out of spec because of an EMI-induced failure, then a measure of EMI degradation can be defined.

Some examples of malfunctions and hypothetical scoring calibrations are indicated in Table 18-1. There is a gray area of marginal function, as most performance malfunctions are not two-state (i.e., go/no-go) in nature but represent a degree of degradation or mission compromise. Note, however, that in the last device, engine overheat alarm relay, there is no margin for malfunction; the alarm is either activated, or it is not. This situation is often the exception rather than the rule. Usually scoring criteria representing the shades of gray or marginal performance are required to define and evaluate malfunction response.

System Exercise and Loading. To realistically test the EMC of any system, it must be tested under real-life situations. This is "mission profile" or

Table 18-1. List of hypothetical scoring criteria for EMI sub-system malfunctions. (Courtesy of Interference Control Technologies.)

System or subsystem	Hypothetical scoring criteria		
	No malfunction	Marginal	Malfunction
Navigation (DME)	< 0.3 km	0.3 to 0.8 km	> 0.8 km
Aircraft control surface movement	< 1 mm	1 to 4 mm	> 4 mm
Digital computer bit error rate	< 10^{-7}	10^{-7} to 10^{-6}	> 10^{-6}
Voice communication word intelligibility	> 95%	85 to 95%	< 85%
Engine overheat alarm relay	No alarm	No alarm	Alarm

"mission scenario" testing. It must be emphasized that all systems are to be "worked" during the EMC testing runs. For example, under mechanical or aerodynamic load, motors, sensors, relays, servos, and actuators will draw more current then under no-load conditions. The loading of the power system may increase the duty cycle of the power supply and can, in the case of a switching power supply, cause an increase in electrical noise in the system. Or, the voltage may drop momentarily (an intrasystem EMI problem due to poor regulation), and systems may be prone to kicking out. In the case of rotating machinery, the increased loading causes an increase in brush-commutator sparking, and thus more noise. Synthetic or artificial loading may be required if real loading cannot be tested, as in the case of a missile. It is important to operate the system in accordance with the mission scenario.

Mission scenarios, designed to test the object, must be done quickly and at low cost. The best way to do so is to establish a schedule for testing. The schedule would be written to simulate the turning on and off of systems in accordance with what the system would do in an actual operational mode. A program like this would also provide for artificially induced operator errors that might occur.

During the testing, malfunction data would be collected and compared against the system operating at the time, providing evidence that would correlate EMI malfunction with the culprit.

Instrumenting the Test Plan. The malfunctions may be sensed and recorded by using personnel to man the stations of output displays such as panel meters, recorders, scope displays, digital presentations, and the like, where the outputs are intended for human interpretation. Here, malfunctions may be scored by personnel using log sheets in accordance with previously defined scoring criteria.

Malfunctions may also be monitored automatically by sensors recording a current or voltage that is an analog of the sensed parameter (2). This technique is especially useful when events are occurring too quickly for human observation, or when the tests involve a large system that must be tested as quickly as possible and/or as economically as possible. The system under test may still be operated by humans in the mission scenario, but in order to record events faithfully, a multichannel recording of events versus time is employed.

Finally, certain auxiliary instruments may be used during the monitoring of malfunctions versus mission scenario to help identify EMI culprits. Such instruments may be used as current probes around certain buses or wiring harnesses, or one or more small loops and/or electric-field rods may be used to probe an area. For these data recordings, plots of peak field amplitude versus time are made to permit all data to be presented on an event-reaction versus time basis. This allows cause-and-effect analysis to be made.

MIL-E-6051D-Tests

This specification is regarded as a special case of the system-level EMC testing discussed above. It also is concerned with other topics. MIL-E-6051D is captioned "Electromagnetic Compatibility Requirements" and is mandatory for use by all departments and agencies of the DoD.

MIL-E-6051D outlines the overall requirements for a system's EMC, including control of system EME, lightning protection, static electricity, bonding, and grounding. It is applicable to complete systems, including associated subsystems and equipment. The emphasis is on intrasystem compatibility whether or not any equipment within the system conforms to other applicable specifications. The test procedures in MIL-E-6051D are general, and do not specify individual equipment or detailed procedures to be followed. Instead, they require the preparation and use of a detailed test plan by the contractor.

The MIL-E-6051D tests are supposed to serve as a final check on the EMI characteristics of a complete system, and are designed to test EMI susceptibility of the actual system in its actual operational environment. Because the tests include all manner of equipment, they require very lengthy preparation and implementation.

The test procedures outlined in the standard are very general. The contractor performing the tests is responsible for preparing a detailed test plan, and there are three basic approaches that can be followed in setting up the procedures for the compatibility demonstration tests required by the standard. These approaches, which may be used either singly or collectively to fulfill the specification requirements, are as follows:

1. Inject emissions into the system at critical points. The injected conducted emission must be at a level that is 6 dB higher than predetermined levels created by the system. Appropriate system test points must then be monitored for malfunction indication.
2. Increase the sensitivity level of the system so that its susceptibility level to interference is increased by the required 6 dB. With this approach, it is also necessary to monitor appropriate system points to determine if a malfunction(s) occurs. This increase in sensitivity by 6 dB is often difficult or impossible to achieve; so this approach is seldom used.
3. Measure the interference susceptibility of key subsystems and system circuits and compare these measured susceptibility levels with existing noise emission levels to determine whether the required 6 dB margin exists.

Each of the above approaches has both advantages and disadvantages. The best approach for any particular system will, of course, depend on that system. In some cases it may be desirable to implement a combination of the above approaches.

Basically, the EMC requirements of MIL-E-6051D dictate that the system operate individually and collectively as per performance specs, and that a susceptibility margin exist of at least 6 dB (20 dB for electro-explosive devices) between the susceptibility and the electromagnetic noise environment that is produced by the total system. This means that any and all elements of the system must be designed to operate with twice the noise level that actually exists in the system, regardless of its frequency, duration, modulation, waveform, repetition rates, duty cycle, sequence, time amplitude, or any other characteristic that is used to define noise. The primary concern is with whether the noise affects the systems and with the manner in which it does so.

It is necessary to evaluate existing noise levels in terms of how the system is affected by the noise. In many cases this is done simply by measuring the effect of the noise on functional elements or circuits in the system. However, in order to determine whether the system will operate as designed, with the required safety margin, it is necessary to utilize test methods and techniques that permit close and accurate time and event correlation of all the quantities measured.

A more recent interpretation of MIL-E-6051D is designed to greatly simplify the test methods and procedures. It involves comparing all receptor output indications with allowable departures; where an EMI cause–effect is observed, it is confirmed that it exists within one-half of the allowance (6 dB), or one-tenth of that (20 dB) for EEDs. The interpretation of 0.5 and 0.1 safety margins (see "Malfunction Indication") is innate to this approach, especially in the realization that many devices operate with nonlinear outputs.

Another interpretation of MIL-E-6051D involves measuring each receptor response to system scenario excitation and comparing these responses to one-half of that (-6 dB) allocated as a susceptibility criterion. If the response is below one-half of that permitted, then the 6 dB safety margin criterion are considered to be met. This will not work for go/no-go situations (e.g., a relay) unless analog voltages or currents are also monitored. It also requires further interpretation for nonlinear system. For a 20 dB safety margin, which is often reserved for squibs or EEDs, a squib simulator output is measured.

CONDUCTED AND RADIATED EMI TERMS

This section discusses the terminology for conducted voltage and current measurement. It also presents electric and magnetic fields in the near and far field. Antenna factors for emission or susceptibility measurements are discussed. Discussions of transients, bandwidth, narrowband and broadband emissions (coherent and incoherent), and detector functions are presented in subsequent sections of this chapter (6).

Conducted EMI Terms

Conducted Voltage Reference. In many electronic and related technical disciplines the fundamental unit of signal or noise amplitude is power, P. To facilitate discussion of large ranges of power, the decibel (dB) system is used, in which the reference is the watt (1 W = 1 joule/sec). Thus, expressing power in dB above 1 watt (see Chapter 2):

$$P_{dBW} = 10 \log_{10} P_W \text{ dBW, for } P_W = \text{power in watts} \qquad (18\text{-}1)$$

(All logarithms will be understood to be to the base 10 unless otherwise specified.)

In many applications the milliwatt (mW or simply m) is used as the power reference. Some examples include signal-generator output calibrations, receiver sensitivities, path loss calculations, and the like. In units of dB, the mW is related to the watt as follows:

$$1 \text{ mW} = 10^{-3} \text{ watt} \qquad (18\text{-}2)$$

or:

$$1 \text{ mW} = 0 \text{ dBm} = -30 \text{ dBW} \qquad (18\text{-}3)$$

Thus:

$$1 \text{ W} = 0 \text{ dBW} = +30 \text{ dBm} \qquad (18\text{-}4)$$

Consequently, the power in P_{dBm} is:

$$P_{dBm} = P_{dBW} + 30 \text{ dB} \qquad (18\text{-}5)$$

EMI specifications, limits, calibrations, and measurements rarely use power as a reference. One reason for this is that signal and noise amplitude measurements are stressed, and, if broadband, they could be either coherent or incoherent. Direct power measurements do not identify coherence, and significant errors could result in some tests involving translation of bandwidth. In any event, the EMI community uses voltage, V, as the basic reference unit for conducted measurements.

The voltage is derived from power:

$$P = \frac{V^2}{R} \text{ watts} \qquad (18\text{-}6)$$

where:

V = circuit voltage in volts
R = circuit impedance in ohms across which V is measured

Hence, combining equations (18-1), (18-5), and (18-6) yields:

$$P_{dBW} = 10 \log (V^2/R) \qquad (18\text{-}7)$$

or:

$$P_{dBm} = 10 \log (V^2/R) + 30 \text{ dB} \qquad (18\text{-}8)$$

The voltage ratio, V_r, of either two networks or one network under different conditions is obtained from their power ratios, P_r:

$$P_r = \frac{P_1}{P_2} = \frac{V_1^2/R_1}{V_2^2/R_2} = \frac{V_1^2 R_2}{V_2^2 R_1} \qquad (18\text{-}9)$$

$$P_r = 10 \log (P_1/P_2) = 10 \log (V_1/V_2)^2 + 10 \log (R_2/R_1) \qquad (18\text{-}10)$$

Thus, V_r as a ratio in dB equals P_r only when $R = R_1 = R_2$:

$$V_r = 10 \log (V_1/V_2)^2 = 20 \log (V_1/V_2) \qquad (18\text{-}11)$$

When $V_2 =$ the reference voltage in volts, equation (18-11) becomes:

$$V_{dBV} = 20 \log V_1 \qquad (18\text{-}12)$$

The EMI community uses the microvolt, μV, as the basic unit of reference voltage:

$$1 \, \mu V = 10^{-6} \, V = 0 \, dB\mu V \qquad (18\text{-}13)$$

Substituting equation (18-13) into equation (18-12) yields:

$$V_{dB\mu V} = 20 \log (10^{-6})V = 20 \log V - 120 \text{ dB} \qquad (18\text{-}14)$$

To convert from units of P_{dBm} to $V_{dB\mu V}$, equations (18-8) and (18-14) are used:

$$P_{dBm} = 10 \log [(10^{-6}V_{\mu V})^2/R] + 30 \text{ dB}$$
$$= -120 \text{ dB} + 20 \log V_{\mu V} - 10 \log R + 30 \text{ dB} \qquad (18\text{-}15)$$

$$P_{dBm} = V_{dB\mu V} - 90 \text{ dB} - 10 \log R \qquad (18\text{-}16)$$

$$P_{dBm} = V_{dB\mu V} - 107 \text{ dB, for } R = 50 \text{ ohms} \qquad (18\text{-}17)$$

or

$$V_{dB\mu V} = P_{dBm} + 107 \text{ dB, for } R = 50 \text{ ohms} \qquad (18\text{-}18)$$

Conducted Current Interference. Several EMI conducted specification limits (10, 12) are given in units of current and in terms of the microampere. This might suggest that a small resistor is added in series with the test lead and the voltage drop across the known resistor, R, is measured with an EMI receiver to determine the unknown current.

$$I_{\mu A} = \frac{V_{\mu V}}{R} \qquad (18\text{-}19)$$

or:

$$I_{dB\mu A} = V_{dB\mu V} - 20 \log R \qquad (18\text{-}20)$$

The above practice is rarely followed, since it disturbs the test items and is impractical when many wires are involved. Consequently, a current probe is used.

Transfer Impedance and Current Probes. Transfer impedance is important when discussing a measuring device (transducer) called a current probe (see Chapter 19). A current probe is essentially a current-to-voltage transformer that operates by converting the current in the sample wire into a voltage at the output terminals of the probe. Thus, it is useful to describe the characteristics of the current probe by a quantity called the transfer impedance, Z_T, which is the ratio of the output voltage of the probe to the input current:

$$Z_T = \frac{V_{out}}{I_{in}} \qquad (18\text{-}21)$$

where V_{out} = the output voltage across the current probe when terminated in 50 ohms (the EMI receiver), and I_{in} = unknown current flowing in the wire(s) around which the probe is placed.

The manufacturer furnishes the transfer impedance, Z_{dBohms}, of the current probe as a function of frequency as the transducer calibration, and the operator measures the voltage, $V_{dB\mu V}$. Subtracting the transfer impedance (in dB) at the frequency of interest from the measured voltage (in dB) at that frequency yields the unknown current (in dB).

Radiated EMI Terms

Near and Far Fields. The previous section discussed reference units for conducted voltage and current used by the EMI community. Now electromagnetic radiation in units of electric and magnetic field intensity will be discussed, but first it is necessary to review the concept of the near field and the far field. (the latter sometimes referred to as plane waves or a plane wave field). Additional information on field theory may be found in Chapter 10.

The power flux flow or simply the power density, P_D, in units of watts/meter^2 is:

$$P_D = E \times H \text{ W/m}^2 \tag{18-22}$$

where

E = electric field intensity in V/m
H = magnetic field intensity in A/m

The electric and magnetic field intensities are related by the wave impedance, Z, in ohms:

$$Z = \frac{E}{H} \text{ ohms} \tag{18-23}$$

As shown in Chapter 10, $Z = 120\pi$ ohms or 377 ohms for far field, free-space conditions only. Thus, for plane waves, the electric and magnetic fields are uniquely related by 377 ohms. Such a unique relation does not exist in the near field. In fact, it will be shown that Z may assume any value [e.g., a small fraction of 377 ohms for magnetic (low impedance) fields, i.e., less than 377 ohms; or many times 377 ohms for electric (high impedance) fields]. The value for Z in the near field is also complex in nature; that is, it contains an imaginary term. Note that equation (18-23) is still valid in the near field. With any time-varying electric field there exists an associated magnetic field and vice versa. This is a direct result of Maxwell's equations.

The near/far field interface distance, r, depends on the wavelength, λ, of the field compared to the dimension, D, of the field source (whether wire or aperture). When the size, $D \ll \lambda$, the near/far field interface dimension, r, is defined by (see Chapter 10):

$$r = \frac{\lambda}{2\pi}, \text{ for } D \ll \lambda \tag{18-24}$$

When $D \geq \lambda$, the interface distance is:

$$r = \frac{D^2}{2\lambda}, \text{ for } D \geq \lambda \tag{18-25}$$

Note that when $D = \lambda/2$, equation (18-25) becomes $r = \lambda/8$, which nearly equals equation (18-24).

Electric Field Reference.   An approach similar to that previously discussed regarding power versus voltage is also developed here. Power density, P_D, previously defined in equation (18-22), is used as the basic unit of power flux flow by the microwave community. Here, far field conditions generally apply at higher frequencies owing to the inverse relationship of frequency and wavelength. The EMI and broadcast communities, however, prefer the field intensity term to power density. The EMI community also states its EMI specification limits in units of field intensity. The two are related as follows:

$$P_D = \frac{E^2}{Z} \text{ watts/m}^2 \qquad (18\text{-}26)$$

$$P_{\text{dBW/m}^2} = 20 \log E_{\text{V/m}} - 10 \log Z \qquad (18\text{-}27)$$

or:

$$P_{\text{dBm/m}^2} = E_{\text{dBV/m}} - 10 \log Z + 30 \text{ dB}$$
$$P_{\text{dBm/m}^2} = E_{\text{dB}\mu\text{V/m}} - 90 \text{ dB} - 10 \log Z \qquad (18\text{-}28)$$
$$= E_{\text{dB}\mu\text{V/m}} - 116 \text{ dB, for } Z = 377 \text{ ohms}$$

or:

$$E_{\text{dB}\mu\text{V/m}} = P_{\text{dBm/m}^2} + 116 \text{ dB, for } Z = 377 \text{ ohms} \qquad (18\text{-}29)$$

where:

E = electric field intensity in V/m
$E_{\text{dB}\mu\text{V/m}} = E$ in terms of dBµV/m
Z = wave impedance = 377 ohms, far field conditions only

Equation (18-29) is tabulated in Table 18-2 for various units. Note that equations (18-16) and (18-28) are nearly identical; the former involves conducted voltages, and the latter applies to radiated electric fields.

Magnetic Field Reference.   From equation (18-23) the magnetic field intensity in A/m is defined in terms of the electric field:

$$H_{\text{A/m}} = \frac{E_{\text{V/m}}}{Z_{\text{ohms}}} \qquad (18\text{-}30)$$

or:

$$H_{\mu\text{A/m}} = \frac{E\mu_{\text{V/m}}}{Z_{\text{ohms}}} \qquad (18\text{-}31)$$

$$H_{\mu\text{A/m}} = E_{\text{dB}\mu\text{V/m}} - 20 \log Z \qquad (18\text{-}32)$$

$$H_{\mu\text{A/m}} = E_{\text{dB}\mu\text{V/m}} - 52 \text{ dB for free space} \qquad (18\text{-}33)$$

The EMI community uses units of magnetic flux density, B, rather than magnetic field intensity:

$$B = \mu \times H \text{ Tesla } (1 \text{ T} = 1 \text{ weber/m}^2 = 10^4 \text{ gauss})$$
$$= (4\pi \times 10^{-7})\mu_r \text{ henry/m} \times H_{\text{A/m}} \qquad (18\text{-}34)$$

Table 18-2. Field intensity and power density relationships (Related by free space impedance = 377 ohms).

Volts/m	dBμV/m	Watts/m^2	dBW/m^2	Watts/cm^2	dBW/cm^2	mW/cm^2	dBm/cm^2
10,000	200	265,000	+54	27	+14	26,500	+44
7,000	197	130,000	+51	13	+11	13,000	+41
5,000	194	66,300	+48	6.6	+8	6,630	+38
3,000	190	23,900	+44	2.4	+4	2,390	+34
2,000	186	10,600	+40	1.1	0	1,060	+30
1,000	180	2,650	+34	.27	-6	265	+24
700	177	1,300	+31	.13	-9	130	+21
500	174	663	+28	.066	-12	66	+18
300	170	239	+24	.024	-16	24	+14
200	166	106	+20	.011	-20	11	+10
100	160	27	+14	27×10^{-4}	-26	2.7	+4
70	157	13	+11	13×10^{-4}	-29	1.3	+1
50	154	6.6	+8	6.6×10^{-4}	-32	.66	-2
30	150	2.4	+4	2.4×10^{-4}	-36	.24	-6
20	146	1.1	0	1.1×10^{-4}	-40	.11	-10
10	140	.27	-6	27×10^{-5}	-46	.027	-16
7	137	.13	-9	13×10^{-6}	-49	.013	-19
5	134	.066	-12	6.6×10^{-6}	-52	66×10^{-4}	-22
3	130	.024	-16	2.4×10^{-6}	-56	24×10^{-4}	-26
2	126	.011	-20	1.1×10^{-6}	-60	11×10^{-4}	-30
1	120	27×10^{-4}	-26	27×10^{-8}	-66	2.7×10^{-4}	-36
0.7	117	13×10^{-4}	-29	13×10^{-8}	-69	1.3×10^{-4}	-39
0.5	114	6.6×10^{-4}	-32	6.6×10^{-8}	-72	66×10^{-4}	-42
0.3	110	2.4×10^{-4}	-36	2.4×10^{-8}	-76	24×10^{-6}	-46
0.2	106	1.1×10^{-4}	-40	1.1×10^{-8}	-80	11×10^{-6}	-50
0.1	100	27×10^{-6}	-46	27×10^{-10}	-86	2.7×10^{-6}	-56
70×10^{-3}	97	13×10^{-6}	-49	13×10^{-10}	-89	1.3×10^{-6}	-59
50×10^{-3}	94	6.6×10^{-6}	-52	6.6×10^{-10}	-92	66×10^{-8}	-62
30×10^{-3}	90	2.4×10^{-6}	-56	2.4×10^{-10}	-96	24×10^{-8}	-66
20×10^{-3}	86	1.1×10^{-6}	-60	1.1×10^{-10}	-100	11×10^{-8}	-70
10×10^{-3}	80	27×10^{-8}	-66	27×10^{-12}	-106	2.7×10^{-8}	-76
7×10^{-3}	77	13×10^{-8}	-69	13×10^{-12}	-109	1.3×10^{-8}	-79
5×10^{-3}	74	6.6×10^{-8}	-72	6.6×10^{-12}	-112	66×10^{-10}	-82
3×10^{-3}	70	2.4×10^{-8}	-76	2.4×10^{-12}	-116	24×10^{-10}	-86
2×10^{-3}	66	1.1×10^{-8}	-80	1.1×10^{-12}	-120	11×10^{-10}	-90
1×10^{-3}	60	27×10^{-10}	-86	27×10^{-14}	-126	2.7×10^{-10}	-96
700×10^{-6}	57	13×10^{-10}	-89	13×10^{-14}	-129	1.3×10^{-10}	-99
500×10^{-6}	54	6.6×10^{-10}	-92	6.6×10^{-14}	-132	66×10^{-12}	-102
300×10^{-6}	50	2.4×10^{-10}	-96	2.4×10^{-14}	-136	24×10^{-12}	-106
200×10^{-6}	46	1.1×10^{-10}	-100	1.1×10^{-14}	-140	11×10^{-12}	-110
100×10^{-6}	40	27×10^{-12}	-106	27×10^{-16}	-146	2.7×10^{-12}	-116
70×10^{-6}	37	13×10^{-12}	-109	13×10^{-16}	-149	1.3×10^{-12}	-119
50×10^{-6}	34	6.6×10^{-12}	-112	6.6×10^{-16}	-152	66×10^{-14}	-122
30×10^{-6}	30	2.4×10^{-12}	-116	2.4×10^{-16}	-156	24×10^{-14}	-126
20×10^{-6}	26	1.1×10^{-12}	-120	1.1×10^{-16}	-160	11×10^{-14}	-130
10×10^{-6}	20	27×10^{-14}	-126	27×10^{-18}	-166	2.7×10^{-14}	-136
7×10^{-6}	17	13×10^{-14}	-129	13×10^{-18}	-169	1.3×10^{-14}	-139
5×10^{-6}	14	6.6×10^{-14}	-132	6.6×10^{-18}	-172	66×10^{-16}	-142
3×10^{-6}	10	2.4×10^{-14}	-136	2.4×10^{-18}	-176	24×10^{-16}	-146
2×10^{-6}	6	1.1×10^{-14}	-140	1.1×10^{-18}	-180	11×10^{-16}	-150
1×10^{-6}	0	27×10^{-16}	-146	27×10^{-20}	-186	2.7×10^{-16}	-156

where:

$$\mu_r = \text{relative permeability of the medium} = \frac{\mu}{\mu_0}$$

Thus:

$$B_{dBT} = H_{dBA/m} - 118 \text{ dB} \qquad (18\text{-}35)$$

$$B_{dBT} = H_{dB\mu A/m} + 2 \text{ dB} \qquad (18\text{-}36)$$

Since the Tesla or dBT is a large unit of magnetic flux density, the EMI community has chosen the picoTesla (pT) (10^{-12} T) as its basic unit. Consequently:

$$B_{dBpT} = B_{dBT} + 240 \text{ dB} \qquad (18\text{-}37)$$

and combining equations (18-35) and (18-37) yields:

$$B_{dBpT} = H_{dBA/m} + 122 \text{ dB} \qquad (10\text{-}38)$$

and combining equations (18-30), (18-34), and (18-38) yields:

$$B = \frac{\mu E}{Z} \qquad (18\text{-}39)$$

or:

$$B_{dBpT} = E_{dBV/m} + 70 \text{ dBpT/V/m, for free space} \qquad (18\text{-}40)$$

$$B_{dBpT} = E_{dB\mu V/m} + 190 \text{ dBpT}/\mu V/m, \text{ for free space} \qquad (18\text{-}41)$$

Finally, it is useful to develop a few other magnetic flux density relations, since one may use these conversions on occasion when interfacing with other disciplines. Since 1 T = 10^4 gauss:

$$B_{dBpT} = B_{dBgauss} + 160 \text{ dB} \qquad (18\text{-}42)$$

and since 1 gamma = 10^{-9} Tesla, and 1 milligamma = 1 pT:

$$B_{dBpT} = B_{dBgamma} + 60 \text{ dB} \qquad (18\text{-}43)$$

or:

$$B_{dBpT} = B_{dBmgamma} \qquad (18\text{-}44)$$

Antenna Factor

Antenna factor is a term used by the EMI community to define the antenna transducer calibration relation for making radiated emission measurements.

It is a measure of the efficiency of the antenna in converting a field incident on the elements of the antenna to a voltage at the output terminals of the antenna. Thus, antenna factor, AF, is:

$$AF = \frac{E}{V} \tag{18-45}$$

where:

E = unknown electric field to be measured in V/m (or μV/m)
V = voltage at the output terminals of the measuring antenna in V (or μV).

The unit of AF in equation (18-45) is meter^{-1}. Expressed in dB, equation (18-45) becomes:

$$AF_{dB/m} = AF_{dB} = E_{dB\mu V/m} - V_{dB\mu V} \tag{18-46}$$

$$E_{dB\mu V/m} = V_{dB\mu V} + AF_{dB} \tag{18-47}$$

The antenna factor m^{-1} unit is dropped in the above equations—manufacturers and others do this out of either carelessness or poetic license. One then measures the voltage $V_{dB\mu V}$, and adds the antenna factor in dB corresponding to the measurement frequency, as shown in Fig. 18-3.

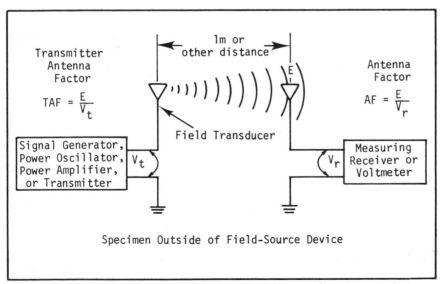

Fig. 18-3. Antenna factors (emission and susceptibility). (*Courtesy of Interference Control Technologies*)

Antenna factors are also a measure of the figure of merit of the antenna transducer. As seen from equation (18-45), if a given E produces a big V, the transducer is sensitive, and the antenna factor is small. On the other hand, if a small voltage is developed, then the transducer is insensitive, and AF is large. The typical range of antenna factors for passive antennas is from about 0 dB to $+60$ dB. Table 18-3 lists antenna factors for typical antennas used in measuring radiated EMI emissions.

Another definition of antenna factor involves the effective height of the antenna and its efficiency in delivering the induced voltage to the antenna load:

$$AF = \frac{1}{h_e A_e} \qquad (18\text{-}48)$$

or:

$$AF_{dB} = -h_{edB} - A_{edB} \qquad (18\text{-}49)$$

where:

h_e = antenna electrical height in meters

A_e = antenna efficiency or ratio of V_1/V_0

V_0 = antenna induced voltage (i.e., available or open circuit voltage)

V_1 = antenna induced voltage delivered to a load (i.e., a 50-ohm EMI receiver)

The term electrical height means the effective height (or length) of the antenna in coupling to an electric field. It does not mean the height of the antenna above earth, although this height also affects the reception.

Table 18.3. Range of Antenna Factors of Receiving Antennas

Antenna	Below/MHz	1–30MHz	30–200MHz	.2–1GHz	1–10GHz
Capacitive Probe		NA	NA	NA	NA
Passive 41" Rod	58–35 dB	35–22dB	NA	NA	NA
Passive 81" Rod	52–29 dB	29–16dB	NA	NA	NA
Active 41" Rod	+6 dB	+6dB	NA	NA	NA
Active 6" Probe	+23 dB	+23dB	+23dB	NA	NA
Tunable Dipole	NA	NA	–2to+14dB	14–28dB	NA
Broadband Dipole	NA	NA	0 – +18dB	NA	NA
Bi-Conical	NA	NA	7 – 18dB	NA	NA
Conical Log Sp.	NA	NA	NA	17–26dB	25–48dB
Ridged Guide	NA	NA	NA	11–18dB	21–40dB

The effective height of a rod or whip antenna above a ground plane is equal to one-half its mechanical or physical height because the induced voltage increment is zero at the base and increases linearly to ΔV, the increment per unit length at the top. Integrated over the mechanical height of the rod then, the electric field induces a voltage of $(V/2)h = V(h/2)$. Thus, a 1-meter rod has an $h_e = 0.5$ meter. The h_e for a half-wave dipole is λ/π meters.

Substituting for the antenna efficiency in equation (18-48) and then substituting into equation (18-45) gives the following expression:

$$AF = \frac{E}{V_1} = \frac{1}{h_e A_e} = \frac{1}{h_e V_1/V_0} = \frac{V_0}{h_e V_1} \qquad (18\text{-}50)$$

or:

$$E_{\mathrm{dB\mu V/m}} = V_{\mathrm{odB\mu V}} - h_{e\mathrm{dB\ meter}} \qquad (18\text{-}51)$$

$$V_{\mathrm{odB\mu V}} = E_{\mathrm{dB\mu v/m}} + h_{e\mathrm{dB\ meter}} \qquad (18\text{-}52)$$

Equation (18-53) is the expression for antenna induced voltage given in some of the earlier EMI specifications such as MIL-I-618D and MIL-I-26600 (8, 11). It is the same as the electric field intensity corrected by the height of the antenna. Thus, in those specification limits it is necessary to use the correct (i.e., specified) antenna.

The antenna factor may be developed either theoretically or experimentally using techniques described in Section 3.1 for Volume 4, DWCI Handbook Series (7). Either approach, however, is based on far field conditions and applies to antennas loaded by a 50-ohm receiver. The experimental approach for determining AF at a 1-meter distance from the test sample is based on SAE report ARP-958. It is not valid below about 50 MHz, since far field conditions then do not exist. The theoretical antenna factor is given in reference 7.

$$AF_{\mathrm{dB}} = 20 \log \frac{9.7}{\lambda \sqrt{G_r}} \qquad (18\text{-}53)$$

where

G_r = antenna power gain.

NARROWBAND AND BROADBAND EMI TERMS

There are two uses of the terms narrowband and broadband emissions: (1) the shop talk or colloquial version and (2) the measurement version. These two usages can become confusing, since they are quite different. Only the accepted measurement version is discussed here.

Narrowband Emissions and Identification

The term narrowband emission means that the emission bandwidth is narrower than or less than some reference bandwidth. Here, the reference bandwidth may be that associated with a potentially susceptible victim receptor. More specifically, the reference bandwidth is that of an EMI measurement receiver. Thus an emission source is narrowband when its 3 dB bandwidth is smaller than that of the EMI measurement receiver 3 dB bandwidth. As discussed in a later section, the definition of broadband emission is just the reverse.

To illustrate, consider a signal that is not continuous wave (CW) and does not exhibit a great bandwidth. For convenience of discussion, a single voltage pulse emission having carrier frequency f_c is selected, as shown in Fig. 18-4. The $\sin x/x$ distributing centered about f_c is recognized with the frequency axis crossings at $f_c \pm 1/t$, $f_c \pm 2/t$, and so on. Its greatest amplitude at f_c is $A \cdot t$ volt-seconds or V/Hz.

Figure 18-4 also shows an EMI receiver RF selectivity response having a bandwidth, B_n, centered about f_c. By inspection of the figure, one concludes that the pulse spectrum is narrowband because its 3 dB bandwidth, B_e, is less than the receiver bandwidth. In other words, it is a narrowband emission because $B_e < B_n$.

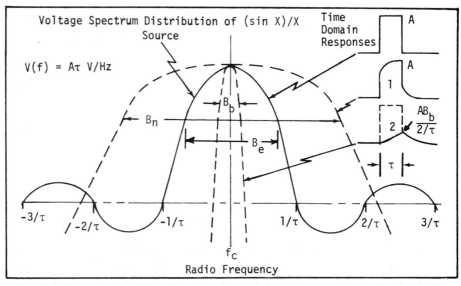

Fig. 18-4. Narrowband and broadband transients relative to the measuring receiver bandwidth. (*Courtesy of Interference Control Technologies*)

To obtain useful information when conducting EMI emissions measurements and recognition testing, there must be one or more tests for broadband or narrowband. One such test is the tuning test, and another is the bandwidth change test.

Narrowband Tuning Test.   If the receiver, with bandwidth B_n, in Fig. 18-4 is tuned up or tuned down in frequency by an amount equal to its own 3 dB bandwidth, and if the output level, ΔV, changes by more than 3 dB, then the emission source is narrowband. If there is a very large change (say 20 dB or more), then the emission is very narrowband. If the change is about 3 dB, however, the emission source has a transitional bandwidth; that is, a transition between narrow- and broadband is occurring. Thus, the tuning test will indicate narrowband, NB, when:

$$NB_{\text{tuning}}: \Delta V > 3 \text{ dB} \tag{18-54}$$

Narrowband Bandwidth Test.   The second test available for narrowband emission identification involves a change in the bandwidth of the EMI receiver. If a second bandwidth in the receiver with bandwidth B_n, in Fig. 18-4, is selected such that the new bandwidth is at least $2B_n$, and if the output level, ΔV, changes by an amount less than 3 dB, then the emission source is narrowband. A further increase in bandwidth would result in little or no perceptible change in output. If the change had been about 3 dB, the emission would be transitional. Thus the bandwidth test will indicate narrowband when:

$$NB_{\text{bandwidth}}: \Delta V < 3 \text{ dB} \tag{18-55}$$

Broadband Emissions and Identification

The term broadband emissions means that the emission bandwidth is broader or greater than some reference bandwidth. Here, the reference bandwidth may be that associated with a potentially susceptible victim receptor. More specifically, as discussed earlier, the reference bandwidth is that of an EMI measurement receiver. Thus, an emission source is broadband when its 3 dB bandwidth exceeds that of the 3 dB EMI receiver bandwidth.

Figure 18-4 is again used to illustrate the tuning and bandwidth test criteria. The EMI receiver now has an RF response as shown with bandwidth B_b, centered about the carrier f_c. By inspection of the figure, one concludes that the pulse emission is broadband, since its 3 dB bandwidth B_e is greater than the receiver bandwidth, or $B_e > B_b$. The preceding two tests will again be applied.

Broadband Tuning Test. If the receiver with bandwidth B_b is tuned up or down in frequency by an amount equal to its own 3 dB bandwidth, and the output level ΔV changes by less than 3 dB, then the emission source is broadband. If there is no perceptible change at all, the emission is extremely broadband relative to the receiver bandwidth. There again exists a transitional region when the change is about 3 dB. Thus, the tuning test will indicate broadband, BB, when:

$$BB_{\text{tuning}}\text{:}\ \Delta V < 3\text{ dB} \qquad (18\text{-}56)$$

Broadband Bandwidth Test. The second test available for determining if a signal is broadband or narrowband involves a change in the bandwidth of the EMI receiver. If a second bandwidth is selected that is at least half as large as the original, or $B_b/2$, and if the output level changes by an amount greater than 3 dB, then the emission source is broadband. If the change had been about 3 dB, then the emission would be transitional. Thus, the bandwidth test indicates broadband emission when:

$$BB_{\text{bandwidth}}\text{:}\ \Delta V > 3\text{ dB} \qquad (18\text{-}57)$$

Summary of Narrowband and Broadband Test

The foregoing may be summarized in a simple matrix as illustrated in Table 18-4.

Special situations may occur when the change in the test receiver output is approximately 3 dB, namely, (1) transitional situations or (2) broadband incoherence. Not much can be done for transitional situations although their impact in terms of identifying the emission type for EMI specification limit compliance can be significant. For example, if an emission is identified as either narrowband or broadband (i.e., a transitional situation), it may be

Table 18-4. Summary of narrowband and broadband emission identification criteria. (Courtesy of Interference Control Technologies.)

Type of test \ Type of emission	Narrowband	Broadband	Notes
RF tuning	>3 dB	<3 dB	(1)
Bandwidth change	<3 dB	>3 dB	(2)(3)

Notes: (1) Up or down frequency change equal to EMI receiver bandwidth.
(2) For narrowband test, increase to new bandwidth by at least two times.
(3) For broadband test, decrease to a new bandwidth of at least one-half.

within specifications for one situation and out of spec for another. While the tester could theoretically select the more favorable situation (if his objective were simply to pass the test sample), a better approach would be to select a different receiver bandwidth and perform the RF tuning test.

In the second special situation mentioned above, it is determined that an emission may be, or indeed is, broadband, if the receiver output level, ΔV, in the bandwidth change test, changes by any of the following:

(1) $\Delta V = 20 \log (B_{new}/B_{old})$, for coherent BB $\qquad$ (18-58)

(2) $\Delta V = 10 \log (B_{new}/B_{old})$, for incoherent BB $\qquad$ (18-59)

(3) $\Delta V =$ somewhere in between Eqs. (18-58) and (18-59)   (18-60)

For example, if B_{old} in Fig. 18-4 were B_b, and if B_{new} were $B_b/2$, then equation (18-58) would indicate a 6 dB change (BB coherent emission) while equation (18-59) would indicate a 3 dB change (BB incoherent emission that could be mistaken as transitional). These special situations are discussed in the following sections.

Coherent and Incoherent Broadband Emissions

A signal or emission is said to be coherent when neighboring frequency increments are related or well defined in both amplitude and phase. For broadband situations, neighboring amplitudes are both equal and in phase. Examples of coherent broadband emission sources are transients and pulsed sources such as computer clocks, radar, and pulse code modulation (PCM) telemetry, as long as the conditions of broadband emissions are confirmed.

A signal or emission is said to be incoherent when it is not coherent, that is, when neighboring frequency increments are random or psuedo-random (bandwidth limited) in either amplitude or phase. Examples of incoherent broadband emission sources are gas lamps (DC energized), noise diodes, black bodies including internal receiver noise, and corona discharge from high voltage sources.

Coherent Broadband Emissions.   In examining the spectrum of a rectangular pulse with pulse duration τ and amplitude step-change Δe, the amplitude is flat within 1.3 dB for frequencies between 0 and $(1/\pi\tau)$, and there is no phase change at all from $f = 0$ to $f = (1/\tau)$. Thus, this qualifies as a coherent emission. Consequently, the voltage developed under the $(\sin x/x)$ curve, V_{BC}, within the receiver bandwidth, B, for $f < (1/\pi\tau)$ is simply the area under the curve:

$$V_{BC} \simeq 2\tau\Delta eB, \quad \text{for } f \leq 1/\pi\tau \qquad (18-61)$$

or:

$$V_{dBV} = 20 \log 2\tau\Delta e + 20 \log B \tag{18-62}$$

where:

τ = pulse duration
Δe = amplitude step-change

All coherent broadband voltages are proportional to their receiver bandwidths, and any change in bandwidth yields a corresponding change in EMI receiver voltage level equal to that presented in equation (18-58).

Incoherent Broadband Emission.   When the voltage term is random from neighboring frequency increment to increment, the incremental voltages do not add in phase, but add in an RMS fashion. This incoherent emission within a bandwidth B is then:

$$V_{BI} = \sqrt{\int_0^B (2\tau\Delta e)^2 \, dB} \quad \text{for } f_B < \frac{1}{\pi\tau} \tag{18-63}$$

$$V_{BI} = \sqrt{(2\tau\Delta e)^2 B} = 2\tau\Delta e \sqrt{B} \tag{18-64}$$

and:

$$V_{dBV} = 20 \log 2\tau\Delta e + 10 \log B \tag{18-65}$$

All incoherent BB voltages are proportional to the square root of their bandwidths, and any change in bandwidth yields a corresponding change in EMI receiver voltage level equal to that presented in equation (18-59). This is further illustrated by thermal noise or bandwidth limited white noise in receivers. Receiver noise power, N, is calculated as:

$$N = FKTB \text{ watts} \tag{18-66}$$

since:

$$V = \sqrt{4RN} = \sqrt{4RFKTB} \tag{18-67}$$

where:

F = Noise figure of receiver
K = Boltzmann's constant (1.38×10^{-23} J/°K)
T = temperature in °Kelvin
B = bandwidth

Again, incoherent broadband noise voltage is proportional to the square root of bandwidth.

DETECTOR FUNCTIONS

The subject of detector functions is reviewed in this section with emphasis on the types of detector functions important to EMI measurement instruments. These types include: peak detection, slide-back detection, quasi-peak detection, average detection, RMS detection, and amplitude probability distribution (APD) detection. Some of these processes are peculiar to EMI measuring instruments and are not found in other receivers.

One function of the detector (second detector for superheterodyne receivers) is to remove the carrier and recover the baseband signal or emissions. While carrier removal is a very elementary and simple process, the detection network may alter the baseband signal in a number of ways, either intentionally or unwittingly (by errors in the detection process by either the designer or the user). The types of detection encountered in EMI measurements are:

1. Peak detection. Peak detectors will stretch pulses, and if no dumping is used over a measurement interval for which stretch decay is negligible, then the greatest upper bound of the baseband pulse-to-pulse emission is indicated.
2. Slide-back detection, which is an indirect type of peak detection and employs no stretching. It measures the peak by means of an adjustable back-bias detector set at the peak threshold.
3. Quasi-peak detection, which uses different rise- and fall-times to weight the circuit in such a manner as to simulate the impact of EMI on the ear for AM broadcast receivers.
4. Average detection, which develops the average level of the baseband modulation amplitude by weighting the emission with equal rise- and fall-times.
5. RMS detection, which is a true thermal detection process in which the equivalent noise temperature or equivalent power of the emission is measured.
6. Amplitude probability distribution detection, which organizes detected emissions over a time interval into an activity count at discrete levels for plotting cumulative probability distributions.

Each of these detection processes will now be discussed in detail.

Peak Detection

The detection and readout or display of the peak of an emission is particularly important in EMI measurements. Because so many potential EMI sources are impulsive in nature, and because an increasing number of potential EMI victims use pulse and digital circuits, the peak of an emission versus frequency per unit bandwidth becomes the most important measure of potential

EMI. For this and other reasons, several EMI specifications including MIL-STD-461A have their limits based upon peak values.

With certain exceptions, direct-reading types of peak detectors existing in EMI receivers produce errors ranging from 0 dB to about 20 dB, depending on the following:

1. The instrument manufacturer.
2. The IF bandwidth selected (larger bandwidths are more prone to larger errors due to faster rise-time requirements).
3. The pulse repetition rate.
4. The signal duty cycle, which is determined by both (2) and (3).
5. Whether a linear or logarithmic amplifier is used.
6. The time constant of the output display or recording device.

To make a peak detector function with not more than 1 dB error for all situations, from single shot transient or impulse to any pulse repetition frequency (PRF < 0.5 IF bandwidth) up to a CW signal of the same envelope amplitude, these conditions must apply when the following procedure is used: (1) logarithmic IF amplifiers, (2) the broadest IF bandwidth, and (3) X-Y plotters. The peak detection procedure is:

1. Choose the operating point of the transistors and type of IF logarithmic circuitry such that the gain is independent of the duty cycle the signal. Automatic gain control should not be used, since the loop gain is intentionally a function of the duty cycle, thus leading to errors in the peak detection circuitry.
2. The impedance of the charging source plus that of the forward diode resistance R_c, and capacitor C, shown in Fig. 18-5, represents a rise-time constant that should be the lesser of the following quantities:
   (a) One-tenth the reciprocal of the largest IF bandwidth.
   (b) The reciprocal of the carrier frequency of the IF if the ratio of IF carrier frequency to bandwidth is five or less. This is a carrier follower when only a few cycles of IF exist per pulse.
3. The decay time constant is on the order of 100 seconds, since a logarithmic response reduces the discharge time constant required for a slowly moving plotter pen to come up to full response.
4. Bootstrap, self-dumping of the detector output is used rather than a selectable dump interval to avoid truncating lower-level plotted signals.

Slide-Back Detection

The slide-back detector, shown in Fig. 18-6, is a special type of peak detector that is a carryover from older EMI receivers. The back-bias voltage V is

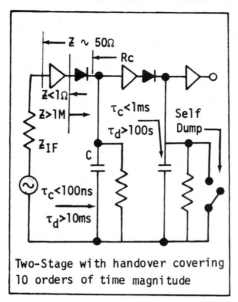

Fig. 18-5. Modern peak detector with less than 1 dB error. (*Courtesy of Interference Control Technologies*)

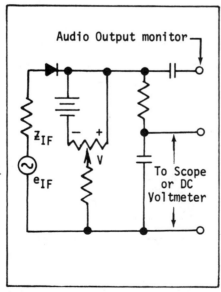

Fig. 18-6. One type of slide-back detector. (*Courtesy of Interference Control Technologies*)

increased manually until the pulse output, as heard in headphones or seen on an oscilloscope, is reduced to zero. At this point, the diode is just cut off, and V is equal to the positive peak of an incoming broadband signal. With the slide-back position preserved, an impulse generator is hooked up to the EMI receiver and the output of the generator adjusted until the reading corresponds to the value of the original signal. The EMI signal amplitude is then read from the impulse generator.

The back-bias detector works satisfactorily if the impulsive signal is both repetitive and of the same amplitude per pulse. However, it is virtually worthless for the single shot transients or in testing in which the test specimen must be restimulated each time to allow the operator to "zero in" on the peak. It does not permit automating EMI tests, since there is no stretching of the output to hand over to a slower recording instrument. In short, for other than certain tests, the slide-back detector is archaic.

Quasi-Peak Detection

The quasi-peak detector is a carryover of EMI instrumentation used many years ago. At that time most of the potential EMI victims were AM broadcast receivers, and potential EMI sources were both man-made impulsive and steady-state in nature. It was observed that by weighting the peak detector rise- and fall-time circuitry in a certain manner, the impact of EMI noise on the broadcast receiver/human ear combination could be more directly correlated with EMI receiver readings. Thus, a data base was built up in

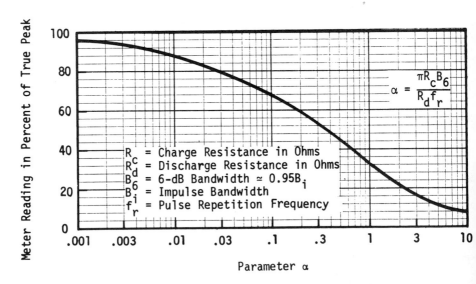

Fig. 18-7. Ratio of quasi-peak to peak detector outputs vs. receiver parameters and impulse repetition frequency. (*Courtesy of Interference Control Technologies*)

terms of ANSI and CISPR specifications to be levied on noise suppression of EMI emitters.

With the wide-scale introduction of TV receivers in the 1950s, the quasi-peak detector appeared to lose its correlation with the effects of noise. After all, the human eye was now involved, which is more sensitive than the ear to EMI. The TV broadcast receiver is approximately nine octaves greater in bandwidth than the AM broadcast receiver, and the RF spectrum involved is many octaves higher, with the result that the quasi-peak detector was modified somewhat, and the bandwidth of the EMI receiver increased by over an order of magnitude.

The quasi-peak detector is distinguished from the peak detector by its time constants. The charge time is much greater (1 msec) than that of the peak detector, and the discharge time constant is less (160 msec for MF and HF EMI receivers and 550 msec for VHF receivers). Figure 18-7 shows the relation between the quasi-peak detector and true peak response versus receiver parameters. As α becomes very small (e.g., a smaller receiver bandwidth, a smaller ratio of charge-to-discharge time constants, and/or a greater number of pulses per second), the quasi-peak output approaches that of the true peak detector. Figure 18-8 shows the reduction of quasi-peak detector output (and other detector functions) versus true peak readings as a function of impulse repetition rate. All the outputs tend to approach the peak as the repetition rate approaches the reciprocal of the bandwidth. ($B_{6\,\text{dB}} = 120\ \text{kHz}$.)

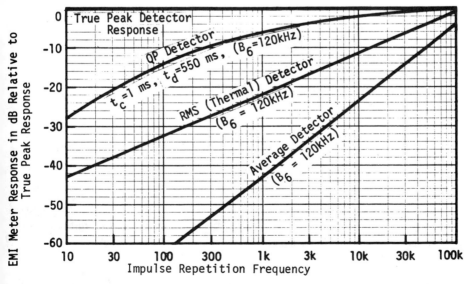

Fig. 18-8. Ratio of different detector outputs vs. impulse repetition frequency. (*Courtesy of Interference Control Technologies*)

Average Detection

Some types of EMI receptor victims are more affected by the average of the culprit EMI emission than by the peak or quasi-peak. Examples include those devices and equipments that have long integration time constants (on the order of 1 sec) for their output displays, such as aircraft cockpit avionics, mobile vehicle and ship instruments, and most other status indicators in use in industry and the commercial world. Ordnance containing EEDs appears to be especially sensitive to short-term averaging (1–10 msec), as do certain biomedical instruments such as EKG recorders. Were one to develop susceptibility standards to protect such devices and equipment, average emissions from potential sources would be of more concern than peak emissions.

The average detector, sometimes called the field-intensity (FI) detector, consists of an envelope detector followed by an averaging network. This circuit is simply a low-pass filter or integrator/capacitor smoothing network consisting of a series resistance and shunt capacitance. Thus, the average of the envelope modulation is produced. For example, a CW signal will develop an output proportional to its peak. When it is 100% amplitude-modulated with either a sinusoidal tone or a square wave, the average remains unchanged although the peak is doubled (i.e., increases by 6 dB). Conversely, a pulse modulation, such as a radar pulse train having a duty cycle of 0.0001, will produce an average signal that is 60 dB below its peak. Figure 18-8 shows the relation between the average of a signal and the pulse repetition rate for the indicated CISPR receiver. This relation is the duty cycle (δ), which is basically:

$$\delta = \tau f_r = f_r/B_i \tag{18-68}$$

where:

δ = duty cycle
τ = responding IF amplifier pulse width
f_r = pulse repetition rate
B_i = receiver IF amplifier impulse bandwidth

Thus, when $f_r \simeq B_i$, the peak and average detectors read about the same.

RMS Detection

Root-mean-square (RMS) detectors are not ordinarily found in EMI instruments. RMS measurements are desirable because of their mathematical convenience. If the RF output is random noise power $P_n(f_0)$, the mean square value of the emission referred to the input at the tuned RF, f_0, is:

$$V_0{}^2 = P_n(f_0)G^2(f_0) \int_0^\infty [G^2(f)/G^2(f_0)] \, df, \text{ for random noise} \tag{18-69}$$

Table 18-5. EMI receiver indication calibrated to read RMS of an input sine wave.

Emission Input \ Detector Type	Peak Detector	Envelope Average	Quasi–Peak $t_c = 1$ msec $t_d = 600$ msec	RMS Detector
Sine Wave RMS Value = V_0	V_0	V_0	V_0	V_0
Impulse Spectral Intensity = $A(f)$	$\sqrt{2}\,AB_i$	$\sqrt{2}\,Af_r$	$\sqrt{2}\,AP(\alpha)B_i$	$A\sqrt{2B_i f_r}$
Random Noise; Power Spectral Density = P_n		$1.25\sqrt{P_n B_{ep}}$	$2.57\sqrt{P_n B_{ep}}$	$\sqrt{2P_n B_{ep}}$

For nonoverlapping repetition pulses, $V_0{}^2$ is calculated as follows:

$$V_0{}^2 = (\tau \Delta e)^2 (f_0) G^2 (f_0) B_{ep} f_r \qquad (18\text{-}70)$$

$V_0{}^2$ as a function of f_r for the indicated receiver characteristics is shown in Fig. 18-8. Its square is a true thermal detector and includes bolometers and thermistors used in microwave measurements.

EMI receivers are sometimes calibrated to read the RMS value of an input sine wave regardless of which detector is being used. An approximate factor is used with each detector function to account for the differences in output voltage with each detector, as shown in Table 18-5.

Amplitude Probability Distribution

Amplitude probability distribution is not really a detection process. Rather, it is a statistical technique of organizing an amplitude–time varying function and presenting it in a graphic manner useful in making certain types of decisions. For example, mobile communication is affected by radiations from automobile engine-ignition noise. Since many automobiles may be involved simultaneously, each with different levels of radiation, emissions level versus distance dependence, speeds, and so on, an APD organization of the impulses is needed. Actually, a cumulative probability distribution is used to describe the peak of field intensities in terms of the number of pulses per second (or relative percent of pulses) exceeding the indicated intensity.

To make APD detection possible, a number of peak level detectors, each back-biased at a different level, are used to drive a similar number of matching counters. At the end of the time sample interval, the counter reading becomes the abscissa on the probability graph, and the corresponding field intensities become the ordinate. The APD technique is also useful to describe transient distributions on a DC or AC power bus, especially when digital or IC equipment may be connected to it.

REFERENCES

1. Audone, B., and Bolla, L. "An Approach to Aircraft Ordnance Test Requirements of MIL-E-6051D." *Record of the 1976 IEEE International Symposium of Electromagnetic Compatibility*, Volume 76-CH-1104-9 EMC, July 13–15, 1976, Washington, DC.
2. Boode, C. N., and Burgan, P. G. "Automated MIL-E-6051 Testing." *Record of the 1975 IEEE International Symposium of Electromagnetic Compatibility*, Volume 75-CH-1102-5 EMC, October 7–9, 1975, San Antonio, TX.
3. Brummett, E. P. "System Level Testing for Electromagnetic Compatibility." *Record of the IEEE International Symposium of Electromagnetic Compatibility*, Volume 76-CH-1104-9 EMC, July 13–15, 1976, Washington, DC.
4. Carr, T. J. "The SAE Specification—A Living Document." *Record of the IEEE International Symposium of Electromagnetic Compatibility*, Volume 76-CH-1104-9 EMC, July 13–15, 1976, Washington, DC.
5. Haber, F., Kocker, C. P., and Forest, L. A. "Space Shuttle Electromagnetic Environment Measurement." *Record of the IEEE International Symposium of Electromagnetic Compatibility*, Volume 76-CH-1104-9 EMC, July 13–15, 1976, Washington, DC.
6. White, Donald R. J. *EMI Methods and Procedures*, Volume 2, EMI/EMC Handbook Series. Gainesville, VA: Don White Consultants, Inc., 1980.
7. White, Donald, R. J., *EMI Instrumentation*, Volume 4, EMI/EMC Handbook Series. Gainesville, VA: Don White Consultants, Inc., 1971.
8. MIL-I-6181D, "Interference Control Requirements, Aircraft Equipment," 25 November 1959, appended 22 June 1965.
9. MIL-E-6051D, "Electromagnetic Compatibility Requirements, Systems," 7 September 1967.
10. MIL-I-16910C, "Interference Measurement, Electromagnetic, Methods and Limits," 26 October 1964.
11. MIL-I-26600, "Interference Control Requirement, Aeronautical Equipment," 2 June 1958, appended 23 April 1959.
12. MIL-STD-461B, "Electromagnetic Interference Characteristics, Requirements for," April 1980.
13. MIL-STD-462, "Electromagnetic Interference, Measurement of," 31 July 1967.
    MIL-STD-826 and 826A, "Electromagnetic Interference Test Requirements and Test Methods," 20 January 1964, appended 30 June 1966.
14. MIL-STD-1541, "Electromagnetic Compatibility for Space Systems," 15 October 1973.
15. Federal Communications Commission, Rules and Regulations, Part 15, RF devices, October 1982.
16. Federal Communications Commission, Rules and Regulations, Part 18, Industrial Scientific and Medical Equipment, October 1982.

Chapter 19
A SUMMARY OF EMI AND
RELATED INSTRUMENTS

This chapter surveys the EMI test areas, including open fields and shielded enclosures, and EMI sensors and exciters, including antennas, current probes and line impedance stabilization networks, and conducted injectors. Also covered are EMI receivers, spectrum analyzers, power-line monitors, and susceptibility testing sources, such as impulse or spike generators and electrostatic discharge (ESD) testers.

EMI TEST AREAS AND ENCLOSURES

This section reviews the electromagnetic and physical environment surrounding the test sample configuration. The object is to establish an isolation barrier between (1) the test sample with its associated test instrumentation and (2) the outside world electromagnetic ambient environment. Chapter 12 discusses shielded cabinets and equipment housing as well as the shielding effectiveness provided by buildings of different types.

The goal of using an EMI test area is to make EMI measurements without interference from ambient emissions from the environment disturbing or confusing the results. At the same time, susceptibility tests can be performed on the test sample without causing electromagnetic interference to the outside world (5). The test sample environment—facilities, instruments, and techniques—must all be carefully considered as potential problem areas.

There are basically two kinds of test areas involved: (1) open field test areas generally associated with far-field antenna pattern measurements, large system test samples, intrasystem EMI testing, and regulatory (FCC, CISPR, and VDE) testing (7), and (2) test areas confined within an enclosure of some type, usually involving EMI measurements on smaller test samples. There are exceptions to both cases as discussed below. In either case, the test environment or surrounding area can invalidate the test results unless a number of precautions are taken.

Open-Field Radiated Emission Measurements

Some types of EMI radiated tests can be performed only in an open field, such as emission measurements pursuant to MIL-STD 449D and FCC/CISPR tests. Tests of the emission spectrum from a device or system include the combined effects of the transmitter, its antenna, and usually the surrounding terrain. For antennas not intended to clear the first Fresnel zone, the terrain is a very important contributor to the far-field antenna pattern which is a part of the emission spectrum.

To satisfy far-field requirements, the test intercept site may have to be as far away as 1 km, in accordance with the (D^2/λ) criteria, where D is the antenna element size and λ the wavelength at the measuring frequency, which specifies minimum distance. To assure validity, meaningful EMI measurements should be performed in the open field at the actual site of installation, at a site as similar to the actual site as possible, or at an approved test site (for FCC/CISPR tests).

Figure 19-1 illustrates a few propagation phenomena contributing to the measured emission spectra that develop misleading results. As shown, the direct path of the radiation and the ground reflection path result in a combined wave that yields an interference pattern. The reflected wave undergoes a 180° phase reversal. When these two path-length differences result in values at or near multiples of one wavelength, λ, the field intensities tend to cancel. When they are at or near values of $n(\lambda/2)$ (n = odd integer), the field intensities are additive. Thus, the local heights of the test specimen and test antennas, the terrain flatness, and the path-length difference all play a major role in the measured emission spectrum amplitude profiles.

Figure 19-1 also shows other phenomena that tend to spoil or invalidate test results, insofar as MIL-STD-449D, RE03 of MIL-STD-462, and FCC/CISPR/VDE open-field specification testing is concerned. Reflections from buildings, towers, fences, and other significant off-axis obstacles may result in spoilage of antenna pattern measurements, since they appear as artificial, phase-coherent sources of secondary emission. Trees or other deciduous vegetation (especially in summer) in the path of the measuring system tend to absorb some of the ground-reflected energy, thus spoiling the combined emission spectrum data. If this is an actual site of installation, the intervening vegetation can affect test measurements; in that case, another test radial may have to be chosen. Finally, Fig. 19-1 shows emission interference emanating from a radar or other transmitter at or near the test sample's frequency or at harmonics thereof. Testing must provide for discriminating against this outside emission source, such as conducting an ambient site survey to determine the level of such emissions.

Most of the above undesired open test area situations can be either avoided or mitigated by selecting another radial for the test instrumentation setup

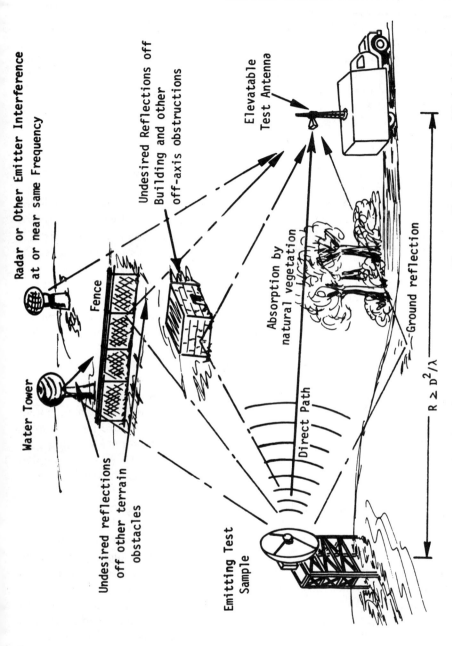

Fig. 19-1. Terrain factors influencing emitter spectrum measurements. (*Courtesy of Interference Control Technologies*)

that has no significant terrain obstructions within $\pm45°$ of the receiving test antenna or ±2 antenna beamwidths, whichever is less. If possible, at the chosen test site, any interfering radars or other nontest emitters will be blocked by a hill or otherwise out of the line of sight. If this is not possible, such emitters may have to be shut down, or interference blankers used. Off-axis vegetation is desirable in that it will partially absorb undesired reflections from the terrain and obstacles.

Shielded Enclosures

Field intensities on the order of 10 V/m (140 dBµV/m) at VHF TV and FM frequencies are not uncommon in some areas where the transmitter is located about 300 meters away. Higher field intensities are observed from UHF TV and radar transmitter frequencies at similar distances. Corresponding REO2 radiated emission narrowband specification limits for MIL-STD-461 vary from 29 dBµV/m at 100 MHz to about 45 dBµV/m at 1 GHz. Therefore, in order to make RE02 radiated emission tests in which the ambient is to be 6 dB below the limits, the required attenuation (offered by intervening obstacles, etc.) is 117 dB at 100 MHz and 101 dB at 1 GHz. In this situation, the shielded enclosure becomes necessary.

The shielded enclosure or shielded room (or, as it is still sometimes called, the screen room because early shielded enclosures were made of copper screen) has been in use for many years for performing electronic measurements where a low electromagnetic ambient is required, or where potentially disturbing emissions must be contained (8, 17). Its use has spread to nonmeasurement applications, such as protecting personnel working near high power radar sites containing certain industrial RF emission sources and protecting sensitive equipment such as biomedical instruments and computers.

The main advantage of the shielded enclosure used for performing EMI measurements is that it provides RF isolation from and to the outside world. Its use allows meaningful emission measurements to be made, both conducted and radiated, in higher ambient locations where such testing would not ordinarily be possible. However, some residual electromagnetic ambient will still exist inside a shielded room, since the room attenuates rather than eliminates outside world emissions (8, 11). In most situations where the outside environment is not abnormally high, a modern shielded room provides attenuation sufficient to reduce all outside emissions to levels below the sensitivity of typical receivers with their normal antennas. Magnetic field shielding at and below ELF is the exception. For example, the enclosure walls offer little attenuation to 60 Hz fields, and input power lines allow easy emission entry. To reduce 60 Hz magnetic fields and their first few harmonics, it is necessary to use extended frequency range shielded enclosures. (This is a term that has been adopted by the EMI community for shielded enclosures to imply significant attenuation to magnetic fields at power-line frequencies and sometimes better performance at UHF and SHF.)

Enclosure Types and Sizes

Some electronic systems are physically too large to employ shielded enclosures for testing. The size of shielded enclosures has no theoretical limit, and they have been built to enclose aircraft hangars. Some structures are lined with absorbent material to form an anechoic chamber. Shielded enclosures are also constructed in mobile configurations that are roving electronic laboratories (viz., shielded enclosures constructed in a trailer or van) (10). Additional discussion of shielded enclosures may be found in Chapter 12.

ANTENNAS USED IN EMI TESTING

This section presents a survey of emission antennas used for conducting RFI/EMC radiated emission testing from 20 Hz to 20 GHz (2, 6, 14, 16). These antennas are mainly employed for open field testing although they may also be used for measurements inside of shielded enclosures, provided certain precautions are taken, especially with active antennas, and at lower test frequencies. Thus, the capacitive probe, loop antennas and magnetic probes, active and passive rod antennas, dipoles and biconical, log spiral, log periodic, and ridged guide antennas are presented.

With some exceptions, these antennas are not useful for radiated susceptibility testing. A number of illustrations will portray typical antenna types that are available together with representative performance data (i.e., typical antenna factors). Readers desiring more comprehensive coverage, including both theoretical and practical aspects of EMI antennas, are referred to reference 16.

Present governmental regulatory emissions standards, such as FCC, CISPR, VDE, and the family of MIL-STD or MIL-Interference specifications, require that radiated emission measurements be performed. Some MIL-STD testing requires that the frequency range extend to as low as 100 Hz. Most military EMI specifications, however, use 10 kHz as the lower frequency limit. FCC and CISPR regulations require radiated testing from 30 MHz to 1000 MHz. VDE specifications require that radiated emissions measurements be made down to 10 kHz using a magnetic loop antenna and converting the magnetic field to an equivalent E-field. Susceptibility testing is required under MIL-STD-461A, which determines requirements for radiated susceptibility (RS) and conducted susceptibility (CS) for systems.

EMI Measurement Antennas: Field-to-Voltage Conversion

To correctly obtain a transfer of field energy (the measured quantity) to conducted energy (to be input to a receiver), a match between the impedance of the field and the input impedance of the receiver is necessary. The impedance of an electric field in the near field at low frequencies, such as from 20 Hz to 50 kHz, is very high. For other than DC conditions, a magnetic field is associated with each electric field. For a predominately electric field source in the near field ($R_f < \lambda/2\pi$) (e.g., monopole), the associated magnetic field component is small, and the impedance ($Z = E/H$) is large ($Z \gg$ 377 ohms.). For a predominately magnetic field source (e.g., loop), the associated electric field component is small, and the impedance is small ($Z \ll 377$ ohms). As the distance increases, the impedance of the electric (magnetic) field becomes lower (higher), and the two impedances asymptotically approach 377 ohms in the far field (see Chapter 10).

Figure 19-2 shows a number of antennas and sensors that are used to convert near- and far-field electric and magnetic field quantities to a conducted voltage to be input to an EMI receiver or spectrum analyzer. The key to Fig. 19-2 is shown in Fig. 19-3.

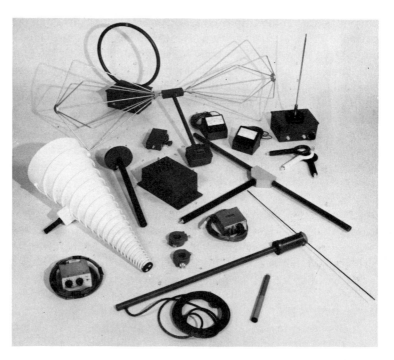

Fig. 19-2. An array of typical antennas and sensors used for EMI testing. (*Courtesy of Electro-Metrics*)

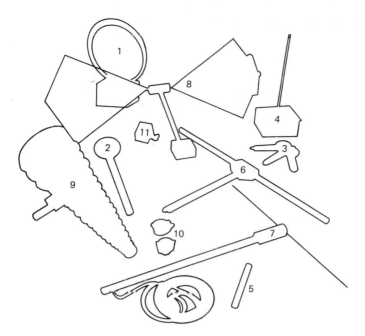

Fig. 19-3. An array of typical antennas and sensors used for EMI testing. (*Courtesy of Electro-Metrics*)

Key

Antenna or sensor no.	Description	Frequency range
(1)	H-field loop	10 kHz–30 MHz
(2)	H-field sensor	20 Hz–100 KHz
(3)	H-field sniffer	10 kHz–230 MHz
(4)	41″ active rod	10 kHz–30 MHz
(5)	E-field sniffer	10 kHz–1 GHz
(6)	Broadband dipole	20 MHz–200 MHz
(7)	Tunable dipole	200 MHz–1000 MHz
(8)	Biconical	20 MHz–200 MHz
(9)	Conical log spiral	200 MHz–1000 MHz
(10)	Current probes	20 Hz–110 MHz
(11)	Current probe amplifier	20 Hz–50 kHz

Magnetic Field Antennas and Probes, 20 Hz to 50 kHz

The loop antenna is used in the near field to intercept the magnetic field associated with an emission and to substantially discriminate against any electric field component. Since magnetic fields exhibit low impedance properties, the loop antenna must present similar characteristics. Because it is comprised primarily of a number of turns of wire, the loop antenna exhibits a low impedance and is an effective coupler of magnetic fields. A nonferrous external tubing provides electric field shielding. A loop antenna is pictured as Antenna #1 in Fig. 19-2, and Fig. 19-4 shows typical antenna factors for the loop antenna.

In addition, a magnetic field sensor, such as Antenna #2, Fig. 19-2, is used to measure the magnetic field in the frequency range from 20 Hz to 100 kHz. The impedance of this particular sensor is in accordance with MIL-STD-461A. A conversion factor is given in chart form as a function of frequency that converts a meter reading in dBμV to dBpT (dB above 1 picoTesla).

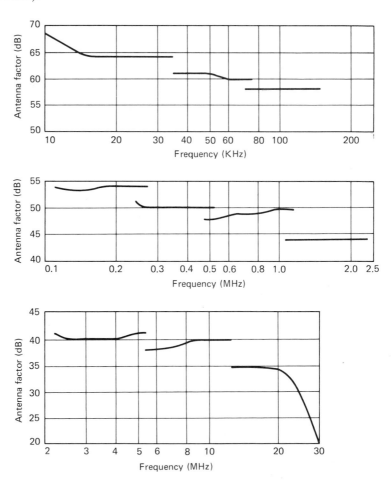

Fig. 19-4. Antenna factors of loop antenna. (*Courtesy of Electro-Metrics*)

Uncalibrated "sniffers" or probes, such as those shown in Fig. 19-2 (Antenna #3), are useful in determining where leaks exist in shielded enclosures and other noncalibrated probing of magnetic fields. Using the three sniffers shown provides a measurement frequency range extending from 10 kHz to 230 MHz.

Electric Field Antennas and Probes

The electric field impedance may be on the order of 1000 megohms at the low frequency end of the spectrum. Accordingly, to measure the electric field, it is necessary that the antenna also provide a very high impedance to the field. At the same time it must exhibit a 50-ohm, 600-ohm, or other low value of output impedance, as required, to match the input impedance of a receiver. This requirement has led to the development and use of the electric field antennas and probes below 50 kHz. In one version, the electric field capacitive antenna employs two short dipole elements that are inputs to electrometer circuitry. The electrometer, whether employing earlier electron tubes or field effect transistors (FETs), exhibits an extremely high impedance at its input terminals and a low impedance output that is coupled to a receiver or other measuring network.

The dipole can be rotated to intercept both horizontal and vertical electric fields. The frequency coverage of this antenna is from 20 Hz to 50 kHz. Portions of the spectrum may be filtered in order to mitigate saturation when used in high fields, and outside electromagnetic ambients such as 60, 120, 180, or 400 Hz from power mains.

The next two antenna types employ the 41-inch or 1-meter rod. These electric field antennas are of the passive and active rod types.

Passive Rod Antennas.   The passive rod antenna, more commonly called a rod or whip, is perhaps the most familiar antenna and most frequently used for radiation emission EMI testing from 14 kHz to 30 MHz. It measures the electric field of an emission in either open field testing or inside a shielded enclosure where it is in the near field below 30 MHz.

The rod element, typically 41 inches long, or about 1 meter (electrical length equals 0.5 meter) for EMI sensor applications, has an equivalent capacitance of about 10 pf at its input terminals. The antenna requires a tuning inductor to be switched on a band-by-band basis to resonate with the rod capacitance. The residual resistance of the LC tuned rod corresponds to values on the order of 10 kilohms at LF. Thus, this antenna, in attempting to drive a 50-ohm receiver, displays a substantial loss due to voltage divider action. Therefore its efficiency is very poor (i.e., the antenna factor, AF, is very high).

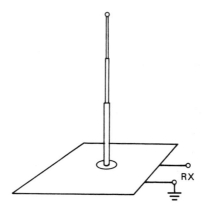

Fig. 19-5. Construction of a typical (passive) rod antenna.

Figure 19-5 illustrates the construction of a typical rod antenna. Figure 19-6 presents typical rod antenna factors used to convert receiver input voltage measurements to electric field intensity equivalents over the several tuned bands from 150 kHz to 30 MHz.

The calibrated passive rod antenna is also available in a 2-meter rod from some suppliers. The greater length is called for in certain older EMI specifications. Another version of the passive antenna is the electric field probe.

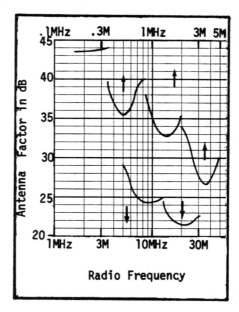

Fig. 19-6. Typical antenna factors of the rod antenna. (*Courtesy of Interference Control Technologies*)

This device is characterized by being small in size, untuned, shielded, and noncalibrated. It is an excellent electric field sniffer for localizing culprit radiation leakage (2, 15).

Active Rod Antennas. The purpose of the active rod antenna is to eliminate most of the deficiencies of the passive rod antenna discussed above, in particular the poor antenna factors associated with passive rod antennas. The active rod antenna utilizes active matching networks to transfer the high input impedance necessary to measure an electric field to the low output impedance necessary to drive an EMI receiver input (50 ohms). An active rod antenna is shown as Antenna #4 in Fig. 19-2.

The notion of the active rod evolved from the earlier cathode followers used in tube versions or the more recent source followers of FETs used in the input circuitry of the rod. These active devices, which are designed for relatively low noise use, typically behave as current transformers in which the high impedance (several megohms) is transformed to a 50-ohm output impedance. The effect is to achieve an efficient antenna factor, typically about 6 dB or better, across the spectrum from 15 kHz to 30 MHz.

The active rod antenna, by virtue of the active devices used in its front end, also tends to exhibit some undesirable characteristics in high field strengths. These limit the measurable field intensities to relatively low values such as generally exist inside of shielded enclosures. Where high field intensities are expected, especially during ambient site surveying, the active devices may saturate, develop harmonics and intermodulation, or generate other undesirable spurious responses due to nonlinear operating conditions. Furthermore, because active rod antennas are wideband untuned devices, they tend to exhibit considerably less dynamic range to broadband signals. Broadband dynamic ranges of 20 to 25 dB are not uncommon. Narrowband dynamic ranges, however, may be on the order of 60 dB or more.

EMI Antennas, 20 MHz to 1 GHz

Higher frequency electromagnetic field measurements are made using tunable dipoles, log periodic dipoles, conical log spiral antennas, biconical antennas, and ridged-guide antennas, which are discussed in the next few sections.

Tunable Dipole. The tunable dipole utilizes either screw-in antenna elements (Antenna #6, Fig. 19-2) or dual telescoping elements (Antenna #7, Fig. 19-2), to vary the overall length of the antenna to one-half of the wavelength corresponding to the frequency under test. This adjustment is made to measure the electric field from 35 MHz to 1 GHz. Below 35 MHz, the tunable dipole becomes too long for use inside shielded enclosures with its dipole elements further extended. Accordingly, this antenna is operated below

35 MHz when adjusted to a length corresponding to 35 MHz, and a modified antenna factor, to correct for the resultant degradation, is employed.

From 35 MHz to 1 GHz the tunable dipole is an excellent standard-gain antenna against which other antennas may be calibrated for both gain and antenna factors. However, since it exhibits relatively narrowband performance, it is not very useful in modern EMI testing involving automatic frequency scanning and recording instruments or systems.

The tunable dipole also exhibits substantial measurement errors at the lower frequencies because its length is about 5.5 meters. When the center of the antenna is only 1 meter away from a MIL-SPEC test sample, the ends may be about 3 meters from the sample, and the substantial curvature of the wave front results in measurement errors. In other words, this antenna is not operating in the far field below approximately 50 MHz. In addition, the ends of the antenna often are near the shielded enclosure walls, which capacitively load and detune the dipole. Accordingly, this antenna is not accurate for measurements performed in shielded enclosures below 50 MHz. Although usable, it has been discontinued in modern EMI MIL-spec testing. Though it is not widely used in industry, the tunable dipole antenna is the preferred antenna for open field CISPR and FCC testing. Other antennas, such as biconicals, conical log spirals, and log periodics, are allowed if their results can be correlated to those that would be obtained with a tunable dipole.

Figure 19-7 shows the antenna factors of a typical tunable dipole. Increased error below 50 MHz is ignored for EMI enclosure measurements. Furthermore, if any of the tunable elements of the dipole come within approximately 1 meter or less of the shielded enclosure walls, detuning will result, and the antenna factors will correspondingly change. This results in an unpredictable error generally unknown to the user and not made available as an additional correction factor.

Biconical Antenna, 20 MHz to 200 MHz. The biconical antenna is perhaps the most widely used for both EMI radiated emission and susceptibility testing in the 20 MHz to 200 MHz portion of the radio frequency spectrum. The design of this antenna is basically a further extension of the development of broadband dipoles. Instead of a simple increase in diameter of the dipole elements, the biconical antenna elements are fanned out from a small diameter near the throat of the elements to a relatively large equivalent diameter at the extremes. The biconical antenna is pictured as Antenna #8, Fig. 19-2.

The antenna directivity is omnidirectional in the H-plane, and bidirectional in the E-plane. The antenna elements may be rotated to measure horizontal (as specified in MIL-STD-462) and vertical polarization, provided that in the vertically polarized mode of operation the lower elements are at least one-half meter above ground so that the antenna is not capacitively loaded at one end sufficiently to change the antenna factors. Figure 19-8

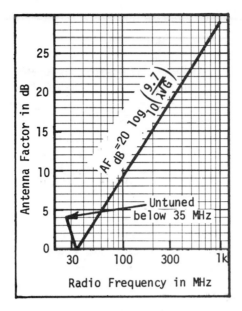

Fig. 19-7. Typical antenna factors for tunable dipole. (*Courtesy Interference Control Technologies*)

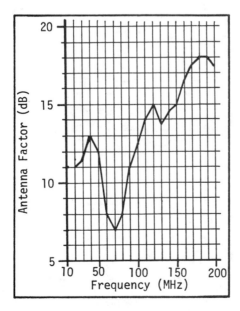

Fig. 19-8. Typical antenna factors of biconical antenna. (*Courtesy of Interference Control Technologies*)

presents the antenna factors for this antenna appearing on page 18 of MIL-STD-461A.

Conical Log Spiral, 200 MHz to 1 GHz.   Until the development of the conical log spiral (CLS), this lower portion of the UHF frequency spectrum was the most difficult to accommodate for radiated emission or suscepti-bility testing inside shielded enclosures. The CLS provides high efficiency resulting in low antenna factors, and since this antenna exhibits circularly polarized properties in the transmitting or receiving mode, it will also ac-commodate electric fields that are either circularly or linearly polarized. Thus, for either horizontal or vertical linearly polarized waves, the circular polarized antenna factors are increased by 3 dB due to the polarization coupling loss.

The conical log spiral antenna is shown as Antenna #9 in Fig. 19-2, and Fig. 19-9 shows representative antenna factors, gain, and VSWR associated with this antenna as presented in MIL-STD-461A.

The antenna factor is greater than might be expected from the antenna gain properties. This is a result of the lossy transmission line that is used to construct the spiral elements. Also note that the VSWR exceeds 3:1 below about 300 MHz. This can cause a 10 dB peak to peak error in measurements.

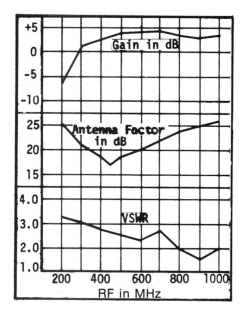

Fig. 19-9. Typical antenna factors of conical log-spiral. (*Courtesy of Interference Control Technologies*)

Log Periodic Antennas 200 to 1000 + MHZ. Usually used for receiving antennas in measurement and broadcast reception (VHF-TV), log periodics play an important role in EMI emissions testing due to their virtually constant input impedance, gain, and radiation pattern over a wide frequency range. The most popular log periodic employs an array of dipoles whose main frequency limitation on the antenna frequency response is the length of the largest dipole, which corresponds to $\lambda/2$. Their light weight makes for extremely sturdy construction and ease of operation. They are usually mounted on an antenna mast that can be moved up and down to determine maximum field strength and rotated to measure vertical and horizontal polarization.

Double Ridged-Guide Antenna, 200 MHz to 2 GHz. The ridged-guide antenna was designed and developed by the U.S. Army Electronics Command in an effort to improve the radiated emission and especially the susceptibility antenna factors of antennas between 200 MHz and 1 GHz. This antenna is basically a microwave horn that employs a double-ridged waveguide to expand the bandwidth from typically less than one octave to over three octaves. Waveguide horns are efficient and generally exhibit good VSWR properties, but they are not ordinarily broadband. A further benefit in using a ridged waveguide is to reduce the physical size that would otherwise result. This antenna is approved for MIL-STD-462 testing for Army procurements.

Figure 19-10 is a diagram of a double ridged-guide antenna, and Fig. 19-11 shows typical antenna factors that accompany this antenna. The antenna factor performance of the ridged-guide antenna is superior to those of both the conical log spiral and the log periodic, although it is also more expensive to manufacture. However, it should permit more repeatable and accurate measurements, and require less RF power for radiated susceptibility tests.

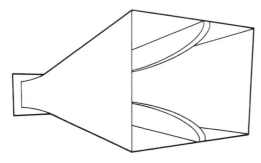

Fig. 19-10. Double-ridged-guide antenna.

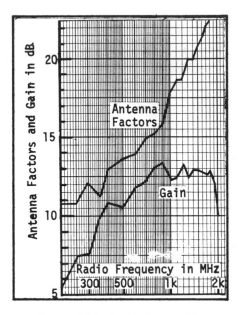

Fig. 19-11. Typical antenna factors of ridged-guide antenna (*Courtesy of Interference Control Technologies*)

EMI Antennas, 1 GHz to 40 GHz

Conical Log Spiral, 1 GHz to 10 GHz.   An additional increase in high frequency measurement performance is obtained by using a smaller version of the conical logarithmic spiral antenna. The physically smaller antenna is capable of covering a decade from 1 GHz to 10 GHz. It exhibits circular polarization and has a 3 dB relative loss to either horizontal or vertical polarization. As a result, the antenna factors are increased accordingly by 3 dB in measuring linearly polarized fields.

The conical log spiral antenna is one of two presently accepted EMI antennas in this frequency region for emission and susceptibility measurements inside shielded enclosures. Other antennas may be used if included and accepted in the EMI Test Plan. The construction properties of the CLS are available from the U.S. Army.

Double Ridged-Guide Antenna, 1 GHz to 12 GHz.   This ridged-guide antenna is similar to that discussed above except that it has been scaled up in frequency (down in dimensions) by nearly a decade. It is a suitable substitute for the conical log spiral, and exhibits a greater efficiency than the CLS because there exists no cable loss in the antenna per se. Its polarization is linear rather than circular.

Again, in comparing this antenna with the conical log spiral covering the same portion of the spectrum, it is seen that this antenna exhibits a somewhat higher gain and lower antenna factors. Accordingly, this more efficient antenna permits the development of a higher field intensity for a given signal generator input RF power for radiated susceptibility measurements.

Parabolic Antennas, 12 GHz to 40 GHz. For RE03/MIL-STD-463 harmonic and spurious output measurements, involving radiated emission properties of test items in the open field, a 45-cm-diameter dish illuminated by a small horn is employed from 12 GHz to 18 GHz. Construction information for this antenna is presented in U.S. Army drawing ES-DL-201090. The typical antenna factor for this antenna is about 19 dB.

A 30-cm-diameter parabolic dish is used with either one of two horn feeds to cover the frequency spectrum from 18 GHz to 26 GHz and from 26 GHz to 40 GHz. The typical antenna factor for these horn-fed dish combinations is about 23 dB.

Antennas for Susceptibility Testing

A number of the antennas mentioned above may also be used for susceptibility testing; that is, they are hooked up to an RF oscillator/power amplifier and used to generate electromagnetic fields. The test specimen is then immersed in the field, and the response of the specimen is then measured. They are also useful in measuring the shielding effectiveness of shielded enclosures and measuring the electrical characteristics of test sites. The antennas that are useful for this type of work include dipoles, biconical antennas, the conical log spiral, and ridged-guide antennas. Because of reciprocity, the frequency ranges for measurement and radiation are alike. The antenna factor in susceptibility testing is a measure of the efficiency of the antenna in converting a given input voltage on the terminals of the antenna to a desired radiated field strength.

To generate uniform magnetic fields for magnetic field susceptibility testing, Helmholtz coil systems are employed. A Helmholz coil system is constructed of two equal-sized current-carrying coils (diameter a) spaced $a/2$ apart and sharing the same axis. A known uniform magnetic field (parallel B-field lines) is produced in an area halfway between the two coils. The test specimen is placed in the uniform magnetic field, and the response of the specimen to the field is measured.

CONDUCTED SENSORS AND INJECTORS

The previous section reviewed antennas that are used to make radiated emission (RE) and radiated susceptibility (RS) measurements for EMI and related applications. The other class of EMI test requirements involves conducted emission (CE) and conducted susceptibility (CS) tests that depend on a measured voltage or current or injecting a current or voltage into the test item. The generic family of devices used to accomplish this is called conducted sensors and injectors, which is the topic of this section.

Conducted Sensors

Conducted emission sensors reviewed in this section are the 10 μf capacitor, current probe and the line impedance stabilization network (LISN), sometimes called an artifical mains.

Figure 19-12 illustrates a hypothetical composite EMI test setup showing the relative locations of the conducted sensors to be reviewed in this section.

In general, either the current probe or the LISN can be used to measure conducted power-line emissions, depending upon which is required. Sometimes both are used together. The current and the voltage probe, on the other hand, are both useful for measuring conducted signal and control-line emissions appearing on equipment interconnecting wires and cables.

Two questions frequently arise: (1) which conducted sensor is the more meaningful to use for CE measurements, and (2) how does one relate a conducted EMI current measurement to a conducted voltage measurement? In partial answer to the first question, it may be said that since the determination of a particular EMI specification limit was made using a specific

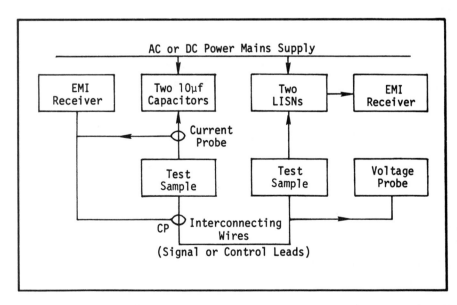

Fig. 19-12. Test measurement setup illustrating conducted emission sensors and associated devices. (*Courtesy of Interference Control Technologies*)

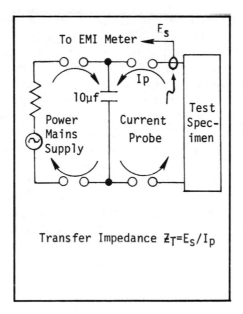

To EMI Meter ← F_s

$10\mu f$

I_p

Power
Mains
Supply

Current
Probe

Test
Spec-
imen

Transfer Impedance $Z_T = E_s / I_p$

Fig. 19-13. Equivalent R-F circuit of 10-μf capacitor. (*Courtesy of Interference Control Technologies*)

device for measurement, either current probe or LISN (artificial mains), subsequent testing should use the same types of measurement equipment.

In response to question (2), the problem with converting EMI conducted current measurements into voltages or vice versa is that the impedance versus frequency of the test sample and that of the power mains feeding it are both unknown. It is not uncommon to assume that the test sample appears as a constant current source to a 50-ohm load (the LISN impedance). Then the equivalent EMI voltage is simply the current probe measurement times 50 ohms. For the reverse situation, the equivalent current is the LISN voltage measurement divided by 50 ohms. The conversion from current to voltage may result in a somewhat higher value, but assures compliance with specification limits. When voltage is converted to current, the result may be somewhat lower than the true value.

None of the EMI measurements describes most installed situations, since the equivalent impedance is different for every power main, varying with frequency, loading, and so on. For these reasons some future EMI specifications are examining the possibility of using power limits in which both voltage and current probe measurements will be made on the test sample driven by a power mains simulating typical installations.

The 10 µf Capacitor. Figure 19-13 shows 10-µf capacitors appearing between a test sample and the AC or DC power mains supply. Used only with a current probe, one capacitor appears across each power-line lead to be tested. (cf. CEO1 and CEO3 of MIL-STD-462). Its purpose is to provide a low RF impedance path at the lowest frequency of interest to any EMI noise appearing on the power mains and/or test sample, as shown in Fig. 19-13. Thus, conducted emissions to be measured by the current probe will originate from the test sample of interest and not from the power mains as long as it appears to the right of the capacitor and satisfies other conditions.

As a practical matter, the 10-µf capacitor cannot perform well at the low RF portion of the frequency spectrum because it must simultaneously not overload an AC power source and still provide a low impedance RF path. For an AC power mains frequency of 60 Hz (or 400 Hz), the capacitive reactance is 265 ohms at 60 Hz (40 ohms at 400 Hz). Since the test sample draws a current, I_s, from the power mains supply voltage, V_s, its impedance is $Z_s = V_s/I_s$. Thus, its impedance approaches that of the capacitance, X_c, at a frequency for which:

$$X_c = Z_s = \frac{V_S}{I_S} = \frac{1}{2\pi f C} = \frac{115 \text{ VAC}}{I_S}, \quad \text{for } V_S = 115 \text{ V} \qquad (19\text{-}1)$$

or:

$$f = \frac{I_s}{2\pi \times 10^{-5} \times 115} \simeq 140 I_s \text{ Hz} \qquad (19\text{-}2)$$

Assuming that the test sample input impedance is constant over the ELF portion of the spectrum, then equation (19-2) indicates the capacitor will start to become effective (3 dB cutoff frequency) at 1.4 kHz for a test specimen load of 10 amperes, for example.

The impedance of the test item is not really constant with frequency. Among other considerations, it probably has an EMI protection input filter. If the filter has an input inductor element, the impedance will rise with frequency, and the 10-uf capacitor becomes more effective. This argument is one reason why the supporters of the LISN or artificial mains believe that the LISN is more effective than the capacitor.

While the purpose of the 10-µf capacitor is very simple, the physical realizability of one is not. The capacitor must be capable of handling at least 50 amperes on its feed-through bus and a line voltage of 230 VAC RMS for general-purpose use. It must behave as a capacitor with no parasitics from DC to 50 MHz for conducted emission (MIL-STD-462 CE01 and CE03) measurements. Thus, any parasitic resonance effects must appear well above 50 MHz. The 10-µf capacitor must conform to the requirements of SAE ARP-936.

EMI Current Probes.   Current probes are convenient to use because they are simply snapped on around one or more wires and clamped in place (usually one wire at a time, although for interconnecting harnesses and for measurement of common-mode quantities, many wires may be tested at once). Two current probes are shown as sensor #10 in Fig. 19-2.

Current probes work on the principle of sampling the magnetic field around a wire (the primary) by secondary turns on a toroidal transformer. The induced voltage in the output of the current probe is proportional to the permeability of the toroid, its cross-sectional area, the number of secondary turns, the current flowing in the wire, and the frequency of measurement. Figure 19-14 shows the transfer impedance for three different current probes increasing with frequency at a rate of 20 dB per decade. Thus, to make a sensitive current probe, especially at lower frequencies, the permeability, area, and number of turns on the toroid should be made as large as practical with a relatively small toroid radius.

To make it functionally useful, the toroid is made in two equal halves. One end of each of the two halves is hinged, and the other ends butt together when clamped. The mating faces at both ends are machined flush in order to reduce the air gaps to an insignificant amount. This ensures that the permeability of the toroid and not the air gap is the limiting factor in concentrating the magnetic field and reducing the overall path reluctance.

The size, number, and deployment of the coupling turns of wire around the toroid determine both the circuit inductance and parasitic capacitance. The maximum usable frequency is limited to a value below self-resonance.

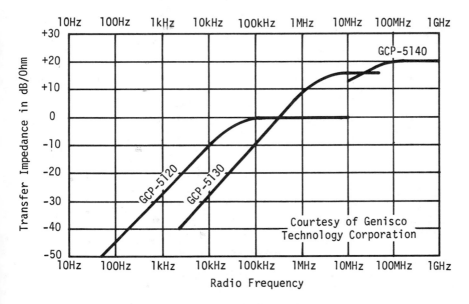

Fig. 19-14. Transfer impedance current probes. (*Courtesy of Genisco Technology Corporation*)

Thus, it is evident that any one current probe can be designed to cover only a limited range in the frequency spectrum.

The current rating capacity of current probes is another limiting factor in design. Should the power source current in the test wire be too large, the resulting magnetic flux density in the toroid would saturate the core, and the transfer impedance would no longer be independent of the level of the conducted emissions being measured. Consequently, it is necessary that current probes designed for EMI use be able to handle the largest currents expected at DC or power mains frequencies. A value of 350 amperes capacity is typical. For larger currents, special probes are obtainable at greater cost. Current probes used by the power companies (public utilities) are not useful in EMI testing because they are rated for 60 Hz measurement only, and not for RF.

Line Impedance Stabilization Networks. The line impedance stabilization network (LISN), sometimes called a line stabilization network (LSN), a power line impedance stabilization network (PLISN), or an artificial mains, is a coupling device. It is used to measure conducted emissions from a test sample's power leads and not that coming from the power mains supply. Such an installation was shown in Fig. 19-12. In other words, the LISN is a buffer network that permits connecting the power leads of the test item to the power mains by (1) passing only DC or AC power to the test sample, (2) preventing the test sample's electromagnetic noise from getting back into the power bus, (3) blocking the power mains RF from coupling into the test sample, and (4) stabilizing the impedance presented to the test sample.

The LISN provides for direct connection of the 50-ohm input terminals of an EMI receiver to the 50-ohm connector provided on the LISN. Figure 19-15 is a schematic diagram of a typical LISN used in FCC 15J testing.

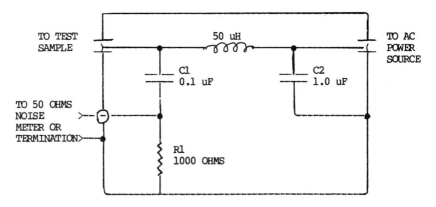

Fig. 19-15. Schematic of 50 μH LISN used for FCC conducted emissions testing.

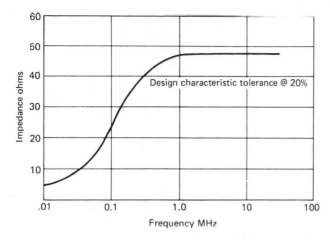

Fig. 19-16. Impedance vs. frequency for the LISN of Fig. 19-15.

Representative impedance versus frequency for this LISN is shown in Fig. 19-16. The LISN is operated over only that frequency portion that corresponds to about 50 ohms at the output connector jack. Two LISN pairs are required to cover the spectrum up to 30 MHz, or three up to 400 MHz for a two-wire power service. Four identical units are required for three-phase, four-wire mains.

In examining the schematic diagram of the LISN, it is noted that the power leads from the power supply mains bus and those from the test specimen are RF isolated by the 50-μH series inductance. It is transparent at 60 and 400 Hz to permit power current to flow to the test item. RF noise originating on the power bus is shunted to ground through the decoupling capacitor, $C2$. Conversely, any RF noise on the test item's power leads is coupled to the 50-ohm connector jack through capacitor $C1$. As it is necessary to place the LISN on each current-carrying conductor, some LISNs employ two such networks in a single package.

Conducted Injectors

Conducted susceptibility tests on AC and DC power supply lines are performed to simulate interference on the line in the form of transients or spikes from various sources such as switching surges and lightning (4). The impulse waveform is to be injected onto the line is shown in Fig. 19-17 for Army procurements. Navy and Air Force specify the same waveshape but with peak voltage equal to 400 V and duration t less than or equal to 5 μs.

Either parallel or series injection, specified in CS06 testing, is performed, depending on the loading of the power lines and the RF impedances of the

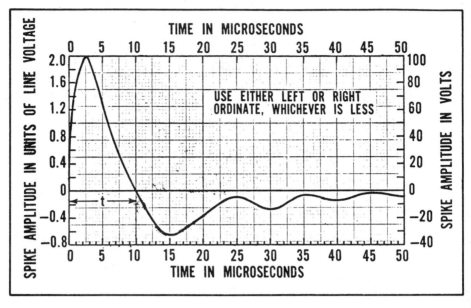

Fig. 19-17. Spike voltage applied to DC or AC power leads of test specimen for CS06 specification limit.

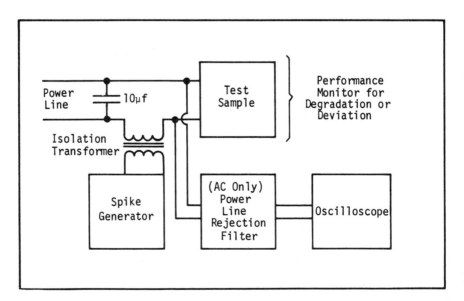

Fig. 19-18a. Test setup for CS06—spike voltage, power-line-conducted susceptibility measurements, series injection.

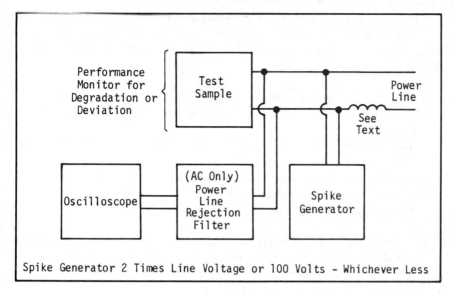

Fig. 19-18b. Test setup for CS06—spike voltage, power-line-conducted susceptibility measurements, parallel injection.

power lines and the test sample. One side of an isolation transformer (turns ratio of 1:1) is inserted in series with the AC or DC power line. The other side is connected to a suitable impulse generator, and the voltage shown in Fig. 19-17 is applied. The test setup for series injection per CS06 requirements is shown in Fig. 19-18a, and the parallel injection setup is shown in Fig. 19-18b.

The specimen is said to have passed the test if no "malfunction, degradation of performance on deviation from specified indications beyond the tolerances indicated in the individual equipment on subsystem specification when the test spike . . . is applied to the AC or DC power input leads" (21).

RECEIVERS AND SPECTRUM ANALYZERS

This section surveys the function and characteristics of EMI receivers and spectrum analyzers used to perform emission measurements. The characteristics of CISPR receivers are also covered.

EMI Receivers

EMI receivers are tunable, frequency-selective audio and RF voltmeters that measure the unknown voltage appearing at the input terminals from a suitable conducted or radiated sensor (as discussed above). The EMI receiver is a superheterodyne receiver with emphasis placed upon attempting to accurately measure signal or noise amplitude. As such, it is usually characterized by the following:

- Built-in impulse generator for calibration by substitution up to 1 GHz.
- Interlocked RF-IF attenuator to mitigate saturation and spurious responses.
- Synchronous tuning of IF to eliminate overshoot and facilitate impulse bandwidth determination.
- Special detector functions to accomplish particular calibration objectives or interpretation of measurements.
- RF-shielded housing to reduce radiated susceptibility.

EMI receivers are commercially available covering the frequency spectrum from 20 Hz to 26 GHz. There are basically two types of these receivers: (1) a separate self-contained complete receiver covering a portion of the spectrum; or (2) a unit consisting of basic RF attenuator, calibrator, IF amplifier, detectors, and meter display with separate plug-in RF heads, each of which covers portions of the spectrum. EMI receivers are to some extent standardized, although the standards differ, and different suppliers incoporate their own features. Instruments in the 15 kHz to 1 GHz spectrum are standardized by the American National Standards Institute (ANSI), whereas the International Electrotechnical Commission (IEC) has set up standards for the 15 kHz to 300 MHz spectrum.

Application.   The very nature of the EMI problem implies that for control to be possible, accurate measurements are necessary. A special calibrated RF voltmeter must be used which contains a number of features to facilitate measurements. EMI emission measurements are basically performed to help accomplish one or more applications, as follows:

1. *Specification tests* are those associated with either first look (development test) or final quality assurance and/or regulatory compliance of production specimens. They involve conducted and/or radiated emission measurements to determine whether the test specimen is within procurement or regulatory specification limits. Most of the MIL-specs involve testing inside shielded enclosures, while CISPR and FCC regulations involve open field tests.
2. *Electromagnetic ambient surveys* are applications that determine the ambient into which equipment is to be installed. They are used to:
   (a) Determine radiation profiles for emission control such as at test ranges.

(b) Determine radiation profiles of CE radiators and man-made noise in order to require that test specimens meet this ambient environment.

(c) Assist in selecting an electromagnetically quiet site from several candidate pieces of real estate.

(d) Assist in identifying relatively quiet portions of the frequency spectrum for colocating proposed new equipment or reassigning frequencies of existing CE equipment.

3. *CE equipment signatures* involve making transmitter measurements at fundamentals, harmonics, their emission sidebands, and other spurious radiations. The calibrated data are needed for procurement compliance of specifications and for EMI prediction and analyses.

4. *Culprit emitter identification* develops from the outcome of (1) or (2) above, in which the sources of EMI are to be identified, often using uncalibrated probes. The object is to identify the culprit source(s) and the nature of the emission. Examples include either interfering or illegal radiators; RF leaky cabinet joints; hot cables or wires; noisy power-line insulators; and disturbing industrial, scientific, and medical equipment.

5. *Peripheral applications* include propagation studies, discrete antenna pattern measurements, filter and shielded enclosure attenuation tests, circuit crosstalk performance measurements, and establishing cable shield and grounding EMC performance.

Bandwidth and Sensitivity Considerations. Receiver sensitivity, S, is defined in terms of its internal noise power, N, referred to the receiver input terminals. One such definition is:

$$\frac{S + N}{N} = 2 \text{ or } S = N \tag{19-3}$$

where:

$N = FKTB$ watts is the internal noise of the receiver $\qquad$ (19-4)
$F = $ noise figure of receiver (also called noise factor)
$K = $ Boltzmann's constant $= 1.38 \times 10^{-23}$ W/°K/Hz
$T = $ thermal temperature of receiver front end in °K
$B = $ receiver bandwidth in Hz

For a receiver front end with typical room temperature ($21°C \simeq 293°K$), equation (19-3) becomes:

$$N \simeq 4 \times 10^{-21} FB \text{ watts} \tag{19-5}$$

$$N \simeq 4 \times 10^{-18} FB \text{ milliwatts} \tag{19-6}$$

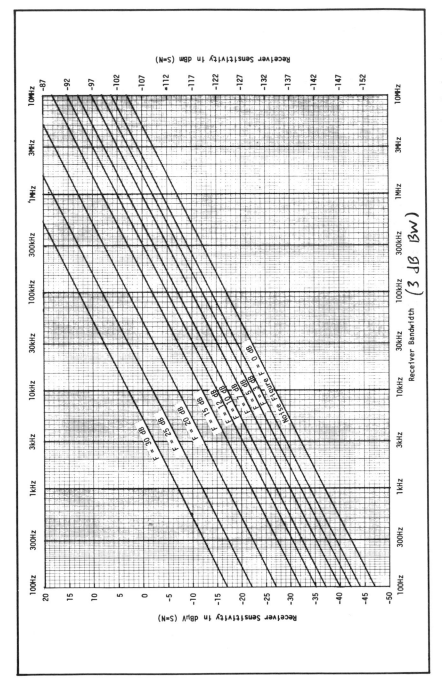

Fig. 19-19. Receiver narrowband sensitivity vs. bandwidth and noise figure. *(Courtesy of Interference Control Technologies)*

Expressed in units of dBm (decimals above 1 milliwatt), equation (19-6) becomes:

$$N_{dBm} = -174 + 10 \log_{10} (FB) \tag{19-7}$$

$$N_{dBm} = -174 + F_{dB} + 10 \log_{10} B_{Hz} \tag{19-8}$$

$$N_{dBm} = -114 + F_{dB} + 10 \log_{10} B_{MHz} \tag{19-8a}$$

Equation (19-8) is useful in quickly determining the receiver noise power in units of dBm (dB above 1 milliwatt) for any specified receiver noise figure and bandwidth. These relations are shown in Fig. 19-19 where the right ordinate is in units of dBm, the x-axis is receiver 3 dB bandwidth, and the parameter is noise figure in units of dB. The left ordinate in Fig. 19-19 is expressed in units of dBµV and corresponds to the right ordinate in units of dBm for a 50-ohm system ($S_{dB\mu V} = S_{dBm} + 107$ dB; see Chapter 18).

The preceding definition of sensitivity is in terms of receiver thermal noise which is incoherent (phase components of adjacent frequency increments are random). This definition holds for either narrowband signals or broadband incoherent noise. When the signal is developed from an impulse-like source such as a transient or an impulse generator, a second definition of sensitivity is useful:

$$S_{dB\ V/MHz} = -7\ dBm + F_{dB} - 10 \log_{10} B_{MHz} \tag{19-9}$$

Functional Description.   Figure 19-20 is a simplified block diagram of a typical modern EMI receiver. It consists of one to three stages of heterodyning. Signal flow in the figure is from the sensors on the upper left, through the receiver stages in the center, to the processing circuitry on the lower right. The input sensors consists of intercept antennas (#1) for radiated emission pickup. Other sensors include current probes (#2) for conducted measurements and line impedance stabilization networks (LISNs, #3). (Previous sections of this chapter discuss these sensors.) The figure also shows an impulse generator (#4) and an amplitude calibrating device for obtaining absolute voltage measurements by substitution when an unknown signal is being measured.

The four inputs to the EMI receiver shown in Fig. 19-20 are in reality a single terminal, since only one of the sensors is connected to the receiver input at one time. An input signal is passed through an RF attenuator (#5), which is a broadband three- or four-decade step attenuator that can be passive or active (FET) networks. The first decade is usually ganged to the IF amplifier (#12) for IF attenuation insertion. The other decade steps exist back at the RF attenuator to reduce the levels of large signal inputs. The output from the RF attenuator is passed through either a low-pass (#6) or a band-pass filter (combination of high- and low-pass filters) in order to reduce strong out-of-band signals. Nearly all EMI receivers have either pre-selectors or tunable

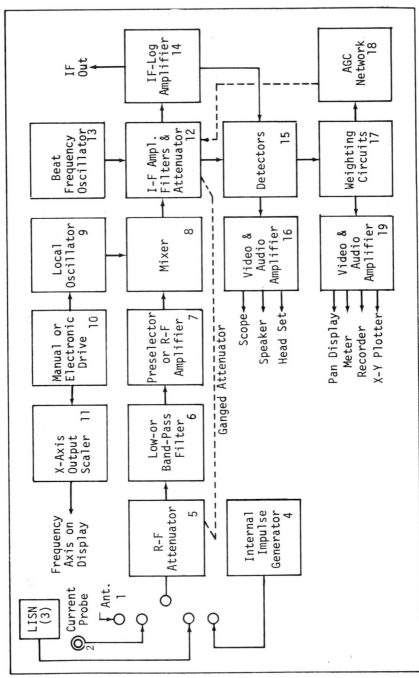

Fig. 19-20. Simplified block diagram of modern EMI receiver. (*Courtesy of Interference Control Technologies*)

RF amplifiers (#7) in order to further attenuate strong out-of-band signals that may otherwise cause intermodulation and other spurious responses.

The output from the pre-selector drives a first converter consisting of a mixer (#8) and a local oscillator (#9). The convertor is generally a down frequency conversion with the exception of LF receivers, which first up-convert. The local oscillator is tuned either manually or by an electronic ramp. If the latter is used, it is usually a voltage-tuned oscillator. The tuning shaft or ramp voltage also drives an x-axis scalar (#11) to either a scope display or an X-Y plotter for frequency.

The mixed output from the converter drives an IF amplifier and attenuator (#12), including bandwidth determining filters. IF response is generally gaussian to avoid overshoot and to facilitate calibration of EMI receivers. To maintain a large dynamic range, the final IF amplifier stages are designed to allow waveforms of large peak amplitude to pass undisturbed. Data on instrument bandshape are essential because the instrument response depends upon both bandwidth and its shape factor. Some EMI receivers have practically constant bandwidth, at least within each tuning band, while others do not.

A beat frequency oscillator (#13) is generally provided to assist in identification of CW signals below about 30 MHz. Its output produces a recognizable tone at audio frequency by beating with the IF signal.

To further compress the dynamic range over which the output of the IF amplifier swings, most instruments have a logarithmic amplifier (#14), whose input may swing over 60 dB and whose output varies by about 20 dB. The log IF amplifier typically provides unity gain to high-level signals near saturation and 60 dB gain to low-level signals. While large dynamic ranges (e.g., greater than 60 dB) are easy to obtain for narrowband signals, they are difficult to realize for broadband signals owing to problems with saturation.

Either the linear or the logarithmic amplifier drives a number of selectable detector functions (#15). One detector output (AM and/or FM) is used to drive a video and audio amplifier (#16), whose outputs may be presented directly on a high-impedance scope, or to a low impedance speaker or head set. Weighting circuits (#17) employed in the detectors are the devices that make EMI instruments substantially different from the conventional superheterodyne instruments. Some detectors read the peak value of an IF signal through either a direct peak detector circuit or a slide-back network. Other instruments use a diode detector in a quasi-peak circuit whose time constants are chosen to give a reading that depends on both the amplitude and time distribution (duty cycle) of the wave form. Practically all EMI instruments contain an averaging detector that measures the average envelope of the IF output. Some EMI receivers include an FM detector for intercepting and processing frequency modulated signals and for AFC (see Chapter 18, section on "Detector Functions").

The output from the weighting circuits drives other video and audio amplifiers (# 19) for a number of display devices, including meters, recorders, pan adaptors, and X-Y plotters. Another output from one of the circuits in some EMI receivers is used to drive an automatic gain control (AGC) network (# 18). The AGC network changes the gain of the final stages of the IF amplifier (# 12) in order to maintain greater dynamic range of signal input. There is a significant limitation and source of error in the use of AGC networks where either peak signals of low duty cycle or transients are employed, since the gain of the circuit and hence calibration is a function of the duty cycle and pulse width of the signal. AGC networks should be avoided for calibrated EMI instrument applications.

Special output jacks are provided for audio, oscilloscope, IF amplifier, remote meter, and X-Y plotter. These features offer interesting capabilities:

1. The oscilloscope output provides means for observing either frequency or time domain signals.
2. The IF output permits connecting to external devices such as panoramic displays and high frequency oscilloscopes.
3. A jack is provided for other meters to facilitate monitoring of equipment readout remote to the EMI instrument.
4. An X-Y plotter output is provided for plotting amplitude versus frequency as a permanent record, which is especially useful in specification testing.

Microprocessor-Controlled EMI Test Receivers. In addition to the above RF measurement features, many newer EMI receivers are designed for increased measurement automation with the incorporation of microprocessors to manage the analog sections, displays, keyboards, and collection of measurement results (6, 12). Direct readout in μV, dBμV, dBm, dBμV/m, and dBμV/m/MHz are possible, and a synthesized local oscillator allows for band switching without recalibration problems. Automated test runs are preprogrammed into ROMs and EPROMs, and give the user a degree of freedom from manual tuning and spectrum search that can increase measurement accuracy and decrease test run times. In addition, antenna factors can be programmed into the receiver as a special function, eliminating manual conversion. IEEE-488 bus interfaces are available that allow the test receiver to be connected to an external programmable controller.

CISPR Equipment Specifications. The CISPR regulations (see Chapter 20) dictate that emissions measurements be conducted using equipment meeting CISPR specifications, specifically regarding the charge and discharge time constants of the quasi-peak detector. Additional requirements are placed on the measurement bandwidths. These specifications are becoming more widespread, with a variety of CISPR-compliant EMI receivers available, as

Table 19-1. Specifications for CISPR EMI receivers. (Courtesy of Interference Control Technologies.)

CHARACTERISTICS	10K–.15MHz	.15–30MHz	30–300MHz	.3–1GHz
6-dB BANDWIDTH, B_6	200Hz	9kHz	120kHz	120kHz
IMPULSE BANDWIDTH, B_i(~7 dB)		9.8kHz	126kHz	126kHz
20-dB BANDWIDTH, B_{20}		9.8 TO 20kHz	126 TO 280kHz	126 TO 280kHz
RESPONSE TYPE		———————SYNCHRONOUS TUNED———		
QP CHARGE TIME CONSTANT	45ms	1mS	1mS	1mS
QP DISCHARGE TIME CONSTANT	500ms	160mS	550mS	550mS
METER TIME CONSTANT	160ms	160mS	100mS	100mS
PRE-DETECTOR OVERLOAD FACTOR	24dB	30dB	43.5dB	43.5dB
POST-DETECTOR OVERLOAD FACTOR	12dB	12dB	6dB	6dB
I-F REJECTION		$\geq$40dB	$\geq$40dB	$\geq$40dB
IMAGE-FREQUENCY REJECTION		$\geq$40dB	$\geq$40dB	$\geq$40dB
OTHER SPURIOUS RESPONSES		$\geq$40dB	$\geq$40dB	$\geq$40dB
SHIELDING EFFECTIVENESS		$\geq$60dB	$\geq$60dB	$\geq$60dB
ACCURACY, VOLTAGE		$\pm$2dB	$\pm$2dB	$\pm$2dB
ACCURACY, FIELD STRENGTH		$\pm$3dB	$\pm$3dB	$\pm$3dB

well as adapters that convert other EMI receivers to the CISPR specification. Table 19-1 shows CISPR specifications for EMI receivers.

Spectrum Analyzers

This section reviews the functions and characteristics of spectrum analyzers, which are used for some conducted and radiated emissions applications by the EMC and related communities. Since the FCC allows spectrum analyzers in determining complicance with emission specifications, they are becoming increasingly useful and versatile. In contrast to EMI receivers, modern spectrum analyzers are generally characterized by untuned front ends, relatively high noise figures, built-in electro-optical (CRT) spectrum amplitude displays with variable persistence, and digital storage. Amplitude calibration of intercepted signals is achieved through narrowband means. Most spectrum analyzers offer little dynamic range to impulsive signals because of wide-open front ends. Although they are designed to service different applications, the principal advantage of spectrum analyzers over EMI receivers is that the former offer flexibility and functional displays for a smaller investment per octave of coverage.

Application. Spectrum analyzers are used primarily for intercepting, displaying, and examining signal and electrical noise activity over all or part of the radio frequency spectrum from about 100 Hz to 200 GHz. This nine-decade frequency span is generally covered by three or four spectrum analyzers offering a wide choice of functions that are useful in signal identification. They are generally of the superheterodyne type with two to four stages of frequency conversion.

Spectrum analyzers have many applications, with EMI use a small percentage of the total. They are often employed for quick-look emission control during development stages of hardware before EMI qualification testing is performed (i.e., for dev testing). They are occasionally used as supporting instruments in the execution of EMI specification testing. Spectrum analyzers are widely used for electromagnetic ambient site surveying and frequency monitoring interference control (FMIC) to examine signal activity and to identify unknown intercepted emissions by modulation class and other parametric signatures.

With certain exceptions, spectrum analyzers do not use tuned receiver front ends—primarily because of both significant additional cost implications and compromise of flexibility due to physical realizability problems. Since tuned front ends are not used, spectrum analyzers have significant spurious responses due to intermodulations and heterodyning products and they exhibit limited broadband dynamic range due to saturation. Because of the economy and flexibility requirements, they generally exhibit poor equivalent receiver noise figures (e.g., about 30 dB in the HF/VHF spectrum and 40 dB in the microwave region). Consequently, their limited sensitivity often restricts their surveillance applications, and it limits the quality assurance of EMI specification compliance testing. Nevertheless, they are functionally useful laboratory test tools with many applications.

Functional Description.   Figure 19-21 is a simplified functional block diagram of a typical RF spectrum analyzer. One RF input port (#1) is used to accept either a matching pickup sensor (#2) or an internal calibrating voltage (#3). The input signal (#1) is passed through a low-pass filter (#4) that attenuates its higher frequency components in order to reduce spurious responses. However, no tunable amplifier or preselector is used. An RF attenuator (#5) is employed to limit the input signal to the useful dynamic range of the instrument.

In order to permit scanning over a substantial frequency range within a band (i.e., more than one octave), and to achieve a relatively flat frequency response, the incoming signals are up-converted in the mixer (#6) by the voltage-tuned oscillator (#7). While the LO may be tuned over only one octave, the input signal coverage may occupy one to two decades. The IF amplifier (#8) contributes significantly to the noise figure because no RF gain is used. It accepts a band of up-converted signals well beyond cutoff of the low-pass filter (#4). The IF amplifier output is down-converted in the mixer (#9) with the fixed tuned LO (#10).

The second mixer output (#9) drives a second IF amplifier (#11), which in turn may be coupled by selection either to a logarithmic IF amplifier

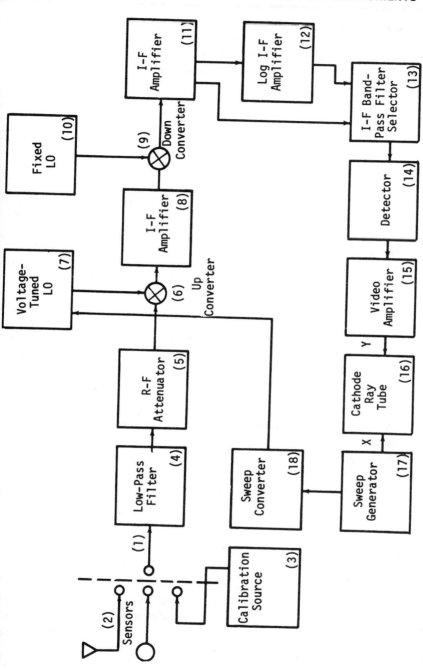

Fig. 19-21. Simplified block diagram of RF spectrum analyzer. (*Courtesy of Interference Control Technologies*)

(#12) or directly to a series of selectable band-pass filters (#13). The log IF amplifier typically permits a 60 dB or more useful display amplitude on the CRT, while a linear display is useful over only about 30 dB. The IF band-pass filters offer a selection of bandwidths over about four orders of magnitude. The output is coupled to the final detector (#14), and is video-amplified (#15) and presented to the Y-axis input of the CRT drive.

Typical Analyzers.   One of the greatest improvements to the spectrum analyzer field was the introduction of the Hewlett-Packard first generation "modern version" of the spectrum analyzer in 1963. The unit included a BWO that scans 2 to 4 GHz. It up-converts the incoming signal below this frequency range to an IF amplifier operated at approximately 2 GHz. Thus, a spectrum from 10 MHz to 2 GHz could be swept out and displayed on the scope at one time. Frequencies from 2 GHz to 10 GHz and higher are obtained by harmonic mixing action. Two serious drawbacks of this analyzer are its limited sensitivity and the number of ambiguous signals presented on the analyzer scope face at one time. The analyzer offered several resolutions (i.e., IF bandwidths), of 1 kHz, 3 kHz, 10 kHz, and 1 MHz. The equivalent noise figure below 2 GHz is about 34 dB.

In order to help reduce the spurious responses, a number of low-pass and interdigital band-pass filters were made available for the spectrum analyzer. However, a later option offered is a pre-selector in the form of a voltage-tuned YIG filter driven by the horizontal output from the display section. The pre-selector, governing 2 to 12 GHz, tracks the tuning of the RF section and reduces undesired responses by about 20 dB + 6 dB/octave beyond 100 MHz off center tuning. Since the pre-selector offers an additional 5 dB of insertion loss, the sensitivity of the spectrum analyzer from 2 to 4 GHz results in an equivalent noise figure ranging from 39 to 51 dB.

A second generation spectrum analyzer, introduced in the late 1960s, incorporated many of the better features of the first generation together with continuing advances in the state of the art.

The newest generation of spectrum analyzers is relying upon the use of microprocessors and digital signal processing techniques to automate the measurement process and increase the dynamic range of the spectrum analyzer to 80 dB and higher. A minicomputer, interfaced with a spectrum analyzer, provides a high level of automation, easing the job of the operator. The microprocessor-controlled analyzers combine automation of digital control and increasingly "user-friendly" displays. A series of programs written by Hewlett-Packard include automated spectrum search, where the spectrum analyzer scans a user-selected frequency band looking for signals, automated harmonic distortion measurement, and a percent amplitude modulation measurement.

For dedicated FCC/CISPR emissions measurement, the Electrometrics ESA-1000 is a relatively low-priced spectrum analyzer specifically designed to perform FCC/CISPR measurements. Conforming to the CISPR receiver specifications shown in Table 19-1, it features peak, CISPR quasi-peak, and average detector functions along with built-in CISPR bandwidths (9 kHz and 100 kHz) in addition to other measurement bandwidths. The frequency range of the ESA-1000 is from 100 kHz to 1000 MHz and, in the 30 to 1000 MHz range, direct E-field strength reading can be obtained by plugging in an optional antenna (i.e., compensation for antenna factor is performed internally). An optional digital memory is available that facilitates measurements and allows measurement comparison between the two channels. A battery pack is available that can power the ESA-1000 for 3.5 hours between charges.

POWER-LINE MONITORING EQUIPMENT

Utility-supplied power mains, as discussed in Chapter 15, are subject to line sags, dropouts, impulsive spikes, and surges. Before sensitive equipment is installed at a site, such as electronic data processing (EDP) and medical equipment, the mains should be tested for the occurrences and severity of the above disturbances. For the most sensitive of equipment, which cannot tolerate power mains disturbances, power conditioners, surge suppression, uninterruptible power supplies (UPSs), or combinations of these devices must be installed on the mains. (See Chapter 15.)

To get an idea of the quality of the power supplied to a site, power-line monitoring equipment should be installed and the variations in power quality observed over the course of days or weeks. This section discusses various aspects of commercially available power-line monitoring equipment.

In general, power-line monitors incorporate some or all of the following features:

- DC, single-, or multi-phase monitoring capability.
- Overvoltage and undervoltage sensing, with user-set threshold.
- Low average and high average, with user-set threshold.
- Number of cycles of occurrence.
- Bipolar impulse amplitudes, 0.5–100 µsec pulse durations.
- Line frequency monitoring.
- Phase voltage unbalance.
- Internal battery to power unit in case of power failure.
- Printout in summary form or on a continuous basis.
- Date and time of occurrence of disturbance.
- Presence of common-mode noise.
- Remote programming and monitoring capability.

Power-line voltage sags or surges (under- or overvoltage) are defined as a change in the line voltage for an extended number of cycles. The recorded over- or undervoltage depends on the threshold set by the user. If the equipment to be installed on the power line is highly sensitive, the threshold is set at a small fraction of the line voltage; less sensitive equipment need not specify a tight threshold. Thus, some information on the power-line susceptibility of the equipment is necessary, which may be available from the manufacturer, or may have to be determined from an analysis by the user.

An important feature of the power-line monitor is the ability to record the quality of the power over a long term, which can aid the design or selection of power conditioning and surge/transient protection equipment.

SIGNAL CALIBRATION AND SUSCEPTIBILITY TESTING SOURCES

The previous sections emphasized conducted and radiated EMI emission measurements. The topic of susceptibility testing will now be discussed, with emphasis on signal sources used to test the susceptibility of equipment. Such sources include impulse and spike signal generators, power oscillators, and power amplifiers, and are used for calibration or for susceptibility testing. In addition, the characteristics of some electrostatic discharge (ESD) testers will be discussed.

Impulse Generators

This section reviews impulse generators (IG) that are designed and used to calibrate EMI receivers over the frequency spectrum from about 200 Hz to 10 GHz. The receiver calibration process is by substitution of a known broadband spectral density below levels of about 100 dBµV/MHz. High voltage impulse generators, having spectral density outputs to 160 dBµV/MHz, are also available. They are particularly useful for irradiating test specimens to determine their radiated transient susceptibility. They are also useful at somewhat lower levels for performing spectrum attenuation tests on filters, circuits, and shielded enclosures up to about 5 GHz (1).

The calibration of the output display meter or plotter of an EMI meter is possible at any frequency and amplitude from the receiver noise level to saturation. Calibration by substitution may also be performed, in which the IG level is adjusted to give the same output reading as a known coherent broadband input signal. In all cases, the receiver gain–bandwidth product comes into play, since the IF uniformly fills the receiver impulse bandwidth,

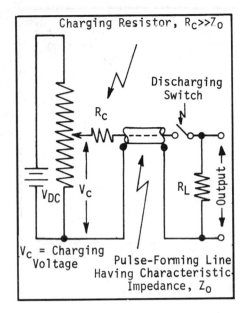

Fig. 19-22. Simplified diagram of an impulse generator. (*Courtesy of Interference Control Technologies*)

as discussed below. This is contrasted with narrowband signal generator techniques of calibration by substitution, involving only the receiver gain. Thus, the calibration technique depends on whether broadband or narrowband signals are being utilized.

Basically, the impulse generator is a device that charges a pulse-forming line to a maximum amplitude of about 200 volts and then rapidly discharges it into a 50-ohm load at the generator output, which is connected to the input of a receiver. This is shown in simplified form in Fig. 19-22. This process, similar to a delay line modulator in a radar, results in a transient spike of one-half of the amplitude of the DC source and an equivalent pulse width of twice the line length, typically on the order of a few tenths of a nanosecond.

The switch shown in Fig. 19-22 is generally a coaxial reed relay with mercury-wetted contacts to avoid contact pitting and permit longer life. The switch may be excited by: (1) a manual single-shot, push-button switch; (2) actuating it synchronously from a 60 Hz power-mains source; or (3) an adjustable internal oscillator from a few Hz to several kHz. Most IGs are operated at 60 Hz from the power line.

The output from an impulse generator is calibrated in terms of the spectral density, generally rated in units of dBμV/MHz. The time domain is shown

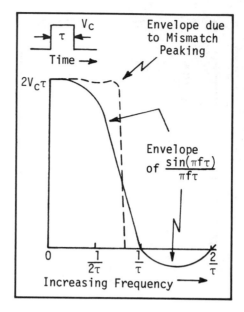

Fig. 19-23. Time and frequency domains of a unit pulse. (*Courtesy of Interference Control Technologies*)

in Fig. 19-23 and corresponds to an amplitude $a = V_c$, where $V_c = $ DC charging voltage.

Upon application of the Fourier integral or Laplace transform to this pulse, the frequency domain, spectrum distribution of the pulse $A(f)$ is defined:

$$A(f) = 2V_c\tau \left| \frac{\sin \pi f \tau}{\pi f \tau} \right| \text{ V/Hz} \qquad (19\text{-}10)$$

where:

$A(f) = $ spectral density in V/Hz
$V_c = $ charging volts
$f = $ frequency in Hz
$\tau = $ pulse duration in seconds

The spectral density function takes on the familiar $(\sin x)/x$ distribution shown in Fig. 19-23. The first null exists at $(1/\tau)$ Hz. Thus, for $\tau = 0.5$ nsec $(0.5 \times 10^{-9}$ sec), $1/\tau = 2$ GHz.

At lower frequencies for which $f \ll 1/\tau$, $\sin (\pi f \tau)/\pi f \tau \simeq 1$, and equation (19-10) becomes:

$$A(f) = 2V_c\tau \text{ V/Hz} \qquad \text{for } f \ll 1/\tau \qquad (19\text{-}11)$$

Thus, the spectral density at low frequencies is essentially flat. This is the basis for the use of impulse generators for broadband calibrators.

Since there is a natural tendency to want to push the limits of instrument performance, relative spectral flatness can be increased by mismatch peaking techniques that compensate for spectral rolloff. This is achieved by deliberately mismatching both the pulse-forming line and the load. The result is to achieve a flatness of better than ± 1 dB, as suggested in Fig. 19-23, out to a frequency of 1 GHz for a 0.5-nsec pulse.

Impulse generators make an excellent secondary standard source for calibration in the field or for a quick calibration of an EMI receiver in the laboratory. Since these calibrations are directly used for broadband signals, they imply a knowledge of the impulse bandwidth of the receiver when narrowband signals are to be calibrated with an IG. The impulse generator is also used to measure and compute the impulse bandwidth of the receiver.

Confirmation of the calibration of the impulse generator output is somewhat difficult to accomplish. The way to do this is to use two or more known (previously measured) standard filters having a gaussian band-pass characteristic. The filter is shock-excited by the IG, and the peak signal level is calibrated on an oscilloscope or other suitable peak reading display. The IG is then removed, and the filter is next driven by a standard CW signal generator whose rms amplitude output is well calibrated and adjusted to give the same oscilloscope peak amplitude display. The ratios of the two generator output levels are then related in a manner that defines the IG calibration:

$$\text{Filter impulse BW} = \frac{\text{CW RMS amplitude in dB}\mu\text{V}}{\text{IG amplitude in dB}\mu\text{V/MHz}} \qquad (19\text{-}12)$$

or:

$$\text{Unknown IG output} = \frac{\text{CW RMS amplitude in dB}\mu\text{V}}{\text{Filter impulse BW in MHz}}$$

Note the 3 dB difference that results from using the RMS amplitude instead of the peak amplitude of the CW generator. While this is a difference of 3 dB (peak-to-RMS), it is the basis established and used by the EMC community, and included in defining broadband specification limits.

A simpler, quicker, but less reliable method of determining that an impulse generator is working properly is to substitute one IG for another to see if an EMI receiver with oscilloscope output will read both impulse generators within 1 dB at several frequencies and amplitude levels. The probability that both are improperly working in such a way as to give the same readings is very small. Another still less conclusive check is to measure the DC charging voltage or supply to the IG as shown in Fig. 19-22. Unless the RF switch is bad, a correct voltage as specified by the manufacturer is often a fair indicator of performance. Yet, most failures are in the RF switch unless the IG is solid state. A poorly performing switch may appear on a scope display as unequal pulse heights in a train, missing pulses, or extraneous pulses.

Spike Generators

The EMI transient generator, better known in EMI parlance as a spike generator, is used as a source for performing transient conducted-susceptibility testing in accordance with the CS06 and CS06.1 test method per MIL-STD-462. Another type of transient generator specification appears in IEEE-STD-472. Electrostatic discharge testers are a special case of spike generators and are designed to simulate the transient phenomena associated with ESD.

The transient shape that must be produced by the MIL-STD-462 spike generator for susceptibility testing is shown in Fig. 19-17. The spike has an equivalent 3 dB bandwidt of about 5 µsec. Therefore, the main spectral energy exists from DC to $1/\tau = 1/5$ µsec $\simeq 200$ KHz. Since little significant transient energy exists above 1 MHz, the required spike is a low frequency transient.

A typical spike generator usually has adjustable peak amplitude from less than 10 volts up to several hundred volts. The spike injected onto a power line can be made to coincide with any point in the voltage cycle—at either positive or negative peaks or at axis crossings. Synchronous and asynchronous operation is available.

Manual or remote operation is usually provided for the injection of a single shot transient in a manner similar to that used on impulse generators. Available output terminals permit either series injection on AC power lines or parallel injection to DC lines carrying up to 50 amperes as called out in applicable specifications.

Other spike or transient generators are available that are used to simulate lightning-induced transients of power and telecommunications lines per ANSI/IEEE Standard C62.41 (formerly IEEE Std 587). These generators produce unidirectional, bipolar, and oscillatory waveforms for the testing of transient protection networks. (See Chapter 15 for a complete discussion of transient waveforms and amplitudes associated with power and telecommunications lines.)

It is highly recommended that such networks be tested against the potential transient environment for which the network was designed. A simplified typical circuit for testing devices and networks against oscillatory lightning-induced transients is shown in Fig. 19-24. With switch SI closed and S2 open, the capacitor C is charged to the specified open circuit voltage by the high-voltage DC generator E (typically 10 kV to 20 kV). The capacitance C is typically 1 µf to 10 µf. When $S2$ is closed, the capacitor discharges through

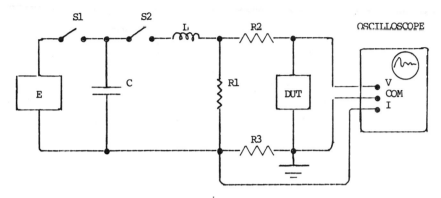

Fig. 19-24. Transient test circuit. (Reference 21)

wave-shaping elements L, $R1$, and $R2$. Depending on L, $R1$, and $R2$, 8×20 or 10×1000 μsec waveforms are produced that are incident on the terminals of the device under test (DUT). The oscilloscope records the response of the DUT to the incident wave, and the clamping voltage of the DUT is checked against the design value (20).

Additional applications of pulse generators are used in NEMP (nuclear electromagnetic pulse) testing. An EMP is produced from the detonation of a nuclear device. The result is a double exponential pulse of high amplitude (50 kV/m) with a rise-time of 3 to 10 nsec. This pulse can couple onto power and telecommunications lines and induce very high energy, fast rise-time transients onto the lines. EMP pin drive sets are available that simulate this type of transient by direct injection onto the line. EMP simulators are also used to drive TEM (transverse electromagnetic wave) cells and parallel transmission line test setups that are used to measure the radiated susceptibility of a piece of equipment to an EMP.

Electrostatic Discharge (ESD) Testers

ESD testers are normally used to test the external enclosures of equipment boxes to verify ESD equipment and cable hardening. They are also useful for developmental susceptibility testing. A number of commercially available ESD testers are available that reproduce the impulsive voltage/current waveform associated with a discharge from a charged human body. The newer types of ESD testers are in the form of hand-held "guns" that contain a capacitor–resistor discharge network with values that approximate those of the "average" human beings (18). The circuit model of a human being who has acquired a charge by walking on a nylon carpet or some other means is shown in Fig. 19-25.

Typically, the resistance, R, values are a few kilohms and the capacitance, C, values are a few hundred picofarads. The charge voltage can reach 20 kV

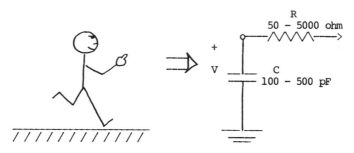

Fig. 19-25. Equivalent ESD circuit of human being.

and higher, with average values of between 4 and 12 kV, and can be of either positive or negative polarity, depending on the charging process and the materials involved. Typical discharge current waveforms have rise-times of 1 nsec, pulse durations of 150 nsec, and peak currents of 5 to 10 amperes. This time behavior extends the frequency content of ESD to $1/\pi\tau_r \simeq 300$ MHz.

Typical ESD testers have an adjustable high voltage supply that slowly charges the capacitor through a several-megohm resistor. Either single or multiple shot operation is available. The discharge is initiated through a metal tip, which may be of several different types and sizes to simulate different discharge conditions (such as holding a key or a tool). The return path of the ESD current is through the capacitance to ground of the EUT, through a low inductance metal strap, and back to the capacitor. In addition, some ESD testers are available with specialized discharge heads that can emphasize either the magnetic field or an electric field of an indirect discharge.

CONTINUOUS WAVE (CW) SUSCEPTIBILITY TESTING SOURCES

This section discusses some aspects of susceptibility testing power sources that produce continuous wave (CW) power to drive antennas to produce radiated fields to check radiated susceptibility of equipment, or that produce signals to be injected via conducted injectors. Currently, the only radiated and conducted susceptibility standards exist in military standards; that is, (except for certain West German and Swedish regulations) commercial products are not required by government regulations to comply with susceptibility limits. As of this writing, there is some movement afoot in the U.S. and Europe to establish susceptibility standards for commercial products.

The antennas discussed in previous sections were, for the most part, used in emissions testing only. Some antennas, however, are used for producing radiated fields, such as some biconicals, conical log spirals, and certain loop antennas. One application of power sources to drive radiating antennas is in the area of Federal Communications Commission Part 15, Subpart J (FCC 15J) radiated emissions compliance testing, which requires that the RF characteristics of a test site be known for it to qualify as an approved test site. The basic procedure is to produce a known radiated field and measure the losses associated with the site. These losses are embodied in a site attenuation curve that is filed with the FCC along with the application for approval of an FCC open field test site. (See Chapter 20.)

A susceptibility power source usually consists of a signal source that drives a wideband power amplifier. Two basic types of the equipment exist: (1) a complete unit, containing a variable frequency oscillator/power amplifier unit; and (2) a separate signal generator/oscillator that drives an external amplifier. The frequency range of the oscillator and amplifier depends on the frequency range of interest, and the output power of the amplifier depends on the field levels that must be generated plus allowances for antenna factors and cable losses, and so on.

The necessary field strengths required by MIL-STD-461 range from 1 V/m (for testing outside of shielded enclosures) to 200 V/m (for aircraft), depending on whether the tests are conducted in a shielded enclosure or not. The use of a shielded enclosure is necessary to prevent the internally generated field from interfering with other equipment, in which case the required field strength is raised to 10 V/m. In order to achieve these field strengths, the power oscillators/amplifiers of susceptibility sources typically are capable of 50 watts to several kilowatts of RF output power at low frequencies and 1 to 10 watts above 1 GHz.

The output power can be rated in terms of maximum, minimum, and nominal (average) power over the operational bandwidth of the amplifier. This is important because the power varies with frequency for any amplifier, and the variance in output power with frequency determines the bandwidth of the amplifier. The user should be aware of how the output power of a given amplifier is measured, especially where swept-frequency measurements are performed. Pulse power is generally difficult to define for a broadband power amplifier because there are many variables involved in rating the pulse power of the amplifier that affect the pulse power, such as duty cycle, repetition rate, and rise-time. In general, a linear amplifier will generate 2 to 6 dB more pulse power than CW power (19).

REFERENCES

1. Andrews, J. R., and Baldwin, E. E. "Baseband Impulse Generator Useful to 5 GHz." *Record of the 1975 IEEE Electromagnetic Compatibility Symposium*, Volume 75CH1002-5, October 7–9, 1975, San Antonio, TX.
2. Bassen, H. I. "A Broadband Miniature Isotropic Electric Fields Measurement System." *Record of the 1975 IEEE Electromagnetic Compatibility Symposium*, Volume 75CH1002-5, October 7–9, 1975, San Antonio, TX.

3. Bronaugh, E. L., Kerns, D. R., and Southwick, R. S. "An Amplitude Probability Distribution Detector System." *Record of the 1975 IEEE Electromagnetic Compatibility Symposium*, Volume 75CH1002-5, October 7–9, 1975, San Antonio, TX.

4. Burnett, R. H. "Transient Generator Requirements for Better Susceptibility Testing." *Record of the 1975 IEEE Electromagnetic Compatibility Symposium*, Volume 75CH1002-5, October 7–9, 1975, San Antonio, TX.

5. Cummings, J. R. "Translational Electromagnetic Environment Chamber, A New Method for Measuring Radiated Susceptibility and Emission." *Record of the 1975 IEEE Electromagnetic Compatibility Symposium*, Volume 75CH1002-5, October 7–9, 1975, San Antonio, TX.

6. Eveleigh, V. W. "Computer Controlled RF Test Facility." *Record of the 1976 IEEE Electromagnetic Compatibility Symposium*, Volume 76CH1004-9, July 13–15, 1976, Washington, DC.

7. Keenan, Kenneth. *Digital Design for Interference Specifications*, Vienna, VA: The Keenan Corporation, 1982.

8. Kesney, E. S. "Shielded Enclosures for EMC and TEMPEST Testing." *Record of the 1975 IEEE Electromagnetic Compatibility Symposium*, Volume 75CH1002-5, October 7–9, 1975, San Antonio, TX.

9. Larsen, E. B., and Andrews, J. R. "Sensitive Isotropic Antenna with Fiber-Optic Link to a Conventional Receiver." *Record of the 1976 IEEE Electromagnetic Compatibility Symposium*, Volume 76CH1004-9, July 13–15, 1976, Washington, DC.

10. Lindgren, E. A. "Demountable RF Shielded Enclosures." *Record of the 1975 IEEE Electromagnetic Compatibility Symposium*, Volume 75CH1002-5, October 7–9, 1975, San Antonio, TX.

11. Nichols, F. J. "Facts and Myths of RF Shielding." *Record of the 1975 IEEE Electromagnetic Compatibility Symposium*, Volume 75CH1002-5, October 7–9, 1975, San Antonio, TX.

12. Ramirez, Ron. "Friendly High Performers—'Intelligent Test Gear.'" *EMC Technology*, Volume 4, Number 1. Gainesville, VA: Don White Consultants, Inc., January 1985.

13. Sugarman, R. H. "Mobile RFI Antenna System." *Record of the 1975 IEEE Electromagnetic Compatibility Symposium*, Volume 75CH1002-5, October 7–9, 1975, San Antonio, TX.

14. Tadeusz, Dr., and Babij, M. "Properties of Wideband Magnetic Field Probes." *Record of the 1976 IEEE Electromagnetic Compatibility Symposium*, Volume 76CH1004-9, July 13–15, 1976, Washington, DC.

15. Trzaska, H., and Babij, T. M. "Superwideband Electric Field Probes." *Record of the 1975 IEEE Electromagnetic Compatibility Symposium*, Volume 75CH1002-5, October 7–9, 1975, San Antonio, TX.

16. White, Donald R. J. *EMI Test Instrumentation*, Volume 4, EMC Handbook Series. Gainesville, VA: Don White Consultants, Inc., 1974.

17. White, Donald R. J. *EMI Test Methods and Procedures*, Volume 3, EMC Handbook Series. Gainesville, VA: Don White Consultants, Inc., 1974.

18. Technical Staff, Keytek Corporation. "Electrostatic Discharge (ESD) Protection Test Handbook." Burlington, MA: Keytek Instrument Corporation, 1983.

19. Technical Staff, Amplifier Research. "Your Guide to Broadband Power Amplifiers," Souderton, PA: Amplifier Research 1984.

20. Technical Staff, GENERAL ELECTRIC CO. "TRANSIENT VOLTAGE SUPPRESSION," GENERAL ELECTRIC CO. ELECTRONIC DATA LIBRARY, Fourth Edition, 1983.

21. MIL-STD-461B, "Electromagnetic Interference Characteristics, Requirements for," 1 April 1980.

Chapter 20
FCC, CISPR, AND VDE
EMISSION STANDARDS
AND COMPLIANCE

The emissions regulations enacted by governments and international regulatory agencies are an essential aspect of electronic equipment manufacture. The ability to market and sell electronic equipment depends upon meeting the specifications embodied in the various regulations. The rule-making organizations of concern in this chapter are the United States Federal Communications Commission (FCC), The International Special Committee on Radio Frequency Interference (CISPR, from the French abbreviation for Comité Special International des Perturbations Radiolectriques), and the Association of German Electrical Engineers (VDE, for Verband Deutscher Elektrotechnicker), which writes (West) German regulations. The German governmental authority responsible for enforcement of the VDE emissions limits is the Minister of Postal and Telecommunications Services, Deutsche Bundespost-Fernmelde-technishes Zentralmant (FTZ-DB). Actually, most European regulations are enforced by the postal agencies of each government, which handle most, if not all, types of regulations concerning communications.

This chapter surveys the regulations enacted by these bodies to reduce the occurrence of harmful interference.

FEDERAL COMMUNICATIONS COMMISSION LIMITS, TESTING, AND COMPLIANCE

The United States Federal Communications Commission (FCC) *Rules and Regulations*, Part 15 Subpart J (FCC R & R Part 15 Subpart J or FCC 15J) deals in unintentional emissions (EMI) from equipment that uses digital techniques and generates and/or uses timing signals or pulses of frequencies in excess of 10 kHz, or has a pulse rate of 10,000 pulses per second or higher. These specifications grew out of public and industrial complaints to the Commission from users of televisions and other radio receivers that were being interfered with by radiation from digital devices operated nearby (3).

They were first embodied in Docket #20780 and later became regulations under Part 15 of the *Rules and Regulations.*

Radio transmitters and receivers that are subject to an emanation requirement elsewhere in the regulations are exempt from meeting the emission limits in FCC Part 15. FCC Part 15, Subpart J (FCC 15 J) specifically deals with computing devices. FCC Part 18 deals in emissions from industrial, scientific, and medical (ISM) equipment, which is equipment that generates RF energy as a part of its operation, and includes RF-stabilized arc welders, sonic welding equipment, and diathermy equipment. FCC *Rules and Regulations* may be obtained from the U.S. Government Printing Office, Washington, DC. 20402.

FCC 15J defines two classes of computing devices that must conform to the emissions specifications. Computing devices include computer peripherals: modems, printers, and other I/0 devices. The peripheral takes on the classification of the device to which it is connected (4) and is generally defined as a component or device to which a cable is intended to be attached. Components and subassemblies, such as switching power supplies, are exempted from this definition, except when the component or subassembly meets the definition of a computing device, is configured as a peripheral, and is sold directly to an end user (9). The two classes are:

Class A: "A computing device that is marketed for use in a commercial, industrial, or business environment; exclusive of a device which is marketed for use by the general public, or which is intended to be used in the home" (2).

Class B: "A computing device that is marketed for use in a residential environment notwithstanding use in commercial, business and industrial environments." A device that passes Class B limits may be used in a Class A environment.

Table 20-1. FCC radiated emissions limits.			
	Class A Section 15.810		Class B Section 15.830
Frequency		converted to 3 m based on $1/D$ variation	
MHz	µV/m at 30 m	µV/m at 3 m	µV/m at 3 m
30–88	30	300	100
88–216	50	500	150
216–1000	70	700	200

Table 20-2. FCC conducted emissions limits.

Frequency (MHz)	Class A Section 15.812 (μV)*	Class B Section 15.832 (μV)*
0.45–1.6	1000	250
1.6–30	3000	250

* Measured across 50-ohm; 50-μH line impedance stabilization network (LISN). (See below.)

Radiated emission limits for the two classes are given in Table 20-1, and conducted emissions limits are presented in Table 20-2 (2). Readings are obtained using a CISPR quasi-peak meter. Peak readings may be substituted for the quasi-peak, but no allowance for correction from peak to quasi-peak is permitted. (See the discussion of detector functions in Chaper 18 for more information on the relationship between peak and quasi-peak detector.)

The receiver bandwidth used in the 30 to 1000 MHz test range must be greater than or equal to 100 kHz. For measurements in the 0.45 to 30 MHz range, the receiver bandwidth must be at least 9 kHz.

Testing for FCC Compliance

Radiated Emissions Testing.   Measurements of radiated emissions are performed between 3 and 30 meters (m) from the equipment under test (EUT). The limits are given in Table 20-1 above at 3 m and 30 m. A scale factor of $1/D$, where D is the distance from the test object to the receiving antenna, is used to convert the emission levels, in terms of $\mu V/m$, to other distances. A tuned dipole antenna is the preferred antenna type for FCC radiated testing, though other antenna types are permitted. Both horizontal and vertical polarizations must be measured.

The FCC testing procedure is addressed in Appendix A of Part 15, entitled "FCC Methods of Measurement of Radio Noise Emissions from Computing Devices" and also in FCC/OST MP-4, dated December 1983. The radiated emissions test setup utilizes (ideally) an infinitely large area, free of obstructions, conductive bodies, and stray RF signals. This situation cannot be achieved except at very remote sites or simulated inside an anechoic chamber. Thus, an approximation of this setup, called an open field site, is used.

FCC Open Field Test Setup.   FCC Bulletin OST 55 (dated August 1982) outlines the requirements for constructing an open field test site to perform FCC radiated testing (6). The user must submit an application to be approved by the Commission prior to conducting FCC compliance testing on the site. There must be a full description of the site in the application, including a physical description and the electrical characteristics of the site, embodied in a site attenuation curve. The site attenuation curve is a measure of the free-space and ground-reflected radiation losses associated with the site. A theoretical site attenuation curve is provided in OST 55, and the allowed deviation from the curve is ± 2 dB.

Electromagnetic Ambient.   FCC measurement procedures are spelled out in FCC/OST MP-4, entitled "FCC Methods of Measurement of Radio Noise Emissions from Computing Devices," dated December, 1983 (8). Paragraph 4.6.1 deals with ambient radio frequency levels. Since an open field test site is subject to RF fields arising from outside interference (e.g., FM/TV broadcast signals, etc.) and from noise on the power line, the radiated and conducted ambient of the test site should be measured. If the ambient signals plus the emanations from the equipment under test (EUT) do not exceed the specified limits, the EUT is considered to be in compliance. In areas of high outside fields or very noisy power lines, where the ambient exceeds the specification limits at certain frequencies, the following alternative test procedures are considered to be acceptable (8):

- Perform measurements at less than the specified distance; extrapolate the measurements to the specified distance using a linear inverse distance ($1/D$, where D is the distance) correction.
- Perform the measurements during times of day when the ambient fields may be low.
- Measure the emanations in an enclosure or anechoic room if the results can be shown to correlate with the emissions that are measured at an open field site.
- Insert line filters between the power source and the LISN or between the power source and the EUT "as appropriate" for the particular measurement.

Elements of an FCC Open Field Test Site.   The EUT and the antenna (tunable dipole) are placed at the two foci of an elliptical area, separated by distance F. The major diameter of the ellipse is equal to twice the distance between the EUT and the antenna, $2F$, and the minor diameter is equal to $\sqrt{3}F$. The area of this ellipse must be free of conductive bodies (metal fences, test equipment, and anything else that may cause reflections). The test equipment should be installed even with or behind the measuring antenna to reduce the possibility of reflections from the equipment influencing the measurements.

A ground plane is preferably installed beneath this area, especially in areas where ground conductivity is low, and/or seasonal variations in ground conductivity are likely. A solid ground plane is not necessary (i.e., a suitably conductive wire mesh may be used for the ground plane, provided that the largest dimension of the mesh material is much shorter than the smallest wavelength of interest). OST Bulletin 55 approves the use of wire meshes with mesh sizes of 1/2 or 1/4 inch (6). At 1000 MHz, the wavelength is 30 cm; so a quarter-inch wire mesh should be sufficient ($1/4'' \simeq 0.6$ cm $\ll 30$ cm). If the ground plane is constructed of sections of wire mesh, the sections should be continuously bonded along the seams of the mesh. The official FCC 3-meter open field test site in Laurel, Maryland employs a ground mesh constructed of sections of wire mesh or "hardware cloth," with individual mesh "cells" measuring approximately 1/4 inch. These sections are soldered along the seams and laid upon a concrete base. It is preferable that the ground plane not be covered by any material that may influence the RF characteristics of the site (although it is noted that the official VDE test site in Offenbach, Germany employs a ground plane that is covered by asphalt). The test site setup is shown in Fig. 20-1.

FCC Test Procedures. If the EUT is normally operated on a table, it is placed upon a rotatable, nonconductive table approximately 1 meter high;

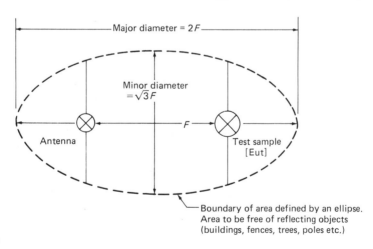

Major diameter = $2F$

Minor diameter $= \sqrt{3}F$

F

Antenna

Test sample [Eut]

Boundary of area defined by an ellipse. Area to be free of reflecting objects (buildings, fences, trees, poles etc.)

Free area $\simeq$

- 6 m X 5 m if test sample is on a rotating table
- Circle of 4.5 m radius around Eut if antenna has to be moved around

At least for 3 and 10 meters test distance, a conductive ground plane (or mesh) must exist to eliminate ground uncertainties

Fig. 20-1. Open field test site. (*Courtesy of Interference Control Technologies*)

if the equipment is normally floor-mounted, it is mounted on a table less than 1/2 meter high. The emissions from each face of the EUT are measured by tuning the antenna to resonance, moving the measuring antenna up and down, and varying lead dress and cable layout of the EUT to achieve the maximum field reading. Both horizontal and vertical polarizations are measured. If the equipment consists of a number of components, the components are arranged in a configuration that yields the maximum radiation. For distances between 3 and 10 meters, the antenna height is varied from 1 to 4 meters; for distances between 10 and 30 meters, the antenna is varied in height between 2 and 6 meters. When the vertical polarization is measured, the edge of the antenna should not come closer than 25 cm to ground; otherwise the antenna may become detuned because of the proximity of the ground plane.

If the EUT is too large to be mounted upon a table, it is placed at the center of a circle of radius D, and the test antenna is moved about the circumference of the circle. The area of a circle of radius $(3/2)D$ from the EUT must be kept free of conductive bodies.

The detector function is set to the CISPR quasi-peak function. If the measuring instrument is not supplied with CISPR weighting circuits, the data may be reported, as observed, on the basis of peak amplitude and without any correction from peak to quasi-peak. If the reading on the meter fluctuates around some amplitude, then the maximum indication on the meter is the measurement of record. The receiver 6 dB bandwidth must be no less than 100 kHz over the range of 30 to 1000 MHz.

For each frequency, the antenna factor and signal cable attenuation (in dB) must be added to the measured field strength (in dB) to obtain the magnitude of the field strength incident on the antenna (in dBμV/m). In some cases the antenna factor is determined by the manufacturer using a specified length and type of cable. Thus, the attenuation of the cable is already accounted for in the antenna factor, provided that the correct cable is used.

Data can be reported in μV/m or dBμV/m in graphical or tabular form, and calibration charts must be supplied with the data if automatic measuring equipment was used to perform the tests. Data may also be presented in the form of photographs of spectrum analyzer displays or the data from chart recorders. A sample calculation showing calibration and antenna factor corrections is required.

The use of an LISN on the power cord of the EUT is recommended. This is to minimize conducted emissions that may couple from the mains into the power cord and possibly radiate, giving false readings. (See the discussion on LISNs below.)

Radiated Emissions Tests at User's Installation.   For large equipment installations, the compliance testing may be performed in situ, that is, at the end user's location. The maximum radiated emissions are to be measured at a radial distance of 30 meters from the EUT, where practical. If testing at 30 meters is impractical, then the measurements can be performed at lesser distances and the results extrapolated to 30 meters. Generally, the results are regarded as being unique to the specific EUT and installation. However, where measurements are made at three or more installations, the results can be considered as representative of all sites for the purpose of meeting the emanation requirements. An LISN must not be used for testing on the premises as the radiated RF emanations are regarded as representative of the specific site.

Conducted Emissions Testing.   There are essentially two reasons for measuring conducted emissions below 30 MHz: (1) the size of antennas suitable for measuring radiation below 30 MHz would make them unwieldy and generally prohibitive; (2) it is believed that above 30 MHz the inductive reactance of wires results in a high impedance, and currents of high frequencies do not flow. The performance of conducted emissions testing is to assure that harmful emissions below 30 MHz do not conduct onto long power lines and eventually radiate.

Conducted emissions measurements are made across a 50-ohm, 50-µH line impedance stabilization network (LISN). An LISN is a network of passive elements designed to present a 50-ohm RF impedance to the EUT, to allow a 50-ohm meter to be connected to measure the interference voltage, and also to isolate the EUT from the power line, so that all emissions that are measured are due to the test object and not incoming power-line noise. A schematic of an LISN is shown in Fig. 20-2 (8).

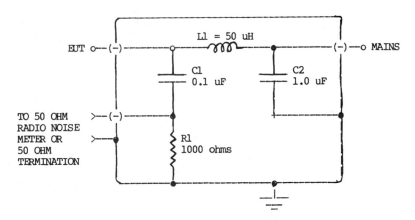

Fig. 20-2. Fifty micro-Henry line impedance stabilization network (LISN).

In the figure, inductor $L1$ is (essentially) a short circuit at power-line frequencies (0.02 ohm), and capacitors $C1$ and $C2$ have high reactive impedance (approximately 26.5 kilohms and 2.65 kilohms, respectively) at power-line frequencies. The function of the LISN is to pass power-line frequencies and to filter out RF energy conducted to and from the AC mains. An RF voltage is developed across the 1000-ohm resistor and input to the meter. This particular LISN has an effective 50-ohm impedance from about 450 kHz to 30 MHz.

Each current-carrying lead must have an LISN inserted in series with the equipment power lines. For example, an EUT powered on a 120 VAC line must have an LISN installed on both the phase wire and the neutral. Three-phase must have one LISN per phase wire and an LISN on the neutral, if the equipment is wye-connected and the neutral is brought out from the equipment.

To test for line-conducted emissions, the equipment under test is placed on a nonconductive support 40 or more cm above a ground plane. The area of the ground plane is a minimum of 2 m². There should not be any conductive bodies within 80 cm of the EUT. The power cord for the EUT is plugged into an LISN, 80 cm away, and any excess cord is folded back and forth into a bundle 30 or 40 cm in length. If the unit is not supplied with a power cord, a 1-meter cord can be used in the test. All power input leads to the EUT are plugged into LISNs. The conducted emissions test setup is shown in Fig. 20-3.

If the EUT is normally grounded through a dedicated terminal, this ground connection must be made to the ground plane. If the ground is supplied through the safety wire of the power cord, then the EUT shall be grounded

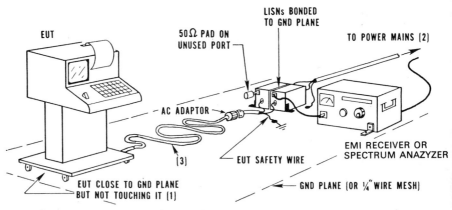

Fig. 20-3. FCC/CISPR typical test setup for a floor standing equipment (shown for single-phase device). (*Courtesy of Interference Control Technologies*)

to the normal utility ground. The LISN housing, measuring instrument case, ground plane, and so on, should be well bonded together so they are at the same RF potential.

It may be necessary to use a simulator with the EUT such that the EUT is run under all modes of operation during the testing.

To make the actual measurements, the receiver/analyzer is connected to one of the 50-ohm LISN ports with a 20-dB pad in series with the connection to protect the receiver/analyzer from line transients. All unused 50-ohm LISN ports should be terminated with a 50-ohm resistive load.

Radio noise measuring instruments shall have the function set to the CISPR quasi-peak setting. The spectrum between 450 kHz and 30 MHz is scanned with the 6-dB bandwidth set to not less than 9 kHz. Measured data are reported in microvolts (μV) or dBμV in tabular or graphical form, as photographs of spectrum analyzer displays, or as recorder charts. Calibration charts must be provided with the data if automatic measuring equipment is to be used.

Conducted Emissions Tests at User's Installation.   For tests performed at the user's installation, both the equipment and the site are considered to be the EUT. Thus, the use of an LISN is not allowed, but another power-line measurement network, connected to the power mains, is employed. This network is shown in Fig. 20-4.

The internal resistance, R, of the meter determines the reactance of the capacitor and inductor, and the resistor in the above network, and these

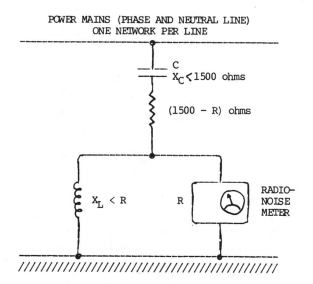

Fig. 20-4. Line probe for conducted test at user's installation. (Reference 8)

values must not exceed the values shown in the figure. In addition, a ground plane is not used as part of the test setup unless it is to be a permanent part of the installation.

FCC Compliance

In order for potentially interfering equipment to be marketed, the equipment must comply with FCC limits. Different types of compliance are provided for in the FCC regulations, as spelled out in FCC R & R 2 and summarized below (1):

Paragraph:

2.903	Type Approval:	Equipment Type Approval is based on equipment examination and test by the FCC. Type Approval attaches to all identical units.
2.905	Type Acceptance:	Equipment Type Acceptance is based on representation and test data for equipment to be used pursuant to a station authorization. Testing is performed by the manufacturer (or his agent); data are not required unless specifically requested by the FCC. Type acceptance attaches to all identical units.
2.907	Certification:	Certification is based upon representations and test data for equipment designed to be operated without individual license under Parts 15 and 18. Testing is performed by the manufacturer (or his agent), and test data are required. Certification attaches to all identical units.

Type Approval means that the FCC conducts the emissions testing at the FCC test site in Laurel, Maryland and utilizes these data to grant or reject applications for marketing and sale of equipment. It is required for marine transmitters, medical diathermy equipment, broadcast modulation monitors, ship radars, microwave ovens, wireless microphones (88–108 MHz), and auditory training transmitters.

Type Acceptance is based upon data supplied by the manufacturer to the FCC. It is required for land mobile transmitters and other transmitters in licensed service.

Certification is based upon data submitted to the FCC and is required for RF devices operating under Part 15, such as TV receivers, broadcast receivers, low power transmitters, RF industrial heaters, computers and other digital devices, and Part 18 ISM equipment.

CISPR REGULATIONS, TESTING, AND COMPLIANCE

The International Electrotechnical Commission (IEC) sponsors the CISPR organization, which is composed of each national committee of the IEC, a United Nations commission, and also a number of international unions, commissions, and committees. The CISPR organization was organized with the intention of setting uniform limits on electromagnetic emissions from equipment so that trade would not be inhibited between member countries as a result of differing emissions specifications. CISPR publications deal with interference for the following items:

- Publication 11: Microwave ovens, with power consumption below 5 kW, and ISM equipment.
- Publication 12: Ignition systems.
- Publication 13: Televisions, FM receivers, and AM receiver power-line susceptibility.
- Publication 14: Conducted and radiated emission of household appliances, portable tools up to 2 kW, office machines, dimmer regulators, and other electrical apparatus.
- Publication 15: Fluorescent lamps.

These documents may be purchased from the IEC or the American National Standards Institute (ANSI).

CISPR Emissions Limits

The rationale behind CISPR limits is to decrease the possibility that interfering sources will disrupt normal radio and television reception. Since EMI contributes to the total noise in the system, the method by which these limits were chosen is as follows: For good, noise-free reception, the signal to noise ratio (S/N) for AM and FM radio must be higher than 15 dB. For good TV reception, $S/N > 35$ dB. The total noise is the combination of the ambient noise plus any EMI. The required field strength for AM, FM, and TV is in the range 48 to 72 dBμV/m. Thus, the maximum interference levels are:

$$48 \text{ dB}\mu\text{V/m} - 15 \text{ dB} = 33 \text{ dB}\mu\text{V/m, for AM and FM radio}$$

$$72 \text{ dB}\mu\text{V/m} - 35 = 37 \text{ dB}\mu\text{V/m, for TV}$$

From these field strengths, a value of 34 dBμV/m at 30 meters was derived for frequencies above 30 MHz. If a piece of equipment meets these limits, but is installed within 30 meters of a radio or TV, the proper S/N ratio will not be met where there is interference on exact radio and TV frequencies, and a noise condition will exist.

Click Interference Relaxation

The CISPR limits make an allowance for discontinuous interference, called "click" interference. This type of interference is usually quasi-random conducted interference generated by relays and switches. The limits are relaxed by an amount according to the number of "clicks per minute." A click is defined as a disturbance that lasts no more than 200 msec and is separated in time by another such disturbance by a minimum of 200 msec. A number of separate impulses may be classified as a single click if the individual impulses are shorter than 200 msec, and the total number of impulses last no more than 200 msec.

The amount of relaxation, in dB, is given in equation (20-1):

$$R_{dB} = 20 \log_{10}(30/N) \quad \text{for } 0.2 \le N \le 30 \tag{20-1}$$

where

N = click rate per minute.

For example, if the average interval between clicks is 5 minutes, then $N = 0.2$. Thus, the allowable relaxation of the CE (conducted emission) limits is:

$$R_{dB} = 20 \log_{10}(30/.2) \simeq 44 \text{ dB} \tag{20-2}$$

The CISPR regulations dictate that no more than 25% of the clicks can exceed the CE relaxation limit.

CISPR Emissions Testing

The test setup for measuring emissions is the same as that shown in Fig. 20-1. A CISPR quasi-peak EMI receiver must be used to measure the emissions. Specifications for CISPR EMI receivers are shown in Table 19-1. The antenna of preference used during the testing is a tunable, half-wave dipole, but others can be used provided that the antenna factors are known and the results can be correlated to the dipole antenna. For measuring of emissions below 30 MHz, the antenna should be an H-field antenna (i.e., loop antenna). The general test procedure is the same as used in FCC testing, which was discussed in the FCC test section of this chapter.

The CISPR Absorbing Clamp (1)

The absorbing clamp is a device used to measure the interference power of household appliances and similar equipment in the frequency range from 30 to 300 MHz. It is clamped on the power cord of the equipment under test. The rationale behind the use of the clamp is that it is assumed that at frequencies above 30 MHz, the interference generated by the equipment is conducted out of the power connection and radiates from the power cord. Thus, the interference power must be limited to reduce harmful emissions occurring via this path.

The structure of the clamp is shown in Fig. 20-5. It is composed of ferrite cores arranged in a split cylindrical fashion, clamped about the power cord. To measure the emissions from a power cord with this clamp, the clamp is placed around the cord, and is moved along the length of the cord until a maximum reading is obtained at the frequency of interest. This procedure checks for maxima due to standing waves on the power line.

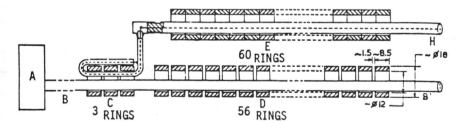

DIMENSIONS IN MILLIMETERS

```
(A) CONSTRUCTION DETAIL
    THE ABSORBING CLAMP CONSISTS OF THREE MAIN PARTS:
    C - FERRITE CURRENT TRANSFORMER.
    D - FERRITE POWER ABSORBER AND IMPEDANCE STABILIZER FOR
        THE EQUIPMENT POWER CABLE.
    E - FERRITE SLEEVE OR ASSEMBLY OF FERRITE RINGS TO REDUCE
        STANDING WAVES ON THE LEAD TO THE MEASURING SET.
```

Fig. 20-5. Construction details of absorbing clamp. (*Courtesy of Interference Control Technologies*)

CISPR Compliance

Compliance with CISPR limits generally varies from country to country, and each country has its own regulations regarding enforcement of the limits. Generally, the government agencies in Europe that are responsible for enforcement of the limits are the postal agencies, which govern all communications (7). Table 20-3 lists the CISPR countries and the various emissions standards to which they subscribe. The reader is directed to Chapter 3 of reference 1 for a breakdown of the various agencies in charge of administering the limits in the CISPR countries.

Table 20-3. Electromagnetic emission standards of the CISPR countries. (Courtesy of Interference Control Technologies).

COUNTRY	IGNITION SYSTEMS	ISM	ELECTRIC APPLIANCES	RADIO AND TV	FLUORESCENT LAMPS AND LUMINARIES	SOLID STATE CONTROLS	COMPUTER PRODUCTS
CISPR	PUB 12	PUB 11	PUB 14	PUB 13	PUB 15		
AUSTRALIA		DR 73117	AS 1044			AS 1054	
AUSTRIA		OVE F67/1957	OVE F60	OVE F60			
BELGIUM		ROYAL DEC. 1960	ROYAL DEC. 1968				
BRAZIL							
CANADA	SOR 75-629	SOR 163-455	CSA C 108.5.4			CSA22.4 VD. 1054	
CZECH.**	CSN 34-2675	CSN 34-2865	CSN 34-2860	CSN 34-2870	CSN 34-2850		
DENMARK*	MPWO 402	MPWO 44	MPWO 377	MPWO 14	MPWO 373	MPWO 213	
FINLAND	PUB T33-65	PUB T33-72	PUB T33-72	PUB T33-72	PUB T33-72		
FRANCE	NORME C91-100 & 91-103	NORME C91-1' & 91-102	NORME C91-100	C91.100 ADD 13	C91.100	C91.100	
E. GERMANY (GDR)**							
W. GERMANY	VDE0879	VDE0871	VDE0875	VDE0872	VDE0871	VDE0871	VDE0871
ISRAEL							
ITALY*	EEC DIR 72/245/CEE(9)						
JAPAN			LAW NO. 234		LAW NO. 234		
NETHERLANDS*	590824/2	IP*			IP*		
NORWAY		NEMKO 662.171 CIR.22/74,13/75	NEMKO 502.167 CIR.23/74	NEMKO 661.174 CIR. 8/75	NEMKO 301.173 CIR. 21/74	NEMKO 665.168 CIR. 13/71	
POLAND**	PN-70/S-76005	PN-71/E-06208	PN-70/E-06008	PN-71/T-05208	PN-68/E-06231	PN-71E-06218	
ROMANIA		STAS 6048/6-71				STAS 6048/7-7	
S. AFRICA	PUB 12	PUB 11	PUB 14	PUB 13	PUB 15		
SPAIN		LINE 20506					
SWEDEN			SIND-FS 1974:1 1975:1			SIND-FS 1974:1	
SWITZERLAND*							
USSR**	(APPEARS TO USE CISPR LIMITS)						
UNITED KINGDOM*	BS 833			BS 905		BS 800 PART 3	
USA	SAEJ551C SAEJ1113	FCC PART 18		FCC PART 15			FCC PART 15j

Notes: (1) * Countries known to follow and are awaiting common market directives (EEC) implementing CISPR requirements.

(2) ** Eastern European countries follow mutual economic AID Council (COMECON) Standards.

VDE REGULATIONS, TESTING, AND COMPLIANCE

The VDE limits are similar to the CISPR limits, with some variations (stricter), and are contained in VDE documents 0871, 0872, and 0875. The VDE regulations may be obtained from the VDE Publishing House (VDE Verlag) with offices in Berlin (1 Berlin 12, Bismarkstrasse 33) or from ANSI. The VDE limits will eventually be referred to as "DIN" standards, which are the German Industrial Norms. The VDE publications dealing with RF emissions limits are:

(1)0871: Regulations for equipment that generate or use RF energy, i.e., ISM equipment and digital devices operating above 10 KHz. All products require a license.
(2)0872: Regulations for radio and TV receivers.
(3)0875: Regulation for household appliances, power tools, power supplies, medical equipment.

The VDE distinguishes three classes of equipment for each publication:

Class A: Industrial or commercial equipment, marketed as such.
Class B: Generally marketed equipment, e.g., residential and other types of equipment. These limits are the same as FCC Class B limits.
Class C: Large equipment installations, where the equipment is installed and then tested for compliance in situ.

In December 1982, the German PTT (Postal and Telecommunications Department) issued Bulletin # 1115/1982 which states that practically all electronic equipment must comply with VDE 0871, Class B. The Class B requirements were also extended down to 10 KHz, from 150 KHz. Prior to this bulletin, it was recommended that EMI limits be met in this frequency range. Now it is *mandatory* that these limits be met.

VDE Emissions Limits

The following figures show the various emission limits as a function of frequency from VDE 0871 and 0875, for the various classes of equipment. Figure 20-6 shows the VDE radiated limits for Class B equipment from 10 kHz to 1 GHz, Fig. 20-7 shows VDE conducted emission limits with the FCC conducted emissions limits included for comparison, the Fig. 20-8 shows the radiated emissions limits for various types of equipment that fall under VDE 0875.

Guidelines for testing equipment are contained in VDE 0877, titled "Procedures for Measurement of Interference Voltages and Field Strengths." In

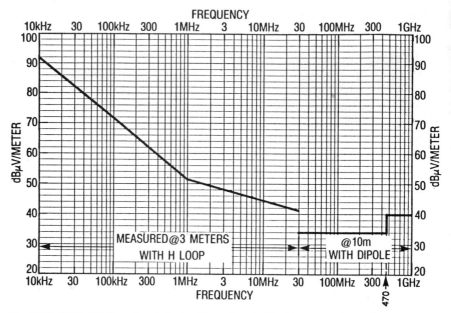

Fig. 20-6. VDE 871-B radiated limits (December 1982). (*Courtesy of Interference Control Technologies*)

general, the VDE test site is essentially the same as the CISPR test site. At radiated measurements of 30 MHz and lower, an H-field antenna is used for measuring radiated H-field strength, and a conversion to E-field strength is performed. Above 30 MHz, the antennas are tunable dipoles and tunable log periodic dipoles.

VDE Compliance

Prior to 1984, certification could be obtained only through the official VDE testing station in Offenbach, Germany, or granted by VDE engineers when they visited the United States three to four times per year. Since 1984, the German government has allowed self-certification as spelled out in PTT (Postal and Telecommunications Department) Bulletin #1078/1984. The bulletin now allows the VDE tests to be carried out by the manufacturer or some third-party testing agency. This relieves the manufacturer (or importer) of the burden of having to send equipment overseas for testing or waiting for VDE engineers to visit the United States.

After compliance has been attained, the device is certified by the VDE, and a label called the Radio Protection Mark is affixed to the product. The type

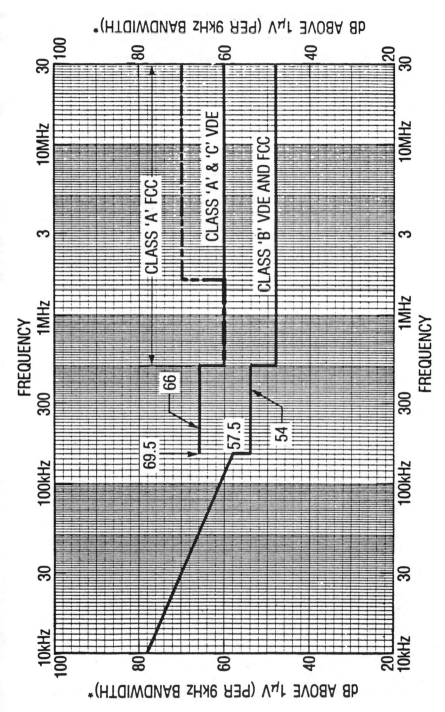

Fig. 20-7. VDE 0871/678 and FCC part 15 limits for conducted emissions. (*Courtesy of Inter-ference Control Technologies*)

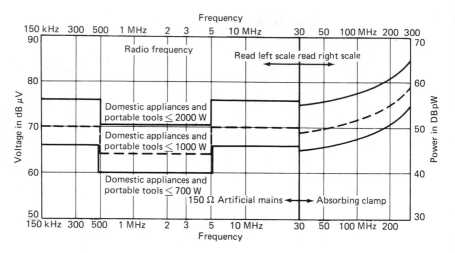

Fig. 20-8. VDE conducted interference limits (from VDE 0875). (*Courtesy of Interference Control Technologies*)

of mark depends upon the class and the type of equipment (either RF or non-RF generating). For instance, a power drill, intended for home use, would fall into the category specifying limits for non-RF equipment generating normal-level emissions and marketed for residential use. In all, there are six types of Radio Protection Marks as shown in Fig. 20-9.

FOR NON-RF EQUIPMENT   (VDE 875)

No license required. Manufacturer *must* show R.P. Mark.
R.P.M. partially superseded by EC conformity label (Household Directive)

FOR EQUIPMENT GENERATING RF   (VDE 871)

All products require a license (Individual or General).
Manufacturer shows R.P.M. (B Limit) or type test (A Limit).

 = HIGH-LEVEL EMISSIONS FOR PRODUCTS EXCLUSIVELY OPERATED IN INDUSTRIAL AREAS

 = NORMAL-LEVEL EMISSIONS FOR PRODUCTS INTENDED FOR USE IN LIVING (RESIDENTIAL) AREAS

 = LOW-LEVEL EMISSIONS FOR SPECIAL CASES IN EMI-SENSITIVE AREAS SUCH AS RADIO RECEIVING STATIONS

 = PRODUCTS WHICH ARE NOT A SOURCE OF EMI

 = HIGH-LEVEL EMISSIONS—USER MUST APPLY FOR A LICENSE

 = GENERAL LICENSE (LEVEL IS 12dB MORE SEVERE THAN A)

Fig. 20-9. Radio protection marks of the VDE. (*Courtesy of Interference Control Technologies*)

REFERENCES

1. Mertel, Herbert K. *International and National Radio Frequency Interference Regulations.* Gainesville, VA: Don White Consultants, Inc., 1978.
2. Federal Communications Commission *Rules and Regulations*, Volume 2, Part 15, July 1981.
3. Keenan, R. K. *Digital Design for Interference Specifications.* Vienna, VA: The Keenan Corporation, 1983.
4. Federal Communications Commission, Office of Science and Technology. "Understanding The FCC Regulations Concerning Computers." Bulletin OST 54, March 1982
5. Straus, Isidor. "Testing Products Correctly Ensures EMI-Spec Compliance." Electronic Design News, Cahners Publishing Company, 1981.
6. Federal Communications Commission, Office of Science and Technology. "Characteristics of Open Field Test Sites." Bulletin OST 55, August 1982.
7. Spiegel, Klaus. "International Standards and Regulations for Power Supplies. *Proceedings of the First Power Sources Conference*, November 27–29, 1984, Boston, MA.
8. Federal Communications Commission, Office of Science and Technology. "FCC Methods of Measurement of Radio Noise Emission from Computing Devices." FCC/OST MP-4 December 1983.
9. Federal Communications Commission, Office of Science and Technology. "Understanding the FCC Regulations Concerning Computing Devices." FCC/OST Bulletin 62, May 1984.

Index